Industrial Chemical
Process Design

To my wife Bettie—
Her countless days of tolerance, understanding,
and never-ending help deserve my expression
of thanks beyond words.

Industrial Chemical Process Design

Douglas L. Erwin

McGraw-Hill

New York   Chicago   San Francisco   Lisbon   London   Madrid
Mexico City   Milan   New Delhi   San Juan   Seoul
Singapore   Sydney   Toronto

Cataloging-in-Publication Data is on file with the Library of Congress

McGraw-Hill

A Division of The McGraw·Hill Companies

1 2 3 4 5 6 7 8 9 0   DOC/DOC   0 8 7 6 5 4 3 2

P/N 137621-6
PART OF
ISBN 0-07-137620-8

The sponsoring editor for this book was Kenneth P. McCombs, the editing supervisor was M. R. Carey, and the production supervisor was Pamela A. Pelton. It was set in Century Schoolbook by North Market Street Graphics.

Printed and bound by R. R. Donnelley & Sons Company.

McGraw-Hill books are available at special quantity discounts to use as premiums and sales promotions, or for use in corporate training programs. For more information, please write to the Director of Special Sales, McGraw-Hill Professional, Two Penn Plaza, New York, NY 10121-2298. Or contact your local bookstore.

This book is printed on recycled, acid-free paper containing a minimum of 50% recycled, de-inked fiber.

Contents

Acknowledgments

The author wishes to acknowledge special appreciation to:

- The American Petroleum Institute, *API Technical Data Book, Third Edition* (API, Washington, D.C., 1976)

- The Microsoft Corporation, Visual Basic 3.0

- Professors R. L. Daugherty and A. C. Ingersoll, PhD, authors of *Fluid Mechanics* (McGraw-Hill, New York, 1954), from which Chap. 10 of this handbook is adapted

- Professors William A. Felsing and George W. Watt, authors of *General Chemistry* (McGraw-Hill, New York, 1951), from which Chap. 11 of this handbook is adapted

- Professor Deane E. Ackers, Robert H. Perry, and staff specialist Don W. Green, editors of *Perry's Chemical Engineering Handbook* (McGraw-Hill, New York, 1997)

- Professor Donald Q. Kern, author of *Process Heat Transfer* (McGraw-Hill, New York, 1950), source of the tables in App. A of this handbook

In the author's worldwide travels as a practicing chemical engineer, countless other colleagues, friends, institutions, and companies have been most encouraging. A sincere and heartfelt thank-you to all.

Introduction: The Quest

At my first job after graduating from college, while fulfilling my routine data-gathering duties in a butadiene-producing chemical plant one day, my boss came running up to me and yelled, "My acid extraction column was totally upset, requiring the plant shutdown." I felt like the underdog who should have an answer. This was one of the most embarrassing moments of my life. My responsibility was clearly to predict—if not prevent—such a happening, or most certainly to be prepared for it. I didn't have the slightest idea what had happened, let alone how to resolve this problem. "You should have known of this happening without my having to tell you," shouted a very red-faced boss.

Years later, I truly believe he had been right; I should have been able to predict this event of extractor flooding, but only later did I know the direct remedy that should have been executed that very hour. The sad part is that my engineering supervisor didn't have a solution either. In fact, I was persuaded that the operator knew more than the both of us that day.

These events led us to the quest for more practical process engineering, and learning how to apply what we learned in college. That's what this book is about. Keep in mind, however, that college has helped you open the door. This book now helps lead you past the door to the feast of useful technology that is our goal, our quest.

With this introduction, let's first consider the practical task items we as process types must and should always be capable of working. The surprise is that only two items are critically needed to fulfill this need completely. First, we must have a workable database through which we can apply the second, which is unit operations. What is this database pointed to, and what is this about unit operations we need yet? After all, we had a heavy helping of unit operations in college and surely the same for hydrocarbon physical data (the database referenced in this book). Yes, but, consider the practical application of this data and the practical application of unit operation selection, design of the sited unit operation, and, finally, resolving unit process operation trouble.

These practical application works only come from the school of hard knocks. With this statement, you have wandered into the very challenge we give you, *the quest for practical process engineering*. This book is written to guide you by providing a critically needed tool for achieving this quest. More simply, PC software for the design of unit operations equipment and unit operations problem analysis and resolution are the principal goals set herein.

The first goal in this quest is to establish a needed database of practical process engineering (PPE). A database is the basic data from which we determine the design and operating conditions of all process equipment. The tools to establish this database were first given to us in college. We used fluid transport properties, density, viscosity, molecular weight, enthalpy, and other tools to determine fluid dynamics, heat-material balances, unit operation design, and much more. Now the big question becomes how we can make practical application of these tools. A good challenge, yes? Of course this is a good challenge, and most certainly an honorable one to win.

Industrial Chemical Process Design offers a good start in meeting this challenge, with nine chapters covering the application of these needed tools, plus yet one more critical need fulfillment, the practical application of our college achievement tools. Chapter 1 offers a good basic database that complements our college building blocks. Next, Chap. 2 covers fractionation, the most practical and the most used process in nearly every unit operation, including the component separation technique, how-to advice, and equipment. Chapters 3 to 7 then follow with needed unit operations and their design techniques, with applications.

Chapter 8 next introduces the cost of equipment. This one chapter is perhaps why many will find this book a vital asset in their daily tasks. It is also perhaps why many will want to own this book, for estimating the investment cost of entire chemical plants, gas processing plants, oil and gas production facilities, and refineries. Chapter 8 establishes a cost method, totally complete within this book, for any single major equipment item, installed and ready to commission. Adding these major cost modular packages together estimates the cost of an entire refinery. Economic analysis of investment return is included, giving the user a complete assay report capability of cost and investment return analysis. A spreadsheet in Excel is included in an example problem and in the compact disc included with this book. The user will find this CD spreadsheet an excellent template for making a cost analysis spreadsheet application for any plant or equipment item desired.

Chapter 9 is another reason why many will want to own this book. Chapter 9 introduces Visual Basic programming to the practicing engineer. This chapter is not only an introduction to programming in Visual Basic, but it demonstrates the common types of skills most engineers need: how to input data into Visual Basic code, how to execute program code in Visual Basic, how to make display screens for input and output data answers in Visual Basic, and how to make and apply data arrays. Many books have been written regarding Visual Basic programming, but none found to the date of this writing have been dedicated fully to the process engineer's needs. Chapter 9 therefore presents a step-by-step walkthrough with actual screen graphic displays. The steps are kept simple, with demonstrated examples. The student merely needs to follow these simple steps for any problem. Countless computer programs applicable to the Microsoft Windows environment may be written by the student simply by using parts of Chap. 9 screens as a template.

Chapters 10 and 11 introduce additional practical unit operations of fluid dynamics and chemistry.

Over 20 computer programs derived in this book are given in executable format in the accompanying CD. The user may apply these programs to countless problems, such as fractionation column design and sizing, heat exchanger design and sizing, air finfan cooler design and sizing, oil electrostatic dehydration pressure vessel design and sizing, and liquid/liquid extractor column design and sizing. The CD program methodologies and the databases are fully covered in the respective chapter text for each of the programs.

Having mentioned the CD computer programs accompanying this book, it is important for the reader, user, and student to realize that these programs have been written for this book and not vice versa. Put in a more simple way, these programs have not been put through an exhaustive beta test. Therefore, if you find that one of these programs stops with a "data error" or other error message, please understand that you have probably entered an area that the program data did not cover, or to which the algorithm did not extend.

You have now stumbled upon another quest of this book, to correct the program through your good ability. After all, I have taken great lengths in all chapters to give you every insight of methodologies, equations, and even a hearty helping of the program code listing. The quest is yours, with my commendations to those who accept the challenge.

Looking back, it has been my sincere intent to give practicing process engineers, both young and old, filler for a gap that has notably existed throughout my entire career. Yes, the day my red-faced boss yelled, "My acid extraction column was totally upset, requiring the plant shutdown," I wished I could have referred to this book. Even more, I wished I could have known some of the basic principles covered herein. Only years later did I find the technology dearly needed to fill this gap. I have found the applications given in this book to be broad, having lengthy experience in most every family type of chemical, refining, and oil and gas plant in our industry. I can state with confidence that I have filled this gap to a reasonable degree. A hearty wish to all that each may find this book a treasure and fulfillment of this quest.

Conventions

[ ]	Indicates a reference listed at the end of the chapter.
*	Indicates multiplication in an equation.
	Blank space indicates multiplication in an equation.
×	Indicates multiplication in an equation.
/	Indicates division in an equation.
—	Indicates division in an equation.
^	Indicates "raised to the power of" in an equation.
( )	Indicate delimiters in equations.

Industrial Chemical Process Design

The Database

Nomenclature

μ	viscosity, cP
°API	gravity standard at 60°F
cP	viscosity, centipoise
cSt	viscosity, centistokes
D	density, lb/ft^3
exp	constant for exponential powers of e base value, 2.7183
MW	molecular weight
P	system pressure, psia
P	pressure, psia
P_C	critical pressure, psia
P_R	reduced pressure (P/P_C), psia
°R	$T + 460$°F
SG 60/60	specific gravity referenced to pure water at 60°F
T	temperature, °F
T_B	true boiling point, °F, of ASTM curve cut component or pure component
T_C	critical temperature, °F or °R
T_R	reduced temperature (T/T_C), °R
V	molar volume, ft^3/lb-mol
Z	gas compressibility factor

To meet this introductory challenge, we must first establish a database from which to launch our campaign. In doing so, consider the physical properties of liquids, gases, chemicals, and petroleum generally in making this application: viscosity, density, critical temperature, critical pressure, molecular weight, boiling point, acentric factor, and enthalpy.

The great majority of the process engineer's work is strictly with organic chemicals. This book is therefore directed toward this database of hydrocarbons (HCs). Only eight physical properties are presented here. Aren't there many others? The answer is yes, but remember, this book is strictly directed toward that which is indeed practical. Many more properties can be listed, such as critical volume or surface tension. Our quest is to take these more practical types (the eight) as our database and thereby successfully achieve our goal, practical process engineering (PPE).

At this point, it is important to present a disclaimer. Many notable engineers could claim the author is loco to think he can resolve all database needs with only the eight physical properties given or otherwise derived in this book. Let me quickly state that many other extended database resources are indeed referenced in this book for the user to pursue. Only in such retrieval of these and many other database resources, such as surface tension and solubility parameters, can PPE be applied. An example is that surface tension and solubility parameters must both be determined before the liquid/liquid software program given herein can be applied. This liquid-liquid extraction program (Chemcalc 16 [1]) is included as part of the PPE presentation. (See Chap. 7.) It is therefore important to keep in mind that many database references are so pointed to in this book—Perry's, Maxwell's, and the American Petroleum Institute (API) data book, to name a few.

Again, why then present only these eight physical properties for our concern? The answer is that we can perform almost every PPE scenario by applying these eight physical properties, which are in most every data source and are readily available. Furthermore, an exhaustive listing would be a much greater book than the one you are reading, such as *Lange's Handbook of Chemistry and Physics* [2]. Incidentally, *Lange's* is a very good reference book which I highly endorse.

Viscosity

Liquid viscosity

The first of these properties is *viscosity*. All principal companies use mainly one of two viscosity units, centipoise (cP) or centistokes (cSt). Centipoise is the more popular. If your database presents only one, say cP, then you may quickly convert it to the other, cSt, by a simple equation:

$$cSt = \frac{cP}{sp\ gr}$$

Specific gravity (sp gr) is simply the density referenced to water, sp gr of water being 1.0 at 60°F. This means that to get the specific gravity of a liquid, simply divide the density of the liquid at any subject temperature of the liquid by the density of water referenced to 60°F. With sp gr so defined, we can subsequently convert cP to cSt or cSt to cP by this simple equation. Thus the conversion is referenced to a temperature. *Viscosity of any liquid is very dependent and varies with the slightest variance of the liquid temperature.*

Viscosity has been defined as the readiness of a fluid to flow when it is acted upon by an external force. The absolute viscosity, or *centipoise*, of a fluid is a measure of its resistance to internal deformation or shear.

A classic example is molasses, a highly viscous fluid. Water is comparatively much less viscous. Gases are considerably less viscous than water.

How to determine any HC liquid viscosity.   For the viscosity of most any HC, see Fig. A-3 in *Crane Technical Paper No. 410* [3]. If your particular liquid is not given in this viscosity chart and you have only one viscosity reading, then locate this point and draw a curve of cP vs. temperature, °F, parallel to the other curves. This is a very useful technique. I have found it to be the more reliable, even when compared to today's most expensive process simulation program. Furthermore, I find it to be a valuable check of suspected errors in laboratory viscosity tests. If you don't have the Crane tech paper (available in any technical book store), then get one. You need it. I have found that most every process engineer I have met in my journeys to the four corners of the earth has one on their bookshelf, and it always looks very used.

The following equations compose a good quick method that I find reasonably close for most hydrocarbons for API gravity basis. Note that the term *API* refers to the American Petroleum Institute gravity method [4]. These viscosity equations are derived using numerous actual sample points. These samples ranged from 10 to 40° API crude oils and products. I find the following equations, Eqs. (1.1) to (1.4), to be in agreement with Sec. 9 of Maxwell's *Data Book on Hydrocarbons* [5].

Viscosity, cP, for 10°API oil:

$$\mu = \exp\,(18.919 - 0.1322T + 2.431\text{e-}04\ T^2) \qquad (1.1)$$

Viscosity, cP, for 20°API oil:

$$\mu = \exp\,(9.21 - 0.0469T + 3.167\text{e-}05\ T^2) \qquad (1.2)$$

Viscosity, cP, for 30°API oil:

$$\mu = \exp\,(5.804 - 0.02983T + 1.2485\text{e-}05\ T^2) \qquad (1.3)$$

Viscosity, cP, for 40°API oil:

$$\mu = \exp(3.518 - 0.01591T - 1.734\text{e-}05\ T^2) \qquad (1.4)$$

where μ = viscosity, cP
T = temperature, °F
exp = constant of natural log base, 2.7183, which is raised to the power in parentheses

You can interpolate linearly for any API oil value between these equations and with extrapolation outside to 90°API. Temperature coverage is good from 50 to 300°F. If outside of this range, use the American Society for Testing and Materials (ASTM) Standard Viscosity-Temperature Charts for Liquid Petroleum Products (ASTM D-341 [6]). The values derived by Eqs. (1.1) to (1.4) are found to be within a small percentage of error by the ASTM D-341 method.

It is good practice to always obtain at least one lab viscosity reading. With this reading, draw a relative parallel curve to the curve family in ASTM D-341. The popular *Crane Technical Paper No. 410* reproduces this ASTM chart as Fig. B-6. If two viscosity points with associated temperature are known, then use the Crane log plot figure, also an API given method (ASTM D-341), to determine most any liquid hydrocarbon viscosity.

Gas viscosity

How to determine any HC gas viscosity.   For most any HC gas viscosity, use Fig. 1.1 (Fig. A-5 in *Crane Technical Paper No. 410*). The constant Sg

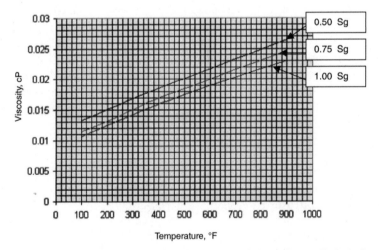

Figure 1.1 Hydrocarbon gas viscosity. *(Adapted from* Crane Technical Paper No. 410, *Fig. A-5. Reproduced by courtesy of the Crane Company [3].)*

is simply the molecular weight (MW) of the gas divided by the MW of air, 29. Note carefully, however, that Fig. 1.1 is strictly limited to atmospheric pressure. The gas atmospheric reading from this figure, or from other resources such as the API *Technical Data Book* [7], is deemed reasonably accurate for pressures up to, say, 400 psig.

In addition to the API *Technical Data Book* and Gas Processors Suppliers Association (GPSA) methods [8], a new gas viscosity method is presented herein that may be used for a computer program application or a hand-calculation method:

For 15 MW gas:

$$\mu = 0.0112 + 1.8e\text{-}05\ T \tag{1.5}$$

For 25 MW gas:

$$\mu = 0.00923 + 1.767e\text{-}05\ T \tag{1.6}$$

For 50 MW gas:

$$\mu = 0.00773 + 1.467e\text{-}05\ T \tag{1.7}$$

For 100 MW gas:

$$\mu = 0.0057 + 0.00001\ T \tag{1.8}$$

where μ = viscosity, cP
          T = temperature, °F

Equations (1.5) to (1.8) are good from vacuum up to 500 psia pressure and temperatures of −100 to 1000°F. Pressures at or above 500 psia should have corrections added from Eqs. (1.9), (1.10), or (1.11). Eqs. (1.5) to (1.8) are reasonably accurate to within 3% of the API data and are good for a pressure range from atmospheric to approximately 400 psig. You may make linear interpolations between temperature-calculated points for reasonably accurate gas viscosity readings at atmospheric pressures.

Many will say (even notable process engineers, regrettably) that higher pressures (above 400 psig) will have little effect on the gas viscosity, and that although the viscosity does change, the change is not significant. *Trouble here!* In many unit operations, such as high-pressure (≤500 psig) separators and fractionators, the gas viscosity variance with pressure is most critical. I have found this gas viscosity variance to be significant in crude oil–production gas separators, even as low as 300 psig. You may make corrections with the following additional equations. These corrections are to be added to the atmospheric gas viscosity reading in Fig. 1.1 or the gas viscosity Eqs. (1.5) to (1.8) [8].

Add the following calculated viscosity correction to Fig. 1.1 or Eqs. (1.5) to (1.8):

Gas viscosity correction for 100°F system:

$$\mu_c = -1.8333\text{e-}05 + 1.2217\text{e-}06\ P + 1.737\text{e-}09\ P^2 - 2.1407\text{e-}13\ P^3 \tag{1.9}$$

Gas viscosity correction for 400°F system:

$$\mu_c = -1.281\text{e-}05 + 1.5484\text{e-}06\ P + 2.249\text{e-}10\ P^2 - 6.097\text{e-}14\ P^3 \tag{1.10}$$

Gas viscosity correction for 800°F system:

$$\mu_c = -1.6993\text{e-}05 + 1.1596\text{e-}06\ P + 2.513\text{e-}10\ P^2 \tag{1.11}$$

where μ_c = viscosity increment, cP, to be added to Fig. 1.1 values or to Eqs. (1.5) to (1.8)

P = system pressure, psia

You may again interpolate between equation viscosity values for in-between pressures. Whenever possible, and for critical design issues, these variables should be supported by actual laboratory data findings.

In applying these equations and Fig. 1.1, please note that you are applying a proven method that has been used over several decades as reliable data. Henceforth, whenever you need to know a gas viscosity, you'll know how to derive it by simply applying this method. You may also use these equations in a computer for easy and quick reference. See Chap. 9 for computer programming in Visual Basic. Applying programs such as these is simple and gives reliable, quick answers.

Now somebody may say here, why should a high-pressure (450 psig and greater) gas viscosity be so important, and, by all means, what is practical about finding such data? Well, this certainly deserves an answer, so please see Chap. 4, page 153, for the method of calculating gas and liquid vessel diameters. Note that Eq. (4.3) in the Vessize.bas program has an equation divided by the gas viscosity. A change of only 10% in the gas viscosity value greatly changes the vessel's required diameter, as may be seen simply by running this vessel-sizing program. Considerable emphasis is therefore placed on these database calculations. *They do count.* Take my suggestion that you prepare for a good understanding of this database and how to get it.

Density

Liquid density

Liquid density for most HCs may be found in Fig. 1.2. This chart is a general reference and may be used for general applications that are not critical for discrete defined components. In short, if you don't have a better way of getting liquid density, you can get it from Fig. 1.2. Note that you need to have a standard reference of API gravity reading to predict the HC liquid density at any temperature.

Generally, you should have such a reading given as the API gravity at 60°F on the crude oil assay or the petroleum product cut lab analysis. If you don't have any of these basic items, you must have something on which to base your component data, such as a pure component analysis of the mixture. If not, then please review the basis of your given data, as it is most evident that you are missing critical data that must be made available by obtaining new lab analysis or new data containing API gravities.

It is important here to briefly discuss specific gravity and API gravity of liquids. First, the API gravity is always referenced to one temperature, 60°F, and to water, which has a density of 62.4 lb/ft^3 at this temperature. Any API reading of a HC is therefore always referenced to 60°F temperature and to water at 60°F. This gravity is always noted as SG 60/60, meaning it is the value interchangeable with the referenced HC's API value per the following equations:

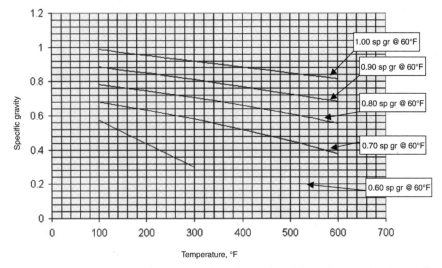

Figure 1.2 Specific gravity of petroleum fractions. *(Plotted from data in J. B. Maxwell, "Crude Oil Density Curves," Data Book on Hydrocarbons, D. Van Nostrand, Princeton, NJ, 1957, pp. 136–154.)*

$$\text{°API} = \frac{141.5}{\text{SG } 60/60} - 131.5 \tag{1.12}$$

$$\text{SG } 60/60 = \frac{141.5}{131.5 + \text{°API}} \tag{1.13}$$

Thus, having the API value given, we may find the subject HC gravity at any temperature by applying Fig. 1.2. Keep in mind that liquid gravities are always calculated by dividing the known density of the liquid at a certain temperature by water at 60°F or 62.4 lb/ft³.

I also find the following equation to be a help (again, in general) in deriving a liquid density.

Liquid density estimation

$$D = \frac{MW}{(10.731 * T_C/P_C) * 0.260^{\wedge}[1.0 + (1.0 - T_R)^{\wedge}0.2857]} \tag{1.14}$$

where D = liquid density, lb/ft³
   MW = molecular weight
   T_C = critical temperature, degree Rankine (°R)
   P_C = critical pressure, psia
   T_R = reduced temperature ratio = T/T_C
   T = system temperature, °R, below the critical point

Let's now run a check calculation to see how accurate this equation is.

n-Octane

$MW = 114.23$

$P_C = 360.6$ psia

$T_C = 1024$°R

SG 60/60 = 0.707

Trial at 240°F, Liquid n-Octane

From Eq. (1.14):

$$D = \frac{MW}{(10.731 * T_C/P_C) * 0.260^{\wedge}[1.0 + (1.0 - T_R)^{\wedge}0.2857]}$$

$D = 38.00$ lb/ft³

From Maxwell, page 140 [9]:

n-octane density at 240°F = 0.625 SG

or $0.625 * 62.4 = 39.00$ lb/ft³

From Fig. 1.2:

$$°\text{API @ 0.707 SG 60/60} = \frac{141.5}{0.707} - 131.5 = 68.64°\text{API}$$

At this API curve, read 0.615 gravity (horizontal line from intersect point) at 240°F, or $0.615 * 62.4 = 38.38 \text{ lb/ft}^3$

Summary of Eq. (1.14) Check

$$\text{Deviation from Maxwell [9]} = \frac{39.0 - 38.0}{39} * 100 = 2.5\% \text{ error}$$

$$\text{Deviation from Nelson [10]} = \frac{38.38 - 38.00}{38.38} * 100 = 1.0\% \text{ error}$$

From the preceding check of n-octane liquid density, we have established that Eq. (1.14) is a reasonable source for calculating n-octane liquid density. Both Nelson and Maxwell data points could also have as much error, 1 to 3%. The conclusion therefore is that Eq. (1.14) is a reasonable and reliable method for liquid density calculations. You may desire to investigate other known liquid densities having the same known variables, T_C, P_C, and MW. You are encouraged to do so.

Gas density

While the density of any liquid is easily derived and calculated, the same is not true for gas. Gas, unlike liquid, is a compressible substance and varies greatly with pressure as well as temperature. At low pressures, say below 50 psia, and at low temperature, say below 100°F, the ideal gas equation of state holds true as the following equation:

$$D = \frac{\text{MW} * P}{10.73 * (460 + T)} \tag{1.15}$$

where D = gas density, lb/ft^3
         MW = gas molecular weight
         P = system pressure, psia
         T = °F

For this low-temperature and -pressure range, any gas density may quickly be calculated. Error here is less than 3% in every case checked. What about higher temperatures and pressures? Aren't these higher values where all concerns rest? Yes, most all process unit operations, such as fractionation, separation, absorption stripping, chemical reaction, and heat exchange generate and apply these higher-temperature and -pressure conditions. Then how do we manage this deviation from the ideal gas equation? The answer is to insert the gas compressibility factor Z.

Add gas compressibility Z to Eq. (1.15):

$$D = \frac{\text{MW} * P}{Z * [10.73 * (460 + T)]} \qquad (1.16)$$

where Z = gas compressibility factor

The question now is how do you derive, calculate, or find the correct Z factor at any temperature and pressure? The first answer is, of course, get yourself a good, commercially proven, process-simulation software program. As these programs cost too much, however, for anyone who works for a living, you must seek other resources. *This is a core reason why this book has been written.* Look at the practical side. After all, who has $25,000 pocket change to throw out for such candor? It is therefore my sincere pleasure to present to you, as the recipient of the software accompanying this book, the following two computer programs.

Z.mak. This is a program derived from data established in the API *Technical Data Book,* procedure 6B1.1 [11]. Please note that Z.mak, although similar, is an independent and separate program from this API procedure. A program listing as in the Z.mak executable file is shown in Table 1.1. Inside the phase envelope, the compressibility factor calculated in Z.mak is more accurate than that calculated in RK.mak (the Redlich-Kwong equation of state). RK.mak is given and discussed later. The Z.mak program may be used with reasonable accuracy, as can the API procedure 6B1.1. Z.mak accuracies range from 1 to 3% error. Most case accuracies are 1% error or less. One caution, however, is necessary, and this is regarding Z values in or near the critical region of the phase envelope.

Important Note: Use Z.mak when at less than the critical pressure and/or in the phase envelope.

The acentric factor is also calculated from the input T_C, P_C, and boiling point. The acentric factor is used in the Z factor derivation. See line 110 in Table 1.1.

Please note that the Z factor so calculated here is to be applied in Eq. (1.16) for calculating the gas density.

The Z factor for butene-1 is now calculated in the actual computer screen display of the Z.mak computer program. (See Fig. 1.3.)

RK.mak. When out of the phase envelope, use this program, the well-known Soave-Redlich-Kwong (SRK) equation-of-state simplified program [12]. The student here may immediately detect the standard SRK

TABLE 1.1   Z.mak Program Code Listing

```
Sub Command1_Click ()
10 'Program for calculating gas compressibility factor, Z
15 'For a Liquid - Vapor Equilibrium Saturation Condition, gas Z
If TxtTC = "" Then TxtTC = 295.6
If TxtPC = "" Then TxtPC = 583
If TxtTB = "" Then TxtTB = 20.7
If TxtP = "" Then TxtP = 200
20 'Data Input lines 30 through 50
  TC = TxtTC: PC = TxtPC: TB = TxtTB: P = TxtP
30 TC = TC + 460          'Critical T in deg R
40 TB = TB + 460          'Atmospheric boiling Temperature in deg R
45 TR = TB / TC
50 PR = P / PC: PR2 = 14.7 / PC 'System P and Reduced PR Calc psia
60 'Calculate Acentric Factor, ACENT
   PRO = 6.629 - 11.271 * TR + 4.65 * TR ^ 2: PR1 = 16.5436
   - 46.251 * TR + 45.207 * (TR ^ 2) - 15.5 * (TR ^ 3)
ACENT = (((Log(PR2)) / 2.3026) - (-PRO)) / (-PR1)
70 'ACENT = .42857 * ((((.43429 * Log(PC)) - 1.16732) / ((TC /
   TB) - 1#)) - 1#
80 'Equations for Z calc follow
90 Z0 = .91258 - .15305 * PR - (1.581877 * (PR ^ 2)) + (2.73536
   * (PR ^ 3)) - (1.56814 * (PR ^ 4))
100 Z1 = -.000728839 + .00228823 * PR + (.217652 * (PR ^ 2))
   + (.0181701 * (PR ^ 3)) - (.1544176 * (PR ^ 4))
110 Z = Z0 - ACENT * Z1
120 'Print ACENT, Z
   TxtACENT = Format(ACENT, "##.######"):
   TxtZ = Format(Z, "##.######")
End Sub
```

SOURCE: Method from data in Calculation method GB1.1, "API Density," American Petroleum Institute, *Technical Data Book,* API Refining Department, Washington, DC, 1976.

equation of state on line 110 and the derivation of coefficients *A* and *B* on line 100 in Table 1.2. This program solves a cubic equation by first assuming a value for *V,* line 80, and then iterating a pressure calculation of *P* on line 110 until the calculated DELV of *V* deviation is less than 0.0001. Thus, this is a unique way to calculate *V* and the density thereof per Eq. (1.16).

Please note that both Z.mak and RK.mak exhibit the same problem for finding the gas density of propane at 100 psia and 200°F. Note also that *Z* calculations from each are appreciably different, 0.86 vs. 0.94 (see Figs. 1.3 and 1.4). Why the difference? *Remember* the previous warning about using the Z.mak program out of the phase envelope? Well, this is a classic example, as these conditions are definitely out of

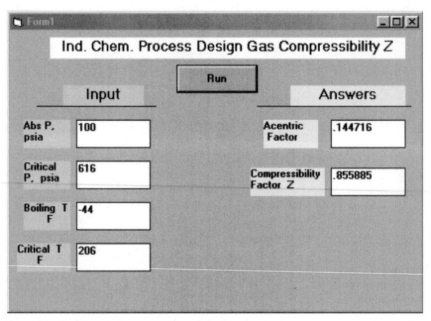

Figure 1.3   Z.mak screen.

propane's phase envelope. Therefore, RK.mak should be correct here, and it is indeed correct. You may verify the answer with any propane pressure-temperature-enthalpy chart. A fluid density of 0.66 lb/ft^3 and a Z factor of 0.94 are correct.

For those of you who are scavengers and are rapidly scanning this book to claim whatever treasure you may find, may I say my good cheers and gung ho (a World War II saying for "go get 'em!"). For those of you who are weeding out every word in careful analysis of what I'm trying to deliver in this book, however, I must share the following thoughts.

I have been a full-time employee in three major engineering, procurement, and construction (EPC) firms, and in each one I had very limited access to these high-priced simulation programs that do almost every calculation imaginable and a few more on top of that. As a practicing process design engineer, I can remember more times I needed these simulation programs on my computer and didn't have one, than I can remember having one when I needed it. Seems these companies always have a young engineer who is indeed a whiz on these simulation programs, the one and only person who runs the simulation program. You need a data set run? Well, you must give the engineer your data in elite form and then wait in queue for the output answers. Uh-oh, you've now got the answers and suddenly realize you didn't cover the entire range critically needed? Do it all over again and wait in queue for your answers, hoping

TABLE 1.2 RK.mak Program Code Listing

```
Sub Command1_Click ()
10 'RK EQUATION OF STATE PROGRAM FOR GAS DENS CALC
20 'Print "   RK EQUATION OF STATE PROGRAM FOR GAS DENS CALC":
   Print : Print
30 'INPUT "  P PSIA, T DEG F ",P,T
40 'INPUT "  PC PSIA, TC DEG F ",PC,TC
   T = TxtT: P = TxtP: TC = TxtTC: PC = TxtPC: MW = TxtMW
50 T = T + 460: TC = TC + 460
60 ' INPUT "  MW OF MIXTURE ",MW
70 'FIRST TRIAL GUESS FOR V, CF PER 1b MOL
DENSITY:
80 V = 10.73 * T * .001 / P
90 Rem A & B CONST CALC
100 B = .0867 * 10.73 * TC / PC: A = 4.934 * B * 10.73 * (TC ^ 1.5)
110 PCA = ((10.73 * T) / (V - B)) - (A / ((T ^ .5) * V * (V + B)))
120 DELP = PCA - P
130 DPDV = (((B + 2 * V) * A / (T ^ .5)) / ((V * (V + B)) ^ 2))
    - (10.73 * T / ((V - B) ^ 2))
140 V = V - DELP / DPDV
160 DELV = (DELP / DPDV) / V
170 If Abs(DELV) > .0001 GoTo 110
180 'PRINT:PRINT USING "  FLUID DENS, 1b/CF = ####.###### "; MW/V
   TxtDen = Format(MW / V, "#####.#####")
190 Z = (P * MW) / (10.731 * T * (MW / V))
   TxtZ = Format(Z, "##.#####")
200 'PRINT: PRINT USING "  Gas Compressibility Factor,
    Z = ###.##### "; Z
210 'End
End Sub
```

SOURCE: Method from O. Redlich and J. N. S. Kwong, *Chem. Rev.* 44:233, 1949.

you got it this time. You have now come to my hit line, *use this book and the software herein to derive your needs!*

The previous example of critically needed density for, say, a hydraulic line sizing or heat exchanger problem is well in order with our modern-day, most advanced, high-priced computer programs. An added thought here is that most medium-sized EPC companies have only one or two keys to run these large computer software programs. Therefore, this book and the accompanying software will help you expedite much of the work independent of these large, costly programs. Just think, you've got your own personal key in this book and software! This book also is a good supplement to these complete and comprehensive simulation programs. As an added plus for you, the major solutions to your problems are given in the CD supplied with this book.

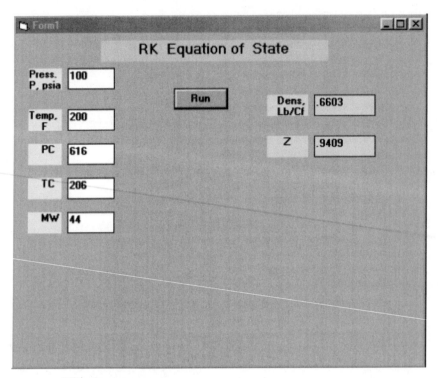

Figure 1.4   RK.mak screen.

Industrial Chemical Process Design is indeed a toolkit offering the user practical process engineering.

Having covered the difficulties of deriving an accurate gas density in Eqs. (1.15) and (1.16), it is important here to understand the practical application of same. First, for hydraulic line sizing, when the pressure of the line is 400 psig or less, consider using a conservative Z factor of 0.95 or 1.0. Look at Eq. (1.16). When Z decreases, the gas density increases, and thus the line size decreases. A conservative approach would be to use a larger Z than calculated or assume $Z = 1.0$ for a safe and conservative design. In most cases no line size increase results, while in some cases only one line size increase is the outcome. I suggest this is good practice.

I have designed many flare systems and performed numerous emergency relief valve sizing calculations applying this $Z = 1.0$ criterion. Herein I suggest you also consider using $Z = 1.0$ for all relief valve and flare line sizing. This is a conservative and safe assumption. In practice, I have found every operating company to admire the assumption even to the point of endorsing it fully.

Critical Temperature, T_C

To this point we have applied the critical temperature to both viscosity and density calculations. Already this critical property T_C is seen as valued data to have for any hydrocarbon discrete single component or a mixture of components. It is therefore important to secure critical temperature data resources as much as practical. I find that a simple table listing these critical properties of discrete components is a valued data resource and should be made available to all. I therefore include Table 1.3 listing these critical component properties for 21 of our more common components. A good estimate can be made for most other components by relating them to the family types listed in Table 1.3.

Also included here is an equation for calculating T_C, °R, using SG 60/60 and the boiling point of the unknown HC. The constants A, B, and C are given for paraffins and aromatic-type families. For naphthene, olefin, and other family-type HCs A, B, and C constants, the process engineer is referred to the API *Technical Data Book*, Chap. 4, Method 4A1.1 [13].

$$T_C = 10^\wedge [A + B * \log (\text{sp gr}) + C * \log T_B] \qquad (1.17)$$

TABLE 1.3   Critical Component Properties

Component	MW	T_B, °F	Sp gr	P_C, psia	T_C, °F	Acentric fraction
Methane	16.04	−258.69	0.3	667.8	−116.63	0.0104
Ethane	30.07	−127.48	0.3564	707.8	90.09	0.0986
Propane	44.10	−43.67	0.5077	616.3	206.01	0.1524
n-Butane	58.12	31.10	0.5844	550.7	305.65	0.2010
Isobutane	58.12	10.90	0.5631	529.1	274.98	0.1848
n-Pentane	72.15	96.92	0.6310	488.60	385.70	0.2539
Isopentane	72.15	82.12	0.6247	490.40	369.10	0.2223
Neopentane	72.15	49.10	0.5967	464.00	321.13	0.1969
n-Hexane	86.17	155.72	0.6640	436.90	453.70	0.3007
2-Methylpentane	86.17	140.47	0.6579	436.60	435.83	0.2825
3-Methylpentane	86.17	145.89	0.6689	453.10	448.30	0.2741
Neohexane	86.17	121.52	0.6540	446.80	420.13	0.2369
2,3-Dimethylbutane	86.17	136.36	0.6664	453.50	440.29	0.2495
n-Heptane	100.2	209.17	0.6882	396.80	512.80	0.3498
2-Methylhexane	100.2	194.09	0.6830	396.50	495.00	0.3336
3-Methylhexane	100.2	197.32	0.6917	408.10	503.78	0.3257
3-Ethylpentane	100.2	200.25	0.7028	419.30	513.48	0.3095
2,2-Dimethylpentane	100.2	174.54	0.6782	412.20	477.23	0.2998
2,4-Dimethylpentane	100.2	176.89	0.6773	396.90	475.95	0.3048
3,3-Dimethylpentane	100.2	186.91	0.6976	427.20	505.85	0.2840
Triptane	100.2	177.58	0.6949	428.40	496.44	0.2568

SOURCE: Data from Table 1C1.1, American Petroleum Institute, *Technical Data Book*, API Refining Department, Washington, DC, 1976.

where   sp gr = SG 60/60

   T_B = normal boiling temperature, °R

Note: Log notation is base 10.

	Constants		
	A	B	C
Paraffins	1.47115	0.43684	0.56224
Aromatics	1.14144	0.22732	0.66929
Olefins	1.18325	0.27749	0.65563

   This equation is acclaimed as good for paraffins up to 21 carbon atoms molecularly, and up to 15 carbon atoms for all others.

   Now for the best resource of all. Table 1.3 has many applications. In recent years new plant designers and plant upgrade designers have chosen many of the components shown in Table 1.3 to represent group compounds meeting the same family criteria. Each grouping may have hundreds of discrete identifiable compounds; however, only one is used to represent the entire group. Such grouping is being found to be acceptable error and most certainly is much better than a rough estimate.

Critical Pressure, P_C

Table 1.3 is also an excellent source for critical pressure P_C. If the particular HC compound or mixture is not listed in this table, consider relating it to a similar compound in Table 1.3. If molecular weight and the boiling points are known, you may find a close resemblance in Table 1.3. Also consider the *API Technical Data Book,* which lists thousands of HC compounds. Grouping as one component per se would also be feasible from Procedure 4A2.1 of the API book. Herein, components grouped together as a type of family could be represented as one component of the mixture. This one representing component may be called a *pseudocomponent.* Several of these pseudocomponents added together would make up the 100% molar sum of the mixture.

   As with T_C, I also present herein a method to calculate P_C applying the molecular group method [6].

$$P_C = \frac{14.7 * \text{MW}}{[(\text{sum DELTPI}) + 0.34]^2} \tag{1.18}$$

where P_C = critical pressure, psia

   MW = molecular weight

   DELTPI = compound molecular group structure contribution

Group Contributions DELTPI

Non-ring-increment group contributions		Ring-increment group contributions	
—CH$_2$	0.227	—CH$_2$—	0.184
—CH$_2$	0.227	—CH	0.192
═CH$_2$	0.198		
═CH	0.198	═CH	0.154
≡CH	0.153	═CH═	0.154

Note: "sum" notation indicates the sum of DELTPI for each group contribution.

Additional molecular group contributions can be found in Reid, Prausnitz, and Sherwood [14]. An example of group contributions is now run for benzene:

sp gr = 0.8844

$T_B = 176.2°F + 460 = 636.2°R$

MW = 78.11

Data taken from Ref. 14

$$T_C = 10^\wedge [A + B * \log (\text{sp gr}) + C * \log T_B] \qquad (1.17)$$

Aromatics $A = 1.14144, B = 0.22732, C = 0.66929$

$T_C = 1013°R$ or $553°F$ \qquad From Table 1.3, $T_C = 552°F$ \qquad Checks okay

$$P_C = \frac{14.7 * \text{MW}}{[(\text{sum DELTPI}) + 0.34]^2} \qquad (1.18)$$

Benzene has 6 ═CH groups at 0.154 each, or 0.924 total, and sum DELTPI = 0.924

$P_C = 719$ psia \qquad From Ref. 14, PC = 710 psia \qquad Checks okay

This example of benzene shows that given the specific gravity at 60/60, the normal atmospheric boiling temperature, and the substance molecular weight, then the T_C and P_C critical properties can be calculated. These exhibited equations, Eqs. (1.17) and (1.18), are within a few percentage points error, up to about 20 carbon atoms for paraffins and 14 carbon atoms per molecular structure for all others.

For determining P_C and T_C from a mixture, having a known P_C and T_C for each component, use molar percentages of each component times the respective P_C and T_C. Then add these P_C and T_C values to get the sum P_C and sum T_C of the mixture.

Molecular Weight

The molecular weight of a discrete component or a group mixture is a very basic and indeed needed data input for defining any component. It is mandatorily defined in any characterization or assay-type hydrocarbon analysis. Molecular weight is indeed a must for solving any fluid-transport design problem. It, together with the subject fluid's boiling point temperature, is the most important data to have or determine. I propose that molecular weight can be determined by means of two methods. Table 1.3 is again the first method proposed. If your component is not specifically listed in Table 1.3, simply estimate using the other similar family-type compounds to secure the MW.

Referring to Table 1.3, please note that molecular weight values are 120 or less for all compounds. The API *Technical Data Book* lists many more HC compounds of 120 MW or less. Compounds of this type should receive MW determinations using these tables, referring to Table 1.3 and Ref. 4.

The second method I propose to determine MW is the crude characterization method. For the past six decades, we have relied on the standard ASTM D86 distillation test to characterize crude petroleum and its products [6]. The next section includes excerpts from the ASTM4 program for crude oil characterization presented in the CD. Please note that there is a proposed MW equation on line 4690. I find this equation to be reasonably accurate, ±3% or less, for most every HC compound or HC pseudogroup above 120 MW. The ASTM4 printout in the next section, in Table 1.5, shows a run for a typical ASTM D86 lab analysis of crude oil. Use this program with caution, however, especially for compounds 100 MW or less. Errors here may exceed 10% in this region.

The ASTM4 program is derived from Fig. 2B2.1 in Chap. 2 of the API *Technical Data Book* [15]. I have derived the equation in line 4690 using a curve-fit math which checks very well with the API figure. This API book is historically a good and reliable source for ASTM crude MW determination. Thus I have included this curve-fitted equation here in the ASTM4 program as the calculation for molecular weight. As seen in ASTM4 line 4690, the API gravity and the average boiling temperature are all that is needed. These two variables, gravity and boiling point, are commonly determined in every lab ASTM analysis run for any crude oil cut, hydrocarbon, or petroleum product. The equation I have presented checks very well within the range it was intended for, ASTM D86–type distillation cuts.

One last MW note I wish to leave with the careful reader: For many years I have been asked to consult on difficult refinery problems concerning naphtha, gasoline, jet fuel, diesel, and gas oil petroleum cuts.

In each and every one of these problems, I have requested immediate petroleum light ends chromatographic and ASTM D86 lab tests from the client's lab services. In answer to half of these requests, I have been handed a document used for the design of the refinery that was at least 15 to 20 years out of date with current operations. While being handed these archaic wonders, I have been told, "We don't get the ASTM test you requested from our lab, and they don't get the proper samples to run the test you requested." I have been even further moved by the fact that these laboratory marvels didn't have the proper apparatus nor the personnel experience in their facilities to run such simple ASTM tests.

Well, I must now say that the light ends, C1, C2, C3, C4, and so on, demand a gas sample–type chromatograph laboratory test. I strongly encourage that a well-experienced and reputable lab service company be contracted to run not only the light ends but also the full ASTM distillation test involving the heavier crude for C6 and the higher boiling point cuts. As a normal service, these professional labs include the cut gravity and the boiling points of each distilled cut logged in the ASTM test.

Boiling Point

The boiling point is the last data herein sought out; however, it is indeed the most important data to secure for a discrete pure component or a pseudo–crude cut component. Since the discrete pure components are generally a known type of molecular structure, their boiling points may readily be obtained or estimated from data sets such as Table 1.3. The crude oil components are left, unfortunately, undefined. Therefore, this section is dedicated to defining the boiling points of crude oil and its products.

Over the past six decades, the petroleum industry has defined the many components making up crude oil and derived products simply by defining boiling range cuts. Each crude oil cut is generally held to a 20°F or less boiling range increment of the total crude oil sample or the crude oil product sample. Each of these boiling ranges is defined by a pseudocomponent. This is not a true single component, but rather a mixture of a type of components. Each grouped pseudocomponent mixture is treated throughout calculation evaluations as a single component. These pseudocomponents thus become the key database defining all petroleum processing equipment and processing technologies. There are two principal types of crude oil boiling point analysis. These are the American Standard Testing Method (ASTM) and true boiling point (TBP) test procedures [6].

First, the ASTM boiling analysis is actually two tests, the D86 and the D1160 [6]. The D86 is performed at atmospheric pressure, wherein the sample is simply boiled out of a container flask and totally condensed in a receiving flask. As the D86 involves cracking or molecular breakdown of the crude sample at temperatures above 500°F, this test method is extremely inaccurate at temperatures of 450°F and above. Thus, a second test method has been added to the D86, the D1160, which uses a vacuum for the same sample distillation at temperatures 450°F and above. The D1160 uses the same setup as the D86, only with an overhead vacuum, usually 40 mmHg absolute pressure. A good complete ASTM analysis of a normal virgin crude oil sample would thus involve starting with a D86 at atmospheric pressure and finishing with a vacuum D1160 when the D86 boiling temperature reaches 450°F.

In the 1930s and earlier, the D86 ASTM–type test was discovered to be inaccurate regarding defining the true boiling ranges of these so-defined pseudocomponents. This inaccuracy is totally due to the fact that every pseudocomponent boiling mixture has boiling range components from its adjacent pseudo cuts. How can this test problem be solved? The answer is simple. Just run a TBP. How can this be done? Use a fractionation-type separation lab setup, refluxing the overhead boilout, which produces a more truly defined boiling cut range pseudocomponent. Thus, for each of the cuts, 20°F or less, a more accurate database of pseudocomponents is so defined.

TBP-type lab tests are more the current-day standard of reputable labs. The Bureau of Mines of the U.S. government uses the Hempel TBP method. The ASTM commission adapted a method called the D2892. Both methods are similar, starting with overhead atmospheric pressure and finishing with vacuum, 40 mmHg or less. Both use trays or internal packing which is refluxed with overhead condensing. The fractionating column is maintained at a stabilized temperature as the temperature profiles of the column increase per distillation boilout progress. The outcome of this test is the true boiling point pseudocomponent definition.

The D2892 lab test is a rather difficult test to run, requiring extensive laboratory work and considerable specialized equipment. It is therefore most apparent that the simpler ASTM tests D86 and D1160 are preferred. No fractionator reflux is required. These ASTM distillations, compared to the more rigorous TBP tests, are more widely used. This is because the ASTM tests are simpler, less expensive, require less sample, and require approximately one-tenth as much time. Also, these ASTM tests are standardized, whereas TBP distillations vary appreciably in procedure and apparatus.

In earlier years, API set up calculation procedures to convert these more easily run ASTM D86 and D1160 boiling point curves to the sought TBP curve data. This book presents a unique program, named ASTM4, which receives ASTM curve inputs, both D86 and D1160 data, and converts them to TBP data. API 3A1.1 and API 3A2.1 methods are referenced. Referencing ASTM4 starting at line 1660 (Table 1.6 in the next subsection), the ASTM 50% points are converted to the TBP 50% points. Then TBP point segment differences with the ASTM D86 and D1160 segment points are established. The end result is that the user may input ASTM D86 and 1160 data into the program and derive answers as though the input were TBP data.

ASTM4 was written by the author and is based on derived mathematical algorithms which simulate the method in Chap. 3 of the API *Technical Data Book* [7]. The reader is referred to the API book for a more detailed derivation of the results from the API curves. Curvefitted equations simulating these API curves are installed in the ASTM4 program and give reasonably close or identical results. This was close to an exhaustive method of deriving a program for generating the needed crude oil pseudocomponent data. But, in due respect to all of those very finite, reliable, accurate, and costly simulation programs, ASTM4 does not equal their perfection. However, as the quest of this book is practical process engineering, ASTM4 produces reasonable and similar results as these high-end programs produce. I, the author, do bow to their excellence, but also imply that ASTM4 earns the right to be compared and in some cases may even produce equal results for the more complex and difficult crude oil runs. Please note we have been comparing and reviewing the API crude oil characterization method. There are several other crude characterization methods, including those of Edmister and Cavit. These however have all been compared, and the API data method upgraded ever closer to perfection as time has allowed.

ASTM4.exe is a PC-format computer program for the ASTM D86 and D1160, or the TBP curve point input. The TBP method is that of the API group, having at least 15 theoretical stages and at least a 5-to-1 reflux ratio or greater. A typical example of ASTM4 input is given in Table 1.4, followed by a refinery gasoline stabilizer bottoms cut having a 10,000-barrel-per-day (bpd) flow rate in Table 1.5. Please note that there are nine points input. Each point is given an ASTM boiling point from a D86 curve and the volume percent that has boiled over into the condensate flask and the accumulative °API gravity reading of the boiled-over fluid.

Please note that nine ASTM curve points are input. Also note that this ASTM D86 lab test is extended well over the 550°F temperature, which is the component cracking and degrading temperature. It would have been much more prudent to have stopped the atmospheric distillation at 550 and used a vacuum procedure such as the previously discussed D1160. A disclaimer, however, is made here for this being an actual lab test in which the lab indeed made a D1160 test for all components above the 500°F temperature and converted all the results to a simple nine-point single ASTM D86 test result as shown. The result

TABLE 1.4   ASTM4 Input

No.	°F	Vol, %	API	Dist, mmHg
1	290	10	50	760
2	380	20	40	760
3	450	30	36	760
4	510	40	32	760
5	575	50	30	760
6	640	60	29	760
7	710	70	24	760
8	760	80	23	760
9	840	99	18	760

NOTE: ASTM curve points input = 9; mid-BP curve point = 5; bpd rate = 10,000.

TABLE 1.5   ASTM4 Output Answers

No.	mol/h	T_C, °R	P_C, psia	MW	Acent	API	Vol BP, °F
01	1.6400e+02	1019	475	116.2	0.29779	49.2	220
02	1.7352e+02	1042	462	120.32	0.31021	47.4	240
03	1.6921e+02	1066	449	124.65	0.32179	45.6	260
04	1.6496e+02	1090	436	129.18	0.33258	43.8	280
05	1.6077e+02	1114	423	133.93	0.34262	42.0	300
06	1.5663e+02	1137	411	138.91	0.35197	40.2	320
07	1.7803e+02	1159	403	144.65	0.37321	39.2	340
08	1.7171e+02	1180	389	150.73	0.38808	38.4	360
09	1.6557e+02	1200	375	157.1	0.40321	37.5	380
10	1.5962e+02	1221	361	163.78	0.41868	36.7	400
11	1.6776e+02	1241	348	170.75	0.4345	35.8	420
12	1.6186e+02	1261	336	177.96	0.4506	34.9	440
13	1.5618e+02	1281	324	185.46	0.46729	34.0	460
14	1.5074e+02	1300	312	193.23	0.48471	33.0	480
15	1.4554e+02	1319	302	201.28	0.50298	32.1	500
16	1.4203e+02	1336	289	210.15	0.5236	31.6	520
17	1.3640e+02	1353	276	219.46	0.54539	31.1	540
18	1.3102e+02	1369	264	229.14	0.5683	30.6	560
19	1.2588e+02	1385	253	239.19	0.59248	30.2	580
20	1.2028e+02	1400	241	250.04	0.61847	29.8	600
21	1.1513e+02	1415	229	261.59	0.64604	29.6	620
22	1.1023e+02	1429	217	273.63	0.67509	29.4	640
23	1.0555e+02	1442	206	286.17	0.70577	29.1	660
24	1.0143e+02	1456	197	298.47	0.73896	28.5	680
25	9.8444e+01	1472	191	309.8	0.77647	27.4	700
26	9.5679e+01	1486	186	321.13	0.81838	26.2	720
27	9.3139e+01	1501	181	332.36	0.86551	25.0	740
28	1.1644e+02	1514	176	343.76	0.9183	24.0	760
29	1.1212e+02	1524	168	357.68	0.97295	23.7	780
30	1.0806e+02	1534	159	371.85	1.03346	23.4	800
31	1.0424e+02	1544	151	386.23	1.10091	23.1	820
32	1.4121e+02	1555	147	397.81	1.18605	22.1	840
33	1.3862e+02	1565	144	408.15	1.28789	21.0	860
34	2.0792e+02	1574	142	417.9	1.40839	19.9	880

Totals: mol/h = 4.7499e+02; lb/h = 1.0974e+05; MW = 231.0

is therefore a smooth curve with nine points taken. The points above the cracking temperature, 500°F, are actually tested under a vacuum and converted to the shown ASTM D86 test. No cracking has occurred in this test. Aren't professional labs wonderful?

ASTM4 has supplied you with pseudocomponent characterization, molecular weights, acentric factors, critical constants, boiling points, and pseudocomponent gravity. With this database you are prepared to resolve most any crude oil and products database problem, deriving calculated needed results.

Crude oil characterization— brief description of ASTM4

Some may ask why I've named this program ASTM4. It is indeed a computer computation method to define or simply characterize petroleum and its products. But why the *4?* Because this is my fourth-generation upgrade of the program. The following will give the reader a brief ASTM4 program walkthrough:

Making the distillation curve and API gravity curve.   ASTM4 is set up for an ASTM D86 distillation curve input. Although the TBP curve input could be set up as an option, this has not been done. The user could, however, make this option input if desired. The DOS version of ASTM4 does indeed offer this option of TBP input. I suggest that the user, if desired, may follow the same pattern as shown in the DOS ASTM4 version, adding a few option steps.

The first lines of the program, to line number 2350, convert the ASTM data points to a TBP database. (See the program code listing in Table 1.6.) As seen in these line codes, curve-fitted equations are applied extensively. The bases used are API *Technical Data Book* Figs. 3A1.1, 3A2.1, 3B1.1, 3B1.2, 3B2.1, and 3B2.2. Code lines 2370 through 2400 correct the ASTM distillation for subatmospheric pressures. Instead of inputting 760 mmHg as shown in the example, you could also input 0 for the same results. The reason for a 0 input is line 2370, where this correction is bypassed. The reference for this subatmospheric pressure correction is the API Fig. 3B2.2 routine. The equations shown are curve-fitted. In fact, this could be a D1160 distillation conversion to the TBP database. If the user desires to make an optional TBP direct database input in place of an ASTM database input, then skip from line 1650 to 2360. The DOS ASTM4 version in the CD disk offers this option.

The next lines of the program, to line number 2480, establish curve-fitted equations. Refer to code lines 2420 through 2490. (See the program code listing in Table 1.7.) Note that linear equations between each of the given ASTM data points are made, giving an overall ASTM

TABLE 1.6   ASTM4 Program Code, Lines 1040 to 2350

```
40 MCT = 0: WCT = 0
1040 For I = 1 To N2
1610 If B(I, 1) <= 475 GoTo 1640
1620 DELTA = Exp(.00473 * B(I, 1) - 1.587)
1630 B(I, 1) = B(I, 1) + DELTA
1640 Next I
1650 If AA1$ = "TBP" GoTo 2360
1660 Rem ASTM50 CONVERSION TO TBP50, API 3A1.1
1670 ASTM50 = B(NI2, 1)
1680 If ASTM50 < 400 Then TBP50 = ASTM50 + (.04 * ASTM50 - 16)
1685 If TBP50 = ASTM50 + (.04 * ASTM50 - 16) GoTo 1730
1690 If ASTM50 < 600 Then TBP50 = ASTM50 + (.06 * ASTM50 - 24)
1695 If TBP50 = ASTM50 + (.06 * ASTM50 - 24) GoTo 1730
1700 If ASTM50 < 700 Then TBP50 = ASTM50 + (.088 * ASTM50 - 40.8)
1705 If TBP50 = ASTM50 + (.088 * ASTM50 - 40.8) GoTo 1730
1710 If ASTM50 < 800 Then TBP50 = ASTM50 + (.155 * ASTM50 - 87.7)
1715 If TBP50 = ASTM50 + (.155 * ASTM50 - 87.7) GoTo 1730
1720 If ASTM50 < 900 Then TBP50 = ASTM50 + (.237 * ASTM50 - 153.3)
1730 Rem ASTM POINT SEGMENT DIFFERENCES, DELA(I)
1740 For I = 1 To N2 - 1
1750 DELA(I) = B(I + 1, 1) - B(I, 1)
1760 Next I
1770 Rem CALC OF DELB(I),TBP POINT SEGMENT DIFFERENCES, API 3A1.1
1780 For I = 1 To N2 - 1
1790 If DELA(I) > 20 GoTo 1850
1800 If B(I, 2) <= 30 Then DELB(I) = 2 * DELA(I)
1805 If DELB(I) = 2 * DELA(I) GoTo 2240
1810 If B(I, 2) <= 50 Then DELB(I) = 1.7 * DELA(I)
1815 If DELB(I) = 1.7 * DELA(I) GoTo 2240
1820 If B(I, 2) <= 70 Then DELB(I) = 1.5 * DELA(I)
1825 If DELB(I) = 1.5 * DELA(I) GoTo 2240
1830 If B(I, 2) <= 90 Then DELB(I) = 1.4 * DELA(I)
1835 If DELB(I) = 1.4 * DELA(I) GoTo 2240
1840 If B(I, 2) <= 100 Then DELB(I) = 1.165 * DELA(I)
1845 If DELB(I) = 1.165 * DELA(I) GoTo 2240
1850 If DELA(I) > 40 GoTo 1920
1860 If B(I, 2) <= 10 Then DELB(I) = 1.385 * DELA(I) + 12.6
1865 If DELB(I) = 1.385 * DELA(I) + 12.6 GoTo 2240
1870 If B(I, 2) <= 30 Then DELB(I) = 1.3 * DELA(I) + 13!
1875 If DELB(I) = 1.3 * DELA(I) + 13! GoTo 2240
1880 If B(I, 2) <= 50 Then DELB(I) = 1.25 * DELA(I) + 9!
1885 If DELB(I) = 1.25 * DELA(I) + 9! GoTo 2240
1890 If B(I, 2) <= 70 Then DELB(I) = 1.235 * DELA(I) + 5.6
1895 If DELB(I) = 1.235 * DELA(I) + 5.6 GoTo 2240
1900 If B(I, 2) <= 90 Then DELB(I) = 1.135 * DELA(I) + 5.3
1905 If DELB(I) = 1.135 * DELA(I) + 5.3 GoTo 2240
```

TABLE 1.6   ASTM4 Program Code, Lines 1040 to 2350 (Continued)

```
1910 If B(I, 2) <= 100 Then DELB(I) = 1.04 * DELA(I) + 2.4
1915 If DELB(I) = 1.04 * DELA(I) + 2.4 GoTo 2240
1920 If DELA(I) > 60 GoTo 2080
2020 If B(I, 2) <= 10 Then DELB(I) = 1.25 * DELA(I) + 18!
2025 If DELB(I) = 1.25 * DELA(I) + 18! GoTo 2240
2030 If B(I, 2) <= 30 Then DELB(I) = 1.02 * DELA(I) + 24.2
2035 If DELB(I) = 1.02 * DELA(I) + 24.2 GoTo 2240
2040 If B(I, 2) <= 50 Then DELB(I) = 1.05 * DELA(I) + 17!
2045 If DELB(I) = 1.05 * DELA(I) + 17! GoTo 2240
2050 If B(I, 2) <= 70 Then DELB(I) = 1! * DELA(I) + 15!
2055 If DELB(I) = 1! * DELA(I) + 15! GoTo 2240
2060 If B(I, 2) <= 90 Then DELB(I) = .975 * DELA(I) + 11.5
2065 If DELB(I) = .975 * DELA(I) + 11.5 GoTo 2240
2070 If B(I, 2) <= 100 Then DELB(I) = 1.1 * DELA(I)
2075 If DELB(I) = 1.1 * DELA(I) GoTo 2240
2080 If DELA(I) > 80 GoTo 2150
2090 If B(I, 2) <= 10 Then DELB(I) = 1.225 * DELA(I) + 19.5
2095 If DELB(I) = 1.225 * DELA(I) + 19.5 GoTo 2240
2100 If B(I, 2) <= 30 Then DELB(I) = .825 * DELA(I) + 36.2
2105 If DELB(I) = .825 * DELA(I) + 36.2 GoTo 2240
2110 If B(I, 2) <= 50 Then DELB(I) = .91 * DELA(I) + 25.2
2115 If DELB(I) = .91 * DELA(I) + 25.2 GoTo 2240
2120 If B(I, 2) <= 70 Then DELB(I) = .9 * DELA(I) + 21!
2125 If DELB(I) = .9 * DELA(I) + 21! GoTo 2240
2130 If B(I, 2) <= 90 Then DELB(I) = .91 * DELA(I) + 15.4
2135 If DELB(I) = .91 * DELA(I) + 15.4 GoTo 2240
2140 If B(I, 2) <= 100 Then DELB(I) = 1.3 * DELA(I) - 12!
2145 If DELB(I) = 1.3 * DELA(I) - 12! GoTo 2240
2150 If DELA(I) > 120 GoTo 2200
2160 If B(I, 2) <= 30 Then DELB(I) = .8575 * DELA(I) + 33.4
2165 If DELB(I) = .8575 * DELA(I) + 33.4 GoTo 2240
2170 If B(I, 2) <= 50 Then DELB(I) = .975 * DELA(I) + 28!
2175 If DELB(I) = .975 * DELA(I) + 28! GoTo 2240
2180 If B(I, 2) <= 70 Then DELB(I) = .9125 * DELA(I) + 20.5
2185 If DELB(I) = .9125 * DELA(I) + 20.5 GoTo 2240
2190 If B(I, 2) <= 90 Then DELB(I) = .9375 * DELA(I) + 13!
2195 If DELB(I) = .9375 * DELA(I) + 13! GoTo 2240
2200 If B(I, 2) <= 30 Then DELB(I) = 1.1333 * DELA(I) - 1!
2205 If DELB(I) = 1.1333 * DELA(I) - 1! GoTo 2240
2210 If B(I, 2) <= 50 Then DELB(I) = 1.1333 * DELA(I) - 4!
2215 If DELB(I) = 1.1333 * DELA(I) - 4! GoTo 2240
2220 If B(I, 2) <= 70 Then DELB(I) = 1.1433 * DELA(I) - 8.8
2225 If DELB(I) = 1.1433 * DELA(I) - 8.8 GoTo 2240
2230 If B(I, 2) <= 90 Then DELB(I) = 1.155 * DELA(I) - 14.4
2240 Next I
2250 B(NI2, 1) = TBP50: TDELB = 0
```

TABLE 1.6   **ASTM4 Program Code, Lines 1040 to 2350 (Continued)**

```
2260 Rem ASTM B(I,1) CONV TO TBP B(I,1)
2270 For I = (NI2 - 1) To 1 Step -1
2280 TDELB = TDELB + DELB(I)
2290 B(I, 1) = B(NI2, 1) - TDELB
2300 Next I
2310 TDELB = 0
2320 For I = NI2 + 1 To N2
2330 TDELB = TDELB + DELB(I - 1)
2340 B(I, 1) = B(NI2, 1) + TDELB
2350 Next I
```

curve. Herein every set of two data point inputs generates a linear equation, as shown in code lines 2430, 2440, 2460, and 2470. Please note that lines 2460 and 2470 generate the API curve. The other two, 2430 and 2440, of course generate the ASTM distillation curve. The user should keep in mind to be generous with curve points input whenever the curve has a sharp deviation. This will help the program produce a better accuracy. In fact, the more the ASTM curve points input, the better the accuracy.

The next lines, 2500 through 2800, make the pseudocomponent TBP curve cuts. (See Table 1.8.) A 20°F segment cut is assumed for each component. Thus, each of the pseudocomponents are herein defined, applying a 20°F segment cut for each. See code line 2520 where the 20-degree segment is first started. The user may optionally install other segments such as a 15- or 10-degree segment cut by replacing each of these 20s as a variable and inputting the variable on form1. Please note that the ASTM4 example run has 34 pseudocomponents gener-

TABLE 1.7   **ASTM4 Program Code, Lines 2410 to 2490**

```
2410 For I = 1 To (N2 - 1)
2415 J = I + 1
2420 Rem CALC CONSTANTS FOR ASTM CURVE, ACx+C=y
2430 AC(I) = (B(I, 2) - B(J, 2)) / (B(I, 1) - B(J, 1))
2440 C(I) = ((B(I, 1) * B(J, 2)) - (B(I, 2) * B(J, 1))) /
     (B(I, 1) - B(J, 1))
2450 Rem CALC CONSTANTS FOR API CURVE, ADx+D=y
2460 AD(I) = (B(I, 2) - B(J, 2)) / (B(I, 3) - B(J, 3))
2470 D(I) = ((B(I, 3) * B(J, 2)) - (B(I, 2) * B(J, 3))) /
     (B(I, 3) - B(J, 3))
2480 If J = N2 Then N3 = ((B(J, 1) - B(1, 1)) / 20) - 1
2490 Next I
```

TABLE 1.8   ASTM4 Program Code, Lines 2500 to 2800

```
2500 J = 1: AT = 0
2510 Y1 = 80
2520 Y1 = Y1 + 20
2530 If B(1, 1) >= Y1 And B(1, 1) < Y1 + 20 Then A(1, 1) = Y1
2535 If Y1 <> A(1, 1) GoTo 2520
2540 'GoTo 2520
2550 For I = 1 To N3
2560 Rem CALC VBP
2570 A(I, 1) = (A(1, 1) - 20) + (I * 20)
2580 If A(I, 1) >= B(J + 1, 1) Then J = J + 1
2590 Rem CALC API GRAVITY, A(I,2), OF PSEUDO COMPONENT
2600 Y2 = AC(J) * A(I, 1) + C(J)
2610 A(I, 2) = (Y2 - D(J)) / AD(J)
2620 Rem CALC 20 DEG F VOL% SEGMENTS, A(I,3)
2630 Y1 = AC(J) * (A(I, 1) - 10) + C(J)
2640 If I = 1 Then Y1 = AC(J) * B(1, 1) + C(J)
2650 If Y1 < 0 Then Y1 = 0
2660 Y3 = AC(J) * (A(I, 1) + 10) + C(J)
2710 If Y3 < 0 Then A(1, 1) = A(1, 1) + 20
2712 If Y3 < 0 Then J = 1
2715 If Y3 < 0 GoTo 2550
2760 A(I, 3) = Y3 - Y1
2770 If A(I, 3) < 0 Then A(1, 1) = A(1, 1) + 20
2702 If A(I, 3) < 0 Then J = 1
CRUDE OIL CHARACTERIZATION
2705 If A(I, 3) < 0 GoTo 2550
2780 AT = AT + A(I, 3)
2790 If I = N3 Then A(I, 3) = 100 - AT - B(1, 2)
2795 API2 = API2 + A(I, 2) * A(I, 3) / 100
2800 Next I
```

ated, because 34 calculated in the program is the variable N3. See code line 2480. If you do choose to go to this higher pseudocomponent count, don't forget to up the variable array dimensions in the Module1.bas component listings. Only 9 ASTM data points are used, as this is reasonably close to being a linear curve.

ASTM4 T_C method

The critical temperature T_C of each pseudocomponent is calculated using the API *Technical Data Book,* Eq. 4D1.1 [16]. This equation is shown here in Table 1.9, code lines 3260 through 3360. Normally, this equation is good for most all types of hydrocarbons, having an error estimation of $\pm6°F$. This equation has been noted to have a maximum

TABLE 1.9   ASTM4 Program Code, Lines 3260 to 4780

```
3260 For I = 1 To N3
3320 VBP = A(I, 1): API = A(I, 2): VL = A(I, 3)
3330 Rem CALC OF TC,DEG F
3340 SG = 141.5 / (131.5 + API): DELTA = SG * (VBP + 100)
3350 TC = 186.16 + 1.6667 * DELTA - .0007127 * (DELTA ^ 2)
3360 A(I, 4) = TC + 459.69
3370 Rem CALC OF MOLAR BP & MEAN BP,OR MOLBP & MNBP, DEG F
3380 M0 = -11981 + 1804.15 * Log(TC)
3390 M1 = -8467.099 + 1313.4 * Log(TC)
4400 M2 = 66.603 * Exp(.0022612 * TC)
4410 M3 = 60.7447 * Exp(.00244804 * TC)
4420 M4 = 55.2457 * Exp(.00267255 * TC)
4430 M5 = 57.712 * Exp(.0027004 * TC)
4440 If API <= 10 Then MOLBP = M0 + ((API / 10) * (M1 - M0))
4445 If MOLBP = M0 + ((API / 10) * (M1 - M0)) GoTo 4500
4450 If API <= 20 Then MOLBP = M1 + (((API - 10) / 10) * (M2 - M1))
4454 If MOLBP = M1 + (((API - 10) / 10) * (M2 - M1)) GoTo 4500
4460 If API <= 30 Then MOLBP = M2 + (((API - 20) / 10) * (M3 - M2))
4465 If MOLBP = M2 + (((API - 20) / 10) * (M3 - M2)) GoTo 4500
4470 If API <= 40 Then MOLBP = M3 + (((API - 30) / 10) * (M4 - M3))
4475 If MOLBP = M3 + (((API - 30) / 10) * (M4 - M3)) GoTo 4500
4480 If API <= 50 Then MOLBP = M4 + (((API - 40) / 10) * (M5 - M4))
4485 If MOLBP = M4 + (((API - 40) / 10) * (M5 - M4)) GoTo 4500
4490 If API > 50 Then MOLBP = M5 + (((API - 50) / 10) * (M5 - M4))
4495 If MOLBP = M5 + (((API - 50) / 10) * (M5 - M4)) GoTo 4500
4500 MNBP = (MOLBP + VBP) / 2: T = MNBP
4510 Rem CALC OF PC,PSIA
4520 P0 = 3738.38 - 515.8401 * Log(T)
4530 P1 = 3442.78 - 481.66 * Log(T)
4540 P2 = 3093.76 - 437.39 * Log(T)
4550 P3 = 2800.98 - 400.02 * Log(T)
4560 P4 = 2538.3 - 365.56 * Log(T)
4570 P5 = 2299.38 - 333.611 * Log(T)
4580 P6 = 2090.985 - 305.666 * Log(T)
4590 If API <= 10 Then PC = P0 - ((API / 10) * (P0 - P1))
4595 If PC = P0 - ((API / 10) * (P0 - P1)) GoTo 4660
4600 If API <= 20 Then PC = P1 - (((API - 10) / 10) * (P1 - P2))
4605 If PC = P1 - (((API - 10) / 10) * (P1 - P2)) GoTo 4660
4610 If API <= 30 Then PC = P2 - (((API - 20) / 10) * (P2 - P3))
4615 If PC = P2 - (((API - 20) / 10) * (P2 - P3)) GoTo 4660
4620 If API <= 40 Then PC = P3 - (((API - 30) / 10) * (P3 - P4))
4625 If PC = P3 - (((API - 30) / 10) * (P3 - P4)) GoTo 4660
4630 If API <= 50 Then PC = P4 - (((API - 40) / 10) * (P4 - P5))
4635 If PC = P3 - (((API - 30) / 10) * (P3 - P4)) GoTo 4660
4640 If API <= 60 Then PC = P5 - (((API - 50) / 10) * (P5 - P6))
4645 If PC = P5 - (((API - 50) / 10) * (P5 - P6)) GoTo 4660
```

TABLE 1.9   ASTM4 Program Code, Lines 3260 to 4780 (Continued)

```
4650 If API > 60 Then PC = P6 - (((API - 60) / 10) * (P5 - P6))
4660 A(I, 5) = PC
4670 Rem CALC OF MOLECULAR WEIGHT, MW
4680 TR = T + 460
4690 MW = 204.38 * (Exp(.00218 * TR)) * (Exp(-3.07 * SG))
     * (TR ^ .118) * (SG ^ 1.88)
4700 A(I, 6) = MW
4710 Rem CALC OF ACENTRIC FACTOR,ACC
4720 ACC = (3 / 7) * (((Log(PC) / Log(10)) - (Log(14.7) /
     Log(10))) / (((TC + 460) / (VBP + 460)) - 1)) - 1
4730 A(I, 7) = ACC
4740 Rem CALC COMP LB/HR,WC, AND MOL/HR,MC
4750 WC = (VL * BPDF / 100) * 42 * 8.34 * SG / 24: WCT = WCT + WC
4760 MC = WC / MW: A(I, 8) = MC
4770 MCT = MCT + MC
4780 Next I
```

$22°F$ deviation on a few component findings. These deviations are deemed acceptable and justified.

API Limitations of the 4D1.1 T_c Equation

T_C	550 to 1000°F
P_C	250 to 700 psia
API gravity	85 to 11

I have found these limitations as set by API to be very conservative. Having accomplished several crude characterizations applying this method, I have found an extended valid range. I therefore submit the following changes as replacements.

Proposed Limitations of the 4D1.1 T_c Equation

T_C	350 to 1000°F
P_C	200 to 900 psia
API gravity	8 to 85

ASTM4 P_C method

This critical pressure P_C method is a curve fit of the API Fig. 4D4.1 [17]. Rather than just one curve-fit equation, however, the reader will soon realize there are seven curve-fit equations, code lines 4520 through 4580. Also, please realize that the mean average boiling point of each pseudocomponent is first calculated in code lines 3370 through

4510. The API figures referenced are 4D3.1 and 4D3.2. Here the six equations, code lines 3380 through 4430, calculate the molal average boiling point. The mean average boiling point is then calculated on line 4500. Using this calculated mean average BP for each pseudo component, P_C is next calculated in code lines 4510 through 4660. The API Fig. 4D4.1 is the chosen basis wherein seven equations are applied to derive a very close curve fit duplicating most every API figure answer with rigorous equations.

It is important to note that all of the equations applied here are selected by windowing the API values in code lines 4440 to 4495 and 4590 to 4650.

The API recommended limitation of this method is simply do not exceed the Fig. 4D4.1 limits.

API Limitations of the 4D4.1 PC Figure

API gravity	5 to 90
Mean average BP	100 to 1200

Acentric Factor

ASTM4 acent method

Having a TC, PC, and a volumetric boiling point for each pseudocomponent, the respective acentric factor of each component is next calculated using the well-known equation from Edmister, shown in code line 4720 [18].

Enthalpy

Please follow this unique method of enthalpy derivation, as no other such methodology has been presented to date, per the best of my research.

The last database of the eight key data items promised is enthalpy. I have broadly used the term *enthalpy* to signify all thermal properties that include specific heat, latent heat, and an absolute enthalpy value, expressed as Btu/lb. This section presents a table which, by interpolation, may be applied to any single component or component mixtures, or to any petroleum characterized component groupings. This enthalpy source table (Table 1.10) may be used conveniently and quickly to derive energy or heat values of both liquids and gases. It is compiled from data in Maxwell (pp. 98 to 127) [5].

The enthalpy source table (Table 1.10) offers data heat value points which are easily interpolated between. First, pure components are listed, and then petroleum fractions are listed, as in Maxwell's book.

TABLE 1.10   Enthalpy Source, btu/lb

Component or petroleum fraction, MW	Two-phase zone				Critical tempera-ture, °F	Single-phase zone			
	Temperature, °F					Pressure, atm @ temperature, °F			
						1.0 @ 600	100 @ 600	1.0 @ 1000	100 @ 1000
	−200	100	200	400					
Methane, 16					−116°F				
Sat. vap.	200	356*	415*	543*		690		1037	
Sat. liq.	0	322†	387†	523†			680		1037
Ethane, 30					90°F				
Sat. vap.	230	336*	377*	491*		623		938	
Sat. liq.	0	200†	300†	440†			600		925
Propane, 44					206°F				
Sat. vap.	220	311	362	470*		597		910	
Sat. liq.	0	171	245	410†			563		889
Butane, 58.1					306°F				
Sat. vap.	217	313	393	465*		594		895	
Sat. liq.	0	160	225	325†			545		870
I-Butane, 58.1					275°F				
Sat. vap.	200	279	315	445*		574		882	
Sat. liq.	0	152	218	367†			557		850
Ethylene, 28					50°F				
Sat. vap.	220	315*	355*	450*		563		830	
Sat. liq.	0	203†	300†	420†			545		817
Propylene, 42					197°F				
Sat. vap.	250	302	360*	455*		568		840	
Sat. liq.	0	170	232†	402†			538		816
Isobutene, 56.1					293°F				
Sat. vap.	215	305	335	403*		573		850	
Sat. liq.	0	160	225	370†			528		825
Pentane, 72.1					387°F				
Sat. vap.	220	306	345	457*		585		888	
Sat. liq.	95	152	215	345†			520		850
I-Pentane, 72.1					370°F				
Sat. vap.	257	293	331	445*		575		882	
Sat. liq.	93	150	212	347†			520		857
Hexane, 86.2					455°F				
Sat. vap.	260	300	343	425		578		882	
Sat. liq.	90	145	207	350			500		847
Heptane, 100.2					513°F				
Sat. vap.	255	295	338	433		570		874	
Sat. liq.	87	140	200	388			487		834
Octane, 114.2					565°F				
Sat. vap.	250	288	332	432		566		867	
Sat. liq.	85	137	195	380			480		820
Benzene, 78.1					551°F				
Sat. vap.	265	290	317	375		477		692	
Sat. liq.	65	107	202	256			372		657
Toluene, 92.1					609°F				
Sat. vap.	254	280	310	376		480		700	
Sat. liq.	63	103	197	253			370		643
200 avg. BP, 95					516°F				
Sat. vap.	170	207	250	337		470		764	
Sat. liq.	0	108	108	242			390		713

TABLE 1.10   Enthalpy Source, btu/lb (Continued)

Component or petroleum fraction, MW	Two-phase zone Temperature, °F				Critical tempera- ture, °F	Single-phase zone Pressure, atm @ temperature, °F			
	−200	100	200	400		1.0 @ 600	100 @ 600	1.0 @ 1000	100 @ 1000
250 avg. BP, 116					570°F				
Sat. vap.	165	203	245	337		415‡		757‡	
Sat. liq.	0	50	105	283			380‡		704‡
300 avg. BP, 128					624°F				
Sat. vap.	162	200	241	335		460‡		750‡	
Sat. liq.	0	50	102	227			380‡		700‡
400 avg. BP, 160					728°F				
Sat. vap.	152	193	235	332		442‡		740‡	
Sat. liq.	0	45	96	217			387‡		689‡
500 avg. BP, 196					828°F				
Sat. vap.	150	185	225	322		440‡		730‡	
Sat. liq.	0	43	94	208			350‡		678‡
600 avg. BP, 245					923°F				
Sat. vap.	142	178	218	315		468‡		719‡	
Sat. liq.	0	43	92	204			339‡		667‡
800 avg. BP, 373					1104°F				
Sat. vap.	130	164	204	299		415‡		693‡	
Sat. liq.	0	42	86	193			323‡		645‡

* Enthalpy value is in the single-phase zone and is at temperature shown in the column at 1.0 atm pressure.
† Enthalpy value is in the single-phase zone and is at temperature shown in the column at 100.0 atm. pressure.
‡ Enthalpy value is in the two-phase zone. In the 1.0-atm column, the value is the saturated vapor enthalpy. In the 100.0-atm column, the value is the saturated liquid enthalpy.

Here the reader is encouraged to observe how easy it is to interpolate between the given points. Thus, the enthalpy of any vapor, and the enthalpy of any liquid, may rapidly be determined.

Maxwell divides the enthalpy of petroleum fractions into charts of mean average boiling points and the fraction's characterization factor. The boiling point is indeed a good identification. Maxwell, however, further considers two values of the petroleum fraction's characterization factor, 11.0 and 12.0. Characterization factor is defined as taking the cube root of the mean average boiling point and dividing this cube root value by the specific gravity of the petroleum fraction. A brief review of Maxwell's enthalpy charts of this type, as shown in Fig. 1.5, indicates very marginal enthalpy differences between these two characterization factors. A 3 to 5% variance of enthalpy is typical between the 11.0 and 12.0 characterization factors, the 12.0 factor having the higher enthalpy. This book exhibits only one, the 12.0 characterization factor enthalpies in Table 1.10. The more practical and the more conservative is the 12.0 characterization factor of

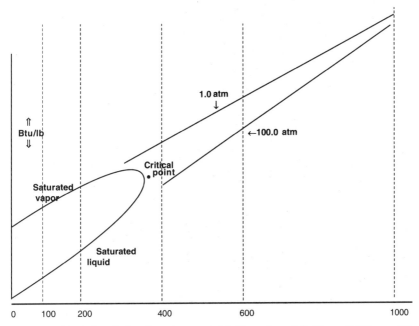

Figure 1.5 Typical enthalpy chart (not to scale). (*Data from J. B. Maxwell, "Thermal Properties,"* Data Book on Hydrocarbons, *D. Van Nostrand, Princeton, NJ, 1957, pp. 98–127.*)

petroleum fractions. As is the case in all of these tabled enthalpy values, linear interpolations may be taken between these boiling point values for any in-between petroleum fraction enthalpy value desired.

Those of you wanting to define a yet closer enthalpy, more accurate for the 11.0 characterization factor, may do so. Take your 12.0 characterization enthalpy reading and subtract a 3% value from it for enthalpies below the critical temperature. Subtract a 5% enthalpy value from readings above the critical point. This simple subtraction at most all points should give you a reasonably close enthalpy reading for the 11.0 characterization factor. Please note, however, that making this correction could be a fruitless exercise, as accuracies here do not merit such detailing.

Figure 1.5 reveals the data curves that Table 1.10 would shape if plotted. Please note the x and y axes are linear. This is a typical replica of the enthalpy charts in Maxwell's *Data Book on Hydrocarbons* [19]. As old as Maxwell's book is, it is still one of the more sought out and technically notable books of our time. Compare Fig. 1.5 to Table 1.10. Observe how Table 1.10 may be used in easy interpolations for any enthalpy reading, liquid or gas state, and pressures from vacuum to 100 atm. Please note in Fig. 1.5 also how the 1.0-atm curve joins the saturation vapor line; same for the 100-atm curve, how it joins the saturated liquid line. Thus, for each component or petroleum fraction, two curves could be realized. Both curves begin at the saturated vapor and liquid low temperature region and end in the upper right of the chart in the supercritical single-phase zone.

If you followed the quest of the previous paragraphs, realize that these referenced Maxwell enthalpy charts and Table 1.10 are near directly and totally governed by temperature alone. Observe how Fig. 1.5 lays all the enthalpies to display two curves plotted as enthalpy vs. temperature. Both curves start at 0°F and end at 1000°F. With the two pressure curves as shown in Fig. 1.5, one can determine any enthalpy value, gas or liquid. You simply need one temperature. Pressure-based interpolation may then be made linearly between the two temperature intercept points of these two curves as shown in Fig. 1.5. Please note that the dashed temperature lines are the same as the column temperatures given in Table 1.10. Thus, Table 1.10 may be used just as if one were using the curve types of Fig. 1.5 to derive enthalpy values. Table 1.10 is proposed as an improved, easier-to-read resource as compared to a curve-plotted chart. The table gives an advanced get-ahead step, giving you the curve points to read to make your interpolation.

Now for those of you who may say, "This table has its place but what about enthalpy values far to the right of this saturated liquid line or far to the left of this saturated vapor line?" Chap. 2 addresses this question most specifically. Please consider the fact that pressure increase has little effect on liquid enthalpy at constant temperature. Similarly, notice how pressure lines tend to converge on the vapor dew point line of the phase envelope. This indicates that at constant temperature, increasing the pressure of vapor tends to approach the enthalpy value of the saturation vapor dew point line of the phase envelope of a P vs. H enthalpy figure. See Fig. 2.1 in Chap. 2.

Interpolation for discrete known components appears easy and obvious in applying Table 1.10. The molecular weight of any component may easily be matched to the nearest tabled component of same molecular weight or interpolated between any two molecular weights. Here molecular weight is stated; however, the boiling points may also be used equally for accurate interpolation.

Mixture enthalpy interpolation

What about mixtures of components? How is interpolation to be performed? Consider a problem:

Example For hydrocarbon mixture enthalpy interpolation: The enthalpy of a mixture as a liquid at 100°F is to be calculated. The enthalpy of the same mixture is to be determined at 1 atm and 500°F. The mixture analysis is as follows:

Composition	Mol fractionation
Ethane	0.10
Propane	0.50
Butane	0.10
Ethylene	0.05
Propylene	0.25
	1.00
MW =	42.7

The molecular weight of the mixture is 42.7. Since the MW of propane is 44, take first the enthalpy value of propane liquid at 100°F, which is 171 Btu/lb. Next take the saturated liquid enthalpy of ethane (30 MW) at 100°F, which is 200 Btu/lb. The following interpolation calculation is made.

solution
$$\frac{44 - 42.7}{44 - 30} \times (200 - 171) + 171 = 173.7 \text{ Btu/lb}$$
$$\times \text{ MW of } 42.7 = 7417 \text{ Btu/lb mol}$$

Maxwell shows a 7480-Btu/lb mol value in his popular *Data Book of Hydrocarbons*. Compared to the Table 1.10 interpolated value of 7417 Btu/lb mol, this is a reasonable check. Please note that while Maxwell computed each component, here only the MW was used to interpolate an enthalpy value.

Next for the liquid at 500°F and 1.0 atm. Using Table 1.10, read the following enthalpies:

	400°F	500°F	600°F
C2	491	557	623
C3	470	534	597

Please note that the enthalpy values under the 500°F heading are interpolated values between the 400 and 600°F values in Table 1.10. Next, perform the similar calculation as previously shown:

$$\frac{44 - 42.7}{44 - 30} \times (557 - 534) + 534 = 536 \text{ Btu/lb}$$
$$\times \text{ MW of } 42.7 = 22{,}887 \text{ Btu/lb mol}$$

Maxwell shows a calculation of 22,400 Btu/lb mol. This is a 2% difference, which for figure curve reading with interpolation is again a reasonable check. I submit therefore that MW or BP interpolation of Table 1.10 is sufficient.

You have now come to the finish of the *Industrial Chemical Process Design* database. I resubmit that these 8 data properties (viscosity, density, critical temperature, critical pressure, molecular weight, boiling point, acentric factor, and enthalpy) are most sufficient to establish the more common and demanding database needed for the practicing process engineer. The following chapters apply these 8 database properties in unique ways, accomplishing the PPE goals.

Now gung ho, and go for those applications!

References

1. Erwin, Douglas L., *Chemcalc 16 Liquid-Liquid Extraction*, Gulf Publishing, Houston,TX, 1992.
2. Dean, John A., and Lange, Norbert Albert, *Lange's Handbook of Chemistry & Physics*, McGraw-Hill, New York, 1998.
3. The Crane Company, *Crane Technical Paper No. 410*, King of Prussia, PA, 1988, Figs. A-3, A-5, A-7, and B6.
4. American Petroleum Institute (API), *Technical Data Book*, API Refining Department, Washington, DC, 1976, Table 1C1.1.
5. Maxwell, J. B., "Crude Oil Viscosity Curves," *Data Book on Hydrocarbons*, D. Van Nostrand, Princeton, NJ, 1957, Sec. 9, pp. 155–165.
6. American Standard Testing Methods (ASTM), *ASTM-IP Petroleum Measurement Tables and Standard Testing Methods*, ASTM, Washington DC, 1952.
7. API, "Crude Oil Characterization," *Technical Data Book*, Chap. 3.
8. Gas Processors Suppliers Association (GPSA), "Physical Properties," *Engineering Data Book*, 9th ed., GPSA, Tulsa, OK, 1972, Figs. 16-25 and 16-26.
9. Maxwell, "Crude Oil Density Curves," Sec. 8, pp. 136–154.
10. API, *Technical Data Book*, Method 6B1.1, API Density.
11. Nelson, W. L., "Determining Densities at High Temperatures," *Oil and Gas Journal*, January 27, 1938, p. 184.
12. Redlich, O., and Kwong, J. N. S., *Chem. Rev.* 44:233, 1949.
13. API, *Technical Data Book*, Method 4A1.1, Critical Temperature; Table 4A1.1
14. Reid, R., Prausnitz, J., and Sherwood, T., *The Properties of Liquids and Gases*, 4th ed., McGraw-Hill, New York, 1985.
15. API, *Technical Data Book*, Fig. 2B2.1.
16. API, *Technical Data Book*, Method 4D1.1.
17. API, *Technical Data Book*, Fig. 4D4.1.
18. Maxwell, "Thermal Properties," Sec. 7, pp. 98–127.
19. Edmister, Wayne C., *Applied Hydrocarbon Thermodynamics*, vol. 2, Gulf Publishing, Houston, TX, 1988, pp. 22–29.

2

Fractionation and Applications of the Database

Nomenclature

A_i	K_i/heavy key K_i (HKK)
D	overhead distribution
D_i	distributed overhead comp i mol frct
HKK	K of IC5
K	equilibrium factor of each component
K_i	component K value
L	reflux rate
L_1	$R_M * R_O * D$
MN	minimum number of theoretical stages
MW	molecular weight
N	number of column theoretical stages
N_m	minimum stages
P	pressure, psia
P_c	critical pressure, psia
P_r	reduced pressure (P/P_c)
Q	feed condition (1.0 for bubble point, 0 for dew point)
R	L_1/D
R	L/D
R_M	minimum reflux ratio (L/D)
R_O	R/R_M
S	$N + 1$
S_m	$N_m + 1$
T	temperature, °F

T_c critical temperature, °F

T_R reduced temperature (T/T_c)

U_i feed comp i mol frct

V_1 $L_1 + D$

X component liquid phase mole fraction, liquid feed

$X(H)$ heavy key mole fraction in distillate

$X(L)$ light key mole fraction in distillate

Y component vapor phase mole fraction, liquid feed

$Z(H)$ heavy key mole fraction in bottoms

$Z(L)$ light key mole fraction in bottoms

Having now established a database in Chap. 1, in all probability the most common application is component separation. Taking raw crude oil and separating it into sellable products is an age-old task. From fuel oil products for running the locomotive steam engines of 1920 to modern-day gasoline products for the automobile, production has been a never-ending task involving distillation. The most common type of distillation is that done in refineries and gas processing plants. Other types of distillation are found in chemical plants and pharmaceutical industries. Even these two types—chemical and pharmaceutical—in many cases are very similar to the refinery-type distillation processes. This book is therefore dedicated to the heart of our distillation or fraction industry, the refinery type. This also involves oil and gas plant fractionation and upstream crude oil and gas production facilities.

Having now established our direction here, one common physical phenomenon merits a discussion: pressure. In almost every refinery fractionation process and gas production fractionation process, the pressure value is seldom above 400 psig. Refinery front-end fractionations are atmospheric crude inlet columns, vacuum stills, 25-psig stabilizers, and 80- to 150-psig overhead liquified petroleum gas (LPG) fractionation trains. These pressure ranges are also common in gas processing plants, seldom ranging over 350 psig. Why refer to these pressure values? What's the point? The point is that component ideal solution modified equilibrium K values can easily be derived most accurately for these separation calculations.

$$K = \frac{Y}{X}$$

where Y = mole fraction of component in vapor phase
        X = mole fraction of component in liquid phase

Edmister, in Chap. 38 of his book *Applied Hydrocarbon Thermodynamics* [1], uses this key principle in generalizing K values of ideal solutions. Here Edmister points out that by using reduced pressure (P/P_c) and Roult's law reduced vapor pressure rather than temperature for the chart parameter, consistent and accurate curve-fitted constants can be made. Applying such curve fitting, the K value can quickly be calculated for any component. Further, Edmister compares these pressure-corrected ideal solution K values to the Gas Processors Suppliers Association (GPSA) data book K values [2], finding excellent agreement.

These generalized K values that Edmister and others have developed and have found to be of good use are installed in the *Industrial Chemical Process Design (ICPD)* program RefFlsh. The uniqueness of these K values is that the input for each component requires only P_c, T_c, and an acentric factor. All three of these database factors are readily derived in Chap. 1. Note here in reference to the preceding paragraphs that refinery fractionation pressures are low values—generally 250 psig maximum. This pressure-deriving K value simplicity is applicable to the majority of refinery and gas plant fractionation unit operations and to most types of component separation processing. This book therefore presents the derivation of these generalized K values for reasons of their small data input requirements and their calculation simplicity.

Equations of State

Before entering into deriving this K generation, I propose the reader fully realize that there are many equation-of-state computer programs that are excellent equilibrium value generators. Peng-Robinson (PR) [3], Soave-Redlich-Kwong (SRK) [4], and Lee-Kesler (LK) [5] are a few equation-of-state choices. Many of these equation-of-state models have been installed in commercial packaged computer program simulators. These commercial software simulator companies have even further refined these equation-of-state models, so that they do not agree with other equation-of-state models of the same type. Also, many equations of the type we now use in these expensive commercial programs do give some erroneous K value results. I have personally experienced the Braun K10 K value model [6] giving excellent results in an aromatic fractionator, compared to very poor results for the Peng-Robinson K values. This does not mean the PR derivations are poor in all cases. In fact, today the PR K derivations are applied more accurately than most other equation-of-state types, especially in or near the critical regions of the phase envelope.

As pointed out earlier, in it would be utopia if we each had a $50,000 computer software simulator continually at our fingertips with which we could solve most every simulation problem. Hey, who's kidding? It would just be nice to have a state-of-the-art computer continually at our fingertips, wouldn't it? Well, we really should at least have a good

Windows PC at our fingertips anyway! Each of us, however, does not have such software luxury, and in many cases we improvise our best solution for our process engineering needs. I have therefore proposed this simple but fully comprehensive and reliable K value derivation that I will call *PPK,* which stands for practical process K equilibrium component values.

PPK Accuracy and Limitations

It is well known and often published that equations of state may produce over 10% error when checked by actual operations. The PR equation of state, for example, ranges from 2.5% average to a maximum 7.5% error for most hydrocarbon systems. PPK should prove equally good, ranging from 2 to 10% maximum error. This assumes, however, a maximum 600-psig pressure. For higher pressures, PPK error can be higher, while the PR equation of state and others like it retain the same error probabilities—2% to 10%. It is therefore good and prudent to place a pressure limitation of 600 psig on the PPK values in this book. In accordance with Edmister's work [1], PPK-derived K values may be equally accurate at much higher pressures, say to 1500 psig. More actual PPK refinery plant operation checks are needed at these higher pressures. Perhaps after the publication of this book, some of you good process people will frontier a better limitation, perhaps finding broader applications. My encouragement goes to you with a hearty gung ho!

The PPK Equation of State

On the CD supplied with this book you will find Visual Basic software for the program RefFlsh. PPK equilibrium K values are calculated for each component in the program RefFlsh. Reference is made in the text to key code lines of this program.

In the RefFlsh program listing, the two key lines for K value calculation are code lines 540 and 590. In the following equations, however, notice Z and X constants must first be calculated. P_r is easy to get, and so is Z. X is simply ln K_r, with ln signifying the natural log of a number of base e. Roult's law K-ratio's K value K_r is much more complex to calculate. K_r brings forth a correction factor for nonideal systems, PRO, which provides a simple way to derive the corrected K values of PPK. In order to derive K_r, it is first necessary to derive the Roult's law reduced vapor pressure PRO. Pitzer [7] first established PRO values up to the critical temperature point, $T_r = <1.0$, with the equation given on code line 250. Stucky [8] made an extrapolation equation (code line 230). This extrapolation is the region above $T_r > 1.0$. After solving one of these equations for PRO, K_r is calculated as shown, dividing PRO by the reduced component system pressure P_r.

$$Z = \ln (P_r)$$

$$P_r = \text{system pressure}/P_c$$

$$P_c = \text{component critical pressure}$$

$$X = \ln (K_r) \quad K_r = \text{PRO}/P_r$$

```
Code Line 230:
PRO = Exp(5.366 * (1 - TR ^ (-1)) + W(I) *
      (5.179 - 5.133 * (1 / TR) - .04566 * (TR ^ (-2))))
Code Line 250:
PRO = Exp(5.366 * (1 - TR ^ (-1)) + W(I) *
      (2.415 - .7116 * (1 / TR) - 1.179 * (1 / TR ^ 2) -
      .7072 * (1 / (TR ^ 3)) + .1824 * (1 / (TR ^ 4))))
Code Line 540:
Y = AA0 * X + A1 * Z + ((1 + A2 * X) * (Exp (X/2)) - 1)
Code Line 590:
Y = AA3 * X + AA4 * X ^ 2 + AA5 * X ^ 3
Code Line 600:
K(I) = Exp(Y)
```

Now, with X and Z constants established for each component, the Y constant is next calculated. (See preceding code lines 540 and 590.) One of these equations will be applied for establishing a Y value. This equation is program-selected by temperature and pressure ranges of other code lines. These code lines window or point out which of these two equations will be used by testing ranges of the calculated values of K_r and P_r and the T_c given data. The last equation for determining the referenced component K value is code line 600, K(I) = Exp (Y). The I here is an array dimension for each component's K value. One K value is calculated for each component at the referenced system's temperature and pressure.

Now that you have an introduction to how generalized K values are calculated, consider next how these sought-out equilibrium values are used. It is first important to note that the program FlshRef is a most valued tool to apply these K values. FlshRef is one of the programs in this book including code listings and executable VB Windows version. Run it a few times to become familiar with it. You will quickly see FlshRef producing these K values from the simple data input of T_c, P_c, and the acentric factor. The molecular weight data input enables the program to display mass liquid and vapor summary answers in pounds per hour. A typical defaulted input data file is supplied for running the program.

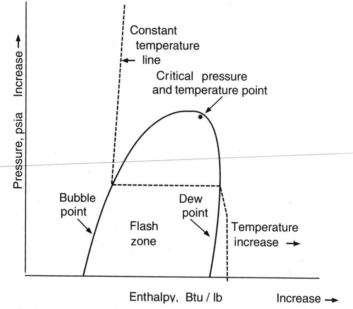

Figure 2.1   Typical pressure-enthalpy phase envelope diagram.

Figure 2.1 presents the simplistic basis upon which all separations are commonly made in our industry. Even membrane separations depend to a large degree upon the vapor pressure and temperature effects shown. A typical temperature dashed line shows how the temperature variance effects a vapor-liquid separation. Notice also the variance for pressure and enthalpy. Inside the phase envelope, the temperature and pressure remain constant while the enthalpy varies. This constant *T* and *P* occur in what is called the *flash zone*.

Bubble Point

The bubble point is defined as that hydrocarbon component condition at which the system is all liquid, with the exception of only one drop (infinitely small) of vapor present. The amount of vapor is specified as a matter of convenience so that the composition of the liquid is the composition of the total system. This means that if we want to find the bubble point of a liquid composition, we simply flash it for a very small amount of vapor specified. You can do this with the RefFlsh program quickly by trial and error. How? Simply try varied temperatures with the pressure held constant. The program RefFlsh will tell you if your temperature is out of the phase envelope by stating SYSTEM IS ALL

LIQUID or SYSTEM IS ALL VAPOR. Make certain you are sufficiently below the critical point for pressure and temperature assumptions. When you find the right temperature, a limited temperature variance will change the system from all liquid to all vapor.

Example   The best way to realize these bubble point facts is to run an example. The following composition is to be run in the program RefFlsh.

Given feed composition

Number	Component	Mol/h
1	C3	4.2
2	NC4	6.8
3	IC4	6.1
4	BP240	10.5
5	BP280	12.4
6	BP320	5.1
7	BP400	1.1

To make this an interesting, realistic problem, let's say the preceding feed analysis is to be fed to a debutanizer column at its bubble point. Find the feed bubble point at an 85-psig pressure, 99.7 psia. Running the program, we find that the system feed is all liquid at 202°F and has these resulting mass compositions at 203°F:

Debutanizer feed bubble point, 85 psig and 203°F

Component	Feed, lb/h	Liquid, lb/h	Vapor, lb/h
C3	1.8522e+02	1.8339e+02	1.8337e+00
NC4	3.9508e+02	3.9336e+02	1.7194e+00
IC4	3.5441e+02	3.5252e+02	1.8866e+00
BP240	1.2632e+03	1.2628e+03	3.7794e−01
BP280	1.6021e+03	1.6018e+03	2.7055e−01
BP320	7.0839e+02	7.0832e+02	6.6924e−02
BP400	1.8018e+02	1.8017e+02	5.1648e−03
Totals	4.6885e+03	4.6823e+03	6.1604e+00

One can see that 6 lb of vapor mass isn't really a drop of liquid. So is this really the bubble point? In truth, it is, because determining to hundredths or tenths of a degree is neither practical nor practiced.

Mathematically, the following must be true for the bubble point:

$$K = \frac{Y}{X}$$

where K = equilibrium factor of each component
Y = component vapor phase mol fraction, liquid feed
X = component liquid phase mol fraction, liquid feed

At bubble point, $\Sigma K * X = 1.0$. This $\Sigma K * X$ is to be the sum of each component in the feed multiplied by its respective K value. The molar flow and K values are:

RefFlsh program component K values

C3	4.2129e+00
NC4	1.8416e+00
IC4	2.2548e+00
BP240	1.2610e-01
BP280	7.1161e-02
BP320	3.9807e-02
BP400	1.2077e-02

Debutanizer feed bubble point, 85 psig and 203°F

Component	Feed, mol/h	Liquid, mol/h	Vapor, mol/h
C3	4.2000e+00	4.1584e+00	4.1582e-02
NC4	6.8000e+00	6.7704e+00	2.9595e-02
IC4	6.1000e+00	6.0675e+00	3.2472e-02
BP240	1.0500e+01	1.0497e+01	3.1417e-03
BP280	1.2400e+01	1.2398e+01	2.0940e-03
BP320	5.1000e+00	5.0995e+00	4.8182e-04
BP400	1.1000e+00	1.1000e+00	3.1531e-05
Totals	4.6200e+01	4.6091e+01	1.0940e-01

solution If you do this calculation for the preceding bubble point problem, you should get:

$$\Sigma K * X = 1.0029$$

where X = mol fraction component, liq. phase

For reasonable accuracy, this answer should be valid for almost any case. The example problem is an actual case in which the debutanizer feed heat exchanger specification will be set at 203°F accordingly. The debutanizer fractionator is designed and specified to receive a bubble point feed. To find such an answer, please note that the computer program simply ran a flash, finding a very small vapor flashing.

Example The *dew point* is defined an all-vapor system except for one very small increment of liquid. Now take the feed again, and consider the fact that this is a flash-off crude still sidestream vapor at 20 psig (34.7 psia). The dew point is to be determined to set the proper sidestream stripper overhead temperature for this desired product. This problem is worked similarly to the bubble point. Simply hold the pressure (34.7 psia) constant, and vary the temperature. Note that when you find the temperature at which just a small amount of liquid is formed, the sign SYSTEM IS ALL VAPOR goes off and the flash component summary appears. Note also that the previous flash summary will remain on the screen (if you had a previous run) until you input a flash zone temperature and click on Run Prog.

The following mass rate summary is found at 313.4°F at 34.7 psia.

Crude column side stripper overhead dew point, 34.7 psia, 313°F

Component	Feed, lb/h	Liquid, lb/h	Vapor, lb/h
C3	1.8522e+02	2.0256e−03	1.8522e+02
NC4	3.9508e+02	8.3977e−03	3.9507e+02
IC4	3.5441e+02	6.5355e−03	3.5440e+02
BP240	1.2632e+03	2.5255e−01	1.2629e+03
BP280	1.6021e+03	5.1950e−01	1.6016e+03
BP320	7.0839e+02	3.7385e−01	7.0802e+02
BP400	1.8018e+02	2.6634e−01	1.7991e+02
Totals	4.6885e+03	1.4292e+00	4.6871e+03

Here, using the program RefFlsh, very little liquid first appears between inputting a temperature of 314°F and one of 313°F. Also, the heading SYSTEM IS ALL VAPOR goes off when 313°F is input after inputting 314°F. This heading signifies you have a temperature with pressure condition out of the two-phase zone. Having an all-vapor or an all-liquid condition, you know to increase or to decrease your temperature input trial to find the dew point or bubble point.

One note about the overhead stripping temperature: 313°F should be observed. This would require superheated steam as a stripping medium. In many stripping still cases this is done, while other cases fix the feed more specifically to the stripping steam available.

Referring to Fig. 2.1, at any point on the dew point curve the summation of each component divided by its respective K value at that point is equal to 1.0. Putting this statement in a mathematical term,

$$\Sigma \frac{Y}{K} = 1.0$$

where Y = Component mol fraction, vapor feed

RefFlsh program component K values at 313.4°F and 20 psig

C3	2.1263e+01
NC4	1.0940e+01
IC4	1.2610e+01
BP240	1.1628e+00
BP280	7.1690e−01
BP320	4.4039e−01
BP400	1.5708e−01

Crude column side stripper overhead dew point, 34.7 psia, 313°F

Component	Feed, mol/h	Liquid, mol/h	Vapor, mol/h
C3	4.2000e+00	4.5932e−05	4.2000e+00
NC4	6.8000e+00	1.4454e−04	6.7999e+00
IC4	6.1000e+00	1.1249e−04	6.0999e+00
BP240	1.0500e+01	2.0994e−03	1.0498e+01
BP280	1.2400e+01	4.0209e−03	1.2396e+01
BP320	5.1000e+00	2.6915e−03	5.0973e+00
BP400	1.1000e+00	1.6260e−03	1.0984e+00
Totals	4.6200e+01	1.0741e−02	4.6189e+01

Now, using the K values of the program displayed, calculate $\Sigma\ Y/K$ of the feed. You should get 1.0005, which is excellent accuracy for most any need.

Herein the overhead temperature of the side stripper may be specified. This also sets the accompanied overhead condenser inlet temperature specification.

$$\Sigma \frac{Y}{K} = 1.0005$$

solution Since you actually found the dew point and bubble point with the program RefFlsh, it is not necessary to calculate the sums $\Sigma\, Y/K$ and $\Sigma\, K * X$. These calculations by hand, if you did them, were simply done to demonstrate that the accuracy of running quick trials of RefFlsh is the same. If you didn't do these calculations, you can indeed trust that the answers in this book merit confidence. The computer program found the first traces of liquid at the dew point and first traces of vapor at the bubble point. You should have now executed these trials, seeing how easily and rapidly you found these two temperatures. Aren't computers and the software with them wonderful?

Reviewing Fig. 2.1 per dew and bubble points. Taking the conditions of feed and using RefFlsh, find the bubble point:

Bubble point at 34.7 psia = 106°F

Dew point at 34.7 psia = 313°F

Referring to Fig. 2.1, notice the vast temperature difference between the bubble point and the dew point. This is 313 − 106 = 207°F. Figure 2.1 displays a single discrete component temperature constant line. A system mixture like the preceding example mixture, debutanizer feed bubble point minus dew point, would yet have a varied-temperature horizontal line running through the phase envelope.

If the temperature were reduced, leaving the dew point or saturated vapor line, the heavier components (BP320+) would first begin to condense into the liquid state. Each mixture component, having different bubble and dew point temperatures at the referenced pressure, makes a temperature range between the dew point and bubble points. Running system bubble and dew point temperatures per se at different pressures, one may plot the phase envelope for this system mixture. The system mixture constant temperature lines as shown in Fig. 2.1 would remain near vertical outside of the two-phase zone. The temperature lines inside the system mixture phase envelope would, however, vary. Our debutanizer example would of course have a varying temperature line beginning at the 106°F bubble point at 34.7 psia liquid and going horizontally on a constant pressure line to the 313°F dew line point at 34.7 psia vapor. Both the saturated liquid and the saturated vapor temperature lines would approach being vertical outside the shown phase envelope.

These vertical constant temperature lines (Fig. 2.1) are the same for a mixture as for a single component. Outside the phase envelope, the temperature lines tend to hold a constant enthalpy value or have very limited variance with variance of pressure. Compare Figs. 2.1 and 2.2. In particular, notice that the constant temperature line approaches verticality outside the phase envelope. This vertical temperature line indicates that outside the phase envelope there is no significant enthalpy variance with pressure variance. Again, outside the phase envelope, temperature variance is most effective to enthalpy. Observe how limited enthalpy is between 1.0 and 100.0 atm pressure.

The simplest way to learn the truth displayed in these two figures is to work a practical problem. Please follow the explanations given in the following problem solution, as many other similar practical everyday problems may be solved the same way.

Example   *Debutanizer Feed Exchanger Heat Cooling Duty Determination:* Take the crude side column stripper overhead at 34.7 psia, 313°F, and determine the heat cooling duty for the shown mixture to condense it and to cool it to its bubble point at 99.7 psia and 203°F.

First find the mole weight, which is:

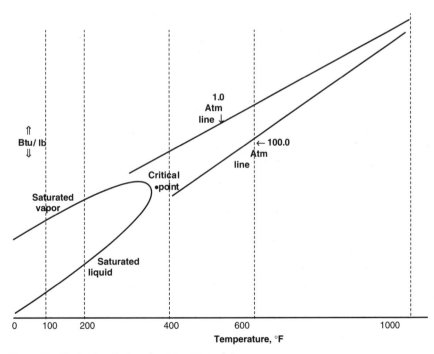

Figure 2.2   Typical enthalpy chart (not to scale).

TABLE 2.1   Interpolation of Table 1.10 Enthalpy Values

Heptane, 200°F	Interpolated enthalpy, 313°F, 34.7 psia	Heptane, 400°F
Saturated vapor, 338 Btu/lb	Dew point, **392 Btu/lb**	Saturated vapor, 433 Btu/lb

Heptane, 200°F	Interpolated enthalpy, 203°F, 99.7 psia	Heptane, 400°F
Saturated liquid, 200 Btu/lb	Bubble point, **202 Btu/lb**	Saturated liquid, 338 Btu/lb

$$\text{Debutanizer feed MW} = \frac{4688.5 \text{ lb/h}}{46.2 \text{ mol/h}} = 101.48$$

Next, note that in Table 1.10 this molecular weight is between those of heptane (100 MW) and octane (114 MW). Also please note the close enthalpy values at the various temperatures of these two components. Here it is expedient and reasonable per our findings to set heptane as the means of reference to determine the debutanizer feed enthalpies. The following calculations are made using the heptane source in Table 1.10 as a reference for all the enthalpy points.

Read the following heptane enthalpies and interpolate as shown in Table 2.1.

solution   Our MW mixture value is near, or for all intents and purposes the same value as, the heptane MW. Mixture 101.5 MW is essentially the same value as for heptane (100 MW). Thus the mixture enthalpy difference is:

$$392 - 202 = 190 \text{ Btu/lb cooling required}$$
for each pound of mixture debutanizer feed

or:

$$190 \text{ Btu/lb} \times 4689 \text{ lb/h} = 890{,}910 \text{ Btu/h cooler duty}$$

We have now found the total cooler duty required for the debutanizer feed exchanger or cooler. In practice, a 1.0 mmBtu/h duty exchanger (10% design safety added) should be installed allowing good judgment and good practice commonly accepted in our industry.

Flash Calculations

Having the K equilibrium constants calculated for each component of the mixture, the FlshRef program applies the equation shown on code line 1010:

```
XL(I) = Z1(I) / (1 + VF3 * (K(I) - 1))
```

You may review the FlshRef program execution in App. B and on the CD.

The VF factor is the flashed vapor to feed molar ratio, which has been calculated in the program in the previous lines. The VF factor is calculated in a unique method using a convergence technique shown in code lines 810 and 890:

```
SK = SK + ((Z1(I) * (1 - K(I))) / (VF1 * (K(I) - 1) + 1))
SK2 = SK2 + ((Z1(I) * (1 - K(I))) / (VF3 * (K(I) - 1) + 1))
```

Applying these equations repeatedly until code line 920 is satisfied for an absolute SK2 value less than 0.00001, the program thereby finds the sum of the flash X_i and sum of the Y_i mole fractions equal to 1.0000. The program then proceeds to display the results, giving both molar and mass flow rates of the component vapor–liquid flash.

Example Continuing with the same debutanizer feed problem, now find the feed analysis of vapor and liquid if 50%, molar, of the feed is to be fed as a vapor to the debutanizer column. Also find the cooler duty for this condition.

Since we already have a total molar flow of the debutanizer feed, 46.2 mol/h, divide 46.2 by 2 to get 23.1 mol/h. Here we must find a temperature at the 85-psig pressure to derive this 23.1 vapor mol/h feed rate. We also know that the bubble point of the feed is 203°F, so here we know to start increasing the flash temperature until this vapor rate value is found.

Debutanizer feed flash at 85 psig and 336°F

Component	Feed, mol/h	Liquid, mol/h	Vapor, mol/h
C3	4.2000e+00	4.3521e–01	3.7648e+00
NC4	6.8000e+00	1.2064e+00	5.5936e+00
IC4	6.1000e+00	9.7084e–01	5.1291e+00
BP240	1.0500e+01	6.5151e+00	3.9849e+00
BP280	1.2400e+01	8.8586e+00	3.5415e+00
BP320	5.1000e+00	4.0540e+00	1.0461e+00
BP400	1.1000e+00	9.9664e–01	1.0336e–01
Totals	4.6200e+01	2.3037e+01	2.3163e+01

Debutanizer feed flash at 85 psig and 313°F

Component	Feed, mol/h	Liquid, mol/h	Vapor, mol/h
C3	1.8522e+02	1.9193e+01	1.6603e+02
NC4	3.9508e+02	7.0092e+01	3.2499e+02
IC4	3.5441e+02	5.6406e+01	2.9800e+02
BP240	1.2632e+03	7.8377e+02	4.7938e+02
BP280	1.6021e+03	1.1445e+03	4.5756e+02
BP320	7.0839e+02	5.6309e+02	1.4530e+02
BP400	1.8018e+02	1.6325e+02	1.6931e+01
Totals	4.6885e+03	2.8003e+03	1.8882e+03

The next objective is to find the cooler duty that will yield this flash result. Since we already have the starting enthalpy value of the debutanizer feed at a dew point of 34.7 psia and 313°F, 392 Btu/lb, find next the enthalpy values of the flashed vapor and the flashed liquid.

This now brings us to a point where I again state to those of you who are scavenging this book, "Skip this paragraph." But for those of you who are attempting to gain every increment of knowledge herein, gung ho! The next portion of methodology is one I and my beloved colleague Wayne Edmister talked about the last time we met. Free energy was the subject. Free energy is a most unique and easy way to resolve many thermodynamic problems. Here is a classic example: we want to find the enthalpy value of a flashed mixture between its bubble point and dew points. We simply take a linear mass direct proportion of the mixture enthalpies between these two points, and this value is the answer.

If you have read my candid comment and agree, the next step is a snap. If you didn't agree, I advise you to run this example on one of those expensive software packages and compare your answer.

Next, using RefFlsh, determine the dew point temperature of the debutanizer flashed feed problem at the 85-psig pressure. The answer found is a 396°F dew point. A rapid reading with a quick mental interpolation of Table 2.1 shows we have a mixture enthalpy of 431 Btu/lb at this dew point temperature. The following calculations derive the cooler/condenser duty.

Using Table 1.5, at 99.7 psia:

$$\text{Dew point mixture enthalpy} = 431 \text{ Btu/lb @ } 396°F$$

$$\text{Bubble point mixture enthalpy} = 202/229 \text{ Btu/lb @ } 203°F$$

Free energy is that portion of energy required to vaporize from the bubble point to the flash of 336°F.

$$\frac{336 - 203}{396 - 203} \times 229 = 158 \text{ Btu/lb}$$

The energy at the stripper overhead was 392 Btu/lb, and the enthalpy at the debutanizer bubble point was 202 Btu/lb, a difference of 190 Btu/lb. This figure was the previous duty multiplier for the cooler calculation. Therefore the free energy not used in this heat balance is the 158 Btu/lb calculated. It appears we need only take the enthalpy difference of 202 and 158, and that is the cooler duty answer. This answer is not feasible, however, as our debutanizer feed starts at 34.7 psia and ends at 99.7 psia. A compressor between these two points would be required. To avoid the expense of a compressor, cool the stripper overhead to a sub–bubble point temperature and then use a pump.

Using a pump to obtain the debutanizer pressure, the cooler is used to obtain a bubble point duty of 1.0 mmBtu/h. Next add a small subcooling duty. Use Table 2.1 and again the heptane liquid line data.

$$\text{Mixture liquid } C_p = \frac{338 - 200}{200°F} = 0.69 \text{ Btu/lb, °F}$$

Subcool 15°: $15 \times 0.69 \times 4689$ lb/h $= 48,530$ Btu/h, say 50,000 Btu/h

Allow a 1.1-mmBtu/h condenser duty.
Next we must add an evaporative heater at 158 Btu/lb latent energy duty, plus the 50,000 sensible heat previously shown and calculated.

$$158 \text{ Btu/lb} \times 4689 \text{ lb/h} + 50,000 = 790,860 \text{ Btu/h}$$

solution   The answer is 1.1 mmBtu/h total condenser, with 870,000-Btu/h duty heater added, and a pump pumping the subcooled liquid to the required debutanizer feed pressure.

Fractionation: A Shortcut Method

Fractionation, or *distillation* as it may be called, is perhaps the most challenging and sought-out separation technique of all the industrial separation techniques. Every refinery, gas plant, and chemical plant, and almost any synthetic processing plant, perform fractionation to some degree. Many books and technical developments have sought the quest of fractionation technology. I also present this same legendary quest, but perhaps in a slightly different approach, a practical methodology: how and what?

First, please consider the many commercial simulation computer programs now available and heavily used by every engineering and operating company in this business. These programs calculate with great precision every detail of fractionation. Product qualities, production rates achievable, fractionation equipment design, and rating are all common achievements. These skillfully developed software computer programs are well proven and accepted. In many cases I have personally performed computations of complex fractionation on these programs, which were immediately accepted as design finalization with approval to fabricate and install. No additional calculations or proving checks were deemed necessary by the client or engineering firm. The program output was the final document, other than the purchase document, to the awarded vendor.

Second, why such confidence in these commercial software computer programs? Fractionation is a critical part of these programs and merits much review and many checks before finalization. In answer, you must realize this confidence grew over many years of field data proving with actual operations. I personally have conducted many such tests of which findings revealed certain equations of state must be used. Particular fractionation designs had to be applied in order to perform checks of proven operating data. Overall, after many such successful field checking accomplishments, our industry has endorsed these proven software programs. In keeping this confidence, the commercially offered programs are currently updated by dedicated staff. I know of two such commercial product software companies that employ over 100 technical personnel, all working each day on these same programs. Constantly updating, generating better database input, making the programs more user-friendly, and applying new applications are a few of their ongoing daily tasks. Their software code is heavily guarded and given to no outsiders. These program codes are known only by a few—very few—within their companies.

Proposed software method

What, therefore, is this proposed practical approach of fractionation technology, seeing such a vast accomplishment of this expertise before us? What possibly could supplement such expertise, having these computer programs at our fingertips? To answer these questions, think about whether you know what you are actually doing when you run one of these programs. If your answer is yes, I have a few items about the practice I would like to discuss with you. If your answer is no, you don't know what in the world is happening in these fractionation computer computations, then I have good news especially for you.

Those of you who have a reasonable knowledge of computer fractionation computations know the uniqueness and usefulness of shortcut fractionation calculations. For those having little or no fractionation knowledge, please accept my heartfelt encouragement to follow the guides in this chapter and to understand the methodology. If you do, you will succeed in many challenges and applications yet to come in your career.

There is yet more to justify our need of a handy shortcut fractionation method. These wonderful commercial software programs previously discussed may not be available. Remember that most companies limit who can run these programs, and you may be required to wait a long time to get your needs met. I again rest my case here that you will have at your fingertips the shortcut method in this book. It is also a computer program. Set up for your particular problem on the shortcut method proposed herein, then run many different conditions, and you will establish expected answers before you can get the answers on the more expensive commercial programs.

The last point of shortcut fractionation justification, and perhaps the most important of all, is the need for determining variable inputs of the more comprehensive tray-by-tray programs. These comprehensive, large computer programs are the major companies' commercial ones discussed previously. Costly to buy and maintain with each year's upgrades, these large, very accurate fractionation programs are the rule of our industry. Many times (I am a user witness) one may spend hours just to get a fractionator column to converge on these large programs. In order to eliminate this unnecessary time spent, run this shortcut fractionation first. You will obtain the correct input, thereby making the large software programs converge the first time. In fact, all of the commercial software packages provide their own shortcut fractionation programs. Therefore, you can run the shortcut first and benefit by knowing the correct input to make the more comprehensive program converge on the first run.

In the years from 1940 through the 1960s, several notable shortcut fractionation methods were published. Of these, one method that included several of these earlier methods has stood out and is today more accepted. Fenske, Underwood, and Gilliland [9–12] are the core of this proposed method. Yet one more entry is added, the Hengstebeck [13] proposed method to apply multicomponent distillation. As these earlier methods pointed out only two component separations (called binary systems), the Hengstebeck added contribution is most important for multicomponent applications.

I will call the proposed shortcut method the Hdist method, which is also the name of the shortcut program provided with this book. Key references will be made to the code listing in this program.

The first objective of a shortcut method is selection of two key components and then setting all conditions around these components. Call the light key component LK and the heavy key HK. The fractionation objective is to separate these two key components per a given specification. The distribution (distillate product analysis and bottoms product analysis) of a multicomponent system is also accomplished in this proposed shortcut fractionator program, Hdist.

The shortcut method consists of (in brief) calculating the minimum number of stages required. Set the tower to a total reflux condition to determine this minimum number of stages. The minimum reflux ratio (R_M = overhead/overhead distillate product) is next calculated using assumed constant relative volatility and constant molal heats of vaporization. These assumed value constants in most fractionation columns tend to be conservative. Hengstebeck stated, "When relative and/or molal heats of vaporization are not constant, their values in the zones of constant composition are the controlling values" [13]. It is therefore reaffirmed here that assuming constant molal heats of vaporization and relative volatility is indeed practical and a conservative design. The final step in the shortcut method is finding the actual number of stages having a given actual reflux setting. The following items describe how this proposed shortcut method accomplishes each of these tasks.

The Fenske-Underwood equation is first used to calculate the minimum number of theoretical stages [14]. Hdist program code lines 210 and 215 show this equation:

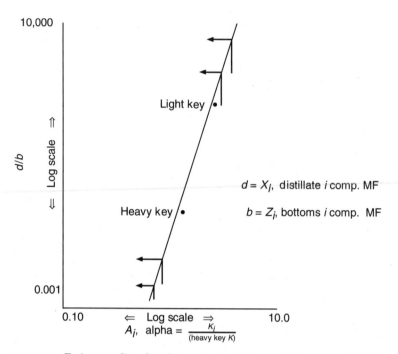

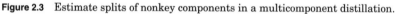

Figure 2.3 Estimate splits of nonkey components in a multicomponent distillation.

```
210 NU = (X(L) * Z(H)) / (Z(L) * X(H))
215 NM = (Log(NU)) / (Log(A(L)))
```

where   X(L)= light key mol frct, MF, in distillate
X(H)= heavy key MF in distillate
Z(L) = light key MF in bottoms
Z(H)= heavy key MF in bottoms
NM = minimum number of theoretical stages

A unique method and result of applying this Fenske-Underwood equation is that, when plotted on log-log graph paper, the components other than key component separations can be determined. Consider Fig. 2.3; notice the straight line drawn through the key component points. The unique thing here is that all other component values are also determined from this straight-line plot.

The equation in Hdist code lines 210 and 215 can also be written:

$$\log \left(\frac{d_i}{b_i}\right) = (N + 1) * \log (A_i) + C_1$$

where N = number of column theoretical stages
C_1= constant

Since we have two of these points given, the constants $N + 1$ and C_1 can be calculated. Then, using these same constants, the split of all other components can be calculated using this same equation. This is accomplished in the program Hdist.

Therefore the distribution of each component in a total reflux operation can be determined by locating the points of the key components on a plot of log d/b vs. log A, then drawing a straight line through the points, and reading off the d/b value for each of the other components at its A. The example worked later in this chapter gives a step-by-step example of how to accomplish this component distribution. The given program Hdist accompanying this book also makes this determination of component distribution, applying the same methodology.

Now that the minimum number of theoretical stages is determined, the minimum reflux is required at infinite theoretical stages. I will reference this reflux as R_M, which is the molar ratio of the tower returned liquid L divided by the outgoing overhead product D. There have been several shortcut methods to derive this minimum R_M. Probably the most popular algebraic method is that of Underwood [11,12], which has been proven over many years and which I personally applied while working in a major EPC company on my first process design job, in 1959.

Underwood's equations were designed for systems with constant relative volatility and with constant molal heats of vaporization. As discussed, these assumptions for most systems prove conservative and therefore indeed merit the application herein of practical process engineering. The Underwood equations for R_M determination are:

$$1 - Q = \sum \frac{A_i * U_i}{A_i - P}$$

$$R_M + 1 = \sum \frac{A_i * D_i}{A_i - P}$$

where Q = feed condition (1.0 for bubble point and 0 for dew point)
         U_i = feed comp i mol fraction
         D_i = dist. overhead comp i mol fraction
         L = reflux rate
         A_i = K_i/heavy key K_i
         R_M = L/D (see previous description)
         P = a constant to be determined

These same equations are applied in the Hdist program.

The previous two equations employ a constant P whose value lies between the As of the keys. The Hdist program first assumes the average key A_i value for P and begins a trial-and-error method until a P value is found that satisfies both equations.

The last step in the Hdist program is the calculation of actual trays with a given actual reflux input. Gilliland's [10] correlation of stages and reflux does this step. Gilliland's equation follows:

$$\frac{S - S_m}{S + 1} = \text{function of } \frac{R - R_M}{R + 1}$$

where $S = N + 1$
N = number of theoretical stages
$S_m = N_m + 1$
N_m = minimum stages
$R = L/D$
L = reflux
D = overhead dist.
R_M = minimum reflux ratio

This equation has been plotted by Gilliland and has been curve-fitted by many. Hdist applies one of these curve-fit equations (code lines 335 through 345) in solving this equation. Thus any value of actual reflux ratio R/R_M may be input into Hdist for deriving an output actual theoretical number of stages. Please note that this Hdist actual number of stages excludes the overhead condenser and reboiler stages. Add 2 to this Hdist exhibited value if you wish to include the condenser and reboiler stages.

Example   *Part 1 of 2, the Column Process Stages and Reflux:* Our best approach to having something practical, and a good template for you to apply, is to work an example problem, designing or rating a debutanizer fractionator. Let's do so by taking the analysis of a 100-mol/h feed rate and set an overhead isopentane specification of no more than 7 molar percent. Do the same for normal butane of no more than 5 molar percent in the bottoms product. In view of the desire to use air finfan overhead condensers throughout this plant, a 120-psia pressure tower operation is to be applied in our design. The feed analysis (taken by chromatograph laboratory analysis and shown in Step 2) is to be used as feed analysis. Using these data, determine the number of theoretical stages required with the tower reflux required. Also establish the feed condition to be used.

Step 1

Establish a mol percent heavy key component in distillate specification and a light key specification in the bottoms product:

HK, IC5 = 7 mol% in overhead product

LK, NC4 = 4 mol% in bottoms product

Step 2

Using these criteria and a feed rate of 100 mol/h, establish a first assumed top and bottoms product component split:

Component	Feed, mol/h	Overhead product, mol/h	Bottoms product, mol/h
C3	5	5	0
IC4	15	14.8	0.2
NC4	25	23.2	1.8
IC5	20	3.4	16.6
NC5	35	2	33.0
	100	48.4	51.6

Hint: Use the log-log plot of d/b vs. alpha and estimate component distributions by drawing a straight line through LK and HK components on the plot. A first alpha estimate can be made simply by assuming a midtower temperature of, say, 190°F. Then run a flash to get K values using the program RefFlsh.

Step 3

Run the overhead analysis for the dew point. At 120 psia and 156.5°F, the following dew point is found running RefFlsh:

	Feed	Liquid	Vapor
C3	5.0000e+00	4.1397e–02	4.9586e+00
IC4	1.4800e+01	2.4298e–01	1.4557e+01
NC4	2.3200e+01	4.8312e–01	2.2717e+01
IC5	3.4000e+00	1.3684e–01	3.2632e+00
NC5	2.0000e+00	9.6186e–02	1.9038e+00
Totals	4.8400e+01	1.0005e+00	4.7399e+01

K values are:

		A_i
C3	2.5284e+00	5.02
IC4	1.2646e+00	2.51
NC4	9.9254e–01	1.97
IC5	5.0338e–01	1.00
NC5	4.1780e–01	0.83

$A_i = K_i/\text{HKK}$

$\text{HKK} = K$ of IC5

K_i = component K value

Step 4

Run the bottoms shown analysis bubble point. At 123 psia and 230°F, the following bubble point is found running RefFlsh:

	Feed	Liquid	Vapor
C3	0.0000e+00	0.0000e+00	0.0000e+00
IC4	2.0000e–01	1.9907e–01	9.2651e–04
NC4	1.8000e+00	1.7930e+00	6.9591e–03
IC5	1.6600e+01	1.6564e+01	3.5548e–02
NC5	3.3000e+01	3.2939e+01	6.1190e–02
Totals	5.1600e+01	5.1495e+01	1.0462e–01

K values are:

		A_i
C3	4.2227e+00	3.90
IC4	2.3481e+00	2.17
NC4	1.9583e+00	1.81
IC5	1.0819e+00	1.00
NC5	9.3600e-01	0.87

Step 5

Run a RefFlsh program flash of the total tower feed at midtemperature, $(156 + 230)/2 = 193°F$. List these K values and calculate the alphas, which are noted as A_is. Midtower K values are:

		A_i
C3	3.2960e+00	4.39
IC4	1.7501e+00	2.33
NC4	1.4199e+00	1.89
IC5	7.4989e-01	1.00
NC5	6.3695e-01	0.85

Step 6

Calculate the average alpha values (K of each component divided by HKK value). The average tower alpha A of each component $= (A_1 * A_2 * A_3)^{1/3}$. These A_is are the tower top, the tower average temperature, and the tower bottom temperature A_is, all of which are used here to calculate the tower average A_is.

Average A_i of each feed component	
C3	4.41
IC4	2.33
NC4	1.89
IC5	1.00
NC5	0.85

Step 7

Start the program Hdist. Input the overhead heavy key mole fraction specification for IC5 (0.07), and the bottoms specification for NC4 (0.05). Also input the number of components (5) and the Q value (1.0) for bubble point feed. Input these key components as the numbers 4 and 5. The first estimated values of recoveries (0.95 for NC4 and 0.80 for IC5) are also required. These figures are the estimates you made in Step 2. Next input a reflux value of 2.0. A 2.0 reflux ratio is chosen economically for reason of tray variance. Using the Step 6 A_i values with the tower feed known, make the following input into the program screen:

Component no.	Tower feed, mol/h	A_i
1,	5,	4.42
2,	15,	2.33
3,	25,	1.89
4,	20,	1.00
5,	35,	0.85

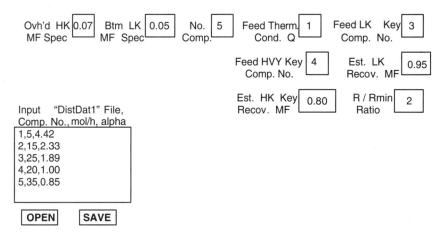

Figure 2.4 Hdist program input.

For your convenience these inputs are in rows and separated by commas. Please note that you must click on the SAVE button in order to establish the file of input you want after finishing all the lines of input. After saving, you must also click on the OPEN button in order to activate this data in the program.

Together with the preceding component data input, your input screen should now appear as shown in Fig. 2.4.

Step 8

Click on the Run Prog. button. The following answers are output:

Component number	Overhead, mol/h	Overhead, mol fraction	Bottom, mol/h	Bottom, mol fraction
1	05.0	0.1039	00.0	0.0001
2	14.6	0.3029	00.4	0.0083
3	22.4	0.4658	02.6	0.0500
4	03.4	0.0700	16.6	0.3205
5	02.8	0.0573	32.2	0.6212
	48.2		51.8	

Compare these answers to the originally assumed values:

Component	Feed, mol/h	Overhead product, mol/h	Bottoms product, mol/h
C3	5	5	0
IC4	15	14.8	0.2
NC4	25	23.2	1.8
IC5	20	3.4	16.6
NC5	35	2	33.0
	100	48.4	51.6

Note these assumed values of overhead and bottoms were never input into the program, but rather were used in estimating the average alphas.

Because we made a good first Step 2 estimate, little if any difference will occur due to running a new set of average alpha values (Step 2 through Step

6). Thus, these answers are valid. This first estimate was made by simply observing a log-log plot of Fig. 2.3. For every case, make a good component distribution estimate on the first trial. Simply draw a straight line on log-log graph paper through the LK and HK components and thereby make your estimates. Please also note that estimating the flash feed alphas at an average tower temperature of 193°F would be a good start in estimating alphas and, subsequently, component distributions for such an estimate. Otherwise, you should take the overhead and bottoms product program calculated analysis and repeat Steps 2 through 6. Also setting new specifications (Step 1) may be required if such is found unreasonable.

solutions

Minimum number of theoretical stages = 5.90

Actual number of theoretical stages calculated for input actual reflux ratio of 2.0 = 9.60 (add two more stages for condenser and reboiler, or 12 theoretical total)

Minimum reflux ratio $R_M = 1.10$

Component number	Overhead, mol/h	Overhead, mol fraction	Bottom, mol/h	Bottom, mol fraction
1	05.0	0.1039	00.0	0.0001
2	14.5	0.3029	00.5	0.0083
3	22.4	0.4658	02.6	0.0500
4	03.3	0.0700	16.7	0.3205
5	02.5	0.0573	32.5	0.6212
MW	58.47		71.39	

Example *Part 2:* Now, having the debutanizer tower answers, material balance, determine the condenser and reboiler duties using Table 1.10. Remember, you may opt to use mixture average molecular weight basis to interpolate enthalpy values in Table 1.10.

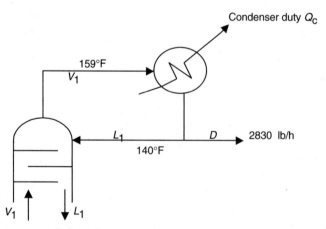

Figure 2.5 Debutanizer top section.

Debutanizer Top-section Equations (see Fig. 2.5)

$$R_O = R/R_M$$

$$R = L_1/D$$

$$V_1 = L_1 + D$$

$$L_1 = R_M * R_O * D$$

R_O = ratio of actual reflux divided by minimum reflux = R/R_M = 2.0

$L_1 = R_M * R_O * 2830 = 1.10 * 2 * 2830 = 6226$ lb/h or 106.48 lb mol/h

$V_1 = L_1 + 2830 = 9056$ lb/h   or   154.88 lb mol/h

$$R = 2.0 * 1.10 = 2.20$$

Energy Balance:

$$Q_C = V_1 * V_1H - L_1 * L_1H - D * D_H$$

159°F   140°F   140°F

V_1H, L_1H, and D_H are respective enthalpy values in Btu/lb at the reference temperatures. Noting all shown streams have a 58.5 MW, refer to Table 1.10 to get enthalpy values shown. Table 1.10 n-butane enthalpy value readings are chosen for MW value and for slightly conservative H values.

$$Q_C = 9056 * 331 \text{ Btu/lb} - 6226 * 186 \text{ Btu/lb}$$
$$- 2830 * 186 \text{ Btu/lb} = 1.31 \text{ mmBtu/h}$$

Condenser duty = 1.31 mmBtu/h

Debutanizer Bottom-section Equations (see Fig. 2.6)

$$B = F + L_1 - V_1 = 5528 + 6226 - 9056 = 2698 \text{ lb/h}$$

Note:
1. Feed at $Q = 1.0$ is bubble point at approximately 178°F.
2. Values of $L_1 * L_1H$ and $V_1 * V_1H$ have been calculated:
   6226 * 186 = 1,158,000 Btu/h
   9056 * 331 = 2,998,000 Btu/h
3. The 71.4-MW isopentane enthalpy value is used for the feed and bottoms product streams (see Table 1.10).

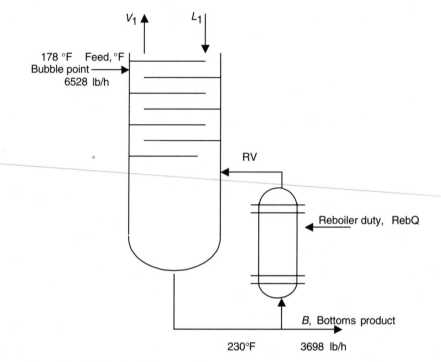

Figure 2.6   Debutanizer bottom section.

solution

$$RebQ = -L_1 * L_1H + V_1 * V_1H + B * B_H - F * F_H$$

$$RebQ = -6226 * 186 + 9056 * 331 + 3698 * 232 - 6528 * 206 = 1.36 \text{ mmBtu/h}$$

$$\text{Reboiler duty} = 1.36 \text{ mmBtu/h}$$

Overview of Example Problem and Validating

Having now completed the proposed shortcut fractionation method with the accompanied energy balance, the final task is validation. I must also include the most important reason and justification for running the shortcut fractionation method. I will use another proven program to make such a check, while also proving the shortcut method reason and justification. What better way to run a check than simply to run these same inputs on a rigorous tray-by-tray computer program? This has been done with one of the current popular, comprehensive, and very rigorous programs. The SRK equation of state is applied with the Jacobian inside-out-tower tray-to-tray convergence technique. This is one of today's most modern programs and is well proven through countless successful similar fractionation problems.

Tray-to-Tray Rigorous Program
Validation Check

Our first task in applying any of these rigorous tray-to-tray programs is to select logical, yet accurate specifications. This is the hit line of why you must first run a shortcut fractionation method. Why? You need and must input accurate specifications such as top and bottom temperatures, heat duties, number of trays, reflux rate, and other such data. These rigorous programs simply will not converge until you input these mandatory and accurate specifications. All of this specification data is derived in the shortcut fractionation method Hdist.

Consider variable degrees of freedom and input the key specifications. In these rigorous programs, key specifications are more than the top and bottom product component specifications. Rather, such key specifications as reflux rate, top and bottom temperatures, heat duties, and many other specifications are in order as derived in Hdist. In this problem I will specify three items and leave two degrees of freedom for the rigorous program to calculate and converge upon. I input the following specifications first.

Input specifications

1. Input 8 theoretical stages, having an additional 2 stages—condenser and reboiler—making a total of 10 stages.

2. Reboiler temperature is specified at 230°F.

3. Specify the overhead product distillate to have 0.07 IC5.

The program is next run and immediately converged in two iterations. Product analysis as in Hdist is achieved; however, the reflux rate is 50% greater and the condenser-reboiler duties are 30% greater than in Hdist. Realize it is much more economical to add two or three more trays vs. this greater exchanger area. Also, three more trays are well justified in reference to the additional energy usage cost alone. The next rigorous program run is made with the following input specifications.

Second-run input specifications

1. Input 10 theoretical stages, having an additional 2 stages—condenser and reboiler—making a total of 12 stages. (Note that Hdist calculated 10 theoretical trays, reboiler and condenser = 12 total.)

2. Reboiler temperature is specified at 230°F (same as Hdist).

3. Specify the overhead reflux ratio, $R = L_1/D$, to be 2.2 (same as Hdist answer).

The rigorous program was run again with these new specifications and immediately converged, again in only three iterations. The following tray-to-tray rigorous program answers were output. These answers are surprisingly very much in agreement with our previous answers in Hdist. These rigorous program answers are recorded in the following for your review and comparison with the previous Hdist answers of this same fractionation problem. The tower is set to 120 psia overhead pressure and given a 3-psi internal pressure gradient and 123-psia bottom pressure. The feed is set at its bubble point, 178°F.

		Mol/h	
	Feed, 178°F	Distributed overhead product, 143°F	Bottom product, 230°F
C3	5	5.0	0
IC4	15	14.7	0.3
NC4	25	22.8	2.3
IC5	20	3.1	16.8
NC5	35	2.6	32.4
Totals	100	48.2	51.8

Tower condenser duty = 1.27 mmBtu/h

Tower reboiler duty = 1.33 mmBtu/h

Top condenser feed temperature = 160°F

Top reflux temperature = 143°F

Bottom reboiler temperature = 230°F

Tray 2 vapor = 154.2 mol/h   Tray 9 vapor = 151.1 mol/h

Tray 2 liquid = 102.5 mol/h   Tray 9 liquid = 203.8 mol/h

Hdist Overview

We have now completed a very comprehensive accuracy review of the program Hdist. Hdist has an excellent answer correctness, but would it yield such correct answers in all problems applying this method? Certainly not. Otherwise Hdist would have replaced these rigorous programs long ago. Rather, the correct usage of Hdist is to run it first, deriving the best estimates and compatible specification program inputs for the rigorous computer programs. Please note that the cost of the rigorous tray-to-tray computer single-user software package is currently over $30,000. Also note that we have again validated Table 1.10 enthalpy values and validated the MW basis usage for its applications. Indeed, this makes Hdist, with the source enthalpy table (Table 1.10) and the program RefFlsh, appear most impressive.

References

1. Edmister, Wayne C., *Applied Hydrocarbon Thermodynamics*, vol. 2, Gulf Publishing, Houston, TX, 1974, Chap. 38.

2. Gas Producers Suppliers Association (GPSA), "Vapor-Liquid Equilibrium," *Engineering Data Book*, GPSA, Tulsa, OK, 1972, Sec. 18.

3. Peng, D. Y., and Robinson, D. B., *I. & E.C. Fund.* 15:59, 1976.

4. Soave, G., *Chem. Eng. Sci.* 27(6):1197, 1972.

5. Lee, B. I., and Kessler, M. G., *AICHE Journal* 21:510, 1975.

6. Braun, W. G., Fenske, M. R., and Holmes, A. S., *Bibliography of Vapor-Liquid Equilibrium Data for Hydrocarbon Systems*, American Petroleum Institute, New York, 1963.

7. Pitzer, K. S., et al., *Thermodynamics*, 2d ed., McGraw-Hill, New York, 1961.

8. Stucky, A. N. Jr., Master of Science thesis, Oklahoma State University, Stillwater, OK, May 1963.

9. Fenske, M. R., *Ind. Eng. Chem.* 24:482, 1932.

10. Gilliland, E. R., *Ind. Eng. Chem.* 32:1220, 1940.

11. Underwood, A. J., *Journal of Petroleum* 32:614, 1946.

12. Underwood, A. J., *Chem. Eng. Prog.* 44:603, 1948.

13. Hengstebeck, R. J., "Stage and Reflux Requirements," *Distillation Principles and Design Procedures,* Robert E. Krieger, Huntington, NY, 1976, Charts 7 and 8.

14. Fenske, M. R., "Fenske-Underwood Equation," *Ind. Eng. Chem.* 32, 1932.

Chapter 3

Fractionation Tray Design and Rating

Nomenclature

ρ_G	gas density, lb/ft^3
ρ_L	liquid density, lb/ft^3
A_A	tray deck active area, ft^2
alpha	equilibrium light component K/heavy component K
AREA	total tower internal cross-sectional area, ft^2
beta	wet tray gas pressure drop factor, accounting for frictional loss
CDC	FPL factor for 2-pass tray, ft
CDCAREA	center downcomer area for 2- or 3-pass tray, ft^2
CDOC	FPL factor for 3-pass tray, ft
cfs	cubic feet per second
DC	side downcomer width, in
DCAREA	side downcomer area, ft^2
DEG	angle of downcomer chord/2, degrees
DIA	tower diameter, ft
D_L	liquid density, lb/ft^3
DP$_{\text{TRAY}}$	total bubble cap tray pressure drop, inches of liquid
DP$_{\text{TRAY1}}$	pressure drop, dry tray condition, inches of liquid
D_V	gas density, lb/ft^3
EOG	one point source referenced to tray efficiency, %
F	packing factor
F_{GA}	dry tray gas rate loading factor
FLD$_2$	bubble cap active area flood

FPL	flow path length of cross-tray liquid flow, downcomer inlet to downcomer outlet, in
G	gas mass velocity, lb/ft^2 * h
gpm	gallons per minute
HDC$_2$	downcomer liquid backup, inches of liquid
H_F	froth height, in
HF$_{10}$, HF$_{45}$, H_F	effective tray froth height, in
HHD	dry tray bubble cap riser pressure drop, inches of liquid
HHDS	height of liquid over downcomer seal for downcomer seal loss, inches of liquid
HHL	height of liquid backup in downcomer, in
HHOW	height of liquid for pressure drop over weir, inches of liquid
H_L	tray liquid height above tray deck surface, in
HUD	head loss under downcomer, inches of liquid
K	mole fraction component in vapor phase/mol fraction component in liquid phase
K_1	0.2 for standard valve trays, 0.10 for low Δ_Ps
KK_1	dry tray press drop derived constant
KK_2	dry tray press drop derived constant
KSB	curve fitted sieve tray loading curves, calculated from Eqs. (3.92) through (3.97)
L	liquid mass velocity, lb/ft^2 * h
LIQM	liquid molar flow rate over subject tray, mol/h
LIQT	tray liquid residence time, s
M	equilibrium curve slope of referenced key component
mmscfd	million standard cubic feet gas rate per day
NG	number of gas phase transfer units
NL	number of liquid phase transfer units
PTCH	dimension from center to center between bubble caps, in
SECTAREA	single bubble cap active area, ft^2
SECTHA	single bubble cap hole area, ft^2
SLOTAREA	total slot area of bubble caps on tray, ft^2
SLTHT	bubble cap slot height, in
SLTNO	number of slots on a single bubble cap
SLTSL	bubble cap slot measure from slot to weir top, in
SLTWD	bubble cap slot width, in
TEFF	key component–referenced tray efficiency, %
theta	angle of downcomer chord/2, radians
TS	tray spacing, in

VAPM	vapor molar flow rate of the subject tray, mol/h
VAPT	froth tray time, s; time of vapor on tray, s
V_{LOAD}	cfs $* [D_V / (D_L - D_V)]^{0.5}$
VLVTH	valve thickness, in; nominally 0.037 in (20-gauge)
V_S	total of actual vapor velocity of caps, ft/s
VUD	velocity of liquid flowing between downcomer outlet and tray deck top, ft/s
WH	weir height, in
X	light key mole fraction in liquid
X	radius – downcomer width, ft
X_1	light key component mole fraction in liquid phase
Y	tower radius, ft
Y_1	M slope ratio–derived Y-axis vapor phase first factor
Y_2	M slope ratio–derived Y-axis vapor phase second factor
Z	downcomer chord/2, ft

Chapter 2 reviewed in detail how to choose methodologies and calculate fractionation column material and energy balances. Now it is time to develop the column internal design patterns and their configurations of good-practice industry design engineering.

First, consider the main objective: selective component phase separation. *Phase separation* here is defined as the separation between vapor and liquid. As there is a large fluid density difference between vapor and liquid, gravity is most useful. Once energy and phase material separation are balanced, the quantities of vapor and liquid are known for each theoretical stage.

A Special Note

For those of you who are scavenging every particle of data from this source, I have a very special data bit and a most needed clarification. For decades our industry has given little thought to fractionation tray efficiencies. Put another way, the general engineering pool of our industry has said efficiency should be applied using proven field operation data. Well, I say this is good, but it leaves us wandering around with many unanswered questions such as, how close is one fractionation operation to another, especially if a certain operation deviates from all other known practiced operations? A fractionation efficiency computer program method called the Erwin two-film (ETF) method is proposed here. ETF has actually been published in a commercially available program [1,2], and is now over 10 years old. Just as ETF appeared 10 years ago, it is presented here in its original code form. In practice it has been applied worldwide, receiving excellent ratings for its accuracy and reliability. Hundreds of accurate fractionation efficiency findings testify to ETF's reliability. Indeed, I am most honored to present it again with encouragement to use it unsparingly.

What does ETF do? ETF simply tells you how many actual trays make up one theoretical stage of a fractionation column. This is an agonizing question I have heard repeated many times. I personally have gained much income by resolving this question as a consultant. (In my case, the IRS has also gained considerable income!) The quest continues even now for a good and reliable answer. My answer is ETF.

Three tray design and rating programs are presented in this chapter. Each program applies the ETF tray efficiency method. You may apply it using hand calculations, as shown in this chapter, or by running one of these programs. To apply it in any of the three computer programs, simply input the T selection at the Frct or Abs (F,T,A) prompt.

Tray Hydraulics

Many books and technical articles deal with our subject. Many of these thoroughly explain beyond any confusion just how any tray-type plate in almost any fractionation column works. Presented here is a slightly different approach that is believed to be a more simple way to understand and apply good practice.

Fractionation, by definition, is simply the mass transfer between a liquid phase and a gas phase in contact with each other. A fractionation column is simply a tall, vertical, cylindrical pressure vessel that contains numerous flat internal metal plates called *trays.* Each tray allows liquid to flow over it, so the liquid flows from tray to tray by the force of gravity. The liquid thus enters the top tray. The liquid portion not vaporized in the column's trays is taken out in the column's bottom liquid accumulation. Gas enters the column's bottom section and flows through each tray to the top section. Entering vapor pressure is its driving force. Gas not absorbed by the liquid exits the column's top section.

The word *gas* is interchangeable with the word *vapor.* Both have the same meaning here, with one exception: most references to fractionation use *vapor* in place of *gas* because the gas phase is in equilibrium or near-equilibrium with the liquid phase. The word *vapor* is generally associated with a liquid phase, meaning gas is being produced from a liquid phase or vice versa. Thus we associate *vapor* with vaporization of part or all of a liquid stream and with condensation of part or all of a vapor stream. Isn't English wonderful? If you are laughing, I join you in amused wonder at how we ever got to where we are.

The gas or vapor passes through holes in the trays, making contact with the liquid as it passes through and over each tray deck. Each tray is configured horizontally in the column so that the liquid tends to flow over the tray while the vapor rises (by pressure flow gradient) through each tray. *Vapor pressure gradient* is defined as the vapor pressure loss across each tray. Vapor pressure loss from tray to tray is summed by two variables, liquid hydraulic gradient and vapor pressure loss through the vapor passage holes. Vapor loss through a hole is readily calculated; however, the added liquid hydraulic variable is more of a challenge. Tray hydraulics is the adjustment mainly of how a liquid passes over each tray and how a constant liquid level is kept on each tray. Liquid froth on each tray at different spray heights is also a program-calculated variable.

The liquid is also flowing by liquid gradient height across each tray in the column. The driving force of the liquid is, of course, gravity. The liquid flows downward from tray to tray through what are called *downcomers*. A downcomer is simply a metal plate placed parallel in the column cylinder, blocking off a portion of the column's diameter. The downcomer traps the liquid in the portion of the column diameter it blocks off. Thus the liquid flows down through downcomers from tray to tray, while vapor flows up through each tray deck.

Now that I've confused you with the preceding description (though I believe it to be very much to the point in a simple way), I present a very easy way to understand tray hydraulics. Only three inputs are necessary to design any fractionation tray:

1. Side downcomer width (as measured along the column radius line)

2. Tower diameter

3. Number of liquid passes

How can defining these three variables set all other dimensional variables? Consider Fig. 3.1. For all systems the following rules apply:

1. Gravity is the motive force that causes liquid to flow.

2. Downcomer areas must be distributed in such an order as to keep liquid flow equally distributed across the tray.

3. Each downcomer of each tray must have the same rate of liquid per downcomer area as all other downcomers on the tray. (*Area* here refers to the top of the downcomer section, in the case of sloped downcomers.)

These three rules are the industry standard for all fractionation systems worldwide. Further, it is not feasible to design any tray outside these rules. Figure 3.1 shows that setting only one of the downcomer widths will truly set the required widths (and areas) of all the subject

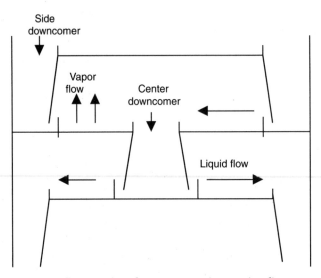

Figure 3.1   Cross section of two-pass tray (conventional).

tray's downcomers. Providing this one downcomer width gives these three *Industrial Chemical Process Design (ICPD)* book programs (Tray10, Tray11, and Tray12), all the necessary input to design or rate a tray. Of course, other data, such as gas and liquid rates with densities and tower diameter, must also be inputted. Inputting all these other variables once on the first run will allow rapid determination of which diameter with downcomer sizing and number of tray passes is optimal.

Because only three values—diameter, downcomer width, and number of passes—define essential tray hydraulics, you can run many trials in a short time. Simply input values into the computer program or execute hand calculations using different widths or passes, seeing if you can get a good design with the chosen diameter. You can even more quickly rate an existing tray, since all trays are designed on the basis of these three rules.

Please notice the two forms in each of the three tray programs: Tray10, Tray11, and Tray12. The first form is called the input form and is just that. Simply input all design tray data in this first form. If you are not sure of the tray data to input, use the default values. How do you input default values? Just click on the Input button and all default values will show. You can do these same calculations by hand with equations shown later in this chapter. Tray10 program form 1 typical default values are shown in Table 3.1.

Notice that certain item inputs are missing. Although default values are provided here so the program will run, you should input proper val-

TABLE 3.1   Valve-Type Tray Data

System factor .95	Weir height, in 3
Tray spacing, in 24	DC clear, in 2
Actual vapor density, lb/ft³ .2	Deck thickness, in .134
Gas molecular weight ■■	Valve thickness, in .06
Liquid density, lb/ft³ ■■	Low tray D_P (Y/N) N
Gas rate, lb/h ■■	Frct. or abs (F,T,A) T
Liquid rate, lb/h ■■	Surf. tens., dyn/cm 10
Slope downcomers (Y/N) N	Liq. visc., cP ■■
Side dcmr. area (opt.), ft² 0	Liq. mol wt. ■■
Active area (opt.), ft² 0	Alpha or K 1.12
Flow path length (opt.), in 0	Liq. frct. lt. key .025

ues in the empty slots. Inputting flow rates with physical properties is mandatory for any case. Liquid surface tension (10), alpha (L_K/H_K) (1.12), and liquid fraction of light key (0.025 mole fraction) default values are shown in Table 3.1. However, each should receive input values, because values vary in each individual case.

If you are using the program supplied on the companion CD, go to the form 2 screen called the Program Run form and notice that there are three remaining items to input: tray diameter, width, and number of liquid passes. You must choose a value for each. Form 2 of the Tray10 program is shown in Table 3.2.

Table 3.3 displays an actual tray design run of the Visual Basic program Tray10. Prior to clicking on the Run Program button, form 1 data were input along with the default values previously shown.

TABLE 3.2   Valve-Type Tray Fractionation Design and Rating

Tower internal diameter, ft	8
Side downcomer dimension, in	10
Number of tray liquid passes (1–4)	4

Program Output Results	
Tray active area flood	72.81%
Tray DC flood	69.16%
Actual tray efficiency	78.83%

Tray spacing, in 24	Sloped downcomers (Y/N) N
FPL, in 10.54	Number of passes 4
Effective weir length, in 233.4	DC backup, in 10.87
Liq. residence time, s 1.582	Gas rate, lb/h 8.4500e + 04
Liq. rate, lb/h 7.4500e + 05	Tray active area, ft² 30.426
Side DC area, ft² 2.777	Center DC area, ft² 7.671
Mid-DC area, ft² 6.613	Center DC width, in 11.41
Mid-DC width, in 11.21	Tray press. loss, in 4.76

TABLE 3.3   Valve-Type Tray Input Data

System factor 0.95	Weir height, in 3
Tray spacing, in 24	DC clear, in 2
Actual vapor density, lb/ft³ 0.2	Deck thickness, in 0.134
Gas molecular weight 21	Valve thickness, in 0.06
Liquid density, lb/ft³ 44	Low tray D_P (Y/N) N
Gas rate, lb/h 8.4500e + 04	Frct. or Abs. (F,T,A) T
Liquid rate, lb/h 7.4500e + 05	Surface tension, dyn/cm 10
Slope downcomers (Y/N) N	Liq. visc., cP 1.5
Side downcomer area (opt.), ft² 0	Liq. mol wt. 56
Active area (opt.), ft² 0	Alpha or K 1.12
Flow path length (opt.), in 0	Liq. frct. lt. key 0.025

Having completed the input form shown in Table 3.3, you may then click on Load Program Run Form. This will display the Program Output Results Form. Three items must next be inputted: column diameter, side downcomer width, and number of passes. Table 3.2 displays these three remaining inputs completed with an 8-ft diameter, 10-in side downcomer width, and four passes. Please notice the closeness of the tray active area flood and the downcomer flood (72.81% and 69.16%, respectively). By trial and error, these flood values were set by varying these three remaining inputs. It is good practice to design for a nearly equal active area flood and downcomer flood percentage.

A special note to those of you who have process engineering responsibilities in fractionation operations or design: for many years I have heard process plant management and college professors alike acclaim, "Let the tray vendors provide their tray designs and operation limits in our plants or our designs." I disagree heartily. First, the process engineer assigned to a plant unit has the duty of maintaining good operation practice. In an EPC company he or she is charged with providing feasible, good-practice fractionation design. Second, to pass the buck to vendors for answers (except to request new equipment quotations) is not logical. I personally represented a well-known worldwide tray fabrication vendor at one time in my career and performed hundreds of tray designs. I can assure you I did not immediately respond to these inquiries (other than valid requests for quotation); in most cases they were thrown into the trash. My most recent communications with the same vendor confirm that nothing has changed. I also assure you no vendor will give you an answer next week—for valid requests for quotation, you might get a response in two to four weeks. These vendors haven't left you in the cold without any resources; after all, you know you are indeed a respectable customer—sometimes. So, all vendors now offer all prospective customers a computer tray program. Most of these programs are of about the same quality as the *ICPD* tray computer programs. However, these vendor programs offer a limited method to derive good design and in some cases can mislead with erroneous results. Of course one may say garbage in, garbage out. I therefore plead with all to take care to make accurate inputs. Input accuracy is indeed a goal of this chapter. Another goal is to do better than most tray vendor programs with Tray10, Tray11, and Tray12 programs. Here you will learn about the code of these programs, their limitations, and how to use good judgment to develop excellent, optimum designs. Operation fractionation tray rating is equally achieved as reliable answer outputs. *ICPD* tray programs also provide excellent tray efficiencies for valve, bubble cap, and sieve-type trays. For reasons of accurate tray efficiency alone, the Tray10, Tray11, and Tray12 programs are indeed the preferred choice compared to public vendor programs.

Conventional tray design

Almost all trays are created on the basis of conventional designs, as shown in Figs. 3.1 and 3.2. *ICPD* tray computer programs apply this conventional design or rating. Figure 3.3 shows a nonconventional design, and Table 3.4 shows the optional data input required in *ICPD* tray programs for this nonconventional design.

The conventional design is the default method of the *ICPD* tray computer programs. All conventional designs have straight, chord-type downcomer areas. (Compare Figs. 3.2 and 3.3.) Note that the non-conventional downcomer areas may have complex weir configurations as compared to the simple, straight weir runs of the conventional downcomers.

All trays must have a balance of downcomer areas. Notice that the areas of the two side downcomers on any tray (straight downcomers, not sloped) are nearly equal area to that of the center downcomer. In other words, the area of the center downcomer tray approximately equals the area of two side downcomers. The logic here is that the liquid backup height in all downcomers must be equal, or otherwise the choked downcomer (the one having a lesser area) will flood, causing tray failure.

Suggestion: Run a few trials of any tray design you wish on the Trayxx program and observe the downcomer areas. Try 1-, 2-, 3-, and 4-pass trays, noting downcomer areas. You will soon realize the importance of keeping a balanced hydraulic system.

The tray program in this book calculates all tray dimensions using your one simple choice of the side downcomer width dimension. Note that the logic given in the preceding discussion is the reason this one and only dimension sets all of the tray downcomer sizes. The two side downcomer areas must equal the center downcomer dimension in the two-pass design shown in Fig. 3.2.

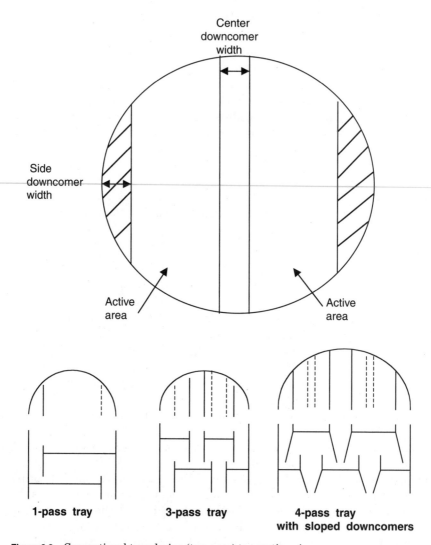

Figure 3.2 Conventional tray design (two-pass) top-section view.

As an example run, use default values in the Tray10 input form. Next input on form 2 an 8-ft column diameter, a 15-in side downcomer dimension, and a 2-pass tray design. Run this Tray10 design. Notice the side downcomers are approximately 4.5 ft² and the center downcomer is 9.0 ft². The program has made slight adjustments for standard valve areas and for flow hydraulics. The hydraulic radii of the center and side downcomers are different and require program conditioning to maintain equal downcomer liquid backup height. This liquid backup height is most critical and should never be allowed to be more than

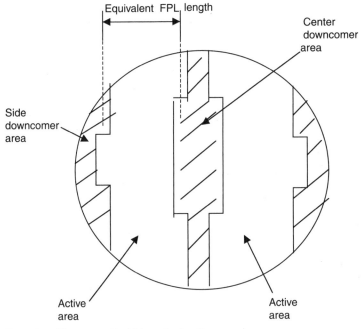

Figure 3.3   Nonconventional tray design (two-pass).

60% of the tray spacing. Please notice you are close to 60%, having a 13.14-in backup. This is an acceptable design at approximately equal flood percentages of the downcomers and tray active area.

A more efficient design for this case is a 4-pass tray. Now, on form 2, input a 10-in side downcomer dimension and 4 passes. Keep the same diameter. Note that the active area flood percentage has dropped in value considerably, and so has the downcomer flood. This example printout is shown in Table 3.4. In many cases you may find advantages to trying different tray pass designs. Of course, if you are rating a tray, you must use the pass design existing in your column.

Tray deck nonconventional active area design

In the last decade, the nonconventional tray added downcomer active area design (Fig. 3.4) has claimed good success. Please note that this design offers a larger tray deck active area than the conventional one in

TABLE 3.4   Nonconventional Tray Design, *ICPD* Tray Program Input

Side downcomer area (optional), ft²	xxx.xx
Active area (optional), ft²	xxx.xx
Flow path length (optional), in	xxxx.xx

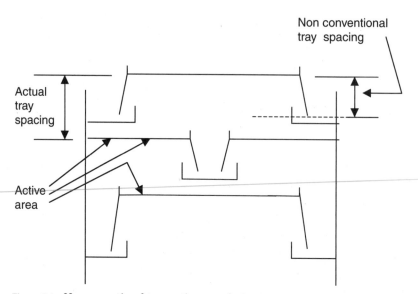

Figure 3.4 Nonconventional tray active area design.

Fig. 3.1. Also note that this nonconventional design has the downcomer outlet area as additional active tray area. This additional active area is the tray deck area under the downcomer having valves, bubble caps, or sieve holes that allow the gas to pass through under the liquid downcomer area of the next tray up. *ICPD* tray programs dealing with the design and rating of sieve, bubble cap, and valve-type trays allow this active area input. This is an option shown in Table 3.1, which is offered in the three tray design/rating computer programs given in this book.

To apply the active area of Fig. 3.4, simply add one-half of the conventional downcomer area to the conventional tray deck area (program-calculated) of one tray to derive the nonconventional tray deck active area for this option. Tray spacing program input should also be reduced; however, keep the same deck spacing for fabrication as was inputted in the conventional design. Derate tray spacing 20 to 30% by simply inputting a tray spacing 20 to 30% smaller than that to be fabricated.

Install any active area calculated by the user per Fig. 3.4 in the Optional active area input offered in the Input form. Please refer to Table 3.1 showing this Optional active area input. This active area input will override the program's conventional calculated active area.

Here Table 3.3 is repeated again, only with these nonconventional input options shown in bold type. Here you will notice all three are inputted as 0, which is also their default values if no values are inputted. Again, inputting any of these three will override all other data calculated in the tray programs and will become the governing factor of your tray design or rating.

Valve-Type Tray Input Data

System factor 0.95	Weir height, in 3
Tray spacing, in 24	DC Clear, in 2
Actual vapor density, lb/ft³ 0.2	Deck thickness, in 0.134
Gas molecular weight 21	Valve thickness, in 0.06
Liquid density, lb/ft³ 44	Low tray D_P (Y/N) N
Gas rate, lb/h 8.4500e+04	Frct. or abs. (F,T,A) T
Liquid rate, lb/h 7.4500e+05	Surf. tens. dyn/cm 10
Slope downcomers (Y/N) N	Liq. visc., cP 1.5
Side dcmr. area (opt.), ft² 0	Liq. mol wt. 56
Active area (opt.), ft² 0	Alpha or K 1.12
Flow path length (opt.), in 0	Liq. frct. lt. key 0.025

Inputting into a tray design program or choosing your tray data. Having a good introduction to tray hydraulics, our next objective is to determine and apply good and valid data. There are three input screens designated as form 1, which are the accompanied programs supplied with this book. As the valve-type input screen has been shown in Tables 3.1 and 3.4, the following input forms are presented as program input screens for bubble caps and sieve trays. Sieve tray data input has nearly the same input format as shown in Table 3.5.

Key data

As shown on the computer program run forms, you must input key data for the program to run, even with the default values automatically inputted by the program itself (see Fig. 3.5). There are three key inputs for all three programs (valve, bubble cap, and sieve programs). Before any of these programs will run, the user must choose a value for these three inputs.

Tower internal diameter, ft	XX.X
Side downcomer dimension, in	XX.X
Number of tray liquid passes	XX.X

TABLE 3.5   Bubble Cap Tray Input Data

System factor 0.95	Weir height, in 2
Tray spacing, in 24	Cap dia, in 5.875
Actual vapor density, lb/ft³ 0.2	Riser O.D, in 4.125
Gas molecular weight 21	Cap pitch (S/T) T
Liquid density, lb/ft³ 44	Cap pitch, in 8.25
Liquid molecular weight 56	Slot height, in 0.75
Gas rate, lb/h 8.4500e + 04	Slot width, in 0.25
Liquid rate, lb/h 7.4500e + 05	No. slots/cap 40
Slope downcomers (Y/N) N	Cap slot seal, in 0.5
Side dcmr. area (opt.), ft² 0	Surf. tens. dyn/cm 10
Active area (opt.), ft² 0	Liq. visc., cP 1.5
Flow path length (opt.), in 0	Frct. or abs. (F,T,A) T
Downcomer clearance, in 2	Alpha or K 1.12
	Liq. frct. key 0.025

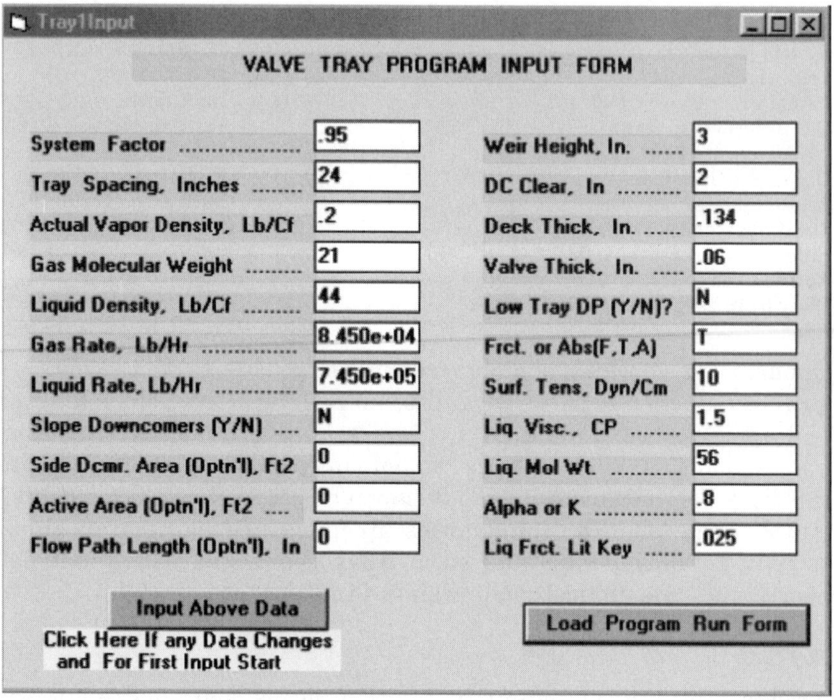

VALVE TRAY PROGRAM INPUT FORM

System Factor	.95	Weir Height, In.	3
Tray Spacing, Inches	24	DC Clear, In	2
Actual Vapor Density, Lb/Cf	.2	Deck Thick. In.	.134
Gas Molecular Weight	21	Valve Thick, In.	.06
Liquid Density, Lb/Cf	44	Low Tray DP (Y/N)?	N
Gas Rate, Lb/Hr	8.450e+04	Frct. or Abs(F,T,A)	T
Liquid Rate, Lb/Hr	7.450e+05	Surf. Tens, Dyn/Cm	10
Slope Downcomers (Y/N)	N	Liq. Visc., CP	1.5
Side Dcmr. Area (Optn'l), Ft2	0	Liq. Mol Wt.	56
Active Area (Optn'l), Ft2	0	Alpha or K	.8
Flow Path Length (Optn'l), In	0	Liq Frct. Lit Key	.025

Input Above Data

Click Here If any Data Changes
and For First Input Start

Load Program Run Form

Figure 3.5 Valve tray program input form.

The smallest feasible tower diameter size is 2 ft, but any size over 2 ft may be chosen. For pilot plant units, however, you may use the special optional inputs shown in Table 3.4. These same inputs are required for any of the three tray types, and allow accurate program answers for diameters smaller than 2 ft.

Choose any side downcomer dimension that conforms to the geometry shown in Figs. 3.1 through 3.5. Set this downcomer dimension to get near-equality between the active area flood and the downcomer flood. This is good design practice and will help to avoid short circuits such as downcomer gas blowthrough. As explained earlier, this side downcomer fixes all other downcomer dimensions. Choose this dimension to keep your downcomer flood less than 90%. By choosing the number of tray passes for any case, you are in fact determining the minimum tower diameter.

Tray Design Data

General tray design methodology for all tray types

As you can see in Tables 3.6 and 3.7, the valve, bubble cap, and sieve trays share many of the same input prompts in the tray design programs supplied on the accompanying CD.

TABLE 3.6   Bubble Cap Tray Data Input

Unit name (20-character limit)		BUBCP1.EXP	
System factor	1.00	Weir height, in	2.00
Tray spacing, in	20	Cap diameter, in	5.8750
Actual vapor density, lb/ft³	2.7500	Riser diameter, in	4.1250
Gas molecular weight	30.00	Cap pitch (S/T)	T
Liquid density, lb/ft³	29.3	Cap pitch, in	8.2500
Liquid molecular weight	40.0	Slot height, in	0.7500
Gas rate, mmscfd	82.4	Slot width, in	0.2500
Liquid rate, GPM	1100.0	Number of slots per cap	40
Sloped downcomers (Y/N)	N	Cap slot seal, in	0.5000
Side downcomer area (optional), ft²	0.0	Surface tension, dyn/cm	20.0
Active area (optional), ft²	0.0	Liquid viscosity, cP	1.00
Flow path length (optional), in	0.0	Frct or abs (F,T,A)	T
Downcomer clearance, in	4.0	Alpha or K	4.050
		Liquid fraction light key	0.10000

Tray key data for calculations. Please notice in Fig. 3.6 that three key data inputs are required: tower diameter, side downcomer dimension, and number of tray liquid passes.

To this point, we have covered mostly the accompanying computer tray program inputs. These programs are on the CD in the back cover of this book. The remaining part of this chapter will exhibit the equations used in these programs and the database resources for them.

System factor

The system factor SF is simply a dimensionless factor relating how foamy your system is. Use Table 3.8 as a guideline [3].

Tray spacing

The range for tray spacing TS is 6 in to any feasible limit. The industrial standard is 24 in. Choose your tray spacing, keeping the active

TABLE 3.7   Sieve Tray Data Input

Unit name (20-character limit)		SIEVE1.EXP	
System factor	1.000	Weir height, in	3.00
Tray spacing, in	20	Downcomer clearance, in	2.00
Actual vapor density, lb/ft³	2.7500	Deck thickness, in	0.1340
Gas molecular weight	30.0	Hole diameter, in	0.1875
Liquid density, lb/ft³	29.3	Hole pitch (S/T)	T
Gas rate, mmscfd	82.4	Pitch, in	0.7500
Liquid rate, GPM	1100.0	Frct or abs (F,T,A)	T
Sloped downcomers (Y/N)	N	Liquid molecular weight	40.00
Side downcomer area (optional), ft²	0.0	Surface tension, dyn/cm	20.0
Active area (optional), ft²	0.0	Liquid viscosity, cP	1.00
Flow path length (optional), in	0.0	Alpha or K	4.050
		Liquid mol fraction, X	0.10000

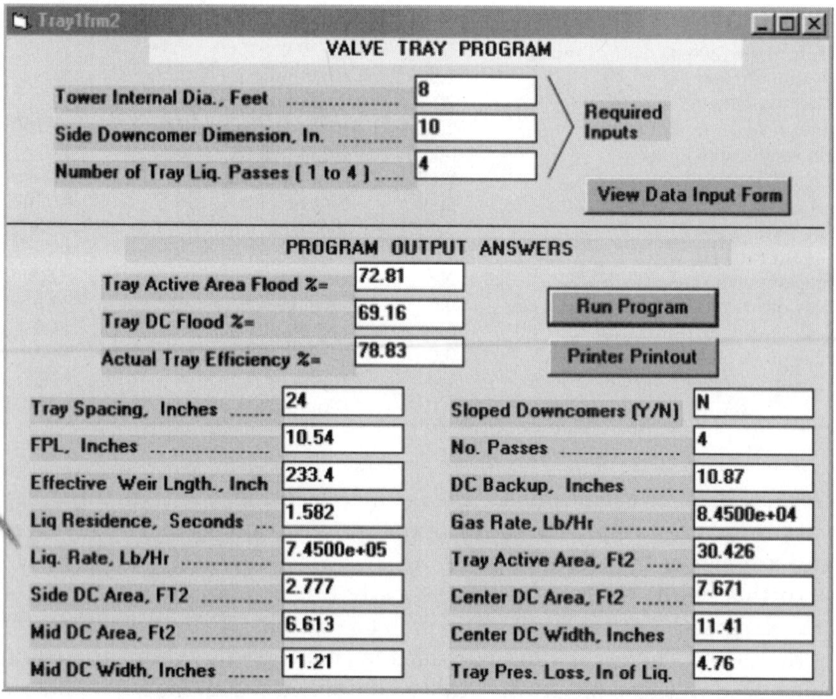

Figure 3.6 Program output answers from Fig. 3.5 program input form.

area tray flood below 90%. Note that some systems work well even to 110% flood, but 90% should be regarded as a safe guideline. A 24-in spacing is recommended for most systems. High jet flood may merit the larger spacing (>24 in). Jet flood is calculated as shown in the following and by the supplied program. Remember that a jet flood condition is very sensitive to tray spacing. Generally, tray spacing is never larger than 48 in and seldom less than 18 in.

TABLE 3.8   System Factors for Various Tray Systems

System service	System factor
No foam (freon)	1.00
Amine acid gas absorbers (heavy foam)	0.70
Amine stripper	0.80
Oil absorbers (moderate foam)	0.85
Glycol absorbers (heavy foam)	0.65
Hydrocarbon fractionators (moderate foam)	0.85

D_L liquid density, lb/ft^3

D_V actual vapor density, lb/ft^3

gpm   liquid rate, gal/min

LMW   liquid molecular weight

mmscfd   gas rate, million standard cubic feet per day

MW   gas molecular weight

GPM normally indicates gallons per minute. Here apply the hot gpm flow condition. If you have the mass rate, then:

$$\text{gpm} = \frac{\text{lb/h}}{500 * \text{SG} \, @ \, T} \quad (3.1)$$

All the preceding prompts are critical factors that require logical and realistic values. The respective value ranges should be governed by your system at actual physical conditions on the particular tray. It is therefore good practice to use mass flow and pounds per hour, and not volume flow, as it is so easy to reference a standard-temperature flowing mass and thereby introduce considerable error. For example, use hot density of flowing liquid and vapor density at actual pressure and temperature. The one exception to this rule is the gas rate, which is normally in units of mmscfd. The following equations are given for your convenience to convert from lb/h or mol/h to mmscfd.

$$\text{mmscfd} = \frac{\text{lb/h}}{\text{MW}} \times 24 \times 379.49 \times (1.0e\text{-}6) \quad (3.2)$$

or

$$\text{mmscfd} = (\text{mol/h}) \times 24 \times 379.49 \times (1.0e\text{-}6)$$

where   mmscfd = million standard cubic feet per day of gas (at 14.7 psia and 60°F)

MW = molecular weight of the gas passing through the tray

International convention is to use mmscfd gas flow in most all calculations and to use it with true flowing gas density, lb/ft^3. Refer to Chap. 1 for gas density calculation.

Sloped downcomers

Conventionally, most downcomers are sloped (see Fig. 3.4). A 70% downcomer slope is normal convention. This means the area of the bottom of the downcomer is 70% of the area of the top of the downcomer. The program supplied with this book makes a conservative calculation, accounting for dead area that cannot be used for active area. This sloping does in fact decrease the active area flood while holding the downcomer flood constant as compared to no slope. Although some tray vendors go to as much as 50% sloping, my experience has shown 70% to be good practice. A minimum tray spacing of 16 in is recommended for sloped downcomers. Excessive downcomer entrainment backup may occur at tray spacings of 14 in or less. You may opt to install more downcomer sloping by inputting an active area value in the following prompt.

Calculating the active area

The first calculation necessary for any tray design or rating is the tray size and configuration. For hand calculations (a major topic in this book), you must first assume a tray size and configuration for a new tray design. Then, by trial, the more optimum tray design is found. Of course, if you are rating an existing tray design, then the tray diameter and downcomer dimension are fixed.

Figure 3.7 displays tray downcomer area geometry. With a given tower diameter size, only one dimension, DC, sets the downcomer area. The following equations show how the downcomer area is determined.

AREA = total tower internal cross-sectional area, ft^2

DC = side downcomer width dimension, in

DCAREA = side downcomer area, ft^2

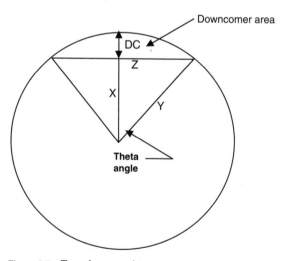

Figure 3.7   Tray downcomer area geometry.

DEG = angle of downcomer chord/2, degrees

DIA = tower diameter, ft

theta = angle of downcomer chord/2, radians

X = radius − downcomer width, ft

Y = tower radius, ft

Z = downcomer chord/2, ft

$$Y = \frac{DIA}{2} \tag{3.3}$$

$$X = Y - \frac{DC}{12} \tag{3.4}$$

$$Z = (Y^2 - X^2)^{0.5} \tag{3.5}$$

$$AREA = 3.1416 * Y^2 \tag{3.6}$$

$$theta = A_{tn} \frac{Z}{X} \tag{3.7}$$

$$DEG = theta * \frac{360}{6.2832} * 2 \tag{3.8}$$

$$DCAREA = \frac{DEG}{360} * AREA - (Z * X) \tag{3.9}$$

Refer to the earlier section of this chapter concerning tray hydraulics. Please note that the balance of the tray hydraulics fixes all downcomer areas when one side downcomer area is determined. This is true for any multitray liquid pass: 2-pass, 3-pass, or 4-pass and greater. Please refer again to Fig. 3.1, which diagrams these multitray liquid passes. The following equations are therefore presented:

For pass = 1:

$$FPL = 12 * DIA - 2 * DC \tag{3.10}$$

For pass = 2:

$$CDCAREA = 2 * DCAREA \tag{3.11}$$

For pass = 2:

$$CDC = \frac{CDCAREA}{DIA} \tag{3.12}$$

For pass = 2:

$$FPL = \frac{12 * DIA - 2 * DC - CDC * 12}{2} \qquad (3.13)$$

For pass = 3:

$$CDCAREA = \frac{0.69}{0.31} * DCAREA \qquad (3.14)$$

For pass = 3:

$$CDOC = \frac{8.63 * CDCAREA}{0.69 * DIA} \qquad (3.15)$$

For pass = 3:

$$FPL = \frac{(DIA * 12) - 2 * DC - 2 * CDOC}{3} \qquad (3.16)$$

For pass = 4:

$$H_5 = 6.78 * \frac{DCAREA}{0.21 * DIA} \qquad (3.17)$$

For pass = 4:

$$H_3 = 6.9 * \frac{DCAREA}{0.21 * DIA} \qquad (3.18)$$

For pass = 4:

$$FPL = \frac{12 * DIA - 2 * DC - H_3 - 2 * H_5}{4} \qquad (3.19)$$

where   CDC = FPL factor for 2-pass tray, ft
     CDCAREA = center downcomer area for 2- or 3-pass tray, ft^2
       CDOC = FPL factor for 3-pass tray, ft
        FPL = flow path length of cross-tray liquid flow from downcomer inlet to downcomer outlet, in
H_3 and H_5 = span factors of 4-pass tray for FPL calculation

Downcomer flood percent calculation

Two factors must be known for downcomer flood calculation: the flow rate of liquid on the subject tray and the downcomer area. We will assume that liquid flow is given in gallons per minute at the hot flowing temperature. Next determine the total downcomer area. Note that each tray pass type has a different total downcomer area.

For 1-pass tray: downcomer area = DCAREA

For 2-pass tray: downcomer area = 2 * DCAREA

For 3-pass tray: downcomer area = 3 * DCAREA

For 4-pass tray: downcomer area = 4 * DCAREA

DCAREA is simply the previously defined side downcomer area in square feet. The procedure for establishing downcomer flood uses Eqs. (3.20), (3.21), and (3.22). Take the smallest VD_s value of the three equations and apply it to Eq. (3.23)[3]:

$$VD_s = 250 * SF \tag{3.20}$$

$$VD_s = 41 * (D_L - D_V)^{0.5} * SF \tag{3.21}$$

$$VD_s = 7.5 * TS^{0.5} * (D_L - D_V)^{0.5} * SF \tag{3.22}$$

$$DCFF = \frac{gpm}{DCAREA} * \frac{100}{VD_s} \tag{3.23}$$

where   DCFF = downcomer flood, %

The importance of percent flood loadings

The downcomer percent flood is a critical tray-loading factor. If it is over 90%, then tray flooding failure is likely. If it is below 20%, tray vapor blowthrough (downcomer side) is likely to happen. It is therefore important to keep the downcomer flood below 90% and above 20%. One more limiting downcomer flood value, the active area flood, is also important. The active area flood calculation is covered in the following section.

Adjust the downcomer flood approximately equal to the tray active area flood. With these two flood values equal or nearly equal, the tray is considered to have a balance of loading between the downcomer loading and the active tray area loading. This balance ensures that the tray will operate efficiently even if it has less than 50% flood loadings. Review these two flood values (downcomer and active area flood values) carefully and make adjustments, especially in new tray design, ensuring that these flood values are close to equal.

Tray active area calculation

Having downcomers configured, the tray active area (ft^2) is calculated next. The tray active area is defined as that area of tray cross section open to vapor-liquid contact. Downcomer areas and their inlets and outlets are not part of the tray active area. Considering this, the following equations determine tray active area A_A (ft^2).

For 1-pass trays:

$$A_A = \text{AREA} - 2 * \text{DCAREA} \tag{3.24}$$

For 2-pass trays:

$$A_A = \text{AREA} - 2 * \text{DCAREA} - \text{CDCAREA} \tag{3.25}$$

For 3-pass trays:

$$A_A = \text{AREA} - 2 * \text{DCAREA} - 2 * \text{CDCAREA} \tag{3.26}$$

For 4-pass trays:

$$A_A = \text{AREA} - 2 * \text{DCAREA} - 0.58 * 2$$
$$* \frac{\text{DCAREA}}{0.42} - 0.5 * 2 * \frac{\text{DCAREA}}{0.42} \tag{3.27}$$

where A_A = tray deck active area, ft^2

Tray V_{LOAD} factor

The V_{LOAD} factor is used to determine what is known as *jet flood*. Jet flood is simply the liquid jetting, causing liquid to recycle from one tray back to the tray above, from which the liquid passed. In some cases jetting can be so severe that it blocks the gas passage with pressure buildup. The following equations calculate V_{LOAD}:

$$\text{cfs} = \frac{\text{mmscfd} * 1,000,000}{379.49} * \frac{\text{MW}}{86,400 * D_V} \tag{3.28}$$

$$V_{\text{LOAD}} = \text{cfs} * \left(\frac{D_V}{D_L - D_V} \right)^{0.5} \tag{3.29}$$

Nonconventional tray active area design

Figure 3.5 displays typical nonconventional tray active area design. Figure 3.4 displays a more unusual nonconventional design. Each of these designs objectively provides more tray active area, as you may observe in these figures.

The following three prompts are options in the three tray design programs accompanying this book. Their purpose is to allow design and/or rating of such configurations as shown in Figs. 3.4 and 3.5. Some tray vendors do conform to these deviations to a limited extent. In such cases, however, you need only input the equivalent side downcomer dimension and these three tray design programs will perform an accurate calculation.

```
SIDE DCMR AREA (OPTIONAL), FT2 ...................... X.X
ACTIVE AREA (OPTIONAL), FT2 ..........................X.X
FLOW PATH LENGTH (OPTIONAL), IN. .....................X.X
```

The next program prompt allows you to install a special side downcomer top area that cannot normally be sized with only the key data input. This input will overwrite the key data input of the main screen. A zero input here voids this option.

```
SIDE DCRM AREA (OPTIONAL), FT2 ......................... X.X
```

Install any special active area condition as shown in Fig. 3.3. This input will replace the active area calculated in the program. Running the program before inputting this option will help verify your input choice. A zero input here voids this option.

```
ACTIVE AREA (OPTIONAL), FT2 .............................. X.
```

For configurations such as that shown in Fig. 3.5, it is helpful to input the equivalent flow path length as shown. The input units are in inches. This option, however, should seldom be used without one or both of the two preceding options, because the equivalent FPL will be calculated (in most cases) as having an accurate side downcomer equivalent dimension input.

```
FLOW PATH LENGTH (OPTIONAL), IN ........................ X.X
```

Tray efficiency data

Data specific to tray type must be established next, but these inputs will be discussed later. The data inputted for the next six prompts are the same for all tray types and are primarily for tray efficiency calculations. If tray efficiency or tray liquid residence time values are not desired, these inputs may be skipped (i.e., remain as zero values). However, for bubble cap and sieve trays, the SURF TENS DYN/CM prompt is for active area tray flood calculation. This value should therefore be inputted.

```
FRCT or ABS (F,T,A) .............. T
```

Enter F or A for the O'Connell method [4]. Enter T for the Erwin two-film method [1].

Select the method of calculation for tray efficiency. Two methods are presented: the O'Connell method and the two-film method. In the programs accompanying this book, you may select the O'Connell method by entering either an F for fractionator or an A for absorbers. In 1946, O'Connell [4] published curves on log-log plots showing both absorber and fractionator efficiencies vs. equilibrium-viscosity-density factored equations. Separate curves for absorbers and fractionators were given. Such data have been curve-fit using a modified least-squares method in conjunction with a log scale setup. The fit is found to be reasonably close to the O'Connell published curves.

For fractionators:

$$X = \text{alpha} * \text{liquid feed viscosity}$$

where   alpha = equilibrium light component K/heavy component K
$\quad\quad\quad\quad$ K = mole fraction component in vapor phase/mole fraction component in liquid phase

For absorbers:

$$X = \text{light key mole fraction in liquid} \times \text{liq. avg. mol wt.}$$
$$\times \text{avg. visc./liq. avg. density}$$

Campbell's series [5] states that many authors recommend, as do I, the O'Connell method for hydrocarbons. You may select the two-film method by entering T. The Fractionation Research Institute (FRI) has for more than 30 years used and proven a much more thorough method called the two-film method, which is equally applicable to both fractionators and absorbers [6]. The FRI method accounts for the actual tray internal configuration and fluid dynamics, making it far superior to the O'Connell method. I have therefore produced a type of two-film method for determining tray efficiency that has been checked to be within 3% accuracy of hundreds of answers calculated with the FRI method. Called the Erwin two-film method, it is recommended for all types of trays and will also calculate the liquid tray residence time in seconds. This method is included in the three tray computer program executable files in the CD accompanying this book.

Input the surface tension in dynes per centimeter. For the program to run, a value must be inputted. If data are not available and the surface tension is not deemed too great, then input 20. A default value is created in the computer programs supplied in the accompanying CD.

SURF TENS DYN/CM .... XX.XX

(Range: for hydrocarbons, 20 dyn/cm)

Input the liquid viscosity in centipoise. The liquid here is that on the subject tray. This is an important factor in determining tray efficiency. The value should be entered accurately and derived from reliable data sources, such as a reliable computer program for tray-to-tray fractionation calculations. Values greater than the normal range may occur, causing lower tray efficiencies.

```
LIQ VISC CP ................... XX.XX
```

(Range: 0.05 to 2.0 cP)

Input the molecular weight of the liquid. If the liquid is a multicomponent system, use a basis of mixed components for an average mol weight.

```
LIQ MOL WT. ................... XX.XX
```

Enter an alpha value if you have chosen F or T for the method. Enter a K value for a light key component if you chose A. Input the factor alpha or K. Alpha is defined as simply the light key K divided by the heavy key K component. The K factor is simply the particular component's vapor phase mole fraction divided by its liquid mole fraction. The alpha value is therefore a ratio of the chosen two key components. These key components should be those that readily point to how well the fractionator is doing its job of separation. For example, for a depropanizer tower, choose propane as the light key component and butane as the heavy key, since you wish to separate the propane from the butane to make a propane product specification. For a multicomponent system, you may try several components to determine a controlling alpha and/or to factor an average tray efficiency.

```
ALPHA or K ................... XXX.XXX
```

This factor should be inputted if you have chosen the two-film method. Enter the value of the chosen light key component liquid phase mole fraction. This value should be obtained from an equilibrium curve, a two-component binary system, or a tray-to-tray computer program printout. The program uses this factor to calculate the slope of the equilibrium curve. Be sure to select the liquid light key mole fraction at the proper curve point or tray condition that corresponds to the K value used.

```
LIQ FRCT LIT KEY ................................... XXXXX
```

Erwin two-film method

The triumph of this chapter is tray efficiency determination. Today we have commercial chemical engineering process simulation programs, which are very expensive but are mandatory for every design company and operating company. These programs do marvelous things concerning material and energy balances. They accurately analyze the most complex fractionation columns. Concerning tray efficiency accuracy, however, most fall short of expectations.

The Erwin two-film (ETF) tray efficiency method is listed in the following four steps. This is a program subroutine copied from the fractionation tray computer program Chemcalc 13 [1]. This tray efficiency method has been used successfully for over a decade and has hundreds of successful applications. It is presented here for you to apply as a supplemental to the other, more expensive computer simulation programs to make them more complete. Key variable nomenclatures follow.

EOG = one point source–referenced tray efficiency, %

H_F = froth height, in

HF_{10}, HF_{45}, H_F = effective tray froth height, in

H_L = tray liquid height above tray deck surface, in

LIQM = liquid molar flow rate over the subject tray, mol/h

LIQT = tray liquid residence time, s

M = equilibrium curve slope of referenced key component

NG = number of gas phase transfer units

NL = number of liquid phase transfer units

TEFF = key component–referenced tray efficiency, %

VAPM = vapor molar flow rate of the subject tray, mol/h

VAPT = froth tray time, s

The ETF tray efficiency principal equations are given as Eqs. (3.30) and (3.31).

$$\text{TEFF} = \frac{100 * \log \{1 + \text{EOG} * [(M * \text{VAPM/LIQM}) - 1]\}}{\log (M * \text{VAPM/LIQM})} \quad (3.30)$$

$$\text{EOG} = 1 - \exp \frac{-1}{1/\text{NG}} + \left(M * \frac{\text{VAPM/LIQM}}{\text{NL}} \right) \quad (3.31)$$

There are two key variables that are difficult to calculate: NG and NL (gas phase transfer units and liquid phase transfer units, respectively). The number of transfer units for each phase is calculated in terms of contact time and vapor rate. The contact times are determined from liquid volumetric holdup on the tray deck and froth in the tray deck bubbling area. In Step 2 in the following text, NG is calculated from froth height H_F and vapor phase holdup time VAPT.

Two empirical equations, Eqs. (3.40) and (3.41) for NG and Eqs. (3.45) and (3.46) for NL, have been derived [1] and applied to hundreds of varied fractionation cases. These equations have now been successfully used for over a decade, and are presented here as unique solutions for accurate determination of NG and NL.

Equations (3.32) and (3.33) display how to calculate molar rates starting with gas in million standard cubic feet per day (mmscfd) and liquid in gallons per minute (gpm).

$$LIQM = \frac{gpm * 60 * D_L}{7.48 * LMW} \tag{3.32}$$

$$VAPM = \frac{mmscfd * 1{,}000{,}000}{24 * 379.49} \tag{3.33}$$

The objective is first to choose one key component that is the light end component or the heavy end key component in the overhead of the column. You may choose any component in the system for the efficiency determination. Your choice, however, should be a specified component in the column overhead or bottom tray section. You may also run this efficiency for several components (one heavy key and one light key per one run at a time), choosing the lowest tray efficiency of all runs for your final answer. The following four steps will guide you through this two-film tray efficiency method.

Step 1. The first step in the procedure is to determine the key component equilibrium curve slope M. This is for one component only. You must choose a light key and a heavy key component for this tray efficiency calculation. Select components that are keys in the fractionator split. A single $K = Y/X$ equilibrium value is to be applied to absorbers and strippers as this M value. The change of Y per the change of X is sought out for the light key component.

For fractionation column tray efficiency determination, use the alpha ratio (the light key K value divided by the heavy key component K value). Please note that the fractionation column involves both the enriching section and the stripping section, as compared to absorber- or stripping-type columns.

You may easily obtain these K values from a process simulation program for your particular case. If you do not have this source, use Chap. 2 of this book or the program RefFlsh, which is included in the computer programs that accompany this book. The following hand-calculation method may be used to determine M, the equilibrium curve slope.

$$Y_1 = \frac{X_1}{1 + X_1 * (alpha - 1)} \qquad (3.34)$$

$$Y_2 = \frac{X_1 + 0.001}{1 + (X_1 + 0.001) * (alpha - 1)} \qquad (3.35)$$

$$M = \frac{Y_2 - Y_1}{0.001} \qquad (3.36)$$

where Alpha = ratio of light key component to heavy key component equilibrium K values
$\quad X_1$ = light key component mol fraction in liquid phase
$\quad Y_1$ = M slope ratio–derived Y axis vapor phase, first factor
$\quad Y_2$ = M slope ratio–derived Y axis vapor phase, second factor

Equations (3.34) through (3.36) are used in the tray computer program supplied with this book.

Step 2. Next the number of gas phase transfer units NG is to be calculated. First froth height in inches H_F must be calculated. If the dry tray pressure drop is 0.5 in or more, then use Eq. (3.37).

$$H_F = 1.559 + 9.048001 * \frac{gpm}{L_{WI}} - 2.64 * \left(\frac{gpm}{L_{WI}}\right)^2 + 0.3234 * \left(\frac{gpm}{L_{WI}}\right)^3$$

$$(3.37)$$

where L_{WI} = weir length

If the dry tray pressure drop is equal to or less than 0.1 in, use Eq. (3.38):

$$H_F = 0.2226 + 0.16065 * \frac{gpm}{L_{WI}} - 0.011993 * \left(\frac{gpm}{L_{WI}}\right)^2$$

$$+ 0.00050725 * \left(\frac{gpm}{L_{WI}}\right)^3 \qquad (3.38)$$

For your calculated dry tray pressure drop DP_{TRAY1} (inches of liquid), you may use linear interpolation of the preceding H_F values calculated to find your froth height case. Vapor froth tray time may now be calculated from Eq. (3.39):

$$\text{VAPT} = \frac{H_F}{12} * \frac{A_A}{\text{cfs}} \tag{3.39}$$

where   VAPT = time of vapor on tray, s

For VAPT = 0.1 s or less, apply Eq. (3.40) for NG calculation:

$$\text{NG} = 0.9272 + 0.035637 * \frac{\text{cfs}}{A_A} + 0.075124 * \left(\frac{\text{cfs}}{A_A}\right)^2 - 0.006525$$

$$* \left(\frac{\text{cfs}}{A_A}\right)^3 + 0.0002109 * \left(\frac{\text{cfs}}{A_A}\right)^4 \tag{3.40}$$

For VAPT = 1.0 s or greater, apply Eq. (3.41) for NG calculation:

$$\text{NG} = 6.683 + 2.15 * \frac{\text{cfs}}{A_A} + 0.16665 * \left(\frac{\text{cfs}}{A_A}\right)^2 \tag{3.41}$$

For VAPT time calculation, you may use linear interpolation to calculate the time case for NG, applying Eqs. (3.40) and (3.41) for their respective times of 0.1 and 1.0 s. If your calculated vapor time is greater than 1.0 s, assume a 1.0-s vapor tray time. Vapor tray time is seldom greater than 1.0 s except for large tray spacings of 48 in or more.

Step 3.   Determine the number of liquid phase transfer units NL in this step. As you did for vapor, first calculate the liquid holdup time on the tray, and then calculate NL from a proven empirical equation.

Calculate the clear liquid pressure loss in flow over the tray H_L (inches of liquid) using Eq. (3.42):

$$H_L = 0.4 * \text{WH} + 0.4 * \left(\frac{\text{gpm}}{L_{\text{WI}}}\right)^{2/3} \tag{3.42}$$

Calculate the volumetric rate CFL (ft³/h) of liquid flow over the tray using Eq. (3.43):

$$\text{CFL} = \frac{\text{gpm}}{7.48 * 60} \tag{3.43}$$

Calculate the liquid on tray holdup time LIQT (s) using Eq. (3.44):

$$\text{LIQT} = \frac{H_L}{12} * \frac{A_A}{\text{CFL}} \tag{3.44}$$

For LIQT = 1.0, apply Eq. (3.45) for NL calculation:

$$NL * (VISC^{3/4}) = 0.34352 + 0.0023188 * \frac{cfs}{A_A}$$

$$+ 0.015256 * \left(\frac{cfs}{A_A}\right)^2 - 0.0008312 * \left(\frac{cfs}{A_A}\right)^3$$

$$+ 0.000015077 * \left(\frac{cfs}{A_A}\right)^4 \tag{3.45}$$

For LIQT = 8.0 s, apply Eq. (3.46) for NL calculation:

$$NL * (VISC^{3/4}) = 0.5826 + 1.1379 * \frac{cfs}{A_A}$$

$$- 0.09721 * \left(\frac{cfs}{A_A}\right)^2 + 0.010381 * \left(\frac{cfs}{A_A}\right)^3$$

$$- 0.0003605 * \left(\frac{cfs}{A_A}\right)^4 \tag{3.46}$$

where   VISC = viscosity of liquid at tray temperature, cP

For LIQT other than 1 or 8 s, use linear interpolation of NL vs. LIQT.

Step 4.   Now, having the NG, NL, M, LIQM, and VAPM variables, calculate the ETF tray efficiency TEFF from Eqs. (3.47) and (3.48):

$$TEFF = 100 * \log \frac{1 + EOG * [(M * VAPM/LIQM) - 1]}{\log (M * VAPM/LIQM)} \tag{3.47}$$

$$EOG = 1 - \exp\left[- \frac{1}{1 / NG + M * (VAPM/LIQM)/NL}\right] \tag{3.48}$$

Equation (3.48) is known as point efficiency, having been given in a number of publications, one notable one being *Distillation Principles and Design Procedures* [7]. Equation (3.48) is the two-film method of predicting the ETF tray one-point efficiency, and refers to a small element of a tray that must be converted to a Murphree efficiency (Eq. 3.47) [8].

Valve-Type Tray Design and Rating

To this point we have covered general tray design for any type of tray—valve, bubble cap, or sieve type. Beginning here, we will review the design and rating of valve-type trays, followed by bubble cap and sieve tray design and rating.

Valve tray capacity factor

Referring to the explanation of liquid jet flooding (see Eq. (3.29), tray V_{LOAD} factor). Jet flooding is accountable in a factor named the capacity factor for valve-type trays CAFO. The following equations derive CAFO. For vapor density D_V (lb/ft^3) equal to or less than 0.170, then:

$$CAFO = \frac{TS^{0.65} * D_V^{0.1667}}{12} \qquad (3.49)$$

where   TS = tray spacing, in

Values of D_V greater than 0.17 lb/ft^3, the following curve-fit equations apply:

CAFO calculation by curve-fit.   The following tray capacity curves have been curve-fit from various tray manufacturing companies, including F. W. Glitsch and Koch [3]. These curve-fitted equations have been reviewed and checked against actual fractionation tray operating data, yielding excellent results for more than a decade. Nomenclature is referenced to that previously given, except CAF48, CAF24, and CAF12 = tray capacity factors for tray spacings of 48, 24, and 12 in, respectively.
For 48-in tray spacing and $D_V \leq 0.023$, use Eq. (3.50a).

$$CAF48 = 0.7923 + 0.06818 * \log (D_V) \qquad (3.50a)$$

For 48-in tray spacing and $D_V \leq 2.3$, use Eq. (3.50b).

$$CAF48 = 0.541 * \exp (-.06138 * D_V) \qquad (3.50b)$$

For 48-in tray spacing and $D_V > 2.3$, use Eq. (3.50c).

$$CAF_{48} = 0.71185 - 0.27787 * \log (D_V) \qquad (3.50c)$$

For 24-in tray spacing and $D_V \leq 0.119$, use Eq. (3.50d).

$$CAF_{24} = 0.65259 * (D_V^{0.16596}) \qquad (3.50d)$$

For 24-in tray spacing and $D_V \leq 3.1$, use Eq. (3.50e).

$$CAF_{24} = 0.46157 * \exp (-.035858 * D_V) \qquad (3.50e)$$

For 24-in tray spacing and $D_V > 3.1$, use Eq. (3.50f).

$$CAF_{24} = 0.71185 - 0.27787 * \log (D_V) \qquad (3.50f)$$

For 12-in tray spacing and $D_V \leq 0.17$, use Eq. (3.50g).

$$CAF_{12} = 0.41925 * (D_V^{0.16622}) \qquad (3.50g)$$

For 12-in tray spacing and DV ≤ 5.3, use Eq. (3.50h).

$$CAF_{12} = 0.31643 * \exp(-0.025506 * D_V) \qquad (3.50h)$$

For 12-in tray spacing and $D_V > 5.3$, use Eq. (3.50i).

$$CAF_{12} = 0.71185 - 0.27787 * \log(D_V) \qquad (3.50i)$$

Calculate CAFO by linear interpolation from CAF_{48}, CAF_{24}, and CAF_{12} if your tray spacing is not 48, 24, or 12 in.

$$CAFO = CAF_{24} \quad \text{or} \quad CAFO = CAF_{12} \quad \text{or} \quad CAFO = CAF_{48} \quad (3.50j)$$

Use linear interpolation if your tray spacing is not exactly 48, 24, or 12 in.

Tray active area flood calculation

Tray active area flood is now calculated after all the preceding factors are calculated. The following equations have long been proven by published programs such as Chemcalc 13 [1,2].

$$FLOOD\% = \frac{V_{LOAD} + (\text{gpm} * FPL/12{,}800)}{(A_A * CAFO * SF)} * 100 \qquad (3.51)$$

$$FLOOD\% = \frac{V_{LOAD}}{AREA * CAFO * 0.775} * 100 \qquad (3.52)$$

Use the larger FLOOD% of the two preceding active area flood equations.

Caution: If your tray flood is greater than 90%, you potentially have a failed fractionator column, absorber, or stripper. However, up to 100% tray flood, a number of successful operations have occurred. These are exceptions. Good-practice engineering design demands a 90% tray flood factor limitation! Higher tray flood percentage operation is not recommended by any tray design vendor or consultant. At 110% flood or greater, be assured you indeed have a failed column.

Valve tray pressure drop calculation

Several factors are required first in order to calculate the tray pressure drop in inches of clear liquid. This liquid pressure unit is that liquid density referred to as liquid density used DI, lb/ft^3. First calculate the weir length L_{WI}:

$$\text{If PASS} = 1, \text{ then } L_{WI} = 2 * Z * 12 \qquad (3.53)$$

$$\text{If PASS} = 2, \text{ then } L_{WI} = 2 * Z * 12 + \text{DIA} * 12 \tag{3.54}$$

$$\text{If PASS} = 3, \text{ then } L_{WI} = 2 * Z * 12 + \left\{ \left(\text{DIA} * \frac{12}{2} \right)^2 \right.$$

$$\left. - \left[\left(\text{DIA} * \frac{12}{2} \right) - \text{FPL} - \text{DC} - H_5 \right]^2 \right\}^{0.5} * 2 \tag{3.55}$$

$$\text{If PASS} = 4, \text{ then } L_{WI} = 2 * Z * 12 + \left\{ \left(\text{DIA} * \frac{12}{2} \right)^2 \right.$$

$$\left. - \left[\left(\text{DIA} * \frac{12}{2} \right) - \text{FPL} - \text{DC} \right]^2 \right\}^{0.5} * 2 + \text{DIA} * 12 \tag{3.56}$$

Select a tray deck thickness VLVTH. Standards are 0.074, 0.104, 0.134, and 0.25 in. The hole area of a current-day valve-type tray is calculated from:

$$A_H = 14 * \frac{A_A}{78.5}$$

where A_H = tray deck full open hole area, ft^2

The dry tray pressure drop for the tray under consideration is first found using the following equation [9]:

$$\text{DP}_{\text{TRAY1}} = \left(1.35 * \text{VLVTH} * \frac{510}{D_L} \right) + K_1 * \left(\frac{\text{cfs}}{A_H} \right)^2 * \frac{D_V}{D_L} \tag{3.57}$$

where DP_{TRAY1} = pressure drop in dry tray condition, inches of liquid
K_1 = 0.2 for standard valve trays and 0.10 for low Δ_Ps
VLVTH = valve thickness (in), nominally 0.037 in (20 gauge)

The height of downcomer liquid is next calculated as an additive for the total valve tray pressure drop summation [9]: wet tray downcomer Δ_P, HDC$_2$, inches of liquid, Eqs. (3.58), (3.59), and (3.60):

$$\text{VUD} = \frac{\text{gpm} / [7.48 * (L_{WI}/12) * (\text{DCC}/12)]}{60} \tag{3.58}$$

where   VUD = liquid velocity flowing between downcomer outlet and tray deck top, ft/s

$$\text{HUD} = 0.65 * \text{VUD}^{2/3} \tag{3.59}$$

where   HUD = head loss under downcomer, inches of liquid

The total tray pressure drop HDC$_2$ (inches of liquid) is calculated with preceding factors installed in Eq. (3.60).

$$HDC_2 = DP_{TRAY1} + HUD + WH + 0.4 * \left(\frac{gpm}{L_{WI}}\right)^{2/3} \qquad (3.60)$$

where HDC_2 = downcomer liquid backup, inches of liquid
          WH = weir height, in

CAUTION: Make sure your downcomer backup HDC_2 is not greater than 60% of the tray spacing. If it is, then you have a potential tray flood, which cannot be tolerated. Your fractionator, absorber, or stripping column will likely fail. If HDC_2 is greater than 80%, rest assured you have a failed column! What can you do about it? First, if you are rating an existing tray, consider backing off some of the liquid and vapor traffic flow rates on the subject tray. If you are designing a new tray, then increase your tray active area A_A and your downcomer area DCAREA. One of these solutions will work. So take caution here. This is a common problem experienced countless times by refinery and chemical plant operations.

Bubble Cap Tray Design and Rating

Bubble cap tray flood calculation

A typical bubble cap design is shown in Fig. 3.8. Similar to valve-type trays, the bubble cap tray has a tray loading factor XSB, which is dimensionless. The following three equations reveal the derivation of XSB:

$$LSB = gpm * 500 * \left(\frac{D_L}{62.4}\right) \qquad (3.61)$$

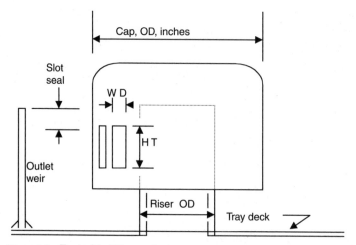

Figure 3.8 Typical bubble cap design.

$$\text{GSB} = \frac{(\text{mmscfd} * 1,000,000/379.49) * \text{MW}}{24} \tag{3.62}$$

$$\text{XSB} = \frac{\text{LSB}}{\text{GSB}} * \left(\frac{D_V}{D_L}\right)^{0.5} \tag{3.63}$$

A bubble cap tray load factor KSB may be calculated from Eqs. (3.64) and (3.65).

For 24-in tray spacing:

$$\text{KSB} = 0.38104 - 0.5748 * \text{XSB}$$

$$+ 0.42074 * \text{XSB}^2 - 0.10669 * \text{XSB}^3 \tag{3.64}$$

For 12-in tray spacing:

$$\text{KSB} = 0.22412 - 0.30904 * \text{XSB}$$

$$+ 0.2178 * \text{XSB}^2 - 0.05416 * \text{XSB}^3 \tag{3.65}$$

If your specific bubble cap tray spacing is, say, 36 or 12 in, use linear extrapolation of Eqs. (3.64) and (3.65) to find your particular KSB. Equations (3.64) and (3.65) are curve-fitted to Fig. 18-10 in the *Chemical Engineer's Handbook* [10,11]. This figure is applicable to bubble cap trays, valve trays, and sieve-type trays. This figure gives flooding gas jetting velocities to $\pm 10\%$, subject to the following restrictions:

1. System is low-flooding.
2. Weir height is less than 15% of plate spacing.
3. Sieve plate perforations are ¼ in or less in diameter.
4. For slot (bubble cap), perforation (sieve), or full valve opening (valve tray) area, the ratio of total hole area A_H to A_A is 0.1 or greater.

Otherwise, the value of KSB should be corrected:

A_H / A_A	KSB corrected multiplier
0.1	1.00
0.08	0.90
0.06	0.80

In addition, correct the KSB for surface tension SURFT, as shown in Eq. (3.66).

$$\text{KSB} = \text{KSB} * \left(\frac{\text{SURFT}}{20}\right)^{0.2} \tag{3.66}$$

where   SURFT = liquid surface tension, dyn/cm

Calculate the bubble cap tray active area A_A from Eqs. (3.67) through (3.72). The tray active area for bubble caps is derated by a factor of 0.92; therefore:

$$A_A = A_A * 0.92 \tag{3.67}$$

Each bubble cap has a certain coverage of the tray active area and a certain hole area for gas passage. These two-dimensional variables are also factored for bubble cap square pitch location or for bubble cap triangular pitch location. The following equations are therefore derived and used:
For square pitch:

$$SECTAREA = \left(\frac{PTCH}{12}\right)^2 \tag{3.68}$$

$$SECTHA = 3.1416 * \left(\frac{RISD}{24}\right)^2 \tag{3.69}$$

For triangular pitch:

$$SECTAREA = \left(\frac{PTCH}{12} * 0.86603\right) * \left(\frac{PTCH/12}{2}\right) \tag{3.70}$$

$$SECTHA = \frac{3.1416 * (RISD/24)^2}{2} \tag{3.71}$$

where   SECTAREA = single bubble cap active area, ft^2
SECTHA = single bubble cap hole area, ft^2

$$SECTNO = \frac{A_A}{SECTAREA} \tag{3.72}$$

where   SECTNO = total number of bubble caps on a single tray

$$RISHA = SECTHA * SECTNO \tag{3.73}$$

where   RISHA = total bubble cap hole area per tray, ft^2

$$V_N = \frac{cfs}{A_A} \tag{3.74}$$

where V_N = vapor velocity through tray hole area, ft/s

$$V_M = \frac{KSB}{[D_V/(D_L - D_V)]^{0.5}} \tag{3.75}$$

where V_M = flood vapor velocity through tray active area, ft/s

$$\text{FLD}_2 = \frac{V_N}{V_M} * 90 \qquad (3.76)$$

where FLD_2 = bubble cap active area flood

Bubble cap pressure drop calculation

Select an L_{WI} equation in accordance with the number of tray liquid passes, Eqs. (3.53) through (3.56). Execute the equation the same as for valve-type trays, determining L_{WI} for the bubble cap tray.

If triangular bubble cap placement is used, then CAPNO = 0.5; for square pitch, CAPNO = 1.0. This is how many bubble caps are in a single section area SECTAREA.

$$\text{SLOTAREA} = \text{CAPNO} * \text{SECTNO}$$
$$* \frac{\text{SLTHT} * \text{SLTWD} * \text{SLTNO}}{144} \qquad (3.77)$$

where   SLOTAREA = total slot area of bubble caps on tray, ft^2
             SLTHT = bubble cap slot height, in
             SLTNO = number of slots on a single bubble cap
             SLTWD = bubble cap slot width, in

$$V_S = \frac{\text{cfs}}{\text{SLOTAREA}} \qquad (3.78)$$

where V_S = total of actual vapor velocity of caps, ft/s

$$KK_1 = 1.2 * \left(\frac{D_V}{D_L - D_V}\right)^{0.2} * \text{SLTHT}^{0.8} * V_S^{0.4} \qquad (3.79)$$

where KK_1 = dry tray pressure drop–derived constant
         KK_2 = dry tray pressure drop–derived constant

$$KK_2 = 1.4791 * \exp\left(-0.8412 * \frac{A_A/0.92}{\text{RISHA}}\right) \qquad (3.80)$$

$$\text{HHD} = KK_1 + KK_2 * \frac{D_V}{D_L} * \left(\frac{\text{cfs}}{\text{RISHA}}\right)^2 \qquad (3.81)$$

where   HHD = dry tray bubble cap riser pressure drop, inches of liquid

$$\text{HHOW} = 0.48 * \left(\frac{\text{gpm}}{L_{\text{WI}}}\right)^{2/3} \qquad (3.82)$$

where   HHOW = height of liquid for pressure drop over weir, inches of
                 liquid

$$\text{HHDS} = \text{SLTSL} + \text{HHOW} + 0.12 * \frac{\text{FPL}}{\text{PTCH}} \qquad (3.83)$$

where   HHDS = height of liquid over downcomer seal for downcomer seal loss, inches of liquid
   PTCH = dimension center-to-center between bubble caps, in
   SLTSL = bubble cap slot measure from slot to weir top, in

$$F_{\text{GA}} = \frac{\text{cfs}}{A_A} * D_V^{0.5} \qquad (3.84)$$

where F_{GA} = dry tray gas rate loading factor

$$\text{beta} = 1 - 1.222 * F_{\text{GA}} + 1.4738 * F_{\text{GA}}^2 - 0.7688 * F_{\text{GA}}^3$$
$$+ 0.1368 * F_{\text{GA}}^4 \qquad (3.85)$$

where   beta = wet tray gas pressure drop factor, accounting for frictional loss

$$\text{HHL} = \text{beta} * \text{HHDS} \qquad (3.86)$$

where   HHL = height of liquid backup in the downcomer, in

$$\text{DP}_{\text{TRAY}} = \text{HHD} + \text{HHL} \qquad (3.87)$$

where DP_{TRAY} = total bubble cap tray pressure drop, inches of liquid

Important Note: Bubble cap HHD factor is equivalent to the DP_{TRAY1} dry pressure drop of valve trays. The bubble cap tray total pressure drop factor DP_{TRAY} is equivalent to the HDC_2 factor of valve-type trays. You may therefore substitute these bubble cap values in the ETF efficiency equations as given for valve trays to determine bubble cap tray efficiency.

Sieve Tray Design and Rating

Sieve tray flood

As with bubble cap flood, sieve tray flood is the ratio of the V_{LOAD} design (notated here as V_N) to the maximum V_{LOAD} of sieve trays. The V_{LOAD} (notated here as V_M) for sieve trays is factored by surface tension and by the actual cubic feet per second throughflow.

Tray active area actual loading:

$$V_N = \frac{\text{cfs}}{A_A} \qquad (3.88)$$

Maximum flood tray loading:

$$V_M = \frac{KSB}{[D_V/(D_L - D_V)]^{0.5}}$$

(3.89)

Sieve tray active area flood%:

$$FLD_2 = \frac{V_N}{V_M} * 90$$

(3.90)

Equations (3.89) and (3.90) equate the tray active area vapor loading V_N to the maximum V_M for determining the gas in liquid entrainment flooding of the tray. The early work of Souders and Brown [12], based on a force balance on an average suspended droplet of liquid, led to the definition of a capacity parameter V_M. Both V_N and V_M here refer to the active area of the tray. This active area is simply the net tower cross-section internal area less the downcomer areas. The downcomer areas include both the downcomer inlets and outlets.

Equation (3.89) is the sieve tray liquid entrainment flood gas loading equation. Equation (3.89) sets the maximum gas rate V_M. At a higher V_M, excess gas–liquid froth buildup would reach the tray above and recycle liquid to it. This liquid recycle would build up to a point at which the liquid would block any vapor passage, resulting in a flooded column and costly shutdown.

Sieve tray jet flooding:

$$FLD_3 = \frac{(cfs/HOLHA)^2 * D_V/D_L}{8.75} * 100$$

(3.91)

Equation (3.91) is the jet flood equation. The chief difference between this equation and the entrainment flood equations, (3.88) through (3.90), is the area references. Equation (3.91) is based on the total sieve tray hole area for gas passage, and Eq. (3.88) through (3.90) are based on the tray active area. Again, the tray active area is simply the tower cross-sectional area less the total downcomer area.

The jet flood equation is also based on the work of Souders and Brown [12]. This equation computes the ratio of the square power of the vapor load (noted in this chapter as V_{LOAD}) to a constant, 8.75, to derive the tray flood. It has been used for over three decades by tray vendors (Koch and F.W. Glitsch [3]) to design and rate sieve-type trays. In many cases, especially for sieve-type tray design, jet flood governs tray flood and thus is the primary sieve tray design and rating equation. In contrast, Eqs. (3.88) through (3.90) are the principal equations for flood determination for both valve- and bubble cap–type trays.

Liquid mass rate passage over tray LSB, lb/h:

$$\text{LSB} = \text{gpm} * 500 * \frac{D_L}{62.4} \tag{3.92}$$

Gas mass rate through tray holes GSB, lb/h:

$$\text{GSB} = \text{mmscfd} * \frac{1{,}000{,}000}{379.49} * \frac{\text{MW}}{24} \tag{3.93}$$

Mass ratio tray loading factor used in Eqs. (3.95) and (3.96) to calculate KSB:

$$\text{XSB} = \frac{\text{LSB}}{\text{GSB}} * \left(\frac{D_V}{D_L}\right)^{0.5} \tag{3.94}$$

The following curve-fitted equations for sieve tray loading refer to Fig. 18-10 in the *Chemical Engineer's Handbook* [10,11].
Use Eq. (3.95) to calculate KSB for 24-in sieve tray spacing.

$$\text{KSB}_{24} = 0.3897 - 0.5516 * \text{XSB} + 0.23386$$

$$* \text{XSB}^2 + 0.05098 * \text{XSB}^3 \tag{3.95}$$

where   KSB = curve-fitted sieve tray loading curves, calculated from Eqs. (3.92) through (3.97)

Use Eq. (3.96) to calculate KSB for 12-in sieve tray spacing.

$$\text{KSB}_{12} = 0.22596 - 0.28475 * \text{XSB} + 0.18179$$

$$* \text{XSB}^2 - 0.042332 * \text{XSB}^3 \tag{3.96}$$

Apply linear interpolation to calculate KSB for sieve tray spacing other than 12 or 24 in.
Correct KSB curve-fitted factor for surface tension SURFT, dyn/cm:

$$\text{KSB} = \text{KSB} * \left(\frac{\text{SURFT}}{20}\right)^{0.2} \tag{3.97}$$

Having now established the main equations for sieve tray flood, Eqs. (3.88) through (3.90), and having calculated a KSB value, we may now calculate the entrainment active area flood of the column. Next, Eq. (3.91) must be used to calculate jet flooding FLD_3. The equation having the highest flood value wins and so governs the tray design or rating. But first, the tray total hole area HOLHA must be determined.

Several equations are applied to calculate sieve tray hole area. Normally sieve tray hole individual diameters are $\frac{3}{16}$ to $\frac{3}{4}$ in. As for bubble caps, sieve tray hole pitch is the measure from center to center between sieve tray holes. The pitch may be square or triangular, each angle of the triangle being 60°. Section areas are the tray deck area of one pitch markoff area, noted as SECTAREA. For square pitch, one complete hole area is measured; for triangular pitch, one-half hole area is measured, noted as SECTHA. The active area divided by SECTAREA equals the number of pitch sections on the tray deck, noted as SECTNO. Equations (3.98) through (3.103) will help clarify these statements.

For triangular-type sieve tray pitch, use Eq. (3.98) to calculate a single triangular tray deck area, noted as SECTAREA. Each leg of the triangle is an equal pitch measure. One triangle area is equal to SECTAREA in ft^2.

$$\text{SECTAREA} = \left(\frac{\text{PTCH}}{12} * 0.86603 \right) * \left(\frac{\text{PTCH}/12}{2} \right) \qquad (3.98)$$

Pitch is simply the measure of the distance from hole center to hole center in inches. Leibson [13] states that the distance of pitch measure should be 2 to 5 times the hole diameter. Leibson also states that the optimum pitch measure is 3.8 times the hole diameter.

For square pitch, apply the following equation:

$$\text{SECTAREA} = \left(\frac{\text{PTCH}}{12} \right)^2 \qquad (3.99)$$

You should recognize that SECTAREA in Eq. (3.99) is simply the area of a square that has equal sides. Each side is one pitch measure. Here also the pitch measure should be 2 to 5 times the hole diameter, and the optimum pitch should be 3.8 times the hole diameter.

Next, calculate the hole area per SECTAREA, noted as SECTHA in ft^2 for triangular pitch, use Eq. (3.100):

$$\text{SECTHA} = \frac{3.1416 * (\text{HOLDIA}/24)^2}{2} \qquad (3.100)$$

Note that each triangle area contains one-half of a single hole area, which is the basis of Eq. (3.100). For square pitch, use Eq. (3.101):

$$\text{SECTHA} = 3.1416 * \left(\frac{\text{HOLDIA}}{24} \right)^2 \qquad (3.101)$$

Note that each square area contains one complete hole area, which is the basis of Eq. (3.101).

Having established the SECTAREA factor area for both square and triangular pitch, next determine how many SECTAREA units are on a complete tray deck. The number of SECTAREAs on a tray deck is noted as SECTNO. Use Eq. (3.102) to calculate SECTNO:

$$\text{SECTNO} = \frac{A_A}{\text{SECTAREA}} \qquad (3.102)$$

Now, having divided the number of section areas by the pitch measure (noted as SECTNO), and having measured the hole areas in each single pitch (noted as SECTHA), the total hole area (noted as HOLHA) may be calculated applying Eq. (3.103):

$$\text{HOLHA} = \text{SECTHA} * \text{SECTNO} \qquad (3.103)$$

Please note that HOLHA is the total hole area in ft^2 on a single tray deck. It is used in Eq. (3.91) to calculate sieve tray jet flood and will be used to calculate sieve tray pressure drop as well.

Sieve tray pressure drop

Sieve tray pressure drop requires factors similar to those used for valve trays, such as weir length L_{WI}. Please refer to Eqs. (3.53) through (3.56) for sieve tray weir length in inches.

The sieve tray dry pressure drop is calculated next, applying the following equations:

$$\text{AHAA} = \frac{\text{HOLHA}}{A_A} \qquad (3.104)$$

$$\text{THDIA} = \frac{\text{DTH}}{\text{HOLDIA}} \qquad (3.105)$$

where   AHAA = A_A ratio of total hole sieve tray hole area HOLHA to sieve tray active area
THDIA = ratio of sieve tray deck thickness to a single sieve tray hole diameter

Using these calculated ratios, AHAA and THDIA, a hole discharge coefficient factor CFCV is calculated from a curve-fitted equation in Fig. 18-14 in the *Chemical Engineer's Handbook* [14]. CFCV is a factor in Eq. (3.112) for calculating the sieve tray dry pressure drop.

If THDIA is equal to 1.2, use Eq. (3.106):

$$\text{CFCV} = 0.8125 + 0.75 * \text{AHAA} \qquad (3.106)$$

If THDIA is equal to 1.0, use Eq. (3.107):

$$\text{CFCV} = 0.7775 + 0.69 * \text{AHAA} \tag{3.107}$$

If THDIA is equal to 0.8, use Eq. (3.108):

$$\text{CFCV} = 0.7 + 0.8 * \text{AHAA} \tag{3.108}$$

If THDIA is equal to 0.6, use Eq. (3.109):

$$\text{CFCV} = 0.6737 + 0.717 * \text{AHAA} \tag{3.109}$$

If THDIA is equal to 0.2, use Eq. (3.110):

$$\text{CFCV} = 0.6446 + 0.686 * \text{AHAA} \tag{3.110}$$

If THDIA is equal to 0.1, use Eq. (3.111):

$$\text{CFCV} = 0.589 + 0.72 * \text{AHAA} \tag{3.111}$$

Use linear interpolation to calculate any THDIA value within or reasonably outside the range presented in Eqs. (3.106) through (3.111).
Calculate the sieve dry tray factor KK_2:

$$KK_2 = 0.186 * \text{CFCV}^2 \tag{3.112}$$

Calculate the sieve tray dry hole gas velocity US, ft/s:

$$\text{US} = \frac{\text{cfs}}{\text{HOLHA}} \tag{3.113}$$

The dry tray pressure loss in inches of clear liquid HHD is now calculated:

$$\text{HHD} = KK_2 * \frac{D_V}{D_L} * \text{US}^2 \tag{3.114}$$

$$\text{HHOW} = 0.48 * \left(\frac{\text{gpm}}{L_{\text{WI}}} \right)^{2/3} \tag{3.115}$$

where   HHOW = height of crest clear liquid over the sieve tray outlet, inches of clear liquid

Next, calculations are made to get the hydraulic gradient HHG, in inches of clear liquid. But first froth with clear liquid velocity across the tray UF, ft/s, must be calculated, and the hydraulic radius of the aerated mass R_H must also be calculated [15].

$$R_H = \frac{TS * 0.4 * [(DIA + 2 * Z) / 2]}{2 * TS * 0.4 + 12 * [(DIA + 2 * Z)/2]} \qquad (3.116)$$

where R_H = mass on tray active area hydraulic radius, ft

$$UF = 12 * \frac{gpm / 448.8}{1.1 * [(DIA + 2 * Z)/2]} \qquad (3.117)$$

where   UF = froth and clear liquid velocity, ft/s

$$HHG = \frac{0.08 * UF^2 * FPL/12}{386.4 * R_H} \qquad (3.118)$$

where   HHG = hydraulic gradient across sieve tray, inches of clear liquid

The height in inches of clear liquid over the sieve tray deck top surface HHDS may next be calculated. Having calculated HHOW in Eq. (3.115), and HHG in Eq. (3.118), HHDS may therefore be calculated in Eq. (3.119).

$$HHDS = WH + HHOW + \frac{HHG}{2} \qquad (3.119)$$

$$F_{GA} = \frac{cfs}{A_A} * D_V^{0.5} \qquad (3.120)$$

Before a total sieve tray pressure drop can be summed, the froth pressure in inches of clear liquid over the active area must be calculated. This froth height actually reduces the HHDS value by a factor called the aeration beta correction. This has been done by Smith, who plotted the aeration factor beta vs. F_{GA} (see Eq. (3.120) for F_{GA}). Equation (3.121) is a curve-fit of Smith's beta curve plot [16]. Generally a beta factor of 0.7 to 0.8 is calculated using Eq. (3.121).

Beta vs. F_{GA} curve fit:

$$beta = 1 - 1.222 * F_{GA} + 1.4738 * F_{GA}^2$$

$$- 0.7688 * F_{GA}^3 + 0.1368 * F_{GA}^4 \qquad (3.121)$$

Correct the HHDS sieve tray hydraulic gradient with the beta factor (froth correction):

$$HHL = beta * HHDS \qquad (3.122)$$

The total sieve tray pressure drop DP_{TRAY} may now be calculated using the dry tray drop HHD from Eq. (3.114) and HHL from Eq.

(3.122). Total sieve tray pressure drop DP_{TRAY} in inches of clear liquid is now calculated in a final equation:

$$DP_{TRAY} = HHD + HHL \qquad (3.123)$$

Table 3.9 lists the code for the sieve tray calculations.

TABLE 3.9   Sieve Tray Code Listing

```
3180 Rem SIEVE TRAY PRESSURE DROP CALC, DPTRAY, INCHES LIQUID
3190 If PASS = 1 Then LWI = 2 * Z * 12
3200 If PASS = 2 Then LWI = 2 * Z * 12 + DIA * 12
3210 If PASS = 3 Then LWI = 2 * Z * 12 + ((((DIA * 12 / 2) ^ 2) - (((DIA * 12 / 2) - FPL -
     DC - H5) ^ 2)) ^ .5) * 2
3220 If PASS = 4 Then LWI = 2 * Z * 12 + ((((DIA * 12 / 2) ^ 2) - (((DIA * 12 / 2) - FPL -
     DC) ^ 2)) ^ .5) * 2 + DIA * 12
3230 AHAA = HOLHA / AA: THDIA = DTH / HOLDIA
3240 AHAA12 = .8125 + .75 * AHAA: AHAA10 = .7775 + .69 * AHAA: AHAA08 = .7 + .8
     * AHAA: AHAA06 = .6737 + .717 * AHAA: AHAA02 = .6446 + .686 * AHAA:
     AHAA01 = .589 + .72 * AHAA
3250 If THDIA <= 1.2 And THDIA > 1! Then CFCV = AHAA10 + (AHAA12 - AHAA10)
     * (THDIA - 1!) / .2
3260 If THDIA <= 1! And THDIA > .8 Then CFCV = AHAA08 + (AHAA10 - AHAA08) *
     (THDIA - .8) / .2
3270 If THDIA <= .8 And THDIA > .6 Then CFCV = AHAA06 + (AHAA08 - AHAA06) *
     (THDIA - .6) / .2
3280 If THDIA <= .6 And THDIA > .2 Then CFCV = AHAA02 + (AHAA06 - AHAA02) *
     (THDIA - .2) / .4
3290 If THDIA <= .2 And THDIA > .1 Then CFCV = AHAA01 + (AHAA02 - AHAA01) *
     (THDIA - .1) / .1
3300 If THDIA <= .1 Then CFCV = AHAA01
3310 KK2 = .186 * (CFCV ^ 2)
3320 US = CFS / HOLHA 'SIEVE HOLE ACTUAL VELOCITY, FT/SEC
3330 HHD = KK2 * (DV / DL) * (US ^ 2)
3340 HHOW = .48 * (GPM / LWI) ^ (2 / 3)
3350 RH = (TS * .4 * ((DIA + 2 * Z) / 2)) / (2 * TS * .4 + 12 * ((DIA + 2 * Z) / 2))'HYDRAULIC RADIUS
3360 UF = 12 * (GPM / 448.8) / (1.1 * ((DIA + 2 * Z) / 2))'FROTH & CLR LIQ FT/SEC
3370 HHG = (.08 * (UF ^ 2) * FPL / 12) / (386.4 * RH)
3380 HHDS = WH + HHOW + HHG / 2
3390 FGA = (CFS / AA) * DV ^ .5
3400 BETA = 1! - 1.222 * FGA + 1.4738 * (FGA ^ 2) - .7688 * (FGA ^ 3) + .1368 * (FGA ^ 4)
3410 HHL = BETA * HHDS
3420 DPTRAY = HHD + HHL
3430 VUD = (GPM / (7.48 * (LWI / 12) * (DCC / 12))) / 60
3440 HUD = .65 * (VUD) ^ 2: WARN$ = "N"
3450 HDC2 = DPTRAY + HUD + WH + .4 * (GPM / LWI) ^ (2 / 3)
3460 If HDC2 > .6 * TS Then WARN$ = "Y": FLD1 = (HDC2 / (.6 * TS)) * 90
3470 Rem TRAY EFFICIENCY CALC, CAMPBELL FIG 11.6
3480 XEFF = KEFF * VISC
3490 If COLMN$ = "F" Or COLMN$ = "f" Then TEFF = Exp(3.9325 - .2477 * Log(XEFF) -
     (.003746 * (Log(XEFF)) ^ 2) - (.0012526 * (Log(XEFF)) ^ 3) - (.0006498 * (Log(XEFF)) ^ 4))
3500 XAEFF = KEFF * LMW * VISC / DL
3510 If COLMN$ = "A" Or COLMN$ = "a" Then TEFF = Exp(3.657 - .2182 *
     Log(XAEFF) - (.016883 * (Log(XAEFF)) ^ 2) - (.0024729 * (Log(XAEFF)) ^ 3) -
     (.00015311 * (Log(XAEFF)) ^ 4))
3520 If COLMN$ = "T" Or COLMN$ = "t" GoTo 3540
3530 Return
```

Tower Random Packing

Any fractionation tray design/rating summary is not complete without a review and general coverage of random tower packing. The remainder of this chapter is dedicated to this subject. Packing-type fractionation tower types have been applied mostly to stripping and absorption operations. Random tower packing-type fractionation may also be applied to any rectifier-stripping-type fractionator. Here we will review typical random packing and the flooding and loading limits applicable to any absorber, stripper, or fractionator. First please notice Fig. 3.9 [17], which contains plotted curves for generalized pressure drop correlation. The equations with nomenclature are:

$$Y = \frac{(G/3600)^2 \, F \, \mu^{0.1}}{32.2 \, \rho_G \, (\rho_L - \rho_G)}$$

$$X = \frac{L}{G} \left(\frac{\rho_G}{\rho_L} \right)^{0.5}$$

where F = packing factor
G = gas mass velocity, lb/ft$^2 \cdot$ h
L = liquid mass velocity, lb/ft$^2 \cdot$ h
μ = kinematic liquid viscosity, cst

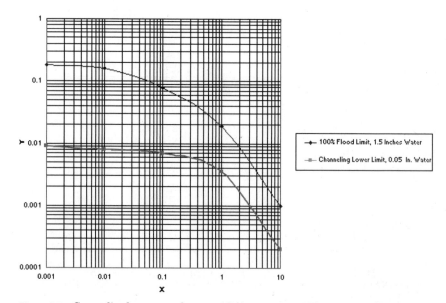

Figure 3.9   Generalized pressure-drop correlation, upper and lower curve limits. (*From J. S. Eckert, "Design Techniques for Sizing Packed Towers,"* Chem. Eng. Progress *57(9), 1961. Used by permission of* Chemical Engineering Progress.)

ρ_G = gas density, lb/ft^3
ρ_L = liquid density, lb/ft^3

In the program section of the software supplied with this book you will find a program called Absorb. Please run this program. You will note that simply clicking on the Run Program button will display both defaulted values and calculated values taken from Fig. 3.9. Please observe that each of the Fig. 3.9 curves are constant packing pressure drop, given in inches of water per foot of packed depth. The key here is noting the values of this pressure drop.

I encourage using vendor-recommended values for the packing factor F. These are readily available via a simple phone request to any vendor or supplier. When you run the supplied PPE program Absorb, a typical listing for one vendor's F value is given in the input side of the Visual Basic form.

Packing flood

A packing pressure drop of 1.5 in/ft is approximately 95% of column packing flood. At 2.0 in/ft of pressure drop D_P, most random packed towers are at the flood point. It is thus prudent and good practice to keep the limiting D_P at 1.5 in/ft or less.

Packing low-load channeling

You will note that the lowest D_P value curve shown in Fig. 3.9 is 0.05 in/ft. This is because lesser D_P values are subject to severe liquid channeling. This D_P value of 0.05 is 40 to 50% of flood value for the packed tower. A lesser value is most surely a liquid channeling condition, meaning the liquid will channel to one side of the packing or will skim only the tower's inside surface. In both of these conditions the needed contact between gas and liquid is severely missing. To avoid this loss of contact, keep your packing D_P at an absolute minimum of 0.05. Some vacuum systems however, are reported to function reasonably well at a D_P lower limit of 0.02.

HETP packing depth efficiency

Numerous equation methods have been presented to determine packing efficiencies through means of determining height equivalent to a theoretical stage HETP. None have proven effective or reliable for all cases. Those that have promise are limited by small window ranges of physical data such as liquid viscosity, liquid surface tension, and mass transfer coefficients. The prudent designer will therefore go by actual proven operating data. However, here we will relate fractionation tray efficiency to actual packing depth required. (See Fig. 3.10 and Table 3.10.) Please note that this is for minimum packing depth.

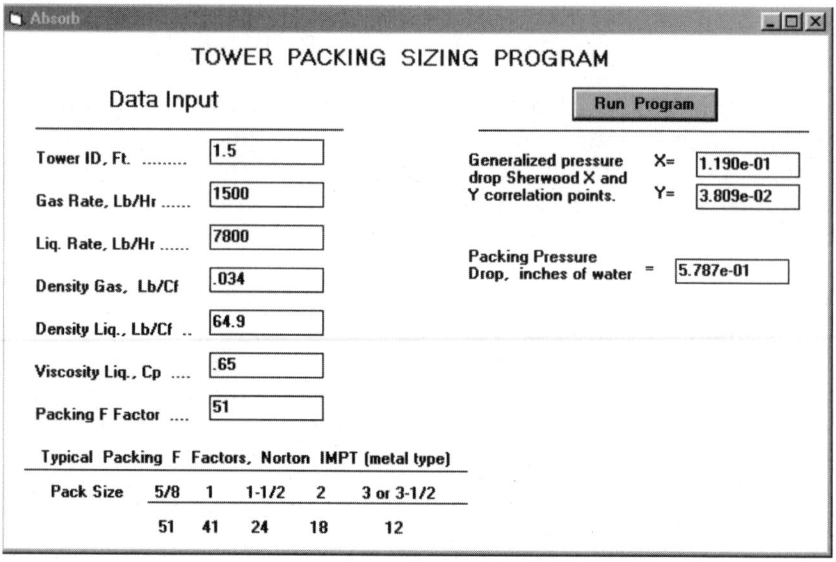

Figure 3.10 Actual screen display in Absorb.exe.

Having presented the preceding for first-time publication, I caution users to confirm final designs with other sources before concluding such designs. From Table 3.10, it appears that a 50% tray efficiency requires only approximately 0.5 ft more of packing than a 90% tray efficiency. In a more simple way, at low-tray efficiency, about 2.5 ft of metal ring packing equals two tower trays. From many years of operation findings, it has been concluded that the packing height equivalent to a theoretical stage HETP ranges from 2 to 3 ft. Table 3.10 affirms this phenomenon one more time.

One last contribution to HETP in this chapter is to establish tray efficiency by simply running one of the tray programs and inputting T for the tray method selection. The program will give you the well-proven tray efficiency two-film method [1]. Then refer to Table 3.10 and estimate your HETP minimum required. I always recommend adding at least 6 in to the HETP for a reasonable, safe design.

TABLE 3.10   Packing Height HETP, ft, Required Related to Tray Efficiency

Metal packing ring size	Tray Efficiency					
	50%	60%	70%	80%	90%	100%
2 in	2.5	2.3	2.2	2.1	2.0	1.9
1 in	2.1	1.9	1.8	1.7	1.6	1.5
½ in	1.7	1.5	1.4	1.3	1.2	1.2

Packing liquid distribution trays and limiting packing height

Having an HETP and the number of theoretical stages required, you are ready to set the packing height in the tower. Please realize that it is critical to ensure that good and effective liquid distribution occurs at the top of each packed tower section. You may ask how many packed sections are necessary. Industry worldwide practice has been to limit the height of each packed section to preferably 6 ft, and to no more than 10 ft in almost every case.

Field test and actual operation findings have revealed that 6 ft is a good practice number, because liquid channeling is very likely in most cases at 70% or less of design liquid-vapor loading. Packed sections 10 ft high or more are referenced here for the 70% design load. Liquid channeling has occurred here even with the best of the tray liquid distributors. Therefore, good practice is to ensure that you select a reliable vendor's liquid distribution tray and keep a 6- to 8-ft packing height in each tower section.

References

1. Erwin, D. L., "Erwin Two-Film Tray Efficiency," *Chemcalc 13 Fractionation Tray Design and Rating*, Gulf Publishing, Houston, TX, 1988, pp. 46–48.
2. Erwin, D. L., *Chemcalc 13 Fractionation Tray Design and Rating* (software), Gulf Publishing, Houston, TX, 1988.
3. Glitsch, F. W., *Ballast Tray Design Manual*, Bulletin 4900, Fritz W. Glitsch & Sons Inc., Dallas, TX, p. 26, 1961.
4. O'Connell, H., "Determination of Actual Trays from Theoretical Trays," *Trans. AICHE* 42:741, 1946.
5. Campbell, J. M., *Gas Conditioning and Processing*, vol. 3, Campbell Petroleum Series, Norman, OK, 1982, p. 11-3.
6. Glitsch, F.W., *Ballast Tray Design Manual*, Bulletin 4900, Fritz W. Glitsch & Sons Inc., Dallas, TX, 1961; Fractionation Research Institute, Alhambra, CA, 1981.
7. Murphee, E. V., *Ind. Eng. Chem.* 17:741, 1925.
8. Hengstebeck, R., "Contacting Efficiencies of Column Internals," *Distillation Principles and Design Procedures*, Robert E. Krieger, 1976, Chap. 10.
9. Erwin, pp. 40, 41.
10. Perry, R. H., and Chilton, C. H., *Perry's Chemical Engineer's Handbook*, 5th ed., McGraw-Hill, New York, 1973, Fig. 18-10.
11. Fair, J. R., and Matthews, R. L., *Petroleum Refining* 37:153, April 1958.
12. Souders, M., and Brown, G. G., *Ind. Eng. Chem.* 26:98, 1934.
13. Perry and Chilton, p. 18-8.
14. Perry and Chilton, p. 18-9.
15. Perry and Chilton, p. 18-7–18-11.
16. Perry and Chilton, Fig. 18-15.
17. Eckert, J. S., "Design Techniques for Sizing Packed Towers," *Chem. Eng. Prog.* 57(9):54–58, 1961.

Chapter 4

Oil and Gas Production Surface Facility Design and Rating

Nomenclature

μ	oil viscosity, cP
μ_g	gas viscosity, cP
A	vertical vessel cross-section area, ft^2
C_d	Percent by weight salt content in dilution water
C_D, CD	drag coefficient, dimensionless
C_m	salt concentration of water mixture, percentage by weight
C_p	salt concentration of production water, percentage by weight
CRE	Reynolds number, dimensionless
d	diameter of vessel, in
D_g	density of vapor at system temperature and pressure, lb/ft^3
D_l	density of liquid at system temperature and pressure, lb/ft^3
d_m	water particle size required for W_c value, μm
D_M, DM	diameter of water, gas, or oil particle, μm
$D_{m1\%}$	water particle size for 1% water cut, μm
DMW	diameter of water droplet in oil, μm
D_p	particle size, ft
DPRT	particle diameter, ft
D_v	the part of the vessel diameter free of liquid, ft
DV	diameter of horizontal vessel, ft
exp	natural log base number 2.71828
FRCT	fraction of vessel cross-section area to be filled with liquid
G	gas rate, lb/h

HOIL	depth of oil, ft
L	effective vessel length for separation of water from oil, ft
ln	natural log of base 2.71828
LRATE	oil throughput rate, lb/h
LWO	required vessel length for water droplet separation from oil phase
MW	molecular weight
OAREA	oil cross-section area, ft^2
PTB	salt in oil at outlet of desalter unit, lb/1000 bbl
Q	total flow of emulsified water and oil
Q_a	actual gas flow rate, ft^3/s
QAO	liquid rate, ft^3/s
Q_o	crude oil flow rate, dry basis, bpd
Q_w	production water rate, bpd
RHOG	gas or vapor density, lb/ft^3
RHOL	liquid or oil density, lb/ft^3
RHOW	water-phase density lb/ft^3
S	total salt content in the production water and dilution water, lb/day
SG	difference in specific gravity of water and oil at system temperature T
SG_m	specific gravity of dilution water and production water mixture at system temperature T
SG_p	specific gravity of production water at system temperature T
T	system temperature, °F
TWSEC	time for water drop in oil phase to fall, s
VISC	viscosity of liquid, gas or oil at system temperature T, cP
VISCL	viscosity of liquid or oil at system temperature T, cP
V_t	terminal velocity of particle, ft/s
VTGO1	particle velocity in oil phase, ft/s
VTWO	terminal velocity of water particle in the oil phase, ft/s
W_c	BS&W of treated desalter outlet crude oil, percentage by volume
W_d	Barrels of dilution water per day mixed with the crude oil fluid upstream of the desalter
$W_{d\%}$	dilution water in crude oil phase, percentage by volume

For over four decades, many process engineers have witnessed and participated in many major crude oil and gas production facility conceptual and detail engineering designs. Some of these designs were the first of their type. However, our main core of surface facility design types have been common, such as the crude oil dehydration gun barrel tank or the electrostatic degasifier and crude oil heater-treater dehydrator. The two-stage or three-stage flash crude oil stabilization train has also been common. In introducing this chapter, there is one important item I wish to impress upon the reader with these common facilities in mind: reliable and accurate field production data.

A Special Note to the Wise

In the early 1970s I was honored to be hired as the principal project manager over all process design for a $4-billion project already in progress. Much of the major equipment fabrication was already awarded to vendors and in progress. Production of oil and gas, 1.2 million barrels per day (bpd) of oil and about 600 million standard cubic feet per day (mmscfd) of gas, was the objective. Specifications of 14.7 psia at 100°F had to be achieved for the oil stabilization. As my first day's task, I personally committed to a thorough review and understanding of the engineering contractor's process flow diagrams with material balances. Well, hold on to your seat, as I found only a one-point design for this entire production field. There was only one case having one crude production flow inlet temperature with an associated gas-oil ratio (GOR), scf/bbl. I was completely puzzled as to how such a giant production field could have only this one case to design all three flash stabilizers and the associated compressor train. Even the smallest production field of several producing wells will have a temperature range with a pressure range and a GOR range.

These three variables will make a *matrix*. In this matrix one of these combinations will be the maximum first-stage gas flash-off in the crude oil stabilization train, which sizes the first-stage separator. Another matrix, totally separate and different, will make the maximum third-stage flash-off gas, which sizes the compressor train. Yet another matrix combination will size the heat exchange and the downstream maximum cooling required. After three days of heart-to-heart talks with my counterparts in the engineering contractors' league, the contractor insisted he was right and had submitted a proper design. He represented one of our nation's largest engineering firms. Did I dare question his actions? In answer, an emphatic *yes!*

After revealing the facts to my management, two meetings followed with me, my management, and the engineering contractor. The final result was that all the engineering contractor's principal staff on this project were replaced, and the engineering contractor's CEO gave my project manager a personal apology. Major equipment already in fabrication was stopped for design changes. This had a very heavy impact on project cost, as well as a critical schedule impact.

In conclusion, be sure you have all the inlet feed production temperature, pressure, and GOR combinations before making your equipment selections and design. I also wish to commend this engineering firm for putting its best people on the project to make corrections. Any engineer would have been honored to participate thereafter.

Crude Oil Production Surface Processes

The main objective in processing crude oil from the production well is to separate it into three phases—gas, oil, and water. Keeping this in mind, all processing is indeed simple. All successful operations are based on making the crude oil vapor pressure acceptable for pipeline transmission and removing sufficient water for the pipeline transmission. Conditioning the vapor pressure or degassing with crude water removal (dehydration) is commonly called *crude oil stabilization.*

Crude oil stabilization is thus the major process quest of all surface facility processing. Crude oil flows into a first-stage separator from the wellhead gathering pipeline. A typical crude oil stabilization process train is shown in Fig. 4.1. Please note that this is a three-stage

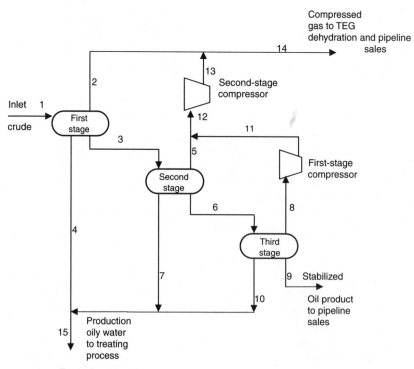

Figure 4.1 Typical crude oil stabilization.

pressure-reduction stabilization train. The third stage should produce approximately atmospheric crude at pipeline transmission temperatures, say 150°F. Assuming the inlet first-stage crude oil GOR ratio (scf gas/bbl crude oil) is 20 or greater, the second-stage pressure should range from 200 to no greater than 300 psig. The third stage should allow whatever wellhead gathering manifold pressure is required, usually 400 psig and greater. If, therefore, inlet crude pressure is 250 psig or less, and GOR >20, only two stages of stabilization are required. If the first-stage inlet pressure is 100 psig or less, only one stage is required. The one-stage scenario is generally a heater-treater-type vessel, which will be covered later in this chapter. The preceding recommended stabilization train stage selections come from much experience and proven good operation practice. One essential reason for these crude oil pressure recommendations is balancing the horsepower requirements of the first and second compressor stages. In some cases, a third compressor may be added in order to match standard vendor compressor units. Horsepower should be nearly equal between these two or three stages of compression. The pressure ranges shown are indeed windows to do this.

First Stabilizer Stage

The first-stage separator is a more critical stage to size, as it must serve not only as a three-stage separator, but also as a liquid-slug catcher. The next question is how big a liquid slug in a pipeline system can become. The answer is to consider surge holdup time. I recommend that any slug catcher vessel receiving any possible liquid slug in a two-phase-flow pipeline be sized to hold a minimum of 5 min of the pipeline's maximum liquid flow design capacity. This 5 min of flowing liquid capacity, please note, is to be added to any occupied liquid volume already normally in the slug catcher vessel. Since the first stage here is serving both as a three-phase separator and a slug catcher, one-half of its volume will be taken at all times for oil-water separation. The remaining half must serve for gas-liquid separation and the added liquid-slug surge, 5 min of maximum pipeline flow.

Considering the GOR, the first-stage vessel diameter may become noneconomic in view of the fact that a second vessel dedicated to slug catching may be added. A slug catcher vessel may be added to serve upstream of the first-stage vessel for a large GOR. Therefore, consider a special slug catcher vessel upstream of the first-stage separator. This vessel can be economically sized to handle the minimum 5-min liquid holdup time. A second and most unique advantage of such a dedicated slug catcher is the fact that a constant feed flow can be achieved in the first stabilizer stage. This advantage is most important when emulsions are present and when large production water rates flow from the well, say 10% water by volume or greater, in the wellhead crude. The level control instrument can use a set point less than three-quarters full, at which the outflowing crude oil fed to the first stage may be flow-controlled at another set point. This first stage feed flow controller should autoreset from a slug catcher high-level signal, say, above one-half full. In summary, the added upstream slug catcher can prove to be economically justified by reducing the first-stage vessel size and by providing a constant feed flow to the stabilizer train.

Crude Dehydration

Crude oil production from the wellhead will always have production water flowing with it. Most of this water is what we call *free water*. That is, it is not dissolved into the crude oil. Soluble water in crude oil has ranged as high as 1000 parts per million by volume (ppmv), and as low as 20 ppmv. No attempt to remove soluble water is ever made in field production facilities. Rather, substantial equipment and processes are dedicated to the removal of free water. This chapter shall therefore address only free water. Refiners request, and in many cases specify, a 1% free water crude content in all pipeline crude oil received. Thus, the crude oil field production facilities must meet this 1% water requirement in treated crude oil to pipeline sales.

Referring to Fig. 4.1, each of the first two stages and perhaps the third stage can be expected to separate out 90% of the respective stage inlet water. In other words, a 10% water cut is passed in each of these stages. This means if 10,000 bpd production water with a 30,000 bpd crude oil is fed to the first stage, 1000 bpd production water will remain in stream 3, feeding the second stage (also called the *intermediate stage*). The second-stage cut will thus be 0.1 * 1000 = 100 bpd fed to the third stage, stream 6 shown in Fig. 4.1. Here note this is less than 0.5 volume percent, already a pipeline spec water remainder cut. What's wrong? This simply doesn't happen in the real world. The first-stage cut is 3.3%, which is very likely a good number. Taking another 90 percent out in a second stage just doesn't happen. Why? First, The free water average droplet size is too small to drop out from the crude in the time allowed in the second stage. This second stage is as shown later in Fig. 4.4, Stokes' law–type gravity separation [1]. Second, oil-water emulsions can hold 3% to approximately 8% free water remainder in the crude oil outlet. More treatment is therefore commonly required in this second, intermediate, stage. Heating and electrostatic treatment for dehydration will be discussed next.

There are two types of dehydration equipment that can make a 0.5% or less water cut. These are the electrostatic treater and the heater-

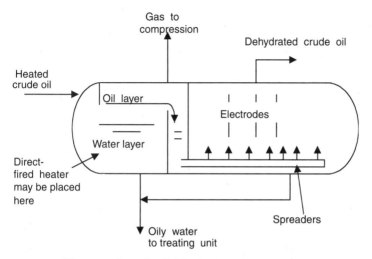

Figure 4.2 Electrostatic crude oil dehydration.

treater type. The electrostatic-type treater is shown in Fig. 4.2. The heater-treater type is reviewed later.

Electrostatic dehydration

The electrostatic separator shown in Fig. 4.2 is normally placed in stream 6, shown in Fig. 4.1. A heater should be installed upstream of this electrostatic treater to control temperature. Here heating temperatures should normally range from 100 to 200°F for light crudes 22°API and above. For heavier crudes, use higher temperatures, such as 250°F for a 14°API crude oil. Some heavier crudes, below 14°API, may require temperatures up to 300°F for adequate electrostatic dehydration. This temperature is to achieve a gravity difference between the crude oil and water of 0.001 or more. The electric power required by these electrogrids is normally 0.05 to 0.10 kVA/ft². The ft² area is the horizontal cross-section area in the electric grid section of the electrostatic dehydrator.

The program ElectSep is provided in the software with this book. Run it with the .exe file and observe the defaulted example run. Please note that the program sizes the vessel to accomplish a 1% water cut or any water cut remainder desired in the treated (dehydrated) crude oil. Vendors guarantee this 1% figure; however, a 0.5% water cut remainder is actually experienced. Removing less water than guaranteed is normal practice, as no vendor desires to barely meet the guarantee made. Figure 4.3 exhibits this defaulted run. You may enter any desired input and run it by clicking on the Run Program button.

ELECTROSTATIC OIL - WATER SEPARATOR

Temperature of Treater, F	300	Spec Grav of oil @ T	0.86
Salt Conc. In Water, lb/bbl	1	Barrels / Day Crude Oil	26310
Viscosity of Oil @ T, Cp	3.5	Prod + Desalting Water, BPD	2690
Oil Time Retention, minutes	30	Water Specific Gravity	.9158
KF Factor of KF*(VISC^.4)	170	Water Cut%	1.0

PROGRAM OUTPUT ANSWERS

Vessel Length Feet	Vessel Dia Inches	Retention Minutes
160	72	30
117	84	30
90	96	30
71	108	30
58	120	30
48	132	30
40	144	30
34	156	30
29	168	30
26	180	30
22	192	30
20	204	30
18	216	30
16	228	30
14	240	30

Water Particle Size, Mcr = 281   Treated Oil Salt, lb/1000 bbl Oil = 10

(a)

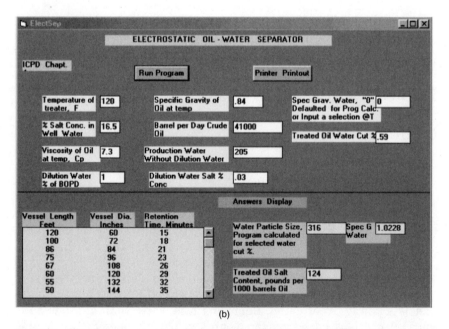

(b)

Figure 4.3 ElectSep program defaulted example runs: (*a*) earlier DOS version, and (*b*) screen capture from the current version. Note that the receiving refiner will apply yet another electrostatic dehydration treatment to reduce the salt content to less than 10 lb per 1000 bbl.

$$D_{m1\%} = K_F \mu^{0.25} \tag{4.1}$$

where $D_{m1\%}$ = water particle size for 1% water cut, μm
$\quad K_F$ = factor for calculating D_m
$\quad \mu$ = oil viscosity, cP

K_F is 200 for heater-treater vessels and 170 for electrostatic vessels. Equation (4.1) is from Arnold and Thro [2].

If you desire to have a 0.5% water cut specification, you may do so by replacing RWBBL = (Q1/.99) * .01 with RWBBL = (Q1/.995) * .005 in the equations shown later. RWBBL is the remaining water in the treated crude oil, bpd. Q1 is the dry crude oil rate, bpd. You shall also need to input another K_F factor of, say, 150 in place of the defaulted 170 K_F factor. I recommend that you also have proven field data and back-calculate more precisely and exactly what K_F factor value is required for a 0.5% water cut. Actual oil operating test runs should be used. A 1-K_F value per water cut is not feasible for all the varied crude oils. The 170 factor used for a 1% water cut is, however, a proven one for crude oils ranging from 12 to over 30°API and for viscosities 15 cP and below at the operating temperature. Verification of this defaulted K_F factor, 170, is recommended for your particular case.

Important note: Referring to Fig. 4.2, please note that the free water is removed in the front section of the electrostatic vessel. ElectSep does not address this front section, but only the later section where oil and water emulsion is present. For this front section you should see the program Vessize, discussed later in this chapter. The ElectSep program herein addresses the precise separation of water from oil. Temperature is an important factor for lowering the oil viscosity. It has been a field-proven concept that the oil viscosity dominates good separation and the achievement of the desired 0.5% water cut. Equation (4.2) governs how large an electrostatic vessel is required.

$$dL = \frac{438 * Q * \mu}{SG * D_M^2} \tag{4.2}$$

where dL = diameter × length of vessel, ft²
$\quad Q$ = total flow of emulsified water and oil
$\quad D_M$ = diameter of water particles, μm
$\quad SG$ = difference in specific gravity of water and oil at T
$\quad \mu$ = oil viscosity, cP

Equation (4.2) is a particle-settling equation that is derived from two equations of force. One force equation is the buoyancy force F_B, and the other is the drag force F_D, produced by the particle movement in the surrounding fluid. As the particle or water droplet velocity increases due to gravitational buoyancy force, this buoyancy force becomes equal to the drag force caused by the particle movement. At this point of $F_B = F_D$, the particle ceases to increase in velocity and stays at this constant velocity, called the *terminal velocity*. This terminal velocity is thus the particle velocity at which all vessel-sizing calculations are based.

Once dL is calculated, the program selects varied diameters and calculates accompanying corresponding diameters. Figure 4.3 displays typical ElectSep program run printouts. Please note the numerous vessel lengths and diameter sizes permissible.

Free water

It is important to note that you should consider removing all free water before attempting to size the electrostatic water separation section. The dL factor calculated is only for this electrostatic section. You should therefore make a good estimate of how much water will pass into the electrostatic section and input this value in the "Prod + Desalting Water, BPD" input block. Inputting all of your production water will seriously err sizing results. Free water should be removed in free water knockout (KO) tanks or vessels upstream. True and needed electrostatic treater sizing may thereby be determined.

Free water is that water which will freely separate from oil in accordance with Eq. (4.1). Time, chemical emulsion breakers, electronic fields, and temperature are the factors that will break an oil-water emulsion. It is prudent to heat the emulsion prior to entering the electrostatic section. Such heating lowers the viscosity, which not only allows more free water removal, but also will enhance the efficiency of the electrostatic-treater section. Hereby, even a 0.5% water cut in treated oil is easily and commonly achieved.

Free water to emulsified water ratios can be as much as 95 parts free water to 5 parts emulsion, while some may even be 10 parts water to 90 parts emulsion. The answer is to lab test and preferably run a field test on actual crude flows. Another concern is free water vs. soluble water, sometimes called *dissolved water*.

Soluble water

Soluble water is not removable by any concept displayed or discussed in this book. Soluble water is molecularly bonded to the oil and may even be a part of oil compound crystal structures, such as hydrate-bonded oil-water molecules. Oil production field equipment never removes soluble water. Soluble water concentrations may range from very low, say 20 parts per million by weight (ppmw), to over 1000 ppmw. In Fig. 4.3, all the water shown entering the electrostatic section is emulsion-type, not soluble water. If it were soluble, 99% would not have been removed as shown. In the oil and gas industries, *water cut,* or water remainder in treated crude oil, always refers to the free or emulsified water remainder, and not to soluble water. Normally, soluble water goes with the product crude unknown as to concentration and never recognized as a water component. More work is respectfully needed here in the future, for when we produce products that are highly sensitive to soluble water content, such as aviation jet fuel, this soluble water content should be known. Why? If for no other reason than jet airliner safety alone. That's enough to merit a world of study and research. My sincere encouragement to all who take this challenge.

Salt removal

A salt–crude oil concentration of 10 pounds per 1000 barrels (ptb) is commonly specified by refiners receiving crude oil. The example run in Fig. 4.3a has derived exactly this figure, 10 ptb. In order to assure this 10 ptb, dilution water should be applied. Having actually experienced this example, I added 3000 bpd of dilution water to the 2690 bpd of production water feeding this electrostatic unit. This, of course, increased my vessel size. For a 12-ft diameter vessel, 4 ft of additional vessel length was required, making a 44-ft length for the electrostatic section. (Please note that the program length and diameter sizes are strictly for the electrostatic section only. Added vessel length is required for the gas-oil separation–free water KO section.) Also, when I added this dilution water I had to hand-calculate the new "Salt conc. in water, lb/bbl" input, 0.5 lb/bbl. This new value replaced the previous salt concentration, 1.0 lb/bbl. The 3000 bpd of dilution water had some salt in it, also making a total salt in water concentration of 0.5 lb/bbl water for the resulting 5690 bpd. Why not run this one yourself and get the feel of conditioning your input?

A final note on the electrostatic dehydrator is how much capacity one of these units will allow. Most vendors guarantee approximately 30 to 90 bbl/ft^2 oil of horizontal cross-section area in the electric grid section. I recommend 60 bbl/ft^2 as a good practice number and one that I find reputable vendors will guarantee. This guarantees a 0.5% water cut remainder in the treated crude oil. Normally, a guarantee of this type will always achieve at least a 0.2% water cut, if not less. Here vendors do not normally make the allowable guarantee, 0.5% water cut, but rather guarantee a 1% water cut instead.

For a continued discussion of electrostatic desalters, see the section on production desalters later in this chapter.

A special note of caution: Be cautious taking these computer runs for your final design criteria. Always confirm with a lab data test or, preferably, actual field operating data from other similar equipment types and with the exact same crude oil and gas source. Once you have such data, these programs should prove to be of great value.

Three-Phase Horizontal Separator

Figure 4.4 displays the common three-phase horizontal separator. Water separation can be as good as 90%, or can be even less than 50%, depending on how much difference there is between the oil and water specific gravities. For this reason, upstream heating can increase this difference, and in many cases heating is installed. Oil-water emulsions are also equally important toward fixing an oil-water separation. Figure 4.4 diagrams a typical stage separator as shown in all three stages of Fig. 4.1.

Vessize program

Run the Vessize program, clicking on the Run Start button. (See Fig. 4.5.) Please note the input data and the output answers. This program sizes the vessel, based on the gas volume rate. You pick a vessel diameter and also input the fraction of the cross-section area you want the vessel to have. With the physical property data and flow rates input as shown, the program calculates a vessel length required to make the average liquid droplet size separation.

For most systems, 150 µm is a good-practice average size for liquids in gas phase for most any liquid-gas separation. In practice, where a fine remainder mist is not critical, a 500-µm droplet size is practical. 25,400 µm = 1 in. A classic example of the 150-µm size is an amine absorber overhead, wherein several hundred thousand dollars annu-

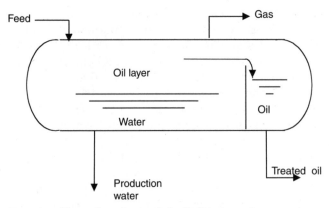

Figure 4.4   Three-phase horizontal cylindrical gravity separator.

```
        THREE PHASE HORIZONTAL VESSEL ANALYSIS CALCULATIONS
           Gas - Oil - Water Separation

     For 5 Foot Horizontal Three Phase Vessel Selection:

Gas Phase Vessel Length Rq'd, ft                        1.38e+01
Oil Phase Water Sep. Vessel Length Rq'd                 2.31e-01
Oil Phase Foam Sep. Vessel Length Rq'd                  2.90e-01

Gas Drag Coef  3.50e+00      Terminal Gas Vel, ft/sec 1.44e+00
Vessel Gas Vel, ft/sec 9.57e+00  Oil Holdup Time, minutes 8.06e-01
Barrels of Oil Volume 1.657e+03  Liquid Drop Fall Time, sec 1.44e+00

         THREE PHASE HORIZONTAL SEPARATOR DATA INPUT

Vessel Dia., ft  5              Gas Rate, lb/hr              40000
Liquid Oil Rate, lb/hr    10000 Liquid Water Rate, lb/hr        10
Water Density, lb/cf      62.4  Gas Viscosity, Cp             .012
Gas Density, lb/cf        .15   Water Drop in Oil, Microns    500
Liquid Drop in Gas, Mcr  150    Gas in Oil Bubble, Microns    200
Liquid Oil Dens, lb/cf    52    Liquid Oil Viscosity, Cp      1.5
Water Depth, inches       10    Oil Depth, inches             25
```

(a)

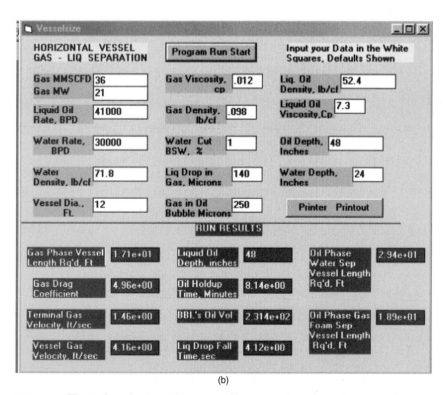

(b)

Figure 4.5 Three-phase horizontal separator Vessize program example runs: (*a*) earlier DOS version, and (*b*) screen display from the current version in Visual Basic format.

ally could be lost in this difference of 150 vs. 500 μm. Another classic example is where a flare KO drum is sized. Here a 500-μm size is preferred, as a fine mist of this type has no effect on a good, clean flare flame. The Vessize program thus offers the user these choices or any range thereof as selectable input.

The equations used in this program are those of Sec. 5 of *Perry's Chemical Engineer's Handbook* [3]. A part of the code is copied as follows:

```
        DV = txtDV ' Horizontal Vessel Diameter, feet
        G = txtG ' Gas rate, lb/hr
        RHOG = txtRHOG ' Vapor Density, lb/cf
        RHOL = txtRHOL ' Liquid Density, lb/cf
        DM = txtDM ' Micron size of Liquid Droplet
        FRCT = txtFRCT ' Fraction of Cross Section Vessel Area
            to be Liquid
        VISC = txtVISC ' Viscosity of Liquid, Cp
        LRATE = txtLRATE ' Liquid Rate, lb/cf

    DPRT = DM * .00003937 / 12
    CRE = 95000000# * RHOG * (DPRT ^ 3) * (RHOL - RHOG) / (VISC
        ^ 2)
    CD = 1.05 * (Exp(6.362 - 1.09352 * (Log(CRE)) + .048396 *
        ((Log(CRE)) ^ 2) - .00050645 * ((Log(CRE)) ^ 3)))
    VT = (4 * 32.2 * DPRT * (RHOL - RHOG) / (3 * RHOG * CD)) ^
        .5
    Y = DV / 2: TAREA = 3.1416 * (Y ^ 2)
If (HOIL + HWATER) > Y GoTo OVERHALF
    XW = Y - HWATER: ZW = ((Y ^ 2) - (XW ^ 2)) ^ .5: WDEG =
        ((Atn(ZW / XW)) * 180 / 3.1416) * 2
    WAREA = ((WDEG / 360) * TArea) - (ZW * XW)
If (HOIL + HWATER) = Y Then OAREA = (TArea / 2) - WAREA:
GAREA = TArea / 2:
        GoTo FinalCalc

    XO = Y - (HWATER + HOIL): ZO = ((Y ^ 2) - (XO ^ 2)) ^ .5:
        ODEG = ((Atn(ZO / XO)) * 180 / 3.1416) * 2
    OAREA = ((ODEG / 360) * TArea) - (ZO * XO)
    If (HOIL + HWATER) < Y Then GAREA = TArea - (WAREA + OAREA)
    GoTo FinalCalc
OVERHALF:
    If HWATER < Y Then XW = Y - HWATER: ZW = ((Y ^ 2) - (XW ^ 2))
        ^ .5: WDEG = ((Atn(ZW / XW)) * 180 / 3.1416) * 2
    If HWATER < Y Then WAREA = ((WDEG / 360) * TArea) - (ZW * XW)

        RHOL = txtRHOL
        DM = txtDM
```

```
FRCT = txtFRCT
VISC = txtVISC
LRATE = txtLRATE

' Water and Oil liquid phases area Calc
If (HOIL + HWATER) > Y GoTo OVERHALF
XW = Y - HWATER: ZW = ((Y ^ 2) - (XW ^ 2)) ^ .5: WDEG =
    ((Atn(ZW / XW)) * 180 / 3.1416) * 2
WAREA = ((WDEG / 360) * TArea) - (ZW * XW)

If (HOIL + HWATER) = Y Then OAREA = (TArea / 2) - WAREA:
    GAREA = TArea / 2: GoTo FinalCalc
XO = Y - (HWATER + HOIL): ZO = ((Y ^ 2) - (XO ^ 2)) ^ .5:
    ODEG = ((Atn(ZO / XO)) * 180 / 3.1416) * 2
OAREA = ((ODEG / 360) * TArea) - (ZO * XO)
If (HOIL + HWATER) < Y Then GAREA = TArea - (WAREA + OAREA)
GoTo FinalCalc
```

CRE is the Boucher [3] Reynolds number calculation to derive a drag coefficient, CD. The CD factor is calculated in the program by a curve-fit equation as shown. The velocity, called the *terminal droplet velocity*, is calculated as the variable VT, also called the *dropout velocity* or the *settling velocity*. All three terms refer to this same variable.

Figure 4.6 shows the cross section of the horizontal vessel, revealing the variables required to calculate the water-phase cross-section area. Similar to the tray downcomer area calculation in Chap. 3, the water phase of the horizontal vessel is calculated here. All the user input that is required is simply the depth of the water phase and the vessel diameter. All the rest is program-calculated. The same is true with similar variables for the oil phase. Note that the user also inputs the oil-phase depth. Having these variables of the vessel cross-section program calculated, the settling equations can next be applied as shown in the following program code listing, derived from Boucher [4].

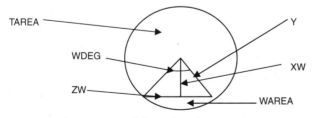

Figure 4.6 Cross section of three-phase horizontal separator.

```
' Calc of Required Vessel length for water drop out of oil

VTWO = .00000178 * ((RHOW / 62.4) - (RHOL / 62.4)) * (DMW ^
   2) / VISCL
TWSEC = HOIL / VTWO: QAO = LRATE / (RHOL * 3600)
LWO = (QAO / OAREA) * TWSEC
txtLWO = Format(LWO, "0.00e+00")

  ' Calc of Required Vessel length for gas bubble rise out of
oil

VTGO = .00000178 * ((RHOL / 62.4) - (RHOG / 62.4)) * (DMGO
   ^ 2) / VISCL
TGSEC = HOIL / VTGO: QAO = LRATE / (RHOL * 3600) ' time,
   sec for gas buble
LGO = (QAO / OAREA) * TGSEC      ' length of vessel rq'd for
   gas-oil separation, oil phase
txtLGO = Format(LGO, "0.00e+00")
```

Gas phase in horizontal separator

For liquid drop separation from the gas phase, the Stokes' law settling equation could be applied. The simplified Stokes' law equation [4] may be written:

$$V_t = \frac{C * g * \text{SG} * D_M^2}{\mu_g} \tag{4.3}$$

where V_t = terminal velocity of droplet
    C = constant
    g = acceleration of gravity
    SG = difference in specific gravity of water and oil
    D_M = diameter of droplet, µm
    μ_g = gas viscosity, cP

V_t is the terminal velocity at which a drop falls from gravity force. From this equation, setting a vessel size appears easily done. This is not the case, however, as a drag force on the liquid oil drop resists this gravity force, and thus the V_t so calculated is incorrect. Here the industry has developed a well-proven drag force coefficient, CD in the Vessize program code. In the following code, terminal velocity VT is calculated by correcting the Stokes' law equation with a drag force correction, CD. CD is calculated by curve-fitting Boucher's Table 5-26 from *Perry's Chemical Engineer's Handbook* [5]. CRE is a Reynolds number calculated for determining the CD correction factor.

API VT, ft/s, Oil Droplet Sphere Terminal Velocity Calculation Method:

```
DPRT = DM * .00003937 / 12
  CRE = 95000000# * RHOG * (DPRT ^ 3) * (RHOL - RHOG) / (VISC
  ^ 2)
  CD = 1.05 * (Exp(6.362 - 1.09352 * (Log(CRE)) + .048396 *
  ((Log(CRE)) ^ 2) - .00050645 * ((Log(CRE)) ^ 3)))
  VT = (4 * 32.2 * DPRT * (RHOL - RHOG) / (3 * RHOG * CD)) ^
  .5
```

where   DM = diameter of oil droplet sphere, μm;

        RHOL = oil density, lb/ft³

        RHOG = gas density, lb/ft³

        VISC = gas viscosity, cP

Please note that the preceding equation for CD calculation has been curve-fitted by the least-squares method [6].

The program next calculates the gas phase area and finally the length of vessel required for the drop separation with the given diameter. The user selects a diameter as a data input. Another key selection by the user is the water and oil depths. Please note that these are discrete vertical measures taken at the center of a cylindrical vessel. The water depth and oil depth, in, is to be added by the program to find the total liquid depth. Herein the user should consider the instrumentation requirements to achieve good level control.

The Vessize program next proceeds to calculate the required horizontal vessel lengths for gas bubble or foam separation from the oil phase. Water separation from the oil phase is also calculated in the following discussion

Oil phase in horizontal separator

As the oil flows within its designated cross-section area through the horizontal vessel, free water droplets form and begin to drop at a terminal velocity rate. The Vessize program calculates this terminal velocity VTWO. As discussed previously in the oil dehydration section, the Stokes' law settling equation is used for the water droplet fall rate VTWO.

Water drops in oil-phase separation

```
' Calc of Required Vessel length for water drop out of oil

  VTWO = .00000178 * ((RHOW / 62.4) - (RHOL / 62.4)) * (DMW ^
        2) / VISCL                                        (4.4)

  TWSEC = HOIL / VTWO: QAO = LRATE / (RHOL * 3600)
```

```
LWO = (QAO / OAREA) * TWSEC
txtLWO = Format(LWO, "0.00e+00")
```

VTWO = Terminal velocity of water particle in the oil phase, ft/s

where   DMW = water droplet size in oil, μm
         RHOW = water-phase density lb/ft^3
         RHOL = oil-phase density, lb/ft^3
         VISCL = viscosity of oil, cP
         HOIL = depth of oil, ft
         OAREA = oil cross-section area, ft^2
         LRATE = oil throughput rate, lb/h
            LWO = required vessel length for water droplet separation
                  from oil phase
         TWSEC = time for water drop in oil phase to fall, s
            QAO = liquid rate, ft^3/s

Assumption: Stokes' law governs and the drag coefficient CD is not significant. From Perry [7]:

```
CD = 18.5 / (CRE^ 0.6) @ CRE > 0.3 & < 1000
```

where   CD = coefficient of drag, dimensionless
         CRE = Reynolds number, dimensionless

Equation (4.4) for terminal particle velocity VTWO assumes a low Reynolds number ranging from approximately 100 to 200. This assumption (CD ~ 1.0) is made in Arnold and Stewart's *Surface Production Operations* [8], from which Eq. (4.4) was derived. In deriving such an equation, the assumption is made that the low Reynolds number places the particle in a laminar flow rate and therefore the drag force coefficient may be neglected. I submit that this is not correct, but yield to the fact that the drag force coefficient value here may be considered unity in many cases. Otherwise, Eqs. (4.5), (4.6), and (4.7) are suggested as the replacement for Eq. (4.4). Equation (4.4) is applied on the basis that terminal velocity occurs when the drag force equals the buoyant force on a sphere, from Archimedes' principle. Equating these two forces yields Eq. (4.4) for VTWO. Arnold and Stewart derive this equation in their book, solving for the terminal velocity of a spherical drop of water in the oil phase.

In my ventures around the world, I have seen that Eq. (4.4) for VTWO constitutes good practice and a conservative approach. I also admit, however, that I believe a more accurate terminal velocity criterion may be derived, using and applying the equations given in this book. Lending more credibility to this more accurate method is Boucher, who writes in Perry's handbook that this method is credible for liquid drops in gas, liquid drops in liquid, and gas bubbles in liquid [9]. I have thus applied this more accurate method to the separation of gas bubbles, or foam, from the oil liquid phase. Since I find few authors who have addressed gas bubble separation from oil, I am very pleased to present what I believe is a new breakthrough of our technology in the following.

Gas bubble in oil-phase separation.   The following Visual Basic line code is copied from the program Vessize.

```
‘ Calc of Required Vessel length for gas bubble rise out of oil
  DPRT = DMGO * .00003937 / 12
```

$$CRE = 95000000\# * RHOL * (DPRT \wedge 3) *$$
$$(RHOL - RHOG) / (VISCL \wedge 2) \tag{4.5}$$

$$CD = 1.05 * (Exp(6.362 - 1.09352 * (Log(CRE)) + .048396 *$$
$$((Log(CRE)) \wedge 2) - .00050645 * ((Log(CRE)) \wedge 3))) \tag{4.6}$$

$$VTGO = (4 * 32.2 * DPRT * (RHOL - RHOG) /$$
$$(3 * RHOL * CD)) \wedge .5 \tag{4.7}$$

$$‘ VTGO = .00000178 * ((RHOL / 62.4) - (RHOG / 62.4))$$
$$* (DMGO \wedge 2) / VISCL \tag{4.8}$$

```
TGSEC = HOIL / VTGO: QAO = LRATE / (RHOL * 3600)
LGO = (QAO / OAREA) * TGSEC
txtLGO = Format(LGO, “0.00e+00”)
```

Please note that two equations, Eqs. (4.7) and (4.8), are given for calculating the gas bubble terminal velocity VTGO. Equation (4.7), having the drag coefficient CD, is the one applied in the Vessize program. CD is calculated again from Perry's curve-fit, least-squares method [5]. This is also the API method, which refers to Perry as the source for this drag coefficient curve [9].

Equation (4.8) is used extensively for strictly Stokes' law applications, as in Eq. (4.2), used for electrostatic water-oil separation. In fact, Eqs. (4.2) and (4.8) are the same equation, with the exception that Eq. (4.8) derives the terminal velocity of Stokes' law, discussed earlier. Equation (4.2) derives the horizontal surface area required for volumetric flow rate and water-oil gravity difference. Equation (4.2) will also be applied later in this chapter for tankage type dehydration of water in oil.

Another method of calculating the terminal velocity is by trial and error. First, assume a velocity VTGO, then calculate the Reynolds number CRE. Next, calculate the drag coefficient CD. Having CD, the VTGO may be calculated using Eq. (4.8). Repeat until the assumed VTGO agrees with the calculated VTGO. The equations shown in Fig. 4.7 with the Visual Basic line code may be used.

The required vessel length LGO for gas bubble separation from the oil phase is calculated next. This is the same as calculated earlier for the water separation. Give it a try.

Settling equations   Note in Fig. 4.7 that the drag coefficient CD is calculated by either of two equations:

```
CD = 18.5 / (CRE^0.6)     for CRE > 0.3          (4.9)
CD = 24 / CRE      for CRE < 0.3                  (4.10)
```

One of the two equations is chosen in the Vessize program by using a Reynolds number test. If the Reynolds number, CRE, is less than 0.3, then Eq. (4.10) is used. It is important to note here that all cases of a water particle settling or a gas particle rising in the oil phase should have a Reynolds number less than 0.3. Why should this be? Please consider the Reynolds number calculated in the standard Reynolds number equation from Fig. 4.7, Eq. (4.11):

```
CRE = 10.325 * DPRT * VTGO1 / VISCL             (4.11)
```

where   DPRT = particle diameter, ft
        VTGO1 = particle velocity in oil phase, ft/s
        VISCL = viscosity of surrounding oil, cP

In Eq. (4.11) the variable DPRT is a very small number, since it is the diameter of a single particle measured in feet. As a particle is, say, 100 to 500 μm, and since 1 in = 25,400 μm, it becomes obvious that the Eq. (4.11) calculation will always have a very small answer, usually 2 to 4 places or more to the right of the decimal point. This fact is supported

```
VTGO1 = 0.001
CRELOOP:
CRE = 10.325 * DPRT * VTGO1 / VISCL   ' Reynolds Number Calc, CRE
CD = 18.5 / (CRE^0.6)                 ' 0.3<CRE< 1000
IF CRE<= 0.3 THEN CD = 24 / CRE
VTGO = (4 * 32.2 * DPRT * (RHOL - RHOG) / (3 * RHOL * CD)) ^ .5
If ABS( VTGO - VTGO1) > 0.01 Then VTGO1 = VTGO1 + 0.001 : GOTO
CRELOOP
```

Figure 4.7 Visual Basic code for another terminal velocity method.

in that VTGO1 is normally greater than 0.1, and VISCL is normally greater than 1.0.

The conclusion here then is that Eq. (4.10) could conveniently be applied to most every gas or water particle separation in the oil phase. If there should be any case in which this is not true—say, for an unusually high VTGO1 value—then consider using the method as shown in Fig. 4.7. This method will apply Eq. (4.9) if CRE is less than 0.3.

One other settling equation method proposed here and strongly endorsed is displayed in Fig. 4.8. This is the recommended method, and it is applied to both horizontal and vertical programs in Visual Basic in the PPE computer programs.

CRE2 is actually the variable of (CD * CRE^2). Here CD is the drag coefficient and CRE is the Reynolds number, both dimensionless. Equation (4.12) is Eq. 5-222 from Perry. Thus plotting CRE2 vs. CD develops the curve of Eq. (4.13). According to Perry, this curve is applicable to gas bubble rise from the oil phase, water drop settling from the oil phase, and liquid drop fallout from the gas phase. Certain variables in Eqs. (4.12) and (4.7) must, however, be changed, such as replacing the liquid oil density RHOL with the gas-phase density RHOG. These replacements are necessary when changing the Eq. (4.13) curve between oil and gas phases.

Now if you are thoroughly confused after a first-time look, please read and try to understand the line code of the Vessize and VessizeV programs. Run these programs by simply clicking on the Program Run Start button. This should help. Also keep in mind that we have covered at least three ways to derive the same answers, namely the terminal velocity of a particle in a fluid, either gas or liquid.

Equation summary. As an attempt to clarify, I offer this further explanation: the curve of Eq. (4.13) will derive the CD drag force constants of both Eqs. (4.9) and (4.10). The Eq. (4.13) CD drag force offers both this same range of Reynolds number coverage, and also extends beyond Eqs. (4.9) and (4.10), especially in the gas-phase zone. Remember also, Eq. (4.13) is applicable in either the liquid or gas zones and is suggested to

```
DPRT = DMGO * .00003937 / 12    ' gas bubble diameter, feet
CRE2 = 95000000# * RHOL * (DPRT ^ 3) * (RHOL - RHOG) / (VISCL ^
  2)                                                        (4.12)
CD = 1.05 * (Exp(6.362 - 1.09352 * (Log(CRE2)) + .048396 *
  ((Log(CRE2)) ^ 2) - .00050645 * ((Log(CRE2)) ^ 3)))      (4.13)
VTGO = (4 * 32.2 * DPRT * (RHOL - RHOG) / (3 * RHOL * CD)) ^ .5
                                                            (4.7)
```

Figure 4.8 Terminal velocity method for gas bubble rise in oil phase.

be the more accurate. Equations (4.2) and (4.3) are the same basic Stokes law equation and should only be applied to water separation from the oil phase. Hopefully, this explanation lightens the confusion. If so, you have done well for a first-time introduction to immersed-body forces.

Gas bubble separation horizontal vessel length. Similar to the water drop separation from the oil phase, the gas bubble separation also requires a vessel length. The greater length required—gas-phase liquid drop, oil-phase water drop, or the oil-phase gas bubble—is to govern the required vessel length. Most all cases are governed by the gas-phase vessel length required. The irony is that in the last decade I have found some (though limited) cases governed by the gas bubble separation length required! Could it be that we have erred in some past designs?

Three-Phase Vertical Separator

Equations

The same principal equations may be used for vertical three-phase separators. Simply, the Stokes law Eq. (4.3) is used for water drop separation from the oil phase. For liquid separation in the gas phase and for gas bubble separation in the oil phase, Eqs. (4.5), (4.6), and (4.7) are applied. These are the key equations in the VessizeV program Visual Basic line code.

Vertical Separators

Coverage has been limited to horizontal three-phase separators up to this point. Considering Fig. 4.9, oil and water must flow vertically downward and gas vertically upward. The same laws of buoyancy and drag force apply. Equation (4.3) may therefore be used in the oil phase for water separation. Equations (4.12), (4.13), and (4.7) (see Fig. 4.8) are applied to the gas phase and oil phase for oil-gas particle separations, as was equally done for horizontal separators. The equations for the horizontal separator from Fig. 4.8 may also be used for the water drop terminal velocity in the vertical separator.

Figure 4.10 displays a typical vertical vessel program run printout. The program calculation output gives a minimum required diameter. This diameter is based on the terminal velocity of liquid drop fall velocity, gas bubble in oil rise velocity, or the water drop fall velocity. The smaller this terminal velocity, the greater the vessel cross-section area required and thus the greater the vessel diameter required. In this example, the oil-phase gas bubble rise terminal velocity is controlling. If you reduce the oil flow to, say, 10,000 lb/h, then the gas-phase liquid

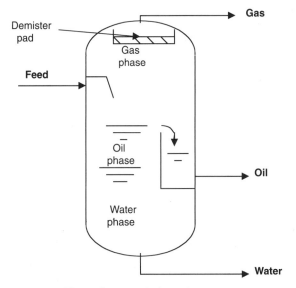

Figure 4.9   Three-phase vertical gravity separator.

oil drop would control the minimum vessel diameter required. Why? Observe the required diameters for each of these three calculated terminal velocities. The greater diameter is calculated from the phase volume flow rate and the particle terminal velocity.

Particle size

Note the three particle sizes given in Fig. 4.10. The liquid drop in gas-phase input is 150 μm. A 500-μm water drop is selected as the oil-phase input, and a 200-μm diameter for a gas bubble in the oil phase is input. These values are intended to be variable inputs reflecting good design practice. The following addresses each of these three particle sizing recommendations in the gas and oil phases. We must also address the justification for selecting these particle sizes. Also please note that the size selections recommended are equally applicable to horizontal vessels, as well.

Particle size note: The particle size inputs are not actual discrete particle sizes. Rather, these three particle size inputs are to be regarded as particle size ranges. For example, the 200-μm diameter input for a gas bubble in the oil phase could have an actual range from as small as 10 μm to over 1000 μm. The range here is considered, however, to be weighted around a 200-μm average. We have not actually field-tested particle sizes, either. These sizes, as input in these theoretical equations, have proven over many field evaluations to be the best

```
           THREE PHASE VERTICAL VESSEL ANALYSIS CALCULATIONS
                      Gas - Oil - Water Separation
     Gas Phase Vessel Diameter Rq'd, ft                     8.08e+00
     Oil Phase Water Sep. Vessel Diameter Rq'd              1.17e+01
     Oil Phase Foam Sep. Vessel Diameter Rq'd               1.41e+01

   Gas Drag Coef  2.40e+01 Terminal Gas Vel, ft/s             1.44e+00
   Barrels of Oil Volume 5.791e+01 Oil Holdup Time, minutes 1.01e+00
   Oil Gas Bub Rise, ft/s                                    3.42e-02
              THREE PHASE VERTICAL SEPARATOR DATA INPUT
```

```
Gas Rate, lb/hr              40000  Liquid Water Rate, lb/hr     10
Liquid Oil Rate, lb/hr     1000000  Gas Viscosity, Cp          .012
Water Density, lb/cf          62.4  Water Drop in Oil, Microns  500
Gas Density, lb/cf             .15  Gas in Oil Bubble, Microns  200
Liquid Drop in Gas, Mcr        150  Liquid Oil Viscosity, Cp    1.5
Liquid Oil Dens, lb/cf          52  Oil Depth, inches            25
Water Depth, inches             10
```

Figure 4.10   Three-phase vertical separator Vessize program example run.

design and rating criteria we have to date. Perhaps some of you who take this study to heart will make further refinements and scientific breakthroughs.

Gas phase.   The gas-phase recommended liquid oil drop particle size is normally 100 to 150 μm. This is a well-tested particle diameter size range, considering countless field-proven cases. The fact is well known that given a 100-μm fallout before reaching a demister pad of common design, particles averaging 100 μm will be removed by the demister pad. When the average liquid particle is larger than, say, 300 μm, most demister pads tend to flood, causing even greater liquid carryover. Thus when you need good liquid removal from a gas stream, a 150-μm or smaller liquid drop particle is recommended.

Regarding the need for absolute removal of all free liquid mist, such as with a gas turbine compressor suction scrubber, I find a 100-μm particle size selection to be good design practice. This particle size fallout before reaching a demister pad will indeed ensure that most any demister pad will remove all remaining free liquid mist.

Caution: I have known a few demister pad vendors to claim that their pads would remove even a 500-μm particle and yet sustain absolute mist-free discharge. I have serious reservations here. Why? First, I have yet to witness myself, or even hear a truly unbiased witness testify, that this is true. Can a 500-μm selection be achieved for absolute mist removal in a demister pad? Second, I leave this question to those who might take up this challenge and prove the case to be unquestionably true or false. In the interim I heartily recommend that all users of the programs presented in this book strictly stay in the 100- to 150-μm range for efficient liquid mist removal in conjunction with a standard demister pad. For a standard demister pad, I propose a stainless steel wire pad of number 20 mesh and a minimum 6-in thickness. If the vessel is less than 5 ft in diameter, I recommend a 4-in minimum thickness.

If there is a question of losing a product such as an amine chemical, say diethanolamine, in an amine gas absorber overhead KO vessel, use this recommended 150-μm liquid particle sizing in the gas phase. I have personally witnessed hundreds of thousands of dollars of annual amine chemical losses in numerous amine gas–treating plants due to poor overhead KO drum design. Spend a few more very well justified dollars at design time and realize a payout of only a few weeks for this added expense! I have witnessed 1.5 lb amine loss per million scf gas processed in an amine absorber overhead KO drum. For a 150-mmscfd gas plant absorber, 355 days per year production, at $1.20 per pound amine chemical cost, this computes to a $95,750 yearly loss. This is not a new discovery, as many an amine absorber installed in the 1950s had several trays in the tower top section dedicated to a water-wash section. I am confident that equal losses can be computed for other chemicals or petroleum products in similar fractionation overheads.

Water drop in oil phase.   Contrary to a 150-μm particle in the gas phase, when absolute free liquid removal is not required, such as with a KO drum upstream of a flare, 500-μm sizing is very common and good-practice engineering. Here free liquid mist, typically 500 μm, will readily burn. Other applications, such as pipeline KO drums, may be of the same order.

The water drop in the oil-phase size is the one for which we have the more abundant data. Please refer again to Fig. 4.1. If each of these stages were the horizontal separator type, as shown in Fig. 4.4, and if a reasonably effective deemulsifier were applied, then each stage should remove 90% by volume of the water in the crude oil feed. This factoring, however, is conditioned upon properly selecting the correct water particle size for the horizontal vessel design. Temperature heating is also a condition, as discussed in the following. Now the key question becomes, "What is the correct water particle size selection in the vessel design basis to make this happen?"

Two physical factors bear heavily upon how well or how completely free water drops can be separated from the oil phase. These are *density* and *viscosity*. The settling equation, Eq. (4.3), is governed mainly by the water-oil density difference and by the water particle size, expressed as a spherical diameter. While we have little control over the difference in density of oil and water, we can control oil viscosity by heating, within limits. Extensive studies have shown that the viscosity of the oil greatly affects the coalescence of the water drop particles.

As these particles may collide, the water particles become larger. The better and sought-out condition here is to have the largest water particles possible. The more these particles collide, the greater the water drop diameter for Eq. (4.3). We find that water particle collisions are greatly affected by, and indeed depend on, the oil viscosity.

The following water particle diameter equation is presented in Arnold and Stewart [10]. Observing five specific oil-water separators for free water removal and checking the evaluation of this particle sizing, I found good agreement. With reasonable deemulsifying chemicals added, 80 to 90% by volume water removal per stage was found, applying the particle size found in Eq. (4.14):

$$D_M = 500(\text{VISC})^{-0.675} \qquad 0.5 < \text{VISC} < 10 \qquad (4.14)$$

where D_M = diameter of water particle drop, μm
         VISC = oil viscosity at system temperature, cP

Please note that Eq. (4.14) is to be considered only for a free water (and emulsion free water)–type KO removal. This is generally the first vessel in the train of stages for oil treating. In Fig. 4.1 it is the first stage. In some applications it can also be applicable to the second stage in Fig. 4.1. There is a common reason why a second stage is required in using Eq. (4.14): first-stage oil residence time. If the first-stage oil residence is less than 10 min, a second-stage D_M input per Eq. (4.14) should be made. This D_M is to be input into the Vessize or VessizeV program. In cases where large oil-water emulsions are present, say, 20% or more by volume oil emulsion, a 30-min first-stage oil residence time is required; otherwise a second stage is required per Eq. (4.14). Please note that both programs, Vessize and VessizeV, calculate and display the oil residence time in minutes.

Now, having a way to calculate a more meaningful D_M for the oil-phase water removal, please review again the vertical vessel run in Fig. 4.10. Note here that a 500-μm D_M variable was initially input. This was the defaulted program value input. Now calculate D_M using the 1.5-cP oil viscosity value. You should get a 380-μm D_M value. Input this new value for water droplet size in oil, and run the program. The oil-water dropout now has become the governing vertical vessel-sizing criterion. Here either two vertical vessels or a horizontal replacement should be used. Vessel fabrication up to 12 ft is generally economic. Above 12 ft, two vessels should be considered with installed cost review. Logistics of transport is also a project cost that must be carefully studied for justification and detail planning.

Equation (4.14) reveals that the lesser the oil viscosity, the larger the allowable water particle drop D_M. The oil viscosity is appreciably lowered by increased temperature. It is thus concluded that temperature should be increased as much as is feasibly and economically practical. This statement, however, is not true for all cases, as increasing the temperature of heavy oil—say, 12°API and lower will decrease the density difference between the oil and water. I have personally witnessed this phenomenon in two projects. For oil 15°API and above, increased temperature causes little if any change in density difference. The red flag here, therefore, is to take caution if the oil gravity is below 15°API; otherwise, apply as high a temperature as feasible.

For downstream oil treatment and completion of dehydration, use the following equation, published by the Society of Petroleum Engineers [2]:

$$D_M = 200(\text{VISC})^{0.25} \qquad \text{VISC} < 80 \text{ cP} \qquad (4.15)$$

Equation (4.15) should be considered the final oil-treatment stage, in which a 0.5 to 1% water cut in treated oil is produced. Here a heater-treater-type horizontal vessel is more commonly used. A direct-fired burner-heater is normally placed in a front section of the horizontal treater. Controlling the system temperature for viscosity is important and may be mandatory for successful water dehydration.

Equation (4.15) applies the third stage shown in Fig. 4.1, the overall flow diagram. Here the free water has been removed, down to near 1% by volume remaining in the third-stage feed. If there is not a near 1% remainder in this feed, then an oil-water emulsion problem is more the case. Oil-water emulsions are hard to break and in many cases require special treatment, such as with electrostatic horizontal treaters or chemically treated tankage. Another equation for electrostatic treaters is discussed later in this chapter. For dehydration tankage (with emulsion breaker chemicals added) Eq. (4.15) is applicable. Dehydration tankage may require long residence time, up to 24 h. Dehydration tankage is also covered later in this chapter.

Gas bubble in oil phase. Little research here has been accomplished, and very little has been published about gas bubble or foam separation from liquid. Herein I offer a good contribution to this technology, along with a plea for more field-proven data. As in the case for liquid droplet fall in the gas phase, I propose that the same equations, Eqs. (4.5), (4.6), and (4.7), be used in the oil media. This is done in these three equations, Eq. (4.7) deriving the gas bubble terminal velocity. We must, however, input a feasible and proven gas particle size D_M, μm. Having accomplished several field-proven foam separation tests, the following D_M determination equation is offered.

$D_M = \exp [5.322 - 0.3446 * \ln \text{VISCL} - 0.0206 * (\ln \text{VISCL})^2]$ (4.16)

where D_M = diameter of gas particle, μm
VISCL = liquid viscosity, cP
ln = natural log of base 2.71828
exp = exponential function of natural log base number

Equation (4.16) examples:

$$\text{VISCL} = 0.71 \qquad D_M = 230$$

$$\text{VISCL} = 6.5 \qquad D_M = 100$$

$$\text{VISCL} = 30 \qquad D_M = 50$$

Equation (4.16) should be limited to an oil-phase viscosity range of 0.5 to 40 cP. Figure 4.11 displays a computer run printout of the VessizeV program. This run is for a vertical vessel requiring a 16-ft diameter. The run was for an actual application in which oil and water were mixed and passed through the vessel as an emulsion mixture.

Note that in Fig. 4.11, the oil defoaming section has set the required vessel diameter, 16.1 ft. Given an oil viscosity of 0.71 cP at the system temperature, the D_M is calculated to be 230 μm in Eq. (4.16). Increase this D_M size in the program and then run it again, clicking the Run Program button. Notice how increasing this D_M size decreases the required vessel diameter. The D_M size we input indeed governs the vertical vessel diameter required.

```
          THREE PHASE VERTICAL VESSEL ANALYSIS CALCULATIONS
                   Gas - Oil - Water Separation
      Gas Phase Vessel Diameter Rq'd, ft                   3.96e+00
      Oil Phase Water Sep. Vessel Diameter Rq'd            1.15e+01
      Oil Phase Foam Sep. Vessel Diameter Rq'd             1.61e+01

Gas Drag Coef          5.91e+00   Terminal Gas Vel, ft/sec 1.76e+00
Barrels of Oil Volume  2.545e+02  Oil Holdup Time, minutes 1.58e+00
Oil Gas Bub Rise, ft/s 7.40e-02

              THREE PHASE VERTICAL SEPARATOR DATA INPUT

Gas Rate, lb/hr             5850   Liquid Water Rate, lb/hr   1000
Liquid Oil Rate, lb/hr   2991000   Gas Viscosity, Cp          .012
Water Density, lb/cf        58.6   Water Drop in Oil, Microns 10000
Gas Density, lb/cf          .075   Gas in Oil Bubble, Microns 230
Liquid Drop in Gas, Mcr      150   Liquid Oil Viscosity, Cp   0.71
Liquid Oil Dens, lb/cf        55   Oil Depth, inches          84
Water Depth, inches           10
```

Figure 4.11 Three-phase vertical separator Vessize program example run.

Similar to the gas phase and the particle drop in the oil phase, the fluid-phase media viscosity is again a controlling factor. The greater the media phase viscosity, the smaller the D_M particle size required, and the larger the diameter of the vertical vessel required.

Please note the gas and oil rates are given in Fig. 4.11 as mass rates. The computer program on the CD accompanying this book presents a more user-friendly input in barrels per day for oil and water and million standard cubic feet per day for gas.

Equation (4.16) is to be considered the calculation of the required gas particle average size for complete foam in oil separation. If you input a larger particle size than that calculated in Eq. (4.16), foam carryover is predicted. How much foam carryover is a good question. Foam carryover is simply the gas entrapment in the oil liquid flowing into the downstream equipment. If some gas flashoff in downstream equipment is tolerable, such as gas venting to flaring, perhaps even an average 500-μm gas in oil bubble value is a good and reasonable input. In contrast, however, when a vessel flashing a large gas volume (say, 100 scf/bbl or greater) is discharging liquid to an atmospheric storage tank, this 500 μm value could be disastrous. Therefore, be cautious and prudent in your designs.

Please note that Eqs. (4.14), (4.15), and (4.16) are equally applicable to horizontal vessels, as well.

Disclaimer: As in all theoretical variable determinations, these equations presented for D_M calculation are subject to field-test verification. Equations (4.14) and (4.16) are not presented as being infallible or able to predict accurately every case of particle size with a given medium viscosity. For example, a crude with a high asphaltene content should be field tested before a final design for construction is issued on the basis of these equations. Small asphaltene crude contents (less than 2%) were used in deriving Eq. (4.16). More tests are needed for foam-liquid separations. Readers and users of this criterion, can perhaps contribute more data, and I indeed solicit such contributions of better methods and data as you may discover.

One last note about particle size: it is an average size of a very wide range, from 1 to over 1000 μm. The number used per Eqs. (4.14) and (4.16) is a representation of the weighted average particle size and is a referenced data source per viscosity of the media in which the particle rises or falls.

Vertical Storage and Treatment Tanks

Vertical storage tanks are most commonly and uniquely used as dehydration vessels as well as desalting vessels for crude oil. The key variable for making a common storage-type tank into an oil dehydration vessel is simply time. Given enough time, most any oil-water mixture will form separate phase layers, normally with oil the top phase and water the bottom phase. This is, of course, the free water in the oil. Soluble water in oil is not addressed in this book, but its content can be anything from a very few to over 1000 ppmv. Free water remainder in treated oil as low as 0.3% by volume (3000 ppmv) can be produced in a storage tank. Tank vendor guarantees or engineering design company guarantees are seldom made for any oil dehydrated values of less than 1% by volume. The reason is the design target values are 0.5% volume, thus allowing a 0.5% margin. A value of 1.0% by volume is commonly guaranteed.

A crude oil residence time range 8 to 30 h or more is the normal dehydration tank design basis to achieve a 1% by volume water remainder specification. Figure 4.12 shows the average heavy crude oil storage time required for this 1% specification. I recommend that you take Fig. 4.12 to heart in selecting your minimum storage time for the 1% quest.

Figure 4.12 shows the results of an actual field storage dehydration tank test by a major oil production, refining, and marketing company. The tank, 150,000 bbl volume, had a daily throughput rate of 160,000 bbl, oil and water mixed. A 6% inlet water cut was fed to the tank with the oil properties of 14°API and 72 cP viscosity, at a tank temperature of 180 to 185°F. 100 ppmv of a deemulsifying chemical was also added. As seen in Fig. 4.12, the 1% treated oil specification was achieved with an

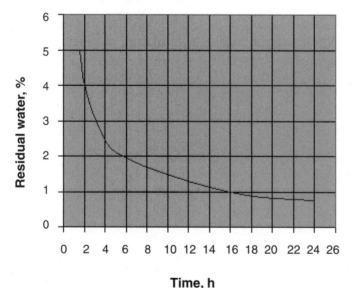

Figure 4.12   Water remainder in wash-tank-treated oil.

approximate 0.8% water remainder. This is a classic but very common test that is used throughout the oil production industry. A lighter oil may have a lesser storage time requirement, say 14 h for a 24°API oil.

The following code line list is taken from the supplied program named TkSep.mak. A storage tank designed to remove water to a 1% dehydration specification is also called a *wash tank*. The name *wash tank* is a proper one because the tank may also be used as a desalting treater and a dehydration tank. Our TkSep program gives you the input callout to input a percentage value for salt concentration in the incoming production well water. In the example run shown in Fig. 4.13, you will see that the salt input is 11%

```
RWBBL = (Q1 / .99) * .01     ' 1% of treated Crude is Water
PTB = RWBBL * SaltConc * 1000 / Q1: txtPTB = Format(PTB,
    "#####")
```

where RWBBL = remaining water in treated crude oil, bpd
       Q1 = treated crude oil flow, bpd
      PTB = salt remaining in treated crude oil, lb/1000 bbl

Refiners normally require an oil mineral content of no more than 10 lb per 1000 bbl. As previously, this is expressed as salt equivalent, NaCl. If you run the program exhibited in Fig. 4.13 and input 2.0 lb salt per barrel inlet feed water (replacing the 1.0 lb shown), you will see the

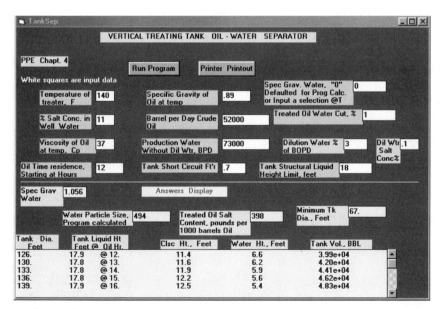

Figure 4.13 Screen from the TkSep.mak program.

treated oil salt content increase to 20 lb salt per 1000 bbl. Thus the key step here to satisfy the 10-lb salt specification is to ensure that your inlet water feed contains no more than 10 lb salt per barrel water. If your water analysis is more than this 1 lb salt per barrel water, then you should add fresh water to your feed to make the total water feed 1.0 lb salt per barrel water or less. A little math on the user's part here may be required. Or perhaps some young and energetic engineers will upgrade this program. Doing so would provide an answer in the program output as to how much additional water and at what allowable salt content will be required.

Following are the main motor code lines of the TkSep program:

```
DM = KF * (VISC ^ .2506): txtDM = Format(DM, "#####")
DiaCalc:
    D = (81.8 / SCF) * (Q1 * VISC / (SG * (DM ^ 2))) ^ .5  ' D
    is calc in inches
    H = TR * Q1 / (.12 * D ^ 2)    ' TR is the retention time
    of oil in the tank, hours
Open "C:\PPEVB\PPEVB44\TankDat" For Output As #1
    For I = 1 To 15
OilHt:
    H = TR * Q1 / (.12 * D ^ 2)    ' H is oil height in inches
    VOLW = Q2 * (TRW / 60) * 5.615 * 1728 / (24)   VOLW is
    water volume in cu. inches
    HW = VOLW / (((D ^ 2) / 4) * 3.1416)  ' HW is water height
    in inches
    If (H / 12) > (HL - (HW / 12)) Then D = D + 12: GoTo OilHt
    HT = H + HW: VOLTK = (((D ^ 2) / 4) * 3.1416 * HT) / (1728
    * 5.615)
```

where   DM = particle size, μm, calculated as shown in the code lines
        KF = factor for calculating DM, as shown in Fig. 4.13
        VISC = oil-phase viscosity, cP

Note: The KF factor is defaulted as a 200 value to derive a 1.0% water remainder specification in the treated oil.) You may find more accurate defaulted values of KF in your company's test. To date, however, this 200 value is proven to be a most accurate and conservative design value to achieve the 1.0% water specification.

Alternate form of Eq. (4.3):

```
D = (81.8 / SCF) * (Q1 * VISC / (SG * (DM ^ 2))) ^ .5
```

The diameter required for the tank is calculated in the preceding modification of Eq. (4.3). Thus, by inputting a starting time TR, the program calculates a varied number of tank diameters, oil height, and water height, with corresponding oil residence time shown. The residence time is increased in the program starting with the user's chosen input time. A number of tank combinations are calculated, as seen in the actual screen exhibited in Fig. 4.13. Each combination of the tank sizes with a residence time shown in Fig. 4.13 is for the 1% water treated-oil specification.

The key issue is how much time is required for the 1% water specification. If you take Fig. 4.12 and let it resolve this issue, then you need 24 h of retention time. See Fig. 4.14 and note the tank design required for the result shown in Fig. 4.13. Here you need a 162-ft-diameter tank with a 65,000 bbl volume for the required 24 h of oil-phase retention time.

Caution: Your company should validate Fig. 4.12, or else produce your own data! A lighter oil, say 34°API, could have a 0.5% water remainder at 12 h retention time, for example. This would require a smaller tank, as shown in the figures. Run a trial for your case in the TkSep.mak program.

Another important TkSep program input is the limiting tank height. In the example runs, Figs. 4.13 and 4.14, we input 18 ft. Allowing 2 ft for freeboard height, this requires a tank height of 20 ft. Tank height is economically governed by the tank wall thickness required. For this

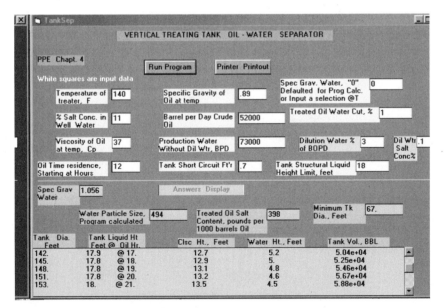

Figure 4.14   Screen from the TkSep.mak program.

reason, tank heights are seldom greater than 40 ft; the range commonly fabricated is 20 to 35 ft in height.

You should now try your own special tank conditions in the TkSep Visual Basic program. Please note carefully that the program does more than determine the oil volume for the residence time shown. It determines the required oil section diameter for the 1% water treated-oil specification. Looking again at the Eq. (4.3) modification, the horizontal surface area is most critical for the water particle separation in the oil phase. Take note of this criterion as you run the program. I am confident you will gain much respect for this program and the reliability of the design answers.

As we started with a simplified process flow diagram in this chapter, we shall end this chapter with a simplified process flow diagram, Fig. 4.15. The first one, Fig. 4.1, gives a diagram for pressure vessel oil-water-gas three-phase separators (gravity separation) and electrostatic three-phase separators. This last flow diagram gives the flow arrangement of atmospheric tanks.

Please note the flows in Fig. 4.15, with the production crude first entering the degasser vessel to flash off the gas and bring it to atmospheric pressure. Then the liquid phase, oil and water, enters the first of two tanks in series flow. All flows shown from the degasser vessel to the oil dehydrator tank are gravity flows. There are no pumps between these vessels and tanks.

Degasser vessel

The degasser does just as its name signifies, it degasses the crude. It not only flashes off the gas but also serves as a surge vessel upstream of the atmospheric tanks. The degasser is generally a vertical vessel, 50

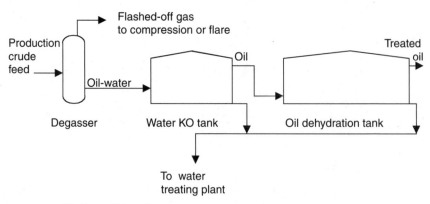

Figure 4.15 Tankage oil treating process.

to 80 ft tall, and has a large open line connecting to the plant flare header. The degasser receives any sudden pipeline surges in both volume and pressure, protecting the downstream tanks. One important design condition of the degasser is to flash out all gas to atmospheric pressure. Thus the VessizeV program offers the methodology to accomplish such a design.

Refer to Fig. 4.11, which shows a classic vertical degasser vessel designed on the basis of removing the gas foam from the liquid phase. If this design basis were ignored, gas carryover into the atmospheric tank could occur, risking mechanical failure of the tank. Ignoring the defoaming concept would have this vertical vessel designed on the basis of liquid particle removal from the gas phase alone. Thus, a smaller diameter, 4 ft, would be the result. Note especially the particle sizes chosen for inputs, which are based on the following data.

1. *Gas-phase liquid particle.* The gas-phase liquid particle size of choice is generally 150 µm or less, which results in excellent gas-phase separation of liquids. For more critical liquid particle removal, such as with turbine blade gas suction, a 100-µm particle size input is recommended with a supporting demister pad in the separator.

2. *Oil-phase water particle.* For the degasser vessel shown in Fig. 4.15, water-oil liquid-phase separation is not required. A very large particle size was therefore input in the example run in Fig. 4.11, 10,000 µm for the water droplet in oil size.

3. *Oil-phase gas particle.* Use 200 to 250 µm here. A gas particle size of 230 µm in the oil phase is therefore input. Experience to date in three heavy oil production fields, 12 to 18°API oil, has shown this gas in oil particle size to be a safe range. A greater gas particle input size in the degasser design may result in severe gas carryover into the downstream atmospheric tank. This input is critically important for the degasser service vessel shown in Fig. 4.15. Repeating, all gas release must be captured in the degasser vessel in this process. Several pounds of gas-release pressure in the downstream atmospheric water KO tank could result in a severe tank-roof rupture. It is therefore urgently and strongly advised to take caution in design here.

The degasser vessel has yet another critical design value—its height. First, it must be tall enough to receive upstream pipeline surges. Designing for a surge equal to the liquid volume of at least 300-ft of the pipeline served is recommended. A second design condition of the height is to pressurize the incoming liquid so it flows into the downstream tank. Remember, both the degasser vessel and the tank are near atmospheric overhead pressure. You may need a height of liquid in the degasser vessel several feet above the top oil level in the downstream tank. This height difference is that required to compensate for the flowing fluid pressure drop in the connecting line between the tanks.

There is one more concern for caution in design. The degasser overhead normally connects to the flare header or to the suction header of a gas compressor train. This could be an error, calling for a pressure of 5 psig or more, which would cause a tank-roof rupture downstream. A downstream tank failure would occur here because the liquid level would be lost in the degasser, causing the gas to release directly into the downstream KO tank. Flare system headers generally operate at 2 to 5 psig and upon emergency relief can reach pressures of 25 psig and higher. A flare header backpressure of 3 psig can cause a very low degasser vessel level and possible loss of any level. Take caution, therefore, that your flare header emergency relief backpressure allows a reasonable degasser minimum level. For the design shown in Fig. 4.15, a flare header backpressure of 2 psig (emergency relief condition) is recommended. Take similar design caution for gas compressor suction headers connected to the degasser vessel. Much larger flare and compressor suction line sizes may be required, but weigh this design investment against the loss of a tank and the accompanied production loss due to a tank rupture.

Water knockout tank

The water knockout tank is designed to separate all free water from the incoming field production oil. It generally uses a 30- to 60-min storage time. In many field cases, the free water is only 30 to 60% of the total water content in the oil. This is due to oil-water emulsions. The oil fed to the downstream dehydration tank (see Fig. 4.15) is treated with an emulsion breaker chemical that is injected into the feed line to the dehydrator tank. As much as 70% of the production may therefore be separated in the downstream dehydrator tank. The KO tank may also receive fresh water for desalting the crude oil through its feed line.

Dehydration tank

The preceding pages cover the dehydration tank shown in Fig. 4.9 in considerable detail. Again note Fig. 4.11, showing the oil-phase residence time in this dehydration tank required to make a 1% or less by volume treated-oil water content specification. You should normally allow 16 h or more residence time for the oil to dehydrate. Again, the point is stressed to make laboratory tests to confirm the residence time required here. Unfortunately, however, our lab tests of the day are indeed approximate and in many cases fail to reveal reliable data. It is therefore prudent to run actual field operating tests, proving without doubt what oil layer residence time is required to make the 1% water specification. If in doubt of the laboratory data (and I have been in numerous cases), apply Fig. 4.11. The 16 h or more residence time you select should be a conservative one for most applications. Once you have selected an oil residence time, run the PPE program TkSep.

TkSep will give you a recommended and calculated tank diameter with calculated height for the oil residence time of your choice. As seen on the program screen in Fig. 4.14, a constant 1% treated-oil water content is maintained for all calculation runs. The program actually aims for 0.5% but displays 1% for a conservative design.

Methodology for Horizontal Separator Sizing and Rating Calculations

Gas-phase liquid particle vessel sizing will first be addressed with design equation methodology in a four-step procedure.

Gas phase

General calculation equations for the gas phase in a four-step sequence are as follows:

Step 1. The first equation of several required is terminal velocity of the liquid particle settling, gas phase, from Eq. 5-212 of *Perry's Chemical Engineer's Handbook* (p. 5-61) [1].

$$V_t = \left(\frac{128.8 \times D_p \times (D_l - D_g)}{3 \times D_g \times C} \right)^{0.5}$$

where V_t = particle terminal velocity, ft/s
D_p = particle size, ft
D_l = liquid density at system temperature and pressure, lb/ft^3
D_g = vapor density at system temperature and pressure, lb/ft^3

Note that a value of 150 μm is used for D_p, per the Gas Producers Suppliers Association (GPSA).

Step 2. The drag factor C is to be determined next, using the following equation (Eq. 5-222 from Perry):

$$C_D(\text{Re})^2 = \frac{(0.95 \times 10^8) \times D_g \times D_p^3 \times (D_l - D_g)}{\mu_g^2}$$

where Re = Reynolds number, dimensionless
μ_g = gas viscosity, cP

The drag coefficient C_D is then read from page 5-64 in Perry, applying $C_D(\text{Re})$ as the x axis.

Step 3. Determine actual gas flow, ft/s:

$$Q_a = \frac{Q_g/(379.49 \times 86,400) \times \text{MW}}{D_g}$$

where Q_a = actual gas flow rate, ft³/s
   Q_g = gas flow, mmscfd
   MW = gas molecular weight

Step 4. Assume a horizontal vessel diameter and solve for horizontal vessel length L, using the following equation:

$$L = \frac{4 \times Q_a}{3.1416 \times V_t \times D_v}$$

where D_v = the part of the vessel diameter free of liquid, ft

For a vertical vessel:

$$A = \frac{Q_a}{V_t}$$

where A = vertical vessel cross-section area, ft²

Having solved the area, calculate the diameter required:

$$A = \frac{\pi D_v}{4}$$

where $D_v = (4A/3.1416)^{0.5}$

If an existing vessel is to be rated, the same equations are applied to determine the maximum gas flow rate, mmscfd.

Normally, all three-phase horizontal separators are designed and maintained to be half full of liquid.

Liquid phase—separating water from oil

Sizing a pressure vessel's liquid-phase section involves four chief factors: area, temperature, chemicals, and oil-water residence time required. The first variable, area, has been well established by the water particle in oil settling equation. This equation is only for *free water*. The second variable, time, addresses the oil-water emulsion problem. Given enough time with a temperature increase and proper chemicals added, most oil-water emulsions will break and free all water. The variable time is an unknown calculation in the industry, however, and has only been determined by actual field-test separators or by full-size equipment operations.

The settling equation has been determined for base, sediment, and water (BS&W) values ranging from 0.05 to 10% [2]. However, actual field tests of all applications are encouraged to determine the correct factor [200 in Eq. (4.17)] which should be derived by test for any one particular case. If field-test data are not available, the following equations below may be used as best estimates.

$$d_{m1\%} = 200 \; \mu^{0.25} \tag{4.17}$$

$$d_m = d_{m1\%} \; W_c^{0.33} \tag{4.18}$$

The settling equation has been derived by Arnold and Stewart [11]. For horizontal vessels, the following equation has been derived:

$$dL = \frac{438 \; Q_o \mu}{\text{SG} \; d_m^2} \tag{4.19}$$

where $D_{m1\%}$ = water particle size for 1% water cut, μm
  μ = oil viscosity, cP
  d_m = water particle size required for W_c value, μm
  W_c = water cut; BS&W remaining in treated oil, percentage by volume
  d = vessel diameter, in
  L = effective vessel length for separation of water from oil, ft
  Q_o = oil flow rate, bopd
  SG = specific gravity difference of oil and water at system temperature

Production Desalters

Equipment design basis

The desalter vessel operates full of liquid and is classified as a two-phase-type separator for oil and water. Nominally, desalters range in capacity from 50 to 80 barrels of oil per day per foot (bopd/ft). The oil-phase rate of flow only is taken as a design criterion, disregarding the water rate because the water falls out in the bottom quadrant. This capacity rating is applied to the horizontal plane of the vessel, the area of which is calculated by multiplying the vessel diameter times the vessel length. For the particular desalter vessels here, 12-ft diameter times 50-ft length equals 600 ft. Thus, for an estimated maximum desalter oil rate capacity limit, say 80 bopd/ft, the expected vendor maximum capacity for this vessel size is 80 × 600, or 48,000 bopd. This is only an estimate and should not be taken as a discrete design for this vessel. This does reveal, however, that the desalter sizes are reasonable for the new 81,000-bopd requirement, or 40,500 bopd for each of two trains.

Methodology

Desalter vessel sizing is similar to sizing a three-phase vessel design (liquid-phase water particles in oil separation), in that both are sized by the settling equation as given in the following procedure. The main difference is that the desalter has an electric field in the oil phase. This electric field tends to accelerate water particle size increase, whereby a much lower BS&W can be achieved as compared to a common vessel.

The following equations show the desalter vessel sizing method. Note that the salt content in crude oil is that salt mineral content which is dissolved in the water phase. To remove salt content, water is to be removed, a process also called *crude dehydration*. Dilution water containing considerably less salt may be mixed with the crude. Thus, injecting dilution water helps to reduce the treated-crude salt content. Equations for determining the amount of salt remainder in desalter treated oil follow.

$$W_d = \frac{W_{d\%} / 100}{1 - (W_{d\%} / 100) \times Q_o} \qquad (4.20)$$

$$S = W_d \times 350 \times \frac{C_d}{100} + Q_w \times 350 \times SG_p \times \frac{C_p}{100} \qquad (4.21)$$

$$C_m = \frac{S}{W_d \times 350 + Q_w \times 350 \times SG_p} \times 100 \qquad (4.22)$$

$$SG_m = \frac{(61.94 \times 0.495 \times C_m) - 0.017 \times (T - 60)}{62.4} \qquad (4.23)$$

$$PTB = 350 \times SG_m \times \frac{C_m}{100} \times \frac{W_c}{100} \times 1000 \qquad (4.24)$$

where W_d = water mixed with the crude oil fluid upstream of the desalter, bpd

$W_{d\%}$ = dilution water in crude oil phase, percentage by volume

Q_o = crude oil barrels per day dry basis

S = total salt content in the production water and dilution water, lb/day

C_d = salt concentration of dilution water, percentage by weight

Q_w = production water rate, bpd

SG_p = specific gravity of production water at T

C_p = salt concentration of production water, percentage by weight

T = system temperature, °F

SG_m = specific gravity of dilution water and production water mixture at T

C_m = salt concentration of water mixture, percentage by weight

PTB = salt in oil at outlet of desalter unit, lb/1000 bbl

W_c = BS&W of treated desalter outlet crude oil, percentage by volume

Note that W_c is selected based upon the electric grid desalter equations that follow.

The preceding five equations reveal that how much salt remains in the treated oil depends very heavily on how much water is removed from the oil in the desalter. Normally, a water remainder of 0.05 to 0.1% by volume is experienced with electric grid desalters. The water particle settling equation shows that this range is obtainable in most cases. Temperature and added deemulsifying chemicals also affect how much water remains in treated oil. The temperature effect is due to variance of oil viscosity with temperature. The greater the viscosity, the less water is removed from the treated oil. Raising the temperature decreases the viscosity, which is desirable. Raising temperature, however, also decreases the difference in specific gravity of water and oil, which is undesirable. The viscosity factor most always governs, however.

The settling equation has been determined for desalter BS&W values ranging from 0.05 to 10% BS&W [2]. However, actual field tests of all applications are encouraged to determine the correct factor [170 in Eq. (4.25)] which should be derived by test for any one particular case. If field test data are not available, the following equations, which determine electric grid desalter sizing, may be used as best estimates.

$$d_{m1\%} = 170 \, \mu^{0.4} \tag{4.25}$$

$$d_m = d_{m1\%} \, W_c^{0.33} \tag{4.18}$$

The desalter settling equation has been derived by Arnold and Stewart [11]. For horizontal vessels, the following equation has been derived:

$$dL = \frac{438\, Q_o \mu}{\text{SG}\, d_m \mu} \qquad (4.26)$$

where $d_{m1\%}$ = water particle size for 1% water cut, μm

μ = oil viscosity, cP

d_m = water particle size required for W_c value, μm

W_c = water cut; BS&W remaining in treated oil, percentage by volume

d = vessel diameter, in

L = effective vessel length for separation of water from oil, ft

Q_o = oil flow rate, bopd

SG = specific gravity difference of oil and water at system temperature

Overview

This concludes our study of oil and gas surface facility design and rating. Chapters 1 to 3 cover gas-liquid separation, while this chapter gives equal coverage to the liquid-liquid phase separations. Indeed this is what surface facility design is all about. It would be a waste, however, to consider the tools presented here to be applicaple only to crude oil surface facility designs. Consider applications to refineries, chemical plants, and gas processing plants as well. For example, the characteristics of water could be replaced with those of certain chemicals in order to derive valued data from these ICPD computer programs. No doubt many engineers of present and future generations will find many more applications.

References

1. Perry, R. H., and Chilton, C. H., "Particle Dynamics" and "Discussion of Stokes Law Definition," *Perry's Chemical Engineer's Handbook,* 5th ed., McGraw-Hill, New York, 1973, pp. 5-60–5-65.
2. Arnold, K. E., and Thro, M. E., "Water-Droplet-Size Determination for Improved Oil-Treater Sizing", SPE Technical Paper No. SPE 28537, Society of Petroleum Engineers Conference, New Orleans, LA, May 1997.
3. Perry and Chilton, "Application of CRE," p. 5-64.
4. Perry and Chilton, Equation 5-215, p. 5-61.
5. Perry and Chilton, Table 5-26, p. 5-64.
6. Lapple and Shepherd, *Ind. Eng. Chem.* 32, 1940. Cited in Perry and Chilton.
7. Perry and Chilton, pp. 5-61, 5-62.
8. Arnold, Ken, and Stewart, Maurice, *Surface Production Operations,* vol. 1, Gulf Publishing, Houston, TX, 1995, p. 105.
9. Perry and Chilton, Fig. 5-78.
10. Arnold and Stewart, p. 174.
11. Arnold and Stewart, p. 172.

Shell/Tube and Air Finfan Heat Exchangers

Nomenclature

Shell/tube exchangers

μ	shell- or tube-side fluid viscosity, cP
μ	fluid viscosity at system temperature, cP $\times$ 2.42
μ_w	fluid viscosity at tube wall temperature, cP $\times$ 2.42
A	tube bundle area, ft^2
A_t	inside tube area, ft^2 $\times$ number of tubes per pass
B	baffle spacing, in
c	fluid specific heat, Btu/lb $\cdot$ °F
C_c	tube clearance, in
D	internal tube diameter, ft; shell-side equivalent diameter, ft
d_o	tube outside diameter, in
F_t	Kern's [1] LMTD correction factor
G	lb/(h) (A_t), tube mass flow per time and cross-sectional area, lb/h $\cdot$ ft^2
GTD	greater temperature difference, °F
h_d	dirt or fouling resistance on tube walls, $1/R_d$
h_{di}	heat transfer coefficient inside tube for dirt fouling factor only, Btu/h $\cdot$ °F $\cdot$ ft^2
h_{do}	heat transfer coefficient outside tube for dirt fouling factor only, Btu/h $\cdot$ °F $\cdot$ ft^2
h_i	coefficient of inside tube wall film resistance to heat transfer, Btu/h $\cdot$ ft^2 $\cdot$ °F

h_{io}	coefficient of inside tube wall film resistance to heat transfer based on outside tube diameter, Btu/h · ft^2 · °F
h_o	coefficient of outside tube wall film resistance to heat transfer, Btu/h · ft^2 · °F
J_h	exp [−4.287 + 0.85 * log (D * G/μ)], curve-fit equation from Kern [1], Fig. 24
K	thermal conductivity of fluid, Btu/(h · ft^2 · °F · ft^{-1})
LMTD	log mean temperature difference, °F
ln	natural log of base 2.7183
LTD	lower temperature difference, °F
P_t	tube pitch, in
Q	overall heat transferred, Btu/h
R_d	fouling factor, $(U_c − U_d)/U_c * U_d$; overall fouling factor, $R_i + R_o$
R_i	tube inside fouling factor, $1/h_{di}$
R_o	tube outside fouling factor, $1/h_{do}$
S	shell-side inside diameter, in
S_p	shell passes
U_c	overall clean heat transfer coefficient for shell tube exchange, Btu/h · ft^2 · °F
U_d	overall dirty heat transfer coefficient for shell tube exchange, Btu/h · ft^2 · °F
W	mass shell-side flow, lb/h

Air finfan exchangers

ϕ	HT correction factor for viscosity, $(U_{VISC}/U_{VISCw})^{0.14}$
ACFM	actual ft^3 per minute volumetric air flow
A_I	tube internal cross-sectional area, in^2
APF	total bare tube external area per foot of fin tube, ft^2/ft (select from Table 5.3 or vendor data)
APSF	external tube OD bare surface area ratio, ft^2, per ft^2 of bundle face area (select from Table 5.3 or vendor data)
AR	ratio of area of fin tube to exterior area of 1-in-OD bare tube (select from Table 5.3 or vendor data)
A_X	extended finned tube surface area, ft^2
CDTM	corrected log mean temperature difference, °F
C_p	specific heat, Btu/lb · °F
DENSA	density of air at 70°F, lb/ft^3
DENSW	density of water at 70°F, lb/ft^3
D_{fan}	fan diameter, ft

D_i	tube internal diameter (normally 0.87 in; 1 in for 16 BWG)
D_O	outside tube diameter, in
D_{PA}	static pressure loss per row of tube cross flow, inches of water
DR	air density ratio; density of dry air at any temperature and elevation pressure divided by density of dry air at 70°F, 14.7 psia, and 0 elevation (sea level), 0.075 lb/ft^3
DTA	first-estimate air temperature rise, °F
DTM	log mean temperature difference, °F
eff	fan motor efficiency (normally 70% or 0.70)
FA	tube bundle horizontal face area, or simply the plot area of the tube bundle, ft^2
FAPF	fan area per fan, ft^2
F_t	log mean temperature correction factor
g	gravity acceleration constant (32.2 ft/s^2)
G_A	air cooler face area air flow, lb/h · ft^2
G_T	tube side mass flow rate, lb/(s · ft^2)
GTD	greater temperature difference, °F
H_A	air-side heat-transfer film coefficient, Btu/h · °F · ft^2
HT	inside tube wall heat transfer film coefficient, Btu/h · ft^2 · °F
J	factor for calculating HT
K	thermal conductivity of fluid, Btu/(h · ft^2 · °F · ft^{-1})
L	tube length, ft
L_D	L/D_i; tube length, ft, divided by tube diameter, in
log	natural log of base e
LTD	lower temperature difference, °F
MW	molecular weight
N_R	modified Reynolds number
NT	number of tubes
OD	tube outside diameter, in (1 in recommended)
P_{force}	static leg + velocity head, $V^2/2 \cdot g$, inches of water
Ptch	tube pitch (select from Table 5.3 or vendor data)
Q	heat transferred, Btu/h
RD_T	allowable tube-side fouling factor
T_1	tube-side inlet temperature, °F
t_1	ambient air temperature, °F
T_2	tube-side outlet temperature, °F (20° approach to t_1 is suggested)
t_2	air cooler outlet air temperature, °F

TA_1	ambient air temperature, °F
U_{VISC}	viscosity of fluid at average temperature, cP
$U_{\text{VISC}w}$	viscosity of the fluid on the tube wall (may be different when temperature of tube wall is at or near temperature of outside tube wall)
U_X	extended tube outside surface basis overall heat transfer coefficient, Btu/h · ft² · °F
UX_2	next U_X assumed value for repeating steps 1 through 13, Btu/h · ft² · °F
W_A	air flow, lb/h
W_T	tube-side flow quantity, lb/h

Probably the most common practical process engineering task is heat exchanger rating. In refineries and field process plants, most of us have encountered countless questions like, "Is the thing fouled?" or "How much of an exchanger do I need for this job?" or "How much process-side temperature cooling do I get with this air cooler when the ambient temperature is 102°F?" Do these questions sound familiar? Certainly they do if you have any exposure at all! This chapter is dedicated to giving you some tools to answer these questions: a good shell/tube exchanger simplified program and a good air finfan simplified computer program. Shell/tube exchanger basics are presented to enable readers (that is, eager, young, enthusiastic readers) to write their own programs, as well as to give the basics to enable everyone to understand the shell/tube program provided with this book. Enough data will be presented for you to do this Visual Basic code writing and to create presentable input-output screens. For examples of these screens, refer to the numerous program input-output screens already written that accompany this book, especially in Chap. 9. Line codes for inputs and outputs are also presented. There is therefore no excuse for an eager engineer not creating such a program. See Chap. 9 for a step-by-step, detailed walkthrough of authoring a typical program.

Nonetheless, for those who don't create a program using the following text, the program STexch, which accompanies this book on a CD, is helpful for all types of exchanger ratings and design. Those who apply STexch should find it a most valued tool. Every process engineer worth his or her pay should have this program or one like it at hand!

Shell/Tube Exchanger Practical Design/Rating Equations

For many years I have carried around with me two sheets of notepaper that contain heat-exchange coefficient equations. These equations calculate the inside and outside tube wall heat-transfer coefficients for every case and every condition of heat transfer. Using them, the overall heat-transfer coefficient may be calculated for any heat exchanger. We should all be familiar with these equations and have them handy for application as the need may arise.

First consider the main equation that governs the heat transferred.

$$Q = U_d \, A \, T \tag{5.1}$$

where A = tube bundle outer surface area, ft^2
        Q = overall heat transferred, Btu/h
        T = log mean temperature difference, °F, denoted LMTD
        U_d = heat transfer coefficient, Btu/h · ft^2 · °F (U_d is the dirty coefficient actually measured, or the design allowable)

Tube side

The most outstanding of these equations that I have used and carried around with me are Eqs. (5.2) and (5.3). These equations actually determine the tube wall heat-transfer film resistance. This is the resistance of significance to any heat-transfer event in a tube. If there were no resistance to heat flow, all materials in contact would immediately reach an equilibrium temperature. We all know this is not the case; for instance, when you pick up (without a rag or gloves) a hot object—such as a hot skillet cooking a porterhouse steak—the small amount of moisture on your skin serves to keep your fingers from being burned severely before you throw the pan down. This holds true for the subject at hand also, as there is a distinct film on both the inside and outside of the tube wall that resists the flow of heat transfer. This film reaches a steady state of resistance to heat flow. Several factors play a role in this film resistance, called h_i for the inside tube film and h_o for the outside tube wall film.

Much could be said here about equation derivations, but I will not try to duplicate the marvelous work of Donald Kern [1]. Any such attempt would be a pale imitation and would also tend to discredit his excellent work. McGraw-Hill has recently republished Kern's book because of its popularity. The following equations are indeed his derivation work. Please allow me to salute one more time a cherished colleague and dear deceased friend, Donald Kern.

Apply the following equation for h_i [2]:

```
hi = (Jh * K/D) * ((c * µ/K) ^ (1/3)) * (µ/µw) ^ .14   (5.2a)
```

where   µ = fluid viscosity at temperature, cP × 2.42
        µw = fluid viscosity at tube wall temperature, cP × 2.42
        c = fluid specific heat, Btu/lb · °F

D = internal tube diameter, ft

G = lb/(h) (A_t), where A_t = inside tube area, ft^2 × number of tubes per pass

hi = inside tube wall film resistance to heat transfer, Btu/h · ft^2 · °F

Jh = exp [−4.287 + 0.85 * [ln (D × G/μ)], curve-fit equation from Kern [3], Fig. 24

K = thermal conductivity of fluid, Btu/(h · ft^2 · °F · ft^{-1})

Equation (5.2a) is valid for any DG/μ value, Reynolds number, turbulent flow zone, or laminar flow zone. First calculate a Reynolds number from DG/μ. Then use Kern's Fig. 24, which appears in App. A as Fig. A.1. You may also derive this value by using Eq. (5.2a) for J_h. This equation is simply a curve-fit to Kern's figure.

Equation (5.2a) may also be written as:

$$h_i = J_h \frac{K}{D} \left(c \frac{\mu}{K} \right)^{1/3} \left(\frac{\mu}{\mu_w} \right)^{0.14} \qquad (5.2b)$$

In Visual Basic (VB) program code line form, this equation would be written as shown in Eq. (5.2a). In this book both are considered interchangeable and may be used indiscriminately.

Please notice that the asterisk (∗) designates the multiplication sign in VB. The up caret (^) designates "raised to the power of" in VB. This is true for all equations in this book.

Substituting the J_h curve-fit Eq. (5.2b) makes:

$$h_i = \left[\exp \left(-4.287 + 0.85 * \ln \frac{D * G}{\mu} \right) \right] * \frac{K}{D} * \left(c * \frac{\mu}{K} \right)^{1/3} * \left(\frac{\mu}{\mu_w} \right)^{0.14} \qquad (5.2c)$$

The term exp denotes that the constant e, 2.7183, is raised to the power of the term or function that follows.

Shell side

Now consider the outside of the tube, where the flowing fluid is on the shell side of the exchanger. For the outside tube wall, the film offering heat-flow resistance is denoted h_o, which has the same units as h_i (Btu/h · ft^2 · °F). The following equation is also from Kern [1].

$$\text{Reynolds number} = D_e * \frac{G_s}{\mu_s}$$

$$J_h = \exp \left[-0.665 + 0.5195 * \ln \left(D_e * \frac{G_s}{\mu_s} \right) \right]$$

where D_e = equivalent shell diameter for fluid flow shell side, ft
        G_s = mass flow per shell flow area

You may use this curve-fit equation to find J_h, or you may use the figure it was taken from, Fig. 28 in Kern [4], which is provided in App. A as Fig. A.2. Using Kern's figure is recommended as more accurate; however, the figure and the curve-fit equation given here have equal accuracies in regard to the final exchanger h_i or h_o calculation [2].

$$h_o = J_h * \frac{K}{D_e} * \left(c * \frac{\mu}{K}\right)^{1/3} * \left(\frac{\mu}{\mu_w}\right)^{0.14} \qquad (5.3)$$

h_o is the outside tube wall film heat transfer resistance in Btu/h · °F · ft². The same units and transport data factors are used as given for the h_i film Eq. (5.2b). An exception is, of course, D_e, the equivalent shell diameter for shell-side fluid flow in feet.
For triangular pitch (see Fig. 5.1) [5]:

$$D_e = \frac{4 \, (0.5P_t \times 0.86P_t - 0.5\pi \, d_o/4)}{6 \, \pi \, d_o} \qquad (5.4)$$

where d_o = tube outside diameter, in
        P_t = tube pitch, in

For square pitch (see Fig. 5.2) [6]:

$$D_e = \frac{4(P_t^2 - (\pi \, d_o^2)/4)}{12 \, \pi \, d_o} \qquad (5.5)$$

G_s is the mass flow per shell flow area. Since the shell fluid must flow perpendicular to the tube configuration, Eq. (5.6) is supplied to calculate G_s [7].

$$G_s = \frac{144 \, W \, P_t}{S \, C_c \, B} \qquad (5.6)$$

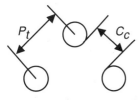

Figure 5.1 Cross-sectional view of triangular tube pitch.

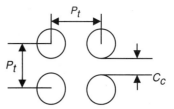

Figure 5.2 Cross-sectional square tube pitch view.

where B = baffle spacing, in (normally 24 in)
      C_c = tube clearance, in
      P_t = tube pitch, in
      S = shell-side inside diameter, in
      W = mass shell-side flow, lb/h

Equation (5.6) applies to both triangular tube pitch and square tube pitch.

Having established all of these added variables for the shell-side h_o film heat-transfer coefficients D_e and G_s, Eq. (5.3) is expanded here, since these variables must be determined for this equation.

$$h_o = \exp\left(-0.665 + 0.5195 * \ln\frac{D_e * G_s}{\mu}\right) * \frac{K}{D_e} * \left(c * \frac{\mu}{K}\right)^{1/3} * \left(\frac{\mu}{\mu_w}\right)^{0.14}$$

All of the preceding shell-side calculations are based on Kern, App. A Figs. A.1 and A.2 for deriving the J_h values. Please notice that the preceding equation form has the J_h value replaced with the curve-fit term derived in this book. You may do the same for the h_i film equation, replacing the J_h variable with the curve-fit term. Here you have the complete equations for the h_i and h_o heat-transfer film coefficients, without the need to resort to any figures. You may apply these equations to a PC program, deriving the heat-transfer coefficient for any exchanger problem. The PC program STexch accompanying this book applies these equations.

Checking the shell-side Reynolds number value.   The lower Reynolds numbers (less than 2100) begin to become uneconomical, since the shell-side flow area increases proportionally as the Reynolds number decreases. At some Reynolds number value—say 1000—the shell-side tube pitch and baffle spacing or even the shell diameter should be reduced to get a Reynolds number of 2100 or greater. I recommend always ensuring a shell-side Reynolds number of 2100 or greater. In most cases, this will ensure a reasonable and economical basis for design. These low Reynolds numbers promote much smaller tube wall film heat-transfer coefficients h_o. This of course means a much larger heat exchanger tube bundle is required to do the job.

$$\text{Shell-side Reynolds number} = D_e * \frac{G_s}{\mu}$$

Using this equation, you should always consider the shell-side Reynolds number. Considerable exchanger bundle cost savings may be gained by making these small but unique exchanger design checks.

Overall heat transfer coefficient

Equations (5.2) and (5.3) derive h_i and h_o tube wall film heat flow resistances. Next a common tube wall film resistance is established. The convention here is to let the outside tube wall be the basis. Therefore the inside tube wall h_i is to be converted to the outside tube wall basis and designated h_{io}. This is a simple multiplication:

$$h_{io} = h_i \times \frac{\text{inner diameter of tube, in}}{\text{outer diameter of tube, in}}$$

The overall heat transfer U_c now may now be defined [8]:

$$U_c = \frac{h_{io} \times h_o}{h_{io} + h_o} \tag{5.7}$$

U_c denotes the clean coefficient. Actually a dirty coefficient U_d governs or controls the overall heat transfer. See Eq. (5.1). U_d is calculated as follows [9]:

$$U_d = \frac{U_c \times h_d}{U_c + h_d} \tag{5.8}$$

where h_d = dirt or fouling resistance on tube walls, $1/R_d$
$\quad\quad U_d$ = overall heat transfer coefficient, dirty, Btu/h $\cdot$ °F $\cdot$ ft^2

Exchanger fouling factor R_d

h_d now becomes a sought-out value because it reveals how well the exchanger is performing in reference to its design. h_d is conventionally given in the term $1/R_d$ as shown in the preceding text. R_d values, known as *fouling factors,* may range from 0.001 to 0.020 for hydrocarbon systems. The smaller the R_d, the smaller the fouling in the exchanger. An R_d of 0.001 is considered indicative of light fouling and is a good indication the exchanger is performing well. An R_d of 0.02 is an indication of massive fouling. However, some exchangers are designed for R_d values of 0.02 and even greater. Considerable exchanger design size increase from these lower R_d values (<0.002) is realized when the exchanger is designed for the higher R_d values (0.02 and greater). Kern's [10] Table 12, reprinted in App. A as Table A.1, reveals expected common fouling R_d factors that are within the 0.001 to 0.005 R_d value range [11].

$$R_d = \frac{U_c - U_d}{U_c * U_d} \tag{5.9}$$

Objective of the fouling factor. In a refinery, chemical plant, or gas-oil production facility plant, exchangers should be checked regularly for the fouling factor R_d. For example, when the R_d value of an exchanger is found to increase by more than 50% in a 2-month or longer time period, consider increasing fluid flow through the controlling fouling tube side or shell side. Exchanger improvements can be determined by applying the previous equations. Savings of thousands, even millions of dollars annually have been realized by keeping R_d in check and making the proper plant material balance adjustments. Cooling water exchangers, hot oil exchangers, and process/process fluid exchangers are classic example systems.

In a later section, "Exchanger Rating," a step-by-step procedure is given for determining R_d and evaluating R_d in existing exchangers in service. This is one of our most valued tools as practicing technical service process engineers. *Use it!*

Log mean temperature difference

The log mean temperature difference LMTD is simply the average or weighted temperature difference between the hot side and the cold side of the exchanger. Use Eq. (5.10) to determine this mean average temperature. It is simply, as the term indicates, the average temperature difference between the tube and shell sides of the exchanger.

$$\text{LMTD} = \frac{(\text{GTD} - \text{LTD})}{\ln(\text{GTD/LTD})} \tag{5.10}$$

where LMTD = log mean temperature difference
       GTD = greater temperature difference
       LTD = lower temperature difference
       ln = natural log of base 2.7183

An easy way to determine LMTD is given in Table 5.1. Please note that it is possible for the LTD to be at the higher-temperature end of the exchanger. Likewise, the GTD could be at the lower-temperature end of the exchanger. Make sure to define correctly the LTD and GTD at the terminal ends of the exchanger.

TABLE 5.1 Determining LMTD: LMTD = (GTD − LTD)/[ln (GTD/LTD)]

Hot fluid			Cold fluid	Difference
T_1		Higher temperature ↑	t_2	$\text{GTD} = T_1 - t_2$
T_2	↓	Lower temperature ↑	t_1	$\text{LTD} = T_2 - t_1$
				$\text{GTD} - \text{LTD}$

It is now important to establish a correction factor F_t due to varied configurations of shell-tube exchangers. Please see Kern's [12] Figs. 18 through 23, which are reprinted in App. A as Figs. A.3 to A.8. These figures are self-explanatory, guiding the user to optimum configurations.

At this point in our review of shell/tube heat exchangers, all values of Eq. (5.1) are fully presented in the preceding text.

Exchanger rating

Now for the highlight of this chapter—exchanger rating. First please briefly review Fig. 5.3, which shows a screen capture from STexch, which is on the CD supplied with this book. Please bring up the executable copy of this program on your CD. Click on the command button Run Program. Here you are presented all the default values for two 2-4–type exchangers, both in series flow. The term *2-4–type exchanger* simply means a 2-pass shell with four or more tube passes.

Example 5.1   Figure 5.3 shows an exchanger rating of an actual exchanger in a lean-oil exchange with a rich-oil stream. Both streams serve a refinery gas absorber column in which the lean oil absorbs certain hydrocarbon components from the absorber-fed gas stream. 84,348 lb/h of a 35°API lean oil is fed to the overhead of this absorber column. Before entering the absorber, this lean oil is fed to the tube side of a shell/tube exchanger called an L-R exchanger. This L-R exchanger receives 86,357 lb/h of rich oil on its shell side from the absorber bottoms. Per the exchanger original design, the lean oil enters the exchanger at 100°F and leaves at 295°F. The rich oil enters at

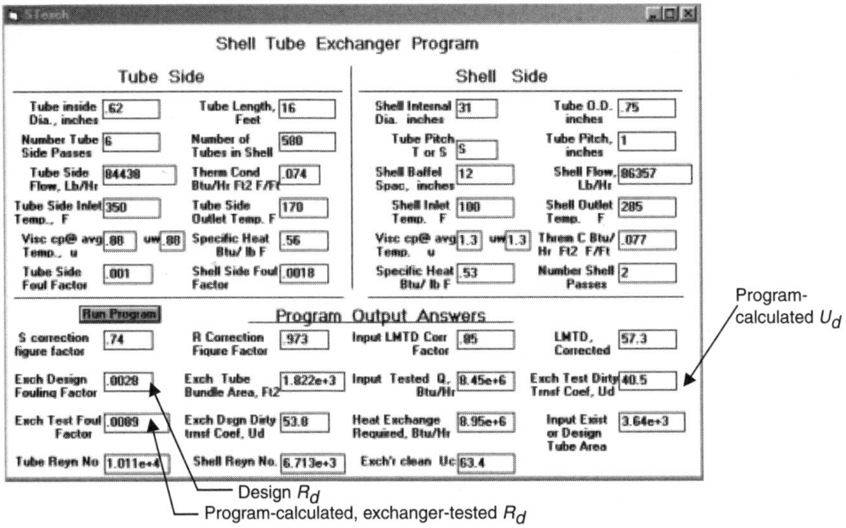

Figure 5.3   Actual refinery lean-oil gas absorber column for lean-oil–rich-oil exchanger.

350°F and leaves at 160°F, also per the original design. The design calls for an exchange rate of 8.95e+6 Btu/h with a fouling factor allowance of 0.0028. As seen in Fig. 5.3, the exchanger does not meet the design criteria.

We will now proceed through a step-by-step procedure to rate the given exchanger. In this example you, the plant engineer, have found the spec sheet showing the exchanger has a 2-4 configuration (see Kern's Fig. A.4 in App. A). The term *2-4* here indicates the exchanger has a 2-pass shell with four or more tube passes. You confirm this, using the plant's specifications that show there are two shells in series. Both have two shell passes and six tube passes. There are 580 tubes 16 ft long in each shell, which comes to 1822 ft² tube area per shell.

solution

Step 1. The heat balance is:

$$Q = 84{,}438 \text{ lb/h} * 0.56 \text{ Btu/lb} \cdot °F * (350 - 170) = 8.51 \text{ mmBtu/h}$$

Some difference is noticed if you calculate the shell-side Q. This difference is mainly due to variations in the accuracies of the specific heat values obtained. For our purpose and for accuracy within 3%, however, the calculated tube-side heat is sufficient.

Step 2. From Kern's [1] Table 8 (Table A.2 in App. A) you can choose a starting U_d value, and from Eq. (5.11) you can calculate a first-estimate bundle exchanger requirement. Choose a value (50 is suggested for this case) having medium organics on both the tube side and the shell side. If you are rating an existing exchanger, go to step 3 and with the corrected LMTD calculate an accurate U_d. The following equation can be used to determine U_d:

$$U_d = \frac{Q}{A * \text{LMTD}} \tag{5.11}$$

where A = tube bundle area from your plant's spec sheet, ft² (in this case, 3644 ft², both shell areas included)
LMTD = 57.3 (step 3)
Q = heat duty (step 1), 8.51 mmBtu/h

In this case, $U_d = 40.8$.

Step 3. Now the LMTD is the only remaining factor needed to make a first U_d estimate. From Table 5.1 and Eq. (5.12), next calculate the LMTD:

$$\text{LMTD} = \frac{\text{GTD} - \text{LTD}}{\ln (\text{GTD/LTD})} = 67.5 \tag{5.12}$$

The exchanger terminal temperatures are:

	Tubes	Shell
GTD =	170 − 100 =	70°F
LTD =	350 − 285 =	65°F
		5°F

$$\text{LMTD} = \frac{5}{\ln (70/65)} = 67.5$$

Referring to Kern's [12] Figs. A.3 through A.6 in App. A, do the following:

$R = 0.973$ $S = 0.74$

1-2 exchanger, F_t = inoperable (Kern, Fig. A.3)
2-4 exchanger, F_t = inoperable (Kern, Fig. A.4)
3-6 exchanger, F_t = 0.725 (Kern, Fig. A.5)
4-8 exchanger, F_t = 0.85 (Kern, Fig. A.6)

$$\text{LMTD corrected} = 0.85 * 67.5 = 57.4$$

Step 4. There are two possible bases for calculating this step:
Step 4.1 basis: rate an existing exchanger
Step 4.2 basis: design a new exchanger

In either case, you will use the following Eq. (5.13) to determine U_d, the heat transfer for step 4.1, and A, the new tube bundle area for step 4.2.

$$U_d = \frac{Q}{A * \text{LMTD}} \text{ or } A = \frac{Q}{U_d * \text{LMTD}} \tag{5.13}$$

If you are rating an exchanger, you have already applied Eq. (5.13) to calculate U_d (see step 2). If you are designing an exchanger, the area A is calculated here with the U_d from step 2 and LMTD is calculated from step 3.

Step 5. This step, which is for designing a new exchanger only, will help you calculate the number of tubes required to satisfy the step 4 bundle area. Please refer to Kern's [13] Table 10 (Table A.3 in App. A). From this table you will determine how many tubes are required for the area calculated in step 4, with an assumed tube length. Please note that Kern's Table 9 (Table A.4 in App. A) should also be used to determine the limiting number of tubes for an assumed shell diameter.

The chosen tube length should be your plant's practice or standard length and should consider the plot land area needed to pull and service the bundle in maintenance turnarounds.

Step 6 (hot fluid, tube side, lean oil). The hot fluid is on the tube side in the existing exchanger you are rating. Please note again here that you have taken operations control room data point readings gathering data that have established the exchangers' terminal temperatures and the flow rates through the tube bundles and shells.

The tubes are ¾-in OD on 1-in square pitch, 580 tubes per shell, 16 ft long each. Two shell passes are in each of the two shells, and six tube passes are in each shell. Each tube has an area of 0.302 in², or, divided by 144, 0.0021 ft².

For $DG / \mu < 2100$ Reynolds number, laminar flow zone, apply the following equation for inside tube wall h_i value:

$$h_i = \left[\exp\left(-4.287 + 0.85 * \ln \frac{D * G}{\mu}\right)\right] * \frac{K}{D} * \left(c * \frac{\mu}{K}\right)^{1/3} * \left(\frac{\mu}{\mu_w}\right)^{0.14} \qquad (5.2c)$$

where μ = fluid viscosity at temperature, cP × 2.42 (0.88)
μ_w = fluid viscosity at tube wall temp., cP × 2.42 (0.88)
c = fluid specific heat, Btu/lb · °F at average temperature (0.56)
D = internal tube diameter, ft (0.0516)
G = lb/h (A_t), where A_t = inside tube area, ft² × number of tubes/pass (use Kern's [13] Table A.3) = 0.0021 ft² * (580/6) = 0.203 ft² (84,438/0.203 = 416,000)
h_i = inside tube wall film resistance to heat transfer, Btu/h · ft² · °F
K = thermal conductivity of fluid, Btu/h · ft² · °F/ft (0.074)

Note: Please refer to Fig. 5.3 for these data values.

Therefore, $h_i = 34.76 * 1.434 * 2.526 * 1.0 = 126$.

Step 7 (cold fluid, shell side, rich oil). Calculate the shell-side film coefficient with the following equations.
    Calculate G_s from:

$$G_s = \frac{S_p 144 \, W \, P_t}{S \, C_c \, B} = 267,400 \qquad (5.14)$$

where B = baffle spacing, in (12)
C_c = tube clearance, in (0.25)
P_t = tube pitch, in (1.0)
S = shell-side inside diameter, in (31)
S_p = shell passes (2)
W = mass shell-side flow, lb/h (86,357)

Equation (5.14) applies to both triangular and square tube pitch.
    For Reynolds number $D_e \, G_s / \mu > 2100$, use the following equation:

$$h_o = \exp\left(-0.665 + 0.5195 * \ln \frac{D_e * G_s}{\mu}\right) * \frac{K}{D_e} * \left(c * \frac{\mu}{K}\right)^{1/3} * \left(\frac{\mu}{\mu_w}\right)^{0.14} \qquad (5.15)$$

Therefore, $h_o = 50.0 * 0.976 * 2.787 * 1.0 = 136$.

where μ = fluid viscosity at temperature, cP × 2.42 (1.3)
μ_w = fluid viscosity at tube wall temperature, cP × 2.42 (1.3)
c = fluid specific heat, Btu/lb · °F at average temperature (0.53)
D_e = equivalent shell pass diameter, ft (0.0789), see Eq. (5.14)
G_s = 267,400, see Eq. (5.14)
h_o = outside tube wall film resistance to heat transfer, Btu/h · ft² · °F
K = thermal conductivity of fluid, Btu/h · ft² · °F/ft (0.077)

    To calculate D_e for triangular pitch, use Eq. 5.4. For square pitch, use Eq. 5.5.

$$D_e = \frac{4\,(0.5P_t \times 0.86P_t - 0.5\pi\,d_o/4)}{6\,\pi\,d_o} \tag{5.4}$$

$$D_e = \frac{4(P_t^2 - (\pi\,d_o^2)/4)}{12\,\pi\,d_o} \tag{5.5}$$

Use the square-pitch equation for D_e calculation in your lean-oil–rich-oil exchanger, since it is designed with square pitch.

$$\text{Square-pitch } D_e = \frac{4 * (1.0 - (\pi * 0.563)/4)}{12 * \pi * 0.75} = \frac{0.558}{28.27} = 0.0789$$

Step 8 (rating the existing exchanger). You have now calculated the h_i and h_o values. Next convert h_i to h_{io}:

$$h_{io} = h_i \times \frac{\text{tube inner diameter}}{\text{tube outer diameter}}$$

For nearly every application, the term $(\mu/\mu_w)^{0.14}$ may be disregarded. It is equal to unity (1) in almost every case.

From the equations in steps 6 and 7, $h_i = 126$, $h_o = 136$, and $h_{io} = 126 * 0.62 / 0.75 = 104$.

Calculate the overall clean heat-transfer coefficient U_c as follows:

$$U_c = \frac{h_{io} * h_o}{h_{io} + h_o} = \frac{104 * 136}{104 + 136} = 58.9 \tag{5.7}$$

$U_d = 40.8$ from step 2.

$$R_d = \frac{U_c - U_d}{U_c * U_d} = 0.0075 \tag{5.9}$$

The reason for the slight differences between these hand-calculated answers and the computer program run (see Fig. 5.3) is that these hand calculations have been rounded off in many instances. The result amounts to some error difference; however, this error is considered to be less than 5%. For those of you who desire to duplicate the computer run, my heart goes out to you! You can only do so by accurately calculating the variables to the fourth-digit place. Please notice how sensitive Eq. (5.7) is to the fourth-digit place of significance. This makes it desirable to use a computer program so as to avoid this small but discrete error.

The preceding equations are equally applicable to designing a shell-tube exchanger. Specifically, apply Eq. (5.8), using a selected R_d value for your particular service. In Eq. (5.8), remember that $h_d = 1/R_d$. Thereby selecting an R_d value, you then calculate U_d, applying Eq. (5.8).

Conclusion.   Having now gone through an eight-step procedure to not only rate a shell-tube exchanger, but also to design a new exchanger, you have a useful tool indeed. Please notice one important item when rating in-service exchangers—the fouling factor R_d. One hears much plant operating staff conversation about exchanger fouling, but next to nothing about R_d. This factor, when calculated as in the preceding steps, is the sum critique that tells how bad or how clean an exchanger really is. *At an early date, establish your plant's shell-tube fouling factors and check them every three months.* You may be surprised. If you check the R_d for a carefully designed exchanger over time, there should be no surprises, since a well-designed exchanger has a stable and reasonably constant R_d between maintenance cleanings.

As seen in Example 5.1, you can very easily calculate your exchanger's fouling factor by taking some operating data and doing a little research on specifications in your company archives. In my experience of nearly 50 years, I have found most exchangers to be within an R_d range of 0.001 to 0.005. I accumulated these data in about 30 refineries scattered over the United States. The listings in Table 5.2 should prove of good value for comparison to your findings in refineries.

Perhaps an exhaustive list of R_d values is not justified, as only a field test as described in Example 5.1 will reveal the truth. You may make good guesses, but only this field test actual operation data can reveal the true R_d. Now that you have an R_d value, what do you do with it? First, consider Eqs. (5.8) and (5.9):

$$U_d = \frac{U_c * h_d}{U_c + h_d} \tag{5.8}$$

$$R_d = \frac{U_c - U_d}{U_c * U_d} \tag{5.9}$$

TABLE 5.2 R_d Values for Various Petroleum Fractions

Oils			
Crude oil, less than 25°API	0.005	Crude oil, greater than 25°API	0.002

Atmospheric Distillation Units			
Bottoms less than 25°API	0.005	Distillate bottoms ≥ 25°API	0.002
Overhead untreated vapors	0.0013	Overhead treated vapors	0.003
Side-stream cuts	0.0013		
Machinery and transformer oils	0.001	Quenching oil	0.004

Water Services			
Brine water (cooling)	0.001	Cooling tower water	0.002
Tempered closed circulating water	0.002	Steam condensate	0.001

Now:

$$\frac{1}{U_d} = \frac{1}{U_c} + R_i + R_o \tag{5.16}$$

where R_i = fouling factor inside tube wall
R_o = fouling factor outside tube wall

$$R_d = R_i + R_o = \frac{1}{h_{di}} + \frac{1}{h_{do}} \tag{5.17}$$

where h_{di} = heat transfer coefficient inside tube for dirt fouling factor only, Btu/h · °F · ft²
h_{do} = heat transfer coefficient outside tube for dirt fouling factor only, Btu/h · °F · ft²
R_d = overall fouling factor sum = $R_i + R_o$
R_i = tube inside fouling factor = $1/h_{di}$
R_o = tube outside fouling factor = $1/h_{do}$

Assume now you have taken the inside and outside readings for an exchanger over a 6-month period. You have also used the program STexch, supplied on the CD, and have the result shown in Fig. 5.3. Reviewing Fig. 5.3, assuming you have run the program, reveals a serious fouling factor problem in this exchanger. From Eq. (5.16) and R_d = overall fouling factor sum = $R_i + R_o$, you may calculate U_d:

$$\frac{1}{U_d} = \frac{1}{U_c} + R_i + R_o; \ U_d = 1/\left(\frac{1}{U_c} + R_d\right) = 40.5 \text{ Btu/h} \cdot \text{ft}^2 \cdot \text{°F} \tag{5.18}$$

Note that this U_d is program-calculated for you in Fig. 5.3.

Please observe in your program run that you have lost 6% of your exchange design duty:

$$\frac{8.96e + 06 - 8.45e + 06}{8.45e + 06} = 0.06 \text{ or } 6\% \text{ loss}$$

This loss has occurred over 6 months. There is still time to consider absorber options, such as air-cooling exchanger fan speed increase or absorber pressure adjustments, to keep your absorber product specs. Perhaps, hopefully, within the next month or so, a maintenance turnaround is planned in which you may consider some surgical alterations. Looking at the 6000 Reynolds number, it appears that you could consider replacements using two shells having a higher shell-side velocity—thus less resistance, lower shell-side heat-transfer coefficient, and less shell-side fouling.

At this point you have been alerted to the problem, have discovered it with excellent data accumulations, and have pointed it out to the refinery management. Submitting a resolving report to management is one more step you should take. Your report should include options of upgraded exchanger replacement, fouling materials origin and elimination, immediate process remedies, and, planning for the next maintenance activities—a good cleaning. Gung ho!

Air Finfan Heat Exchangers

The air-cooling heat exchanger is perhaps the most popular type of heat exchanger throughout the industry. It is used in every refinery and gas processing plant from the smallest to the largest capacities known. A surprising factor about air-cooler design and rating is the limited data available to design or rate any particular unit. I have noted that for over two decades little reference data was available to create any methodology for rating, let alone designing, air finfan coolers. Also, few of our expensive chemical engineering process simulators have air-cooler rating software available. This is bad news. This chapter will help to fill in this large gap of missing technology. First, a discussion of air-cooler configuration will examine cooler components—fans, finned tubes, fan bays, and so on. Second, air finfan cooler thermal design and rating will be covered for sensitive as well as latent heat-condensing-type heat transfer.

Regarding our industry's failure to provide adequate heat-transfer calculation methodologies for air finfan cooler exchangers and the software industry's equal failure, I am most pleased to include on the CD provided with this book two unique air finfan cooler programs named AirClr8 and AirClr9. These programs will rate and design almost any air cooler. The user merely needs to gain an understanding of how these programs work and learn about proper input data. When you execute these programs they will come up in the DOS mode. I have written them to make them look like Visual Basic, however, so nothing here is lost. The reason I have left them as they are in DOS is perhaps some nostalgia on my part. Since writing these programs over 12 years ago, I have used them numerous times, with excellent results indeed. I therefore offer them to you in hopes that you will find their applications a treasure, as I have.

AirClr8 and AirClr9 are the only programs on the accompanying CD that are not in Visual Basic. To get a printout of the program input and answer screens, use the PrintScreen key in DOS mode.

To understand the calculation methodology for air finfan cooling and how to rate these popular coolers is an objective of this chapter. The methodologies presented are those installed in the AirClr8 and AirClr9 programs. If you follow through with the examples and the calculation techniques, I am confident you will go far toward mastering rating and, yes, even design of these most popular cooling exchangers.

Air finfan cooler configurations

Figures 5.4 and 5.5 show typical elevation and plot plan section sketches of horizontal air-cooled exchangers. The essential components are one or more tube sections, called *tube bundles,* and one or more axial flow fans, all enclosed in a structural module made to anchor footings into concrete or to stand on an offshore platform. As shown, one fan may serve more than one tube bundle. A bay may be made up of multiple tube bundles and may also be served by one or more fans. Most important is the fact that one exchanger may be composed of any number of tube bundles. Tube bundles may be arranged in series or in parallel, just like shell/tube bundles.

Air-cooled exchangers are classified as forced-draft when the tube section is located on the discharge side of the fan (see Fig. 5.4). Air coolers are classed as induced-draft when the tube section is located on the suction side of the fan.

Forced-draft or induced-draft. Figures 5.4 and 5.5 show the two kinds of air finfan cooler configurations: forced-draft and induced-draft. The more popular and commonly used is the forced-draft type. The reason for this is that air coolers are normally placed in well-vented and open areas. The main disadvantage of the forced-draft type is its poor ability to disperse the hot air generated. Placing these coolers in open or well-ventilated areas overcomes this deficiency. The advantages and disadvantages of both types of coolers are summarized in the following lists.

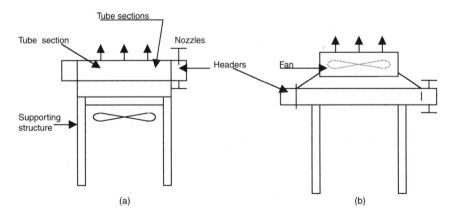

Figure 5.4 Typical side elevations of air coolers: (*a*) forced draft, and (*b*) induced-draft.

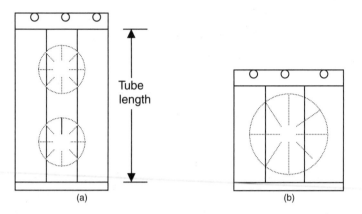

Figure 5.5 Typical plan views of air coolers: (*a*) two fan bays with three tube bundles, and (*b*) one fan bay with three tube bundles.

Advantages of induced-draft coolers

- Much better distribution of air across the plot surface facial area

- Little if any recirculation of hot air

- Rain, hail, and sun protection for the tube bundles, thus better outside film heat-transfer coefficients

- Upon fan failure, increased capacity compared to forced-draft coolers, since the airflow is designed for natural draft

Disadvantages of induced-draft coolers

- Location in the hot airflow requires a higher horsepower.

- Maximum effluent air temperature is limited to 200°F due to the fan motor bearings, fan blades, V-belts, and other fan-associated components.

- For inlet process fluid temperatures 350°F or higher, the fan motor is subject to early failure due to being subjected to high temperatures. Fan blade failures may also occur.

Advantages of forced-draft coolers

- Slightly lower horsepower due to cooler air temperature as compared to induced-draft fan motors. (Fan horsepower is directly proportional to air temperature.)

- Easily accessible for maintenance.

- Better functioning in cold climates.

- Preferred for high process temperature inlet of 300°F and higher because of fan system temperature protection.

Disadvantages of forced-draft coolers

- Poor distribution of air in the air cooler area
- Low air discharge from the tube bundle top, greatly increasing the probability of hot air recirculation with absence of stack
- Upon fan failure, very poor natural draft capability
- Tube bundles exposed to rain, hail, and sun damage

Air cooler sizes

Air cooler fans are normally 14 to 16 ft in diameter. Fan motors are almost always electric, although a few hydraulic and gasoline engines have been used. Fan tip speeds are normally 12,000 ft/min or less. In order to produce this speed, U.S. practice is to use V-belt drives or reduction gearboxes. V-belt drives up to about 30 hp are used, and gear drives at a higher motor horsepower. Single fan motors are usually limited to 50 hp.

Generally, the design basis is two fan bays. If one fan fails, the other keeps reasonable cooling going for the process fluid. Better control of turndown cooling is available when using two or more fan bays. Fan staging is a unique control methodology. Good-practice design is to keep the ratio of fan diameter area to face area of the tube bundles to 0.40 or above. Face area FA, ft^2, is the plan area of the heat-transfer surface available for airflow at the face of the section.

Air coolers are usually 6 to 50 ft long. Tube bundle bays are normally 4 to 30 ft wide. Use of longer tubes tends to cost less as compared to using shorter, wider tubes. Tubes are stacked in horizontal rows. A two-pass tube bundle would have an equal number of tube rows dedicated to each pass, in which the top tube pass realizes a hotter air coming from the bottom section. Likewise, the tube bundle(s) can have multiple tube section passes, where each pass has the full count of tube rows in the air cooler. Here each tube bundle pass has a bottom-to-top tube row in the air cooler. (See Fig. 5.6.)

Tubes with outer diameters (ODs) of 1 in are very common and are used in most cases. An OD range of $\frac{5}{8}$ to $1\frac{1}{2}$ in is design practice. Tension-wrapping is the most common and popular means of fin placement on tubes. Tube fins (normally aluminum) are tube-wrapped 7 to 11 per in and stand $\frac{1}{2}$ to 1 in tall. The great advantage of tube fins is the extended outside surface area provided for heat transfer to air. The tube fins provide from 12 to 25 times the outside surface area of the tube OD area. The tubes are usually placed in triangular pitch and in horizontal rows 3 to 8 high. A four-row tube stack is a very common design. Clearances between fin tips are $\frac{1}{16}$ to $\frac{1}{4}$ in.

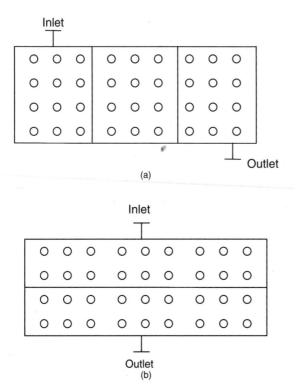

Figure 5.6 Tube pass configurations: (*a*) one-pass and side-by-side tube section passes (use Kern Fig. A.9*a* F_t correction), and (*b*) over-and-under two-pass tube section (use Kern Fig. A.9*d* F_t correction) [14]. See App. A.

Table 5.3 offers considerable needed data to rate-check an existing cooler or to design an air-cooler exchanger. Consider the fact that almost all air coolers are fabricated using 1-in OD tubing. Those whose ODs are not 1 in may still be rated by making the bare-tube OD surfaces equal to the bare-tube 1-in OD surface area and rating as for 1-in OD tubes. Thus Table 5.3 can be used in numerous rating applications. In scenarios where a different tube diameter is to be rated (other than 1-in OD), vary the tube length to get the bare-tube area desired, and vary the tubes per pass to get the tube inner diameter (ID) fluid velocity required. This is done by applying the following rating methodology.

1. Make your 1-in tube-pass cross-sectional area equal to the tube-pass cross-sectional area you are rating. Vary the number of 1-in tubes per pass.

TABLE 5.3   Fin Tube Data for 1-in-OD Tubes

	½-in-high fins, 9 fins per in		⅝-inch-high fins, 10 fins per in		
APF, ft²/ft	3.8		5.58		
AR, ft²/ft²	14.5		21.4		
Tube pitch	2-in Δ	2¼-in Δ	2¼-in Δ	2⅜-in Δ	2½-in Δ
APSF (3 rows)	68.4	60.6	89.1	84.8	80.4
APSF (4 rows)	91.2	80.8	118.8	113.0	107.2
APSF (5 rows)	114.0	101.0	148.5	141.3	134.0
APSF (6 rows)	136.8	121.2	178.2	169.6	160.8

APF, total extended tube external area per foot of fin tube, ft²/ft

APSF, ratio of external tube OD bare surface area to bundle face area, ft²/ft². (Figure 5.6 shows typical face areas of air coolers.)

AR, ratio of fin tube area to exterior area of 1-in-OD bare tube, ft²/ft (bare 1-in tube has 0.262 ft² per foot of tube length)

SOURCE: Natural Gas Processors Suppliers Association (NGPSA), *Engineering Data Book,* NGPSA Refinery Department, 1976, Fig. 10-11.

2. Make your 1-in OD bare-tube area equal to the bare-tube area you are rating. Vary the 1-in tube length to equal the bare-tube OD area you are rating.

A very close rating result will be obtained for your particular tube diameter upon executing these two steps. Thus, Table 5.3 is a good source for rating existing air coolers, as well as designs for 1-in-OD tubes. Vendors can supply other APF, AR, and APSF values. However, getting vendors to reveal their undisclosed databases (which this is) is very uncommon. If you find such a source of data from a vendor, regard it as a valued treasure.

A second note about Table 5.3 is the number of tube rows to be used. Almost every air cooler of any significant heat-transfer value—say, 0.5 MBtu/h or greater—must have three or more tube rows. If there are fewer than three rows, uneconomical fan equipment and structural support will be required, so use three or more rows in your new designs.

Air finfan cooler practical design

The basic equation that must be applied for air-cooling exchangers is:

$$Q = U_X * A_X * \text{DTM} \tag{5.19}$$

where A_X = extended finned tube surface area, ft²
DTM = log mean temperature difference, °F
Q = heat transferred, Btu/h
U_X = overall transfer coefficient, finned tube extended area basis, Btu/h · °F · ft²

The first design factor that will be reviewed is the log mean temperature difference, since it is one of the first considerations in air-cooling design. We normally know the process-side temperatures, the inlet process-side temperature, and the amount of cooling desired. The ambient air temperature is also generally known. If not, 100°F is always a safe and very conservative number for the United States. A complication arises here because the airflow quantity is a variable, making the outlet air temperature an unknown. Not having an airflow quantity results in not knowing an outlet air temperature. To solve these two variables, a trial-and-error assumption is made for the air temperature rise. Equation (5.19) is the first trial assumption for the air temperature rise:

$$\text{DTA} = \frac{U_X + 1}{10} * \left(\frac{T_1 + T_2}{2} - TA_1 \right) \qquad (5.20)$$

where   DTA = first-estimate air temperature rise, °F
T_1 = process inlet temperature, °F
T_2 = process outlet temperature, °F
TA_1 = ambient air temperature, °F

With TA_1 known and the air outlet temperature now estimated $(TA_2 = TA_1 + \text{DTA})$, the log mean temperature DTM may be calculated. Refer to Table 5.1 for a review of DTM calculation. In the case of air coolers, the cold fluid is always the air side, which cools the process-side fluid. T_1 is the process-side temperature and is always the highest of all the shown temperatures. Therefore, $T_1 - t_2$ should most likely be the greater temperature difference GTD. This is not true for every case, especially cases where there is limited airflow and therefore a large air-side temperature rise compared to other cases. You must therefore check to ensure you have chosen the correct temperature difference for these GTD and LTD values.

It is necessary to fix the approach temperature, which is the LTD value. Since we cannot change the ambient air temperature t_1, we must set a T_2 process-side temperature, which becomes a very critical value in the design of the exchanger. The process-side outlet temperature T_2 describes how close the process-side temperature comes to the ambient air-cooling temperature. T_2 should be as low as reasonably possible in most air-cooler cases. In fact, in many cases a process plant is designed with a trim cooler (using cooling water or process cooling stream) to get the final lower process-side temperature desired. If we specify T_2 too close to t_1 in an air cooler, however—say, a 5 to 10°F difference—over one-half of the air-cooler exchanger bundle could be captured to get this last 5°F, making a $T_2 - t_1$ difference of 5 to 10°F.

It is therefore common practice to *set $T_2 - t_1$ never less than 15°F and preferably 20°F or more*. Make sure to keep this temperature approach

15°F or greater in all design cases. Furthermore, you may calculate a 5 or 10°F approach and compare the exchanger size required with that mandated by the 20° approach. Try it!

We can next calculate the DTM from Eq. (5.21):

$$DTM = \frac{GTD - LTD}{\log(GTD/LTD)} \tag{5.21}$$

where   DTM = log mean temperature difference, °F
  GTD = greater temperature difference
  log = natural log of base e
  LTD = lower temperature difference

The next task is to determine the DTM correction factor, if any. The interesting fact here is that all over-under-type tube-pass configured bundles having three or more tube passes have a correction factor of 1.0 (unity). A large number of air-cooler designs incorporate one or two tube passes of this type, however, requiring a DTM correction factor. Refer to Kern's [14] Figs. 16.17a and 16.17d, reprinted in App. A. as Figs. A.9a and A.9d. Use Kern's Fig. A.9a F_t correction for all one-pass and side-by-side tube sections. Use Kern's Fig. A.9d F_t correction for over-and-under two-pass tube sections.

$$CDTM = F_t * DTM \tag{5.22}$$

This equation is used for over-and-under configuration; if there are more than two tube passes, then $F_t = 1.0$.

There are two types of tube passes: side-by-side and over-and-under. The majority of tube bundle passes are of the over-and-under configuration type. Most of these include three or more tube passes. Thus the majority of cases allow a 1.0 F_t DTM correction factor. In cases where a side-by-side configuration is applied, however, use a correction factor taken from Kern's Fig. A.9a in App. A.

Now, having calculated a corrected DTM value CDTM, review again Eq. (5.19):

$$Q = U_X * A_X * CDTM \tag{5.23}$$

The normal procedure for solving Eq. (5.23) is:

- Q is known by the heat balance of process cooling required.
- U_X, the overall heat transfer coefficient, is given a first assumption to start a trial-and-error solving procedure.
- CDTM is a known value calculated from Eq. (5.20) through (5.22).

U_X is given a first assumed value and then calculated from the other inputted values. A trial-and-error process is executed until the assumed U_X value matches the calculated value. Table 5.4 lists several U_X values good for a first-trial assumption.

Design methodology

The following procedure steps are given to calculate the required transfer bare tube surface, the number of tubes required, the number of tube passes required, the required airflow, and the fan horsepower. First, however, certain data are required, which are itemized in the following.

The basic data required for hot fluid are:

- Viscosity of fluid at average temperature U_{VISC}, cP
- Specific heat C_p, Btu/lb · °F
- Thermal conductivity of fluid K, Btu/(h · ft^2 · °F · ft^{-1})
- Heat transferred Q, Btu/h
- Tube-side flow quantity W_T, lb/h
- Tube-side inlet temperature T_1, °F
- Tube-side outlet temperature T_2, °F (suggested 20° approach to the ambient air temperature t_1)
- Allowable tube-side fouling factor RD_T

The basic data required for the air side are:

- APF selection from Table 5.3 or vendor
- Tube pitch selection from Table 5.3 or vendor
- AR selection from Table 5.3 or vendor
- APSF selection from Table 5.3 or vendor

TABLE 5.4 Estimates of U_X

Service	U_X range
Water cooling	6.0–7.5
Glycol-water	5.0–6.5
Hydrocarbon liquid, 0.5–3.0 cP	3.0–4.5 (at 0.5 cP, $U_X = 4.5$)
Hydrocarbon liquid, 4.0–10.0 cP	2.0–0.5
Hydrocarbon gas, 50–300 psig	2.0–3.0 (at 50 psig, $U_X = 2.0$)
Hydrocarbon gas, 400–1000 psig	4.0–5.0
Air and flue gases	1.0–2.0
Steam condensers	5.0–8.0
Hydrocarbon condensers (condensing range 0–60°F)	3.0–6.0 (at 0°F, $U_X = 6.0$)

- Ambient air temperature, °F
- Tube OD selection (1 in recommended)
- Tube pitch, in (triangular)
- Tube bundle passes (three to six over-and-under recommended)
- Tube bundle length, ft (6- to 50-ft lengths are commonly manufactured)

Having now established the basic data requirements, a step-by-step procedure is needed.

Step 1. A trial-and-error method is applied by first assuming a value for U_X. Please refer to Table 5.4 for the first U_X estimate for a particular case run.

Step 2. Calculate the approximate air temperature increase using Eq. (5.20):

$$\text{DTA} = \frac{U_X + 1}{10} * \left(\frac{T_1 + T_2}{2} - TA_1 \right) \tag{5.20}$$

Step 3. Calculate DTM using Eq. (5.21):

$$\text{DTM} = \frac{\text{GTD} - \text{LTD}}{\log\ (\text{GTD/LTD})} \tag{5.21}$$

GTD and LTD are determined from Table 5.1.

Determine the DTM correction factor F_t using Fig. 5.6. Note that in most cases F_t will equal 1.0 since most designs incorporate three or more over-and-under passes. For three or more passes of the over-and-under type, F_t will always equal 1.0. For side-by-side passes, use Fig. 5.6 and the respective F_t (see Fig. A.9 in App. A).

Correct DTM with Eq. (5.22):

$$\text{CDTM} = F_t * \text{DTM} \tag{5.22}$$

Step 4. Calculate the required tube outside extended surface area, A_X, with Eq. (5.24):

$$A_X = \frac{Q}{U_X * \text{CDTM}} \tag{5.24}$$

This A_X value is 10 to 30 times the outside bare tube area. In fact, the AR value given in Table 5.3 is indeed this multiplier factor. In regards to determining the value of Q, note the following:

1. It is important to determine Q. Q is stated in Btu/(h $\cdot$ °F $\cdot$ ft^2). Q is normally determined as a sensible or a latent heat product from calculations. Here you should select which you will use—sensible or latent—and divide your air cooler into section runs if you have both.

2. Q from sensible heat is simply $C_p * W_t * T$. C_p is specific heat of the fluid over the subject temperature range, and W_t is mass rate in lb/h. C_p may be found from Chap. 1, Table 1.10. T is the section process tube side in and out temperature difference in °F.

3. Q from latent heat is $W_t * H$. H is the difference of the inlet fluid gas state enthalpy at its bubble point in Btu/lb, and the liquid state of the fluid outlet enthalpy at its dew point in Btu/lb. Here you may run the bubble point and the dew points of your fluid mixture by using the RefFlsh program discussed in Chap. 2. Then find enthalpies using Chap. 1, Table 1.10.

Step 5. Calculate the air-cooler face area using the APSF factor from Table 5.3 and Eq. (5.25):

$$FA = \frac{A_X}{APSF} \qquad (5.25)$$

In order to select an APSF value from Table 5.3, please note you must also select an APF, AR, tube pitch, and number of rows. All of these are listed in Table 5.3.

Generally any of these values selected will give good results. Simply follow the guides in Table 5.3. For example, for a ½-in fin height with nine fins per inch, at a 2¼-in tube pitch with four rows, a typical selection would be APF = 3.8, AR = 14.5, APSF = 80.8. Therefore FA = A_X / 80.8. Note that an A_X value is calculated in step 4.

Step 6. In this step, the number of tubes NT is calculated. First assume a tube length L in feet, which conforms to the area of the respective placement for the new air cooler. Next calculate the width of the tube bundle using your assumed tube length.

$$Width = \frac{FA}{L}$$

Then round off your calculated width and adjust the tube length if needed. Normally tube lengths are 16 to 50 ft. *It is important here to fix the tube length with the width in order to get full coverage of equal leg squares in FA. The reason is that fans of equal blade diameters will cover the entire FA area. Remember also that a multiple number of fans is the better design for turndown and reliability.*

Next calculate the number of tubes NT using the APF factor chosen in step 5. The APF factor is taken from Table 5.3 and is the square feet of bare tube area per foot of tube length.

$$NT = \frac{A_X}{APF * L} \tag{5.26}$$

Step 7. In this step the modified Reynolds number N_R is calculated. First calculate the tube internal cross-sectional area A_I in in^2.

$$A_I = 3.1416 * \left(\frac{D_I}{2}\right)^2 \tag{5.27}$$

A 16 BWG is common and has an A_I of 0.594 in^2. (BWG is a standard gauge for tube-wall thickness.)

Choose the number of passes NP next and calculate the tube-side mass flow rate G_T in lb/(s · ft^2):

$$G_T = \frac{144 * W_T * NP}{3600 * NT * A_I} \tag{5.28}$$

Note that over-and-under tube passes are recommended when there are three or more tube passes, and that the use of less than three tube passes tends to be uneconomical and high in cost.

The modified Reynolds number N_R is calculated next:

$$N_R = D_I * \frac{G_T}{U_{\text{VISC}}} \tag{5.29}$$

where D_I = tube internal diameter (0.87 in; for 16 BWG, 1 in)
        U_{VISC} = fluid viscosity tube wall condensed liquid viscosity, cP, or viscosity of gas or liquid, cP

Step 8. As in the tube/shell calculations, a J factor is calculated for the tube inside heat-transfer coefficient HT.

If N_R is greater than 17, use Eq. (5.30) to calculate J:

$$J = \exp[-3.913 + 3.968 * (\log N_R) - 0.5444 * (\log N_R)^2 + 0.04323 \\ * (\log N_R)^3 - 0.001379 * (\log N_R)^4] \tag{5.30}$$

If N_R is 17 or less, then apply the following equations and interpolate if needed.

$$L_D = \frac{L}{D_I} \tag{5.31}$$

where L = tube length, ft

If $L_D = 2$ and $N_R \le 17$, use Eq. (5.32) for J:

$$J = \exp[3.91 + 0.2365 * (\log N_R) + 0.04706 * (\log N_R)^2] \quad (5.32)$$

If $L_D = 10$ and $N_R \le 17$, use Eq. (5.33) for J:

$$J = \exp[3.389 + 0.3392 * (\log N_R) + 7.910001e\text{-}04 * (\log N_R)^2] \quad (5.33)$$

If $L_D = 20$ and $N_R \le 17$, use Eq. (5.34) for J:

$$J = \exp[3.1704 + 0.33214 * (\log N_R) + 0.0012123 * (\log N_R)^2] \quad (5.34)$$

If $L_D = 50$ and $N_R \le 17$, use Eq. (5.35) for J:

$$J_{50} = \exp[2.851 + 0.3685 * (\log N_R) - 0.011588 * (\log N_R)^2] \quad (5.35)$$

In order to calculate a correct J if $L_D \le 17$, use the following interpolation equations.

If $L_D \le 10$, then use Eq. (5.36) for J:

$$J = J_{10} + \frac{10 - L_D}{8} * (J_2 - J_{10}) \quad (5.36)$$

If $L_D \le 20$, then use Eq. (5.37) for J:

$$J = J_{20} + \frac{20 - L_D}{10} * (J_{10} - J_{20}) \quad (5.37)$$

If $L_D \le 50$, then use Eq. (5.38) for J:

$$J = J_{50} + \frac{50 - L_D}{30} * (J_{20} - J_{50}) \quad (5.38)$$

If $L_D > 50$, then use Eq. (5.39) for J:

$$J = J_{50} - \frac{L_D - 50}{30} * (J_{20} - J_{50}) \quad (5.39)$$

For Eqs. (5.36) through (5.39), $J_2 = J$ at $L_D = 2$, $J_{10} = J$ at $L_D = 10$, $J_{20} = J$ at $L_D = 20$, and $J_{50} = J$ at $L_D = 50$.

The preceding J factor calculations are curve-fitted equations from the GPSA *Engineering Data Book* [15], Sec. 10, Fig. 10-15.

Step 9. The inside tube wall heat-transfer film coefficient HT is next calculated. Before making this calculation, if this is for steam condensing, the HT value is always 1500 Btu/h · °F · ft². Otherwise, apply Eq. (5.40) [3]:

$$\text{HT} = J * K * \frac{(C_p * U_{\text{VISC}}/K)^{1/3}}{D_I} \tag{5.40}$$

where C_p = tube fluid specific heat, Btu/lb · °F
         D_I = internal tube diameter, in
         K = thermal conductivity of tube fluid,
              Btu/(h · ft² · °F · ft⁻¹)
   U_{VISC} = tube wall viscosity, cP

Equation (5.40) is good and applicable for film heat-transfer coefficients for gas-phase, vapor-phase, and condensing fluids. The C_p, K, and U_{VISC} input values should be accurate for the tube-side fluid being cooled.

The specific heat value C_p should reference the gas phase for a condensing fluid.

The thermal conductivity value K should also reference the liquid film value and the gas phase for a condensing film, averaging the two at the tube wall temperature.

U_{VISC} has conventionally been given a correction factor in Eq. (5.40):

$$\phi = \left(\frac{U_{\text{VISC}}}{U_{\text{VISC}w}} \right)^{0.14}$$

and

$$\text{HT corrected} = \text{HT} * \phi$$

However, ϕ here is always near or equal to 1.0. This correction is therefore optional in this procedure. Applying it here usually gives an HT corrected value that is about 3 to 5% less than HT, a more conservative answer.

Step 10.   The air quantity flow is next calculated, knowing the values of DTA (step 2) and Q (step 4) and using a specific heat air constant of 0.24 Btu/lb · °F. This air specific heat value is applicable to all cases. Equation (5.41) is now used to determine airflow W_A:

$$W_A = \frac{Q}{0.24 * \text{DTA}} \tag{5.41}$$

Step 11.   In this step we calculate the air-cooler face-area airflow G_A (lb/h · ft²) of the face area, knowing FA (step 5) and W_A (step 10). G_A is calculated from Eq. (5.42) as follows:

$$G_A = \frac{W_A}{FA} \qquad (5.42)$$

Step 12. The air-side heat-transfer film coefficient H_A (Btu/h $\cdot$ °F $\cdot$ ft^2) is now calculated, using the G_A value from step 11. Equation (5.43) is a data-fitted equation of several actual H_A values vs. G_A values. Note that the following H_A calculated value is the tube extended surface area outside heat-transfer film coefficient. Equation (5.43) also is in good agreement with the GPSA *Engineering Data Book* [15], Sec. 10, Fig. 10-17.

$$H_A = \exp[-7.1519 + 1.7837 * (\log G_A) - 0.076407 * (\log G_A)^2] \quad (5.43)$$

Step 13. In this step, the initial U_X value, which was first assumed, is now calculated. If the calculated value does not agree with the first assumption, then a new U_X is assumed and steps 1 through 13 are repeated.

$$U_X = \frac{1}{(1/HT) * (AR * D_O/D_I) + RD_T * (AR * D_O/D_I) + (1/H_A)} \quad (5.44)$$

where   AR = ratio of area of outside tube extended surface to bare outside tube surface, ft^2/ft^2
  D_I = tube inside diameter, in
  D_O = tube outside diameter, in
  H_A = outside transfer coefficient (step 12)
  HT = inside tube coefficient (step 9), Btu/h $\cdot$ °F $\cdot$ ft^2
  RD_T = tube inside fouling factor (normally 0.001 to 0.005; however, for severe fouling, could be $\geq$ 0.02)

A good estimate for assuming a new U_X value for the second and subsequent trials is the application of Eq. (5.44):

$$U_{X2} = \frac{Q}{A_X * DTM} \qquad (5.45)$$

where U_{X2} = next U_X assumed value for repeating steps 1 through 13

Using reasonable suggested values for AR, FA, and U_X, seldom are more than two trials required. Usually the first trial calculations are within reasonable tolerance of the assumed U_X value—say 15% error or less. Such error is tolerable if a 10 to 15% excess cooler bare tube area is allowed. Thus, having this tolerance of error, in many cases the first assumed U_X value is close enough for hand calculations and a second pass through steps 1 through 13 is not required. The DTA value first calculated in step 2 should also be adjusted for a second calculation pass. Increase DTA if the calculated U_X is greater than the assumed U_X, and decrease DTA if the calculated U_X is the lesser value.

Step 14. Here we calculate the tube-side pressure loss, applying the hand-calculation procedure given in Chap. 6. You may also apply the Fflw program given on the CD accompanying this book. This is made simple by tabulating flow in only one tube and calculating the pressure loss for each pass in the tube. Adding the pressure loss of each pass, of course, sums the full-bundle tube-side pressure loss.

Finfan cooler tube-side pressure losses of 2 to 10 psi are common, and these losses seldom exceed 15 psi. Greater pressure losses indicate that more parallel tubes are needed and/or a parallel tube bundle is needed.

Step 15. Steps 15 through 20 are dedicated fully to the fan criteria and the required air mass flow determined in the previous steps. How many fans should serve each bundle? The minimum fan area per bundle is an industry standard number of 0.40 ft^2 minimum of fan area per square foot face area FA. Thus:

$$FAPF = \frac{0.40 * FA}{\text{number of fans}} \tag{5.46}$$

where FAPF = fan area per fan, ft^2

Selecting the number of fans requires some judgment of the FA layout. Remember that in step 6 the number of fans was fixed by fixing the FA geometry with a number of equal-leg squares layout.

Step 16. The fan diameter per fan D_{fan} (ft) is next calculated:

$$D_{\text{fan}} = \left(\frac{4 \, FAPF}{\pi}\right)^{0.5} \tag{5.47}$$

Step 17. The air-side static pressure loss DPAT (inches of water) is calculated in this step. First consider the fact that almost all average air-side temperatures are in the range of 90 to 150°F. An air-density ratio of elevated levels to sea level DR is needed, which is to be taken at this average air-side temperature. We can easily determine DR by taking a linear interpolation of Table 5.5.

DR may be calculated from the ideal gas law, referenced to 70°F and 14.7 psia conditions, which is at 0 elevation or sea level. Dry air is the basis. Thus the calculation density for air is:

$$\text{Air density at 0 elevation} = 29 \, MW * \frac{14.7 \text{ psia}}{10.73 * (70°F + 460)} = 0.075 \text{ lb/ft}^3$$

TABLE 5.5   Determining DR

Temperature	DR at 0 elevation	DR at 5000 ft
90°F	0.96	0.80
150°F	0.87	0.72
200°F	0.80	0.66

$$\text{Equivalent ft of air at sea level} = \frac{33.93 \text{ ft/atm} * 62.4}{0.075} = 28,230 \text{ ft for 1 atm}$$

Air density at 3,000 ft and 70°F

$$= \frac{MW * 14.7 \text{ psi} * [(28,230 - 3000)/28,230]}{10.73 * (460 + 70)} = 0.067 \text{ lb/ft}^3$$

$$\text{DR at 3000 ft and 70°F} = \frac{0.067}{0.075} = 0.89$$

All values in Table 5.5 are calculated using this procedure. Table 5.5 also matches up accurately with GPSA [15] Fig. 10-16. As a general rule, I have found that DR is in the 0.85 to 0.90 range in almost every air-cooler fan case. Of course, this assumes elevation is sea level or less than 1000 ft. Using a DR value of 0.87 is a conservative approach and is advised if you want a quick but reasonable answer.

The next part of this step is calculating the pressure loss per tube row D_{PA} (inches of water), having the G_A calculation from step 11, and a DR air-density ratio. Eq. (5.48) has shown good results in determining the air-side pressure loss per tube row pass. Equation (5.48) also matches up accurately with GPSA [15] Fig. 10-18.

$$D_{PA} = \exp (1.82 * \log G_A - 16.58) \qquad (5.48)$$

where D_{PA} = static pressure loss per row of tube crossflow, inches of water

It is simple to multiply D_{PA} by the number of tube rows N_{ROWS} to get the total static air-side pressure loss DPAT (inches of water). First, however, the correction for air density must be factored. Here apply the air-density ratio DR, which is given in Table 5.5. Equation (5.48) is based on a D_{PA} value calculated from an air density at 70°F and at 0 ft elevation. It is therefore necessary to correct this D_{PA} value with a DR value determined at the average air-side temperature T_{avg}.

$$T_{avg} = \frac{DTA}{2} + t_1 \qquad (5.49)$$

where t_1 = ambient air temperature, °F

$$\text{DPAT} = N_{\text{ROWS}} * \frac{D_{PA}}{\text{DR}} \qquad (5.50)$$

where DR = ratio of actual dry air density at average temperature
and elevation to dry air density at 70°F and 0 elevation
(interpolated from Table 5.5)

The DR value used in Eq. (5.50) is interpolated from Table 5.5 for the average air-side temperature and the elevation above sea level. An air temperature of 70°F and an elevation of 0 is the basis of DR. Most air coolers are at elevations from sea level (0) to less than 1000 ft, making the interpolation of a DR value from Table 5.5 very simple.

Step 18.   The actual air volumetric flow ACFM is next calculated. A DR value from Table 5.5 will again be used to calculate the air density at the fan ambient inlet temperature t_1. Use a W_A value calculated in step 10. Please note that DR is always referenced to air at 70°F and 0 elevation, whose pressure is 0.075 lb/ft^3.

$$\text{ACFM} = \frac{W_A}{\text{DR} * 60 * 0.075} \qquad (5.51)$$

Step 19.   The fan approximate discharge pressure P_{force} (inches of water) is next calculated applying DPAT, DR, and ACFM. DR which was used to calculate ACFM in Eq. (5.51), has been interpolated from Table 5.5 at the fan elevation and at the ambient temperature of air.

$$P_{\text{force}} = \text{DPAT} + \left(\frac{\text{ACFM}}{4009 * \pi * D_{\text{fan}}^2/4} \right)^2 \qquad (5.52)$$

where P_{force} = pressure head + velocity head, $V^2/2 * g$, inches of water

Equation (5.52) is Bernoulli's theorem [16], an energy equation that is well known to be equal to the sum of the elevation head, the pressure head, and the velocity head. Here the elevation head has been left out because the fan suction air pressure and the exiting air pressure from the air cooler are both equal.

The constant, 4009, is calculated from:

$$4009 = \left(\frac{2 * g * \text{DENSW} * 3600}{\text{DENSA} * 12} \right)^{0.5}$$

$$4009 = \left(\frac{2 * 32.2 * 62.4 * 3600}{0.075 * 12} \right)^{0.5}$$

where   DENSA = density of air at 70°F, lb/ft^3
         DENSW = density of water at 70°F, lb/ft^3
             g = gravity acceleration constant, 32.2 ft/s^2

Step 20.   The final step in this sequence is calculating the fan motor horsepower B_{HP}.

$$B_{HP} = \frac{(\text{ACFM per fan}) * P_{force}}{6387 * \text{eff}} \qquad (5.53)$$

where eff = fan motor efficiency (normally 70%, or 0.70 in equation)

The constant 6387 is now calculated:

$$6387 = \frac{33,000 \text{ ft} \cdot \text{lb}}{\text{min} * \text{h}} * \frac{12 \text{ in}}{\text{ft}} * \frac{\text{ft}^3}{62 \text{ lb}}$$

Note that the ft^3/62 lb term is the density of water at the assumed air inlet temperature, 100°F.

Having accomplished the review of a 20-step design procedure, the objective now is to proceed with a classic example, working all 20 steps. Here the primary objective is deriving the transfer surface and calculating the geometry and fan horsepower.

Example 5.2   Assume that a typical hydrocarbon naphtha liquid from a fractionation tower-side cut stream is to be cooled to 150°F. The naphtha stream enters the air cooler at 250°F at a flow rate of 273,000 lb/h. The physical tube-side properties at the average temperature of 200°F are:

$K = 0.0766$ Btu/(h · ft^2 · °F · ft^{-1})
$C_p = 0.55$ Btu/lb · °F
$U_{VISC} = 0.51$ cP

The exchanger heat duty is:

$$Q = 273,000 \text{ lb/h} * 0.55 * (250 - 150) = 15 \text{ mmBtu/h}$$

Step 1.   Pick a first-trial value of U_X. Reviewing Table 5.4, U_X is 4.5 at an average viscosity of 0.5 cP and decreases to 3.0 at a viscosity of 3.0 cP. Thus, by interpolating, U_X should be approximately 4.4. This is only an estimation and may require a second and third pass for a proper answer.

Step 2.   Using an ambient air temperature t_1 of 100°F, calculate the approximate air temperature increase with Eq. (5.24):

$$DTA = \frac{U_X + 1}{10} * \left(\frac{T_1 + T_2}{2} - t_1\right) \qquad (5.20)$$

$$DTA = \frac{4.4 + 1}{10} * \left(\frac{250 + 150}{2} - 100\right) = 54°F$$

Step 3. Calculate DTM using Eq. (5.21):

$$DTM = \frac{GTD - LTD}{\log \ (GTD/LTD)} \tag{5.21}$$

GTD and LTD are determined from Table 5.1:

Hot fluid		Cold fluid	Difference
250	Higher temperature	154	$GTD = T_1 - t_2 = 96$
150	Lower temperature	100	$LTD = T_2 - t_1 = 50$
			$GTD - LTD = 46$

$$DTM = \frac{46}{\log \ (96/50)} \quad \text{(log = log base of 2.7183)}$$

$$DTM = 70.5°F$$

Plan to use three or more tube rows in an over-and-under pattern. Therefore, $F_t = 1.0$, so no correction of the 70.5°F DTM is required.

Step 4. The required tube outside extended area A_X is:

$$A_X = \frac{Q}{U_X * CDTM} \tag{5.24}$$

$$A_X = \frac{15e\text{-}6}{4.4 * 70.5} = 48{,}356 \ ft^2$$

Step 5. Calculate the air-cooler face area FA using a selected APSF factor from Table 5.3:

- For this first trial, select four rows of tubes (over-and-under configuration).
- Use 1-in-OD tubes.
- Use ⅝-in-high fins, 10 fins per inch.
- APF = 5.58 ft²/ft².
- Choose a 2¼-in triangular tube pitch.
- APSF = 118.8 ft²/ft².

$$FA = \frac{A_x}{APSF} \tag{5.25}$$

$$FA = \frac{48{,}356}{118.8} = 407 \ ft^2/ft^2 \ (4 \ rows \ assumed)$$

Step 6. The tube bundle size and number are now calculated. First assume a tube length L of 40 ft.

$$Width = \frac{FA}{L} = \frac{407}{40} = 10.17 \ ft$$

Four fans would be placed here. Try a tube length of 30 ft to get fewer fans and usually a better fit in space allocation.

$$\text{Width} = \frac{407}{30} = 13.6 \text{ ft}$$

For simplification, use 14-ft-wide and 30-ft-long tubes. Two fans will thus be required.

$$NT = \frac{A_X}{APF * L} \tag{5.26}$$

$$NT = \frac{48,356}{5.58 * 30} = 289 \text{ tubes}$$

At this point in the calculations, it is convenient to round off these figures, allowing a 50,000-ft² extended fin-tube area:

$$NT = \frac{50,000}{5.58 * 30} = 299 \text{ or } 300 \text{ tubes}$$

Step 7. Calculate the modified Reynolds number N_R. For this trial, a 3-pass tube bundle is chosen, which divides the number of tubes equally into 100 tubes for each pass. A 2-pass arrangement is shown in the bottom of Fig. 5.6. Use the over-and-under arrangement here.

$$A_I = 3.1416 * \left(\frac{D_I}{2}\right)^2 \tag{5.27}$$

$$A_I = 3.1416 * \left(\frac{0.87}{2}\right)^2 = 0.5945 \text{ in}^2$$

$$G_T = \frac{144 * W_T * NP}{3600 * NT * A_I} \tag{5.28}$$

$$G_T = \frac{144 * 273,000 * 3}{3600 * 300 * 0.5945} = 184 \text{ lb/(s} \cdot \text{ft}^2)$$

$$N_R = \frac{D_I * G_T}{U_{\text{VISC}}} \quad N_R = \frac{0.87 * 184}{0.51} = 314 \tag{5.29}$$

Step 8. Calculate the J factor. NR is greater than 17; therefore, Eq. (5.30) will be used for J:

$$J = \exp[-3.913 + 3.968 * (\log N_R) - 0.5444 * (\log N_R)^2 + 0.04323$$

$$* (\log N_R)^3 - 0.001379 * (\log N_R)^4] \tag{5.30}$$

$$J = 2027$$

Step 9. Calculate the tube inside film coefficient HT:

$$HT = J * K * \frac{(C_p * U_{VISC}/K)^{1/3}}{D_I} \qquad (5.40)$$

$$HT = 2027 * 0.0766 * \frac{(0.55 * 0.51 / 0.0766)^{1/3}}{0.87} = 275 \text{ Btu/h} \cdot \text{ft}^2 \cdot {}^\circ\text{F}$$

If the tube wall is corrected here for the viscosity ratio ϕ, then:

$$HT * \phi = 275 * 0.95 = 261 \text{ Btu/h} \cdot \text{ft}^2 \cdot {}^\circ\text{F}$$

Step 10. Calculate the airflow W_A:

$$W_A = \frac{Q}{0.24 * \text{DTA}} \qquad (5.41)$$

$$W_A = \frac{15e\text{-}6}{0.24 * 54} = 1,157,400 \text{ lb/h}$$

Step 11. Calculate the air-cooler face area flow mass rate per ft^2 G_A:

$$G_A = \frac{W_A}{\text{FA}} \qquad (5.42)$$

$$G_A = \frac{1,157,400}{407} = 2844 \text{ lb/(h} \cdot \text{ft}^2 \text{ of face area)}$$

Step 12. The air-side extended tube film coefficient H_A is next calculated applying Eq. (5.43):

$$H_A = \exp[-7.1519 + 1.7837 * (\log G_A) - 0.076407 * (\log G_A)^2] \quad (5.43)$$

$$H_A = 9.0 \text{ Btu/(h} \cdot \text{ft}^2 \cdot {}^\circ\text{F)}$$

Step 13. In this step, U_X, the step 1 assumed value, and the overall extended surface heat-transfer coefficient are calculated:

$$U_X = \frac{1}{(1/HT) * (AR * D_O/D_I) + RD_T * (AR * D_O/D_I) + (1/H_A)} \qquad (5.44)$$

$$U_X = \frac{1}{(1/275) * (21.4 * 1.0/0.87) + 0.001 * (21.4 * 1.0/0.87) + (1/9.0)}$$

$$= 4.44 \text{ Btu/h} \cdot \text{ft}^2 \cdot {}^\circ\text{F}$$

Assume a fouling factor on the inner tube side of 0.001, which corresponds to a light distillate such as naphtha for this case. See the shell tube section, especially Table 5.2, for a further discussion of fouling factor values.

In step 1, a U_X value of 4.4 was assumed. The previous calculation shows this assumption is close enough to require no second pass through steps 1 through 13. Thus all initial calculations and assumptions made in steps 1 through 13 are valid and final. There is only a 1% error between calculated and assumed values. Even an error of 5% would be sufficient, requiring no second pass through steps 1 through 13. Our estimation of U_X in step 1 was excellent.

Step 14. Calculate the tube-side pressure drop in this step, applying the Fflw program discussed in Chap. 6, where this step will be reviewed and worked.

Step 15. The number of fans is easily determined in a facial area of 14 ft wide by 30 ft. Install two fans, each taking up a 14- by 15-ft² area. Now calculate the fan area per fan, with two fans required:

$$\text{FAPF} = \frac{0.40 * \text{FA}}{\text{number of fans}} \qquad (5.46)$$

$$\text{FAPF} = \frac{0.40 * 420}{2} = 84 \text{ ft}^2$$

Step 16. Calculate the fan diameter D_{fan}:

$$D_{\text{fan}} = \left(\frac{4 \text{ FAPF}}{\pi} \right)^{0.5} \qquad (5.47)$$

$$D_{\text{fan}} = \left(\frac{4 * 84}{\pi} \right)^{0.5} = 10.3 \text{ or } 11 \text{ ft}$$

Step 17. Calculate the air-side static pressure loss DPAT (inches of water):

$$T_{\text{avg}} = \frac{\text{DTA}}{2} + t_1 \qquad (5.49)$$

$$T_{\text{avg}} = \frac{54}{2} + 100 = 127°\text{F}$$

where T_{avg} = air-side average temperature, °F
        t_1 = ambient air temperature, °F = 100°F

Static pressure loss of airflow across each single tube row DPA (inches of water) is:

$$G_A = \frac{W_A}{\text{FA}} \qquad (5.42)$$

$$\text{GA} = 2844 \text{ lb/h} \cdot \text{ft}^2$$

from step 11. Therefore

$$DPA = \exp(1.82 * \log G_A - 16.58)$$
$$= 0.122 \text{ inches of water per tube row} \qquad (5.48)$$

Next, the air density is corrected with DR from Table 5.5 for dry air at a tube section average air temperature 127°F. A DR factor of 0.90 is interpolated from Table 5.5 at a 0 elevation.

$$DPAT = N_{\text{ROWS}} * \frac{D_{PA}}{DR} \qquad (5.50)$$

$$DPAT = 4 * \frac{0.122}{0.90} = 0.54 \text{ inches of water}$$

Step 18.   Calculate the fan inlet's actual air volumetric flow ACFM, ft³/min. DR may be interpolated from Table 5.5 again, but at 100°F for fan inlet ambient air temperature. However, DR is calculated as previously shown, applying the ideal gas law equation:

$$DENSA = \frac{29 \text{ MW} * 14.7 \text{ psia}}{10.73 * (100°F + 460)} = 0.0709 \text{ lb/ft}^3$$

$$DR = \frac{0.0709}{0.075} = 0.95 \text{ for 0 elevation}$$

$$W_A = 1,157,400 \text{ lb/h} \qquad \text{(from step 10)}$$

$$ACFM = \frac{W_A}{DR * 60 * 0.075} \qquad (5.51)$$

$$ACFM = \frac{1.157e + 06}{(0.95 * 60 * 0.075)} = 271,000 \text{ (or 135,500 per fan)}$$

Step 19.   Calculate the pound force P_{force} (inches of water) that the fan is required to output:

$$P_{\text{force}} = DPAT + \left(\frac{ACFM}{4009 * \pi * D_{\text{fan}}^2/4} \right)^2 \qquad (5.52)$$

$$P_{\text{force}} = 0.54 + \left(\frac{135,500}{4009 * 3.1416 * 11^2/4} \right)^2 = 0.67 \text{ inches of water}$$

Step 20.   In this final step we calculate the fan horsepower B_{HP}. The fan is assumed to have a hydraulic efficiency of 70%.

$$B_{\text{HP}} = \frac{(ACFM \text{ per fan}) * P_{\text{force}}}{6387 * \text{eff}} \qquad (5.53)$$

$$B_{HP} = \frac{(135{,}500 \text{ per fan}) * 0.67}{6387 * 0.70} = 20.3 \text{ hp}$$

Allow a 25-hp motor for each fan. This add-on 5 hp provides sufficient loss for a speed reducer. A 30-hp motor is suggested for motor efficiency allowance and design safety.

Air Finfan Exchanger Computer Programs

AirClr8 applications

The CD supplied with this book contains two air-cooler programs, AirClr8 and AirClr9. The AirClr8 program is for sensible heat transfer as in Example 5.2 (see Figs. 5.7 and 5.8).

Figures 5.9 and 5.10 show actual screen displays from the AirClr8 program. The problem input is the same as in Example 5.2, which was hand-worked earlier in this chapter. Notice that the program has repeated several loops of steps 1 through 13, deriving an answer showing the process naphtha fluid cooled to 141°F instead of the 150°F initially inputted. Why the difference? Observe that the program corrects the heat transferred in making a balance for Eq. (5.23):

$$Q = U_X * A_X * \text{DTM} \tag{5.19}$$

Steps 1 through 13 are repeated in the program until the calculated U_X matches the assumed U_X of the program convergence tolerance input. (See the convergence tolerance input of 0.05 in Fig. 5.7.) Notice the greater U_X value calculated by the computer program (5.05) vs. the hand-calculated value (4.44). This reveals only that the exchanger will

```
              AIR COOLED EXCHANGER RATING/DESIGN PROGRAM

    First assumption overall transf coef....   4.200
    Tubeside fouling factor.................   0.00100
    Process inlet temp, deg F ..............   250.00
    Process outlet temp, deg F .............   150.00
    Process flow LB/HR .....................   273000.00
    Process heat duty, BTU/HR ..............   15000000.00
    Process visc CP ........................   0.5100
    Process Conductivity BTU/HR/FT2 F/Ft ...   0.07660
    Process Specif Ht, BTU/LB F ............   0.55000
    Convergence tolerence ..................   0.050
    Ambient air temp F .....................   100.00
    Outside tube dia, inches ...............   1.000
    Inside tube dia, inches ................   0.870
    Outside extended tube area, FT2/FT .....   118.800
    Tube length, FT ........................   30.00
    Fintube area, FT2/FT of tube ...........   5.50
    Number of tubside passes ...............   4
    Area ratio fin to bare,(FT2/FT)/(FT2/FT)   21.4000
    Number of tubes ........................   300
         PRESS F10 to validate and calc or Press ESC
           Vary the number of tubes to achieve your conditions
```

Figure 5.7   Example 5.2 problem input, AirClr8 program.

```
HEAT TRANSF COEF , available or actual    =    5.041
HEAT TRANSF COEF , extd surface  required =    5.031
BARE TUBE AREA   ft2                       =    2356.2
HEAT TRANSFER COEF bare tube               =    107.239
NUMBER OF TUBES                            =    300
HEAT TRANSF btu/hr                         =    1.6442E+07.
OUTLET AIR TEMP F    actual                =    152.00
AIR MASS RATE   lb/hr                       =    1.3174E+06
CORRECTED PROCESS OUTLET TEMP  deg F       =    140.50
LMTD                                        =    0.651E+02
   Run another Y/N ?
```

Figure 5.8   Example 5.2 answers outputted by AirClr8 program.

do more cooling than specified. Thus the reason for the 141°F process tube-side outlet temperature is this finer iterated computer calculation of U_X and the associated DTM.

The computer-calculated result found here is more accurate and shows what the excess tube surface area will do. Thus the computer program run in Figs. 5.7 and 5.8 reveals that the specified exchanger will do a better cooling job than initially determined in the hand calculations.

With evidence that we have an overdesign in the given hand calculations, it is good to determine what greater fouling factor RD_T may be allowed. See Figs. 5.9 and 5.10 for a second program run using the same input, but specifying an RD_T of 0.002 in place of the initial value of 0.001.

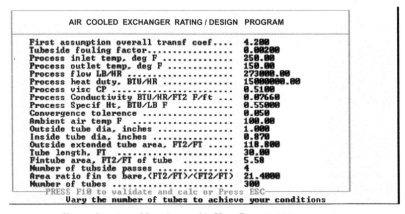

```
        AIR COOLED EXCHANGER RATING / DESIGN  PROGRAM

First assumption overall transf coef.... 4.200
Tubeside fouling factor................. 0.00200
Process inlet temp, deg F .............. 250.00
Process outlet temp, deg F ............. 150.00
Process flow LB/HR ..................... 273000.00
Process heat duty, BTU/HR .............. 15000000.00
Process visc CP ........................ 0.5100
Process Conductivity BTU/HR/FT2 F/ft ... 0.07660
Process Specif Ht, BTU/LB F ........... 0.55000
Convergence tolerence .................. 0.050
Ambient air temp F ..................... 100.00
Outside tube dia, inches ............... 1.000
Inside tube dia, inches ................ 0.870
Outside extended tube area, FT2/FT ..... 118.800
Tube length, FT ........................ 30.00
Fintube area, FT2/FT of tube .......... 5.58
Number of tubside passes .............. 4
Area ratio fin to bare,<FT2/FT>/<FT2/FT> 21.4000
Number of tubes ....................... 300
   PRESS F10 to validate and calc or Press ESC
     Vary the number of tubes to achieve your conditions
```

Figure 5.9   Example 5.2 problem input, AirClr8, $R_{DT} = 0.002$.

```
HEAT TRANSF COEF , available or actual      =      4.408
HEAT TRANSF COEF , extd surface  required   =      4.399
BARE TUBE AREA  ft2                         =      2356.2
HEAT TRANSFER COEF bare tube                =      93.765
NUMBER OF TUBES                             =        300
HEAT TRANSF btu/hr                          =      1.5398E+07.
OUTLET AIR TEMP F    actual                 =      152.00
AIR MASS RATE   lb/hr                       =      1.2338E+06
CORRECTED PROCESS OUTLET TEMP  deg F        =      147.45
LMTD                                        =      0.697E+02
  Run another Y/N ?
```

Figure 5.10   Example 5.2 answers outputted by AirClr8, $R_{DT} = 0.002$.

Now notice in Figs. 5.9 and 5.10 that the output answers match the inputted values much more closely:

	Input	Output
Outlet process temperature	150°F	148°F
Heat transferred	15.0e+06	15.4e+06

The program checks your inputted outlet process temperature and corrects it.

AirClr9 applications

Example 5.3   This example deals with an air-cooler light petroleum product condenser. Please refer to Figs. 5.11 and 5.12 and note the input. The primary objective here is to find the fouling dirt factor RD_T for an existing air-cooler bundle that is part of a larger bundle being served by two fans. The first input (not shown) is an RD_T of 0.001. This is the second line on the input screen. With the 0.001 RD_T, the program run displays the design conditions that would be expected for this exchanger.

RD_T	0.004	0.001
Process outlet temperature, °F	187	164
Air outlet temperature and rate, lb/h	146 and 5.6e+04	142 and 6.1e+04

Here are the steps you need to execute hand calculations and to run the Air-Clr9 program:

1. Determine Q, which is 100% latent heat for condensing-type heat transfer. If this is a partial condenser, make sure you have the enthalpy difference from the exchanger inlet to the exchanger outlet. This enthalpy difference is the correct Q value to use.

2. Determine the tube-side inlet and outlet temperatures on a scale with increasing heat transfer as the liquid condenses and the temperature decreases. This

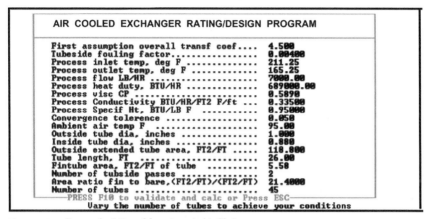

```
AIR COOLED EXCHANGER RATING/DESIGN PROGRAM

First assumption overall transf coef....   4.500
Tubeside fouling factor.................   0.00400
Process inlet temp, deg F ..............   211.25
Process outlet temp, deg F .............   165.25
Process flow LB/HR .....................   7000.00
Process heat duty, BTU/HR ..............   689000.00
Process visc CP ........................   0.5890
Process Conductivity BTU/HR/FT2 F/ft ...   0.33500
Process Specif Ht, BTU/LB F  ...........   0.95000
Convergence tolerence ..................   0.050
Ambient air temp F .....................   95.00
Outside tube dia, inches ...............   1.000
Inside tube dia, inches ................   0.880
Outside extended tube area, FT2/FT .....   118.000
Tube length, FT ........................   26.00
Fintube area, FT2/FT of tube  ..........   5.58
Number of tubside passes ...............   2
Area ratio fin to bare,(FT2/FT)/(FT2/FT)   21.4000
Number of tubes ........................   45
      PRESS F10 to validate and calc or Press ESC
      Vary the number of tubes to achieve your conditions
```

Figure 5.11   Example 5.3 problem input, AirClr9 program.

is normally a linear scale. Record the tube-side outlet temperature T_2, which is required for the cooler duty specified Q.

3. For *rating,* proceed as for the 20 air-cooler steps; however, when making a new estimate of U_X (step 13) by hand or running AirClr9, correct the Q value and use new process-side inlet-outlet temperatures. The process-side outlet temperature must be calculated, matching it to your rating temperature and the assumed U_X to calculated U_X.

4. For *design,* proceed as for the 20 air-cooler steps, but keep the required Q constant and vary the tube number and airflow until the assumed U_X matches the calculated U_X. If the T_2 you recorded earlier is greater than the T_2 calculated by AirClr9, the design is conservative and the number of tubes may be decreased. Otherwise, increase the tube area and rerun the program until the recorded T_2 is equal to or greater than the program-calculated T_2.

```
HEAT TRANSF COEF , available or actual      =   1.348

HEAT TRANSF COEF , extd surface  required   =   1.357

BARE TUBE AREA   ft2                        =   306.3

HEAT TRANSFER COEF bare tube                =   28.933

NUMBER OF TUBES                             =   45

HEAT TRANSF btu/hr                          =   6.8900E+05.

OUTLET AIR TEMP F    actual                 =   146.29

AIR MASS RATE   lb/hr                       =   5.5975E+04

CORRECTED PROCESS OUTLET TEMP   deg F       =   187.10
LMTD                                        =   0.777E+02
     Run another Y/N ?
```

Figure 5.12   Example 5.3 answers outputted by AirClr9.

5. To run AirClr9, execute the first two steps in this list, then vary Q for rating and hold Q constant for design. Note that the program varies the airflow, applying the calculated DTA factor. You may increase the airflow by increasing Q accordingly while keeping U_X constant. Note that the associated tubeside in-out temperatures also are to be adjusted. This is a conservative and safe supplemental way to run AirClr9.

AirClr8 program code listing

The AirClr8 computer program's Basic language code listing is given in Fig. 5.13. This is a Quick Basic or common GW Basic code listing. With the exception of the input-output code, it is also applicable to Visual Basic.

A brief review of the AirClr8 program code should show the similarities between the 20-step hand calculation and the program code in Fig. 5.13. For example, notice the J factor code lines (lines 450–580) in Fig. 5.13 and compare them to Eqs. (5.34) through (5.42). These equations are identical. The program has an advantage, of course, in that countless repeated passes are made through steps 1 through 13 until the calculated U_X matches the assumed U_X. (See code lines 130 and 720–740.) These code lines adjust the process outlet temperature until a match is obtained between the assumed and calculated U_X values. The Visual Basic code used in the AirClr8 program on the CD is the same code as listed in Fig. 5.13.

The PC program AirClr9, included on the CD, is used for air-cooler condenser applications. The code calculates a corrected process outlet temperature T_2. However, the user must adjust the heat input Q in accordance with the T_2 calculated in the program, as discussed previously.

Program summaries

In answer to the question of which program should be run for any definable air-cooler case, the following summaries are provided for clarification.

AirClr8.   This is an executable program for any single-phase fluid to be cooled in an air cooler. A specific heat is given, which the program will use to determine process heat exchanged. The process heat exchanged is calculated by the program on the basis of the user-inputted values for tube number and length, number of passes, and number of tube rows.

This program is also used for condensing steam. A given Q fixes heat transfer. A 1500 Btu/h · ft^2 · °F steam-condensing tube-side film coefficient is assumed. This condensing film coefficient is set by inputting a specific heat value of 0.

```
10 REM AIRCLR8.BAS PROGRAM FOR AIR COOLER RATING VERSION
50 REM INPUT LINES 50-300 FOR PROGRAM VARIABLES
60 UX=4.5    'FIRST ASSUMPTION OF OVERALL TRANSFER COEFFI-
   CIENT
65 RDT=.004    'TUBESIDE FOULING FACTOR
70 T1=211.25    'PROCESS INLET TEMP DEG F
80 T2=120    'PROCESS OUTLET TEMP DEG F
90 WT=7000    'PROCESS FLOW LB/HR
100 Q=689000    'PROCESS HEAT DUTY BTU/HR
110 UVISC=.589    'PROCESS VISC CP
120 K=.335    'PROCESS CONDUCTIVITY BTU/HR/FT2 deg F/FT
125 CP=.95    'PROCESS SPECIFIC HEAT BTU/LB F
130 CONV1=0.05    'CONVERGENCE TOLERANCE use .01 to .5
140 TA1=95    'AMBIENT AIR TEMP deg F
145 DO=1    'OUTSIDE DIAMETER OF TUBE inches
150 DI=.88    'INSIDE DIAMETER OF TUBES inches
160 APSF=118.8 'OUTSIDE EXTENDED TUBE AREA which is FT2/FT2
    of FA
165      'for 3 rows APSF=89.1, 4 rows APSF=118.8, 6 rows
    APSF=178.2
170 L=26    'TUBE LENGTH FT
180 APF=5.58    'TUBE OUTSIDE AREA of FINTUBE AS FT2/FT of
    TUBE in FT2
190 NP=2    'NUMBER OF TUBESIDE PASSES
200 AR=21.4    'AREA RATIO OF FINNED TUBE TO BARE TUBE AS
    FT2/FT / FT2/FT
210 NT=45    'NUMBER OF TUBES
300 'GOSUB 1000 This is the subroutine for the Quick Basic
    Input which is not given here
320 '
330 DTA=((UX+1)/10)*((((T1+T2)/2)-TA1)    'first estimate air
    temp rise
335 UX1=UX
340 TA2=DTA+TA1: GTD1=T1-TA2: LTD1=T2-TA1
350 IF GTD1>LTD1 THEN GTD=GTD1 ELSE GTD=LTD1
360 IF LTD1<GTD1 THEN LTD=LTD1 ELSE LTD=GTD1
370 DTM=(GTD-LTD)/(LOG(GTD/LTD))    'log mean temp deg F
380 IF CP>0 THEN Q=WT*(T1-T2)*CP
390 AX=NT*APF*L
400 FA=AX/APSF
420 AI=3.1416*((DI/2)^2)
430 GT=(144*WT*NP)/(3600*NT*AI)    'tubside mass velocity
    lb/sec/ft
440 NR=DI*GT/UVISC    'modified Reynolds number
450 REM calc of J for HI inside tube film coef
```

Figure 5.13 AirClr8 Basic language code listing.

```
460 IF NR>17 GOTO 580
470 LD=L/DI
480 J2=2.7183^(3.91+.2365*(LOG(NR))+.04706*((LOG(NR))^2))
490 J10=2.7183^(3.389+.3392*(LOG(NR))+7.910001E-04*
    ((LOG(NR))^2))
500 J20=2.7183^(3.1704+.33214*(LOG(NR))+.0012123*
    ((LOG(NR))^2))
510 J50=2.7183^(2.851+.3685*(LOG(NR))-.011588*
    ((LOG(NR))^2))
520 '
530 IF LD<=10 THEN J=J10+((10-LD)/8)*(J2-J10)
540 IF LD<=20 THEN J=J20+((20-LD)/10)*(J10-J20)
550 IF LD<=50 THEN J=J50+((50-LD)/30)*(J20-J50)
560 IF LD>50 THEN J=J50-((LD-50)/30)*(J20-J50)
570 GOTO 600
580 J=2.7183^(-3.913+3.968*(LOG(NR))-
    .5444*((LOG(NR))^2)+.04323*
((LOG(NR))^3)-.001379*((LOG(NR))^4))
590 '
600 REM calc of HT factor
610 CPUK=CP*UVISC/K
620 IF CP=0 THEN HT=1500 ELSE HT=J*K*(CPUK^(1/3))/DI
630 WA=Q/(.24*DTA)
640 REM calc of HA factor
650 GA=WA/FA     'face mass velocity lb/hr/ft2
660 HA=2.7183^(-7.1519+1.7837*(LOG(GA))-
    .076407*((LOG(GA))^2))
670 '
680 REM calc of overall transfer coefficient UX1
685 IF UX2=0 THEN UX2=UX
690 UX1=1/(((1/HT)*(AR*DO/DI)+RDT*(AR*DO/DI)+(1/HA))
693 'IF SET1=1 GOTO 803
694 'IF SET1=2 GOTO 880
700 REM calc of UX1 CORRECTED for initial assumption
705 'PRINT USING " UX1=##.###^^^^   UX2=##.###^^^^ ";UX1,UX2
706 UX2=Q/(AX*DTM)
710 IF ABS(UX1-UX2)<.01 GOTO 800
720 CLS: LOCATE 23,10: PRINT USING " UX1=##.###^^^^
    UX2=##.###^^^^ ";UX1,UX2
725 ' UX2=Q/(AX*DTM)
727 IF ABS(UX1-UX2)>1.0 THEN CONV=1.0 ELSE CONV=CONV1
730 IF UX1>UX2 THEN T2=T2-CONV: GOTO 340
740 IF UX1<UX2 THEN T2=T2+CONV: GOTO 340
745 IF SET1=1 GOTO 803
```

Figure 5.13   (Continued)

```
800 CLS:LOCATE 5,10: PRINT USING "HEAT TRANSF COEF ,
    available or actual  =
####.###";UX1
802 IF SET1=0 GOTO 805
803 UX1=Q/(AX*DTM)
804 LOCATE 7,10:  PRINT USING "HEAT TRANSF COEF , extd
    surface required = ####.###
";UX1
805 AB=3.1416*(DO/12)*NT*L
810 LOCATE 9,10:   PRINT USING "BARE TUBE AREA ft2
    = #######.#";AB
812 LOCATE 11,10:  PRINT USING "HEAT TRANSFER COEF bare
    tube       = ####.###
";Q/(AB*DTM)
815 LOCATE 13,10:  PRINT USING "NUMBER OF TUBES
    = ####### ";NT
820 LOCATE 15,10:  PRINT USING "HEAT TRANSF btu/hr
    = ##.####^^^^.";Q
825 IF SET1=1 GOTO 835
830 LOCATE 17,10:  PRINT USING "OUTLET AIR TEMP F actual
    =  ####.##";TA2
835 LOCATE 17,10:  PRINT USING "OUTLET AIR TEMP F actual
    =  ####.##";TA2
837 LOCATE 19,10:  PRINT USING "AIR MASS RATE lb/hr
    = ##.####^^^^";WA
840 LOCATE 22,10:  PRINT USING "LMTD
    = #.###^^^^";DTM
850 IF SET1=1 GOTO 870
855 IF SET1=2 GOTO 880
860 DTA=Q/(.24*WA): SET1=1: GOTO 340
870 SET1=2
880 QA=.24*WA*DTA
885 IF ABS(QA-Q)<1000 GOTO 910
886 LOCATE 21,10: PRINT USING "Q=##.####^^^^ QA=##.####^^^^
    ";Q,QA
890 IF QA<Q THEN TA2=TA2+.01: DTA=TA2-TA1: GOTO 340
900 IF QA>Q THEN TA2=TA2-.01: DTA=TA2-TA1: GOTO 340
910 LOCATE 21,10: PRINT USING  "CORRECTED PROCESS OUTLET
    TEMP deg F   =
####.##";T2
920 LOCATE 23,10: INPUT " Run another Y/N ";AA$
930 IF AA$="y" GOTO 300
940 IF AA$="Y" GOTO 300
970 END
```

Figure 5.13 (Continued)

AirClr9. This is an executable program for any air-cooler condenser. The inputted Q will be the heat duty transferred. Data inputs for condenser tube-side transport property values of viscosity, thermal conductivity, and specific heat should be determined as for two-phase flow values calculated in Chap. 6. Use the average tube-side temperature for these condensing film transport property values. Weighted average values between gas and liquid should also be determined and applied like that used in the two-phase flow equations in Chap. 6.

Make sure the heat input Q matches the temperature-enthalpy difference calculated from the inlet process temperature/enthalpy minus the outlet process temperature/enthalpy. The program calculates the outlet process temperature; therefore, Q must be adjusted accordingly. This may result in several runs being necessary to get a match of Q input vs. Q allowable per the tube area input.

This completes the discussion of air-cooler design and rating. There are countless applications for which these design tools can be used. They are a priceless treasure for the process engineer designer and plant operations troubleshooter!

Conclusion

This chapter has covered the design and rating of shell/tube exchangers and air-cooler exchangers. These process equipment exchangers are a part of every gas-oil production, petroleum refining, and chemical processing plant. Every process engineer must have a good understanding of these unit operations to be worthy of his or her position and income. This chapter addresses this need. Each user is encouraged to review the articles herein and use them as a template to resolve similar problems or design new exchangers.

The three programs discussed—STexch, AirClr8, and AirClr9—cover every exchanger application in our industry. Use them as supplied on the CD or make your own program applying the step-by-step methodologies given in this chapter. The step-by-step details given are the exact equations and procedures used in the three software programs.

The applications of this chapter are unlimited. Use these tools for designing or rating any exchanger in any application, refinery, gas plant, chemical plant, pipeline, oil-gas production facility, commercial building HVAC, residential HVAC, and so on.

References

1. Kern, Donald Q., *Process Heat Transfer,* McGraw-Hill, New York, 1950. Chaps. 1–7.
2. Kern, Equation 6.15.
3. Kern, Fig. 24.
4. Kern, Fig. 28.

5. Kern, Equation 7.5.
6. Kern, Equation 7.4.
7. Kern, Equations 7.1 and 7.2.
8. Kern, Equation 6.38.
9. Kern, Equation 6.10.
10. Kern, Table 12.
11. Kern, Equation 6.13.
12. Kern, Figs. 18–23.
13. Kern, Table 10.
14. Kern, Fig. 16.17.
15. Gas Processors Suppliers Association (GPSA), *Engineering Data Book,* 9th ed., GPSA, Tulsa, OK, 1972, Figs. 10-15–10-18.
16. Daugherty, R. L., and Ingersoll, A. C., "Bernoulli Theorem," *Fluid Mechanics,* McGraw-Hill, 1954, pp. 72–76.

6

Fluid Flow Piping Design and Rating

Nomenclature

Single-phase flow

ΔP	pipe segment pressure loss, psi
ΔP_d	static leg pressure loss, psi
ΔP_T	total pipe run pressure loss, psi
ε	pipe absolute roughness, ft
θ	time for acceleration, s
μ	fluid viscosity, cP
ρ	fluid density, lb/ft^3
A	Chen's equation constant f, $\varepsilon/(3.7D)$
B	Chen's equation constant f, $5.02/\mathrm{Re}$
C	conversion constant
C_v	constant of water flow rate in a valve at 60°F with a 1.0-psi pressure drop across the valve, gpm
d	pipe diameter, in
D	pipe internal diameter, ft
f	friction factor for internal resistance to flow in pipe, dimensionless
F_p	force of flowing fluid, ft·lb
g	acceleration of gravity, 32.2 ft/s^2
h_L	loss of static pressure head due to fluid flow, ft
h_s	total static leg rise or fall, ft
K	resistance coefficient, dimensionless
L	straight pipe length, ft

$P_{1,2}$ pressure at points 1 and 2, psig

Re   Reynolds number, dimensionless

v fluid velocity, ft/s

W fluid flow rate, lb/h

W_m mass, lb

$Z_{1,2}$ static head of fluid above reference level, ft

Two-phase flow

ΔP_A two-phase flow acceleration pressure loss, psi

ΔP_E pressure loss from elevation changes, psi

ΔP_{ell} 90° standard ell in two-phase flow, psi

ΔP_f pipe friction pressure loss, psi

λ volumetric flow variable, $Q_{LPL}/(Q_{GPL} + Q_{LPL})$, dimensionless

μ_g gas viscosity, cP

μ_l liquid viscosity, cP

ρ_g gas density at flow pressure and temperature, lb/ft^3

ρ_l liquid density at flow pressure and temperature, lb/ft^3

ϕ Flanigan variable for two-phase flow elevation gas effect

D pipe internal diameter, ft

f_o gas-phase friction factor in a pipe run

f_{TP} two-phase flow friction factor in a pipe run

g acceleration of gravity, 32.2 ft/s^2

H_T height of static leg, – for rise and + for fall, ft

L equivalent pipe length, ft

log   natural logarithm of base e, 2.7183

Q_{GPL} volume of gas flow, W_G/ρ_g, ft^3/h

Q_{LPL} volume of liquid flow, W_L/ρ_l, ft^3/h

Re   Reynolds number, dimensionless

W_G gas flow, lb/h

W_L liquid flow, lb/h

Fluid Flow Basics

The most common means of transporting fluid is the pipeline. Every pipe is a long, cylindrical, completely enclosed conduit used to transport gas, liquid, or both from point to point. Mathematical calculations are used to determine the size of pipe, the fluid transport properties, the flow characteristics, and the energy that must be applied to move the fluid. This process is notably called *fluid mechanics.* Unfortunately, only a few special problems in fluid mechanics can be entirely solved by rational, mathematical means. One of these special, solvable problems is that of *laminar flow.* Most every other problem of fluid mechanics, however, requires experimental background work for rational solutions. Many empirical formulas have been proposed for solving problems of fluid flow in pipe; these are often limited and can be applied only when the conditions of the problem closely approach the conditions of the experiments from which they are derived.

In the earlier days of the petroleum age, many pipe experiments were conducted. In the quest for the magic formula, one was found to be the closest to utopia even to this day, called the Darcy formula. The Darcy formula is derived manually from the Bernoulli principle, which simply describes the energy balance between two points of a fluid flowing in a pipe. This energy equation is also applicable to a static condition of no flow between the two points. The classic Bernoulli energy equation [1] is:

$$Z_1 + \frac{144\,P_1}{\rho} + \frac{v^2}{2g} = Z_2 + \frac{144\,P_2}{\rho} + \frac{v^2}{2g} + h_L \qquad (6.1)$$

where $Z_{1,2}$ = static head of fluid above reference level, ft
$\quad P_{1,2}$ = pressure at points 1 and 2 psig
$\quad \rho$ = weight density of fluid, lb/ft^3
$\quad v$ = mean velocity of flow in pipe, ft/s
$\quad g$ = acceleration of gravity, 32.2 ft/s^2
$\quad h_L$ = loss of static pressure head due to fluid flow, ft

All of these variables can be determined with 100% accuracy by mathematical equations, with one exception, h_L. Why? You have found the reason that experimental data must be established to solve for h_L: no one math formula has ever been developed to solve every case of h_L. How smooth is the pipe's internal surface? What is the flow-resisting turbulence due to such roughness? These are questions only experimental data can answer, resolving the true value of h_L.

The h_L variable is named the *friction factor f.* The friction factor is dimensionless and is nomenclated throughout this chapter and in most every other publication the world over as the symbol f. Many mathematical formulas have been derived by experimental data to calculate f, with only two or three formulas having acceptable results.

As of the time of writing, the quest to find this utopian formula for f is still ongoing. All eager young persons full of zest, integrity, and competency (don't leave out wisdom) are encouraged to join this quest.

There are three other factors in the Bernoulli equation:

1. Z, the static head energy force
2. $144\, P_1/\rho$, the pressure head [equivalent to DPAT of Eq. (5.52)]
3. $v^2/2g$, the velocity head equivalent to $[\text{ACFM}/(4009 * \pi * D_{\text{fan}}^2/4)]^2$ of Eq. (5.52)

The Z factor is simply the static head due to elevation height above a referenced elevation point. The $144\, P_1/\rho$ factor is the pressure factor noted as the pressure head. Consider again Eq. (5.52) from Chap. 5. This equation is the basic form of the Bernoulli energy balance equation.

Note that the static head force Z is not included in Eq. (5.52) because, as explained in Chap. 5, the static head of air through the air cooler is negligibly small. DPAT in Eq. (5.52) is the energy of pressure difference of the airflow inlet minus the air pressure outlet. This DPAT factor includes the h_L factor in part of the Bernoulli equation.

The DPAT variable is found by applying Eq. (5.48), the static pressure loss per tube row, a key equation that is to be experimentally determined. The last term, $v^2/2g$, is the velocity head energy of the fan required to move the required airflow through the air cooler. As the air is heated while passing through the fin tubes, it expands, causing an increase in velocity. The air velocity difference, $(v_2 - v_1)^2$, is the velocity head in the Bernoulli equation.

The f factor is seen as the success or failure of fluid flow systems design. It requires experimental input, such as Eq. (5.48), provided for the DPAT term. Equations (5.48), (5.50), and (5.52) are presented again for this review of the Bernoulli equation.

$$D_{PA} = \exp\,(1.82 * \log G_A - 16.58) \qquad (5.48)$$

where D_{PA} = static pressure loss per row of tube crossflow, in of water
$\log$ = natural logarithm of base e

Equation (5.48) is an experimentally determined equation that includes a value for f. The f factor or equivalent D_{PA} factor here accounts for the friction loss of airflow through the fin tubes in crossflow. This equation is emperical and is limited to the specific experimental conditions for which it is derived. It is reasonable to use Eq. (5.48) for most air-cooler design and rating applications. For other applications, however, such as airflow in an HVAC duct, it is not feasible or applicable. Another experimental equation is required for the HVAC-specific applications.

Hydraulic grade line

The term $144\, P_1/\rho$ is called the *hydraulic grade line* [2]. In a fluid flowing in a horizontal straight pipe, if the pressure is measured at both ends, P_1 and P_2, the values will be different, with P_1 being the greater. Why? The friction factor f is the reason. The hydraulic gradient includes the friction factor. How much pressure loss there is between these two pressure points depends on the f value of the particular pipe. Just as in the case described for the air cooler, each row of tubes the air flows through exerts a certain resistance to flow, as calculated in Eq. (5.48) in deriving D_{PA}. Equation (5.48) has been derived experimentally, with data from numerous air-cooler exchangers.

Energy grade line

The last term of the Bernoulli equation is $v^2/2g$, called the *energy grade line* or the *velocity head* of fluid flow. This energy velocity head does not change, assuming a noncompressible fluid. This means that the starting velocity of fluid in the straight horizontal pipe is equal to the ending velocity.

Bernoulli equation terms

Figure 6.1 displays a simple flow of fluid in a pipe with all three Bernoulli terms plus the friction factor h_L. Please note that here the variable h_L represents the friction factor, which is later designated f. Why? Here h_L represents more than f; h_L also represents static head and energy grade. Several important concepts are listed:

- First, observe that the $v^2/2g$ value doesn't change. The energy grade line retains the same energy value from the start to finish. Why? Consider the fact that the velocity of fluid entering the pipe is the same as that exiting. This includes the assumption of a noncompressible fluid. Thus, the horsepower that is required to maintain this velocity of fluid, beginning at 0 velocity, remains constant throughout the pipe. This term may be expressed as foot-pounds per pound of flowing fluid, *horsepower*. Please note that it may also be expressed as feet of fluid, such as for a pump head in feet of fluid pumped.

- Second, the friction loss h_L from pipe entrance to exit may also be expressed as foot-pounds per pound of flowing fluid. This term is more commonly referred to as the *head loss* in feet of fluid. It is loss, because the hydraulic grade line shown in Fig. 6.1 exhibits the h_L loss.

- If the pipe friction were absent, the hydraulic grade line, $144\,P_2/\rho$, would be equal in and out, just as the energy grade line is equal in and out. The hydraulic grade line does, however, show the h_L loss. This grade line may also be expressed in foot-pounds per pound of flowing fluid. It is an energy term, as the other terms are energy expressions. Thus, expression in feet of fluid is equally common, as in the other Bernoulli terms.

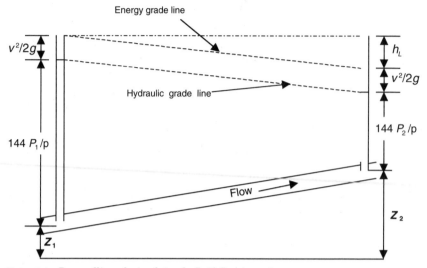

Figure 6.1   Bernoulli analysis of simple fluid flow in a pipe.

- The Z factor in Fig. 6.1 is simply the change of elevation between two points, in feet. Z may also be expressed as an energy term as foot-pounds per pound of mass. Note that here it is pounds of mass rather than of moving fluid, since this is a static measure of force.

Reviewing again the fan-sizing problem of Chap. 5, the following equivalence of the terms in Eq. (5.52) is noted:

$$\frac{144\,(P_1 - P_2)}{\rho} = \text{DPAT}$$

$$\frac{v^2}{2g} = \left(\frac{\text{ACFM}}{4009 * \pi * D_{\text{fan}}^2/4}\right)^2$$

If we regard the fan as a liquid fluid pump, we have the same relationship—the pump head output must equal the forces resisting flow, $144\,(P_1 - P_2)/\rho$, and the mass movement gain, $v^2/2g$. Here we have assumed $v_1 = 0$, wherein all movement of air comes from the fan output, per Eq. (5.52).

The D_{PA} variable was derived in Example 5.2, utilizing the experimental equation Eq. (5.48). The following steps were then made:

$$T_{\text{avg}} = \frac{\text{DTA}}{2} + t_1 \tag{5.49}$$

where T_{avg} = air-side average temperature
t_1 = ambient air temperature, °F

$$\text{DPAT} = N_{\text{ROWS}} * \frac{D_{PA}}{\text{DR}} \qquad (5.50)$$

where DR = ratio of actual dry air density at average temperature
and elevation to dry air density at 70°F and 0 elevation
(interpolated from Table 5.5)

$$P_{\text{force}} = \text{DPAT} + \left(\frac{\text{ACFM}}{4009 * \pi * D_{\text{fan}}^2/4} \right)^2 \qquad (5.52)$$

where P_{force} = pressure head + velocity head, $v^2/2g$, in of water

It was necessary to provide an experimental relationship to friction losses and mass flow in order to calculate the total head in inches of water or foot-pounds force per pound of fluid flow before the problem could be solved. This was accomplished by applying Eq. (5.48).

The same is true for fluid flow in piping. Experimental data must first be established to solve the piping fluid flow problems, specifically the f factor, also called the h_L factor in the Fig. 6.1 analogy. Please note also that f is dimensionless. The f factor is used later in this chapter. The h_L factor is in units of feet; it is a force, requiring a workload, which is the force of friction resisting fluid flow. This friction of resistance is on the internal wall surface of the pipe.

One should note that Fig. 6.1 displays how any pipe experiment can be set up to find this unknown f factor. The h_L value is experimentally determined by taking readings of v and P. The P value is directly read as shown, by reading the liquid column heights. The velocity, however, must be read by adding a venturi tube or orifice flange to Fig. 6.1 for the velocity reading.

Single-Phase Flow Equations

Friction factor f

Having introduced f, the objective is now to find a convenient analogy with which we may calculate any f factor. But first, how is this factor to be applied? Why not just use the simple term h_L and be done? In answer, the h_L factor, if reviewed in Fig. 6.1, is actually a part of the hydraulic grade line factor 144 P_1/ρ. Thus, the correct thing to do here is to put it in this factor. In all equation relationships of fluid flow problem solving, this is done. This simply says that f represents the flow of fluid losses due to friction inside pipe in all equations used.

In order to calculate f, certain dynamic forces of mass flow must first be determined. Knowing these forces resisting flow are along the pipe wall, the work of Osborne Reynolds has shown that certain fluid transport properties compose this friction force. Reynolds proved with numerous tests that these properties are identified as pipe diameter, fluid density, fluid viscosity, and fluid velocity in the pipe. These properties are reasonably constant, considering the fact that the velocity stays constant, the fluid temperature remains constant in flow, and the fluid density remains constant also, or at least the density remains within a negligible variance. Reynolds observed these properties with numerous pipe flow tests and developed what is used even to this day, a dimensionless dynamic force ratio called the *Reynolds number,* Re. The Reynolds number is used to find *f* by having varied pipe sizes and Re values plotted against the *f* factor.

The Reynolds number is calculated as follows [3]:

$$\text{Re} = 123.9 \, \frac{dv\rho}{\mu} \tag{6.2}$$

where d = pipe diameter, in
v = velocity, ft/s
ρ = density, lb/ft^3
μ = viscosity, cP

Several equations have been developed using this Reynolds calculation that have proven accurate for calculating the *f* factor. The first equation proven was the Colebrook equation [4] (also called the Colebrook-White equation):

$$\frac{1}{f^{1/2}} = -2 \log_{10} \left(\frac{\varepsilon}{3.7D} + \frac{2.51}{\text{Re} \, f^{1/2}} \right) \tag{6.3}$$

where D = pipe internal diameter, ft
ε = pipe absolute roughness, ft
$\log_{10}$ = logarithm of base 10

Note that ε, commonly referred to as the *absolute roughness factor,* is in feet of pipe, and ranges from 0.00015 for new smooth pipe to 0.001 for rough old pipe. This factor may also be 0.01 to 0.05 for severely corroded and scaled pipe.

Professor L. F. Moody took the Reynolds work and, combining it with the Colebrook equation (6.3), made a series of curves by plotting constant curve values of ε/D on a plot of *f* vs. Re [5]. The resulting curves perfectly matched *f* experimentally. Both the curves and the Colebrook equation revealed excellent findings for *f*.

The problem regarding the Colebrook equation, however, is the fact that it must be solved by trial and error. Thus, a computer program could easily solve it, such as by applying the Newton convergence method of trial and error. Later in time, however, other well-founded equations were developed to calculate f. The most well-received equation and perhaps the best to date is Eq. (6.4), the Chen equation [6]. For over 17 years, Chen's equation has run the checkout course with excellent findings and wide acceptance. It has a proven record of excellence.

$$\frac{1}{f^{1/2}} = -2 \log \left(A - B \log \left\{ A - B \log \left[A - B \log \left(A - B \log A + \frac{14}{\text{Re}} \right) \right] \right\} \right)$$

(6.4)

where $A = \varepsilon/(3.7\ D)$
   $B = 5.02/\text{Re}$
   $\log$ = natural logarithm of base e

The derivation of the dimensional analysis of fluid flow can be made exact for all but f, but f is a must for experimental values. Please note, for example, that Eq. (6.3) and (6.4) both require that an ε value be used. This value alone is assumed or found experimentally. The ε value normally ranges from 0.00015 for smooth pipe to 0.003 for severely scaled, rough pipe. In recent years we have obtained numerous well-founded and reliable computer programs that calculate the absolute roughness factor ε. One historic finding revealed that an ε factor of 0.04 was required to simulate the actual pressure drop experienced in a certain refinery's cooling tower circulation system piping. Why so high? The reason was that tons of polymer reaction catalyst was accidentally being dumped into the cooling tower system via an improper pipeline connection. The interesting thing here is that we could actually simulate the degree of fouling that took place by using this ε value. Notably, the f factor was calculated from Eq. (6.3) using this ε factor input. Random pressure readings throughout the cooling water system subsequently proved this f calculation to be very accurate.

It is important to realize that Eqs. (6.3) and (6.4) *do not* give a full coverage of the f factor. Why, and what is missing? As to why, note that the Reynolds number is directly proportional to the fluid velocity in the pipe. Therefore, neither of these equations is valid for Reynolds numbers below 4000. In fact, many users of recognition, such as many engineering, procurement, and construction (EPC) companies, note that these equations are invalid for Reynolds numbers below 10,000. How can this problem of finding an accurate f factor be solved? You have now entered into what is called the *laminar flow* regime of fluid flow in pipe. This is where the fluid is moving very slowly, perhaps less than 1.0 ft/s, and the fluid in the center of the pipe is moving faster than the fluid nearer the pipe wall. Equations (6.3) and (6.4) rely on the fluid flow inside the pipe being fully turbulent, or the Reynolds number being 4000 or greater. In principle, turbulent flow is homogeneous, meaning it has the same absolute fluid velocity throughout any cross section of the internal pipe. If the Reynolds number is 2000 or less, the following equation is found to be very reliable for calculating f [7]:

$$f = \frac{64}{\text{Re}} \tag{6.5}$$

f Equation Selection

- Use Eqs. (6.3) or (6.4) for Reynolds numbers greater than 10,000.
- Use Eq. (6.5) for Reynolds numbers equal to or less than 2000.

For Reynolds numbers between these two limits, in the transition zone of 2000 to 10,000, use the more conservative Eq. (6.3) or Eq. (6.5) for the case in question. For example, if the particular case has a limiting pressure drop, and the pressure drop in the pipe is critical (not too large), then consider sizing Eqs. (6.3) and (6.4) for f, as these would calculate larger pipe sizes. Otherwise, use Eq. (6.5).

Fluid flow general equation

Having now established the friction factor f for the general pressure drop equation, the dimensional analysis equation for pipe pressure drop can be derived. A simple equation is now derived from the earlier Bernoulli equation:

$$h_L = \frac{fLv^2}{D2g} \tag{6.6}$$

where h_L = loss of pressure, ft of fluid
$\quad\quad f$ = friction factor
$\quad\quad L$ = straight pipe length, ft
$\quad\quad v$ = fluid velocity, ft/s
$\quad\quad D$ = pipe internal diameter, ft
$\quad\quad g$ = acceleration of gravity, 32.2 ft/s^2

Equation (6.6) is notably named the *Darcy equation* [8]. It is applicable to both laminar flow, Re $\leq$ 2,000, and to turbulent flow, Re $\geq$ 10,000. Two restrictions govern this equation, however:

1. The flowing fluid maintains one constant density, ρ.

2. Only one flowing fluid phase exists. The one phase may be liquid or gas. If liquid, there is no concern for compressibility, as liquid is non-compressible. If gas, however, the fluid is compressible; therefore,

the density factor ρ is constantly changing with the changing h_L increase along the pipe run.

A single-phase flowing fluid must consistently be in one phase throughout the pipe system. The single-phase state is easily simulated, since most cases involve small transport property changes that are nonsignificant. Transport properties are simply fluid density and viscosity. A single-phase gas-flow condition has little change in viscosity, but a very significant variance in density. As the gas flows in a piping system, pressure drop due to fittings and friction losses are significant enough to appreciably decrease the gas pressure. A single-phase flowing liquid, in contrast, does not change in density with decreasing pressure. Nor does it vary in viscosity because of flow conditions in the pipe.

A two-phase flowing fluid is composed of both liquid and gas, generally in an equilibrium state; when the pressure decreases because of flow through the pipe, more of the liquid will flash to vapor. This will lower the temperature, since the removal of latent heat energy taken by the flashed vapor cools the system.

At this point, a review is in order. Please note again Eq. (6.6). The Darcy equation, also known as the Fanning equation, is applicable to any single-phase flowing fluid, liquid or gas. There are seven factors that make up the dimensional analysis of this equation:

1. The h_L factor is an unknown variable to be calculated; it is the pressure loss due to the friction in the pipe. It may be expressed as feet of fluid or pounds per square inch (psi). If expressed as psi, the following equation is used, factoring the density factor ρ:

$$\Delta P = \frac{\rho h_L}{144} \tag{6.7}$$

where ΔP = pressure loss, psi
ρ = fluid density, lb/ft^3

2. The ρ factor is the density of the fluid in pounds per cubic foot (lb/ft^3). It may represent a gas or liquid. If gas, the pressure loss should not be greater than 10% of the pipe run inlet pressure.

3. The f factor is the friction loss factor in any pipe run. This factor is dimensionless and may be calculated from Eq. (6.3), (6.4), or (6.5):

   - If Re > 10,000, use Eq. (6.3) or (6.4).
   - If Re < 2,000, use Eq. (6.5).
   - If 10,000 > Re > 2000, use the most conservative equation for the specific case.

4. The L variable is the straight length of pipe, in feet, to which Eq. (6.6) applies. The L is simply a straight run of pipe having no bends, fittings, valves, or pipe size changes. However, L may also represent the equivalent length of pipe replacing fittings such as elbows, tees, and valves. Some reliable books and reference materials reference these pipe-fitting pressure losses as pipe equivalent length L. Each of these fittings must therefore be added, including the actual straight length of pipe. More coverage of valves and fittings is given in the following section.

5. The velocity v, ft/s, is a variable calculated from Eq. (6.8):

$$v = \frac{0.0509W}{\rho d^2} \qquad (6.8)$$

where v = fluid velocity, ft/s
   W = fluid flow rate, lb/h
   ρ = fluid density, lb/ft^3
   d = pipe internal diameter, in

Using W, lb/h, for fluid flow allows usage of this same equation for gas or liquid; the density term ρ also applies to either. The velocity variable and density transport property factor must stay reasonably near a constant value for the Darcy equation to be valid. For a liquid fluid, ρ and v remain nearly constant. Thus, most any pipeline length may be calculated using a single application of Eq. (6.6).

In conditions where the density of the fluid may vary, such as a gas flowing as a single-phase fluid, computer programs may be applied whereby the pipe lengths L are segmented into short lengths and integrally calculated as straight sections of pipe. These short sections each receive density reduction due to the drop in pressure. Of course, the accompanying fluid velocity is calculated by applying Eq. (6.8).

6. The variable D is the pipe internal diameter in feet. This variable sets the internal surface area for friction resistance to flow and the internal cross-section area flow parameters. A common usage of D is making the equivalent pipe length to diameter ratio L/D for valves and fittings:

$$K = \frac{fL}{D} \qquad (6.9)$$

where K = resistance coefficient, dimensionless

Please note that Eq. (6.9) is actually a part of the Darcy equation, Eq. (6.6). Adding the velocity head gives the following Darcy equation:

$$h_L = \frac{fL}{D} \frac{v^2}{2g} \qquad (6.10)$$

or Eq. (6.6). The term fL/D is noted as the resistance coefficient K. The f variable, friction factor, is the same f as calculated for Eq. (6.6). Each fitting and valve will have a separate and independently calculated K factor. These K factors are to be added, and then make up the entire pressure loss of a pipe run, including all fittings and valves. Thus, an equation such as Eq. (6.11) is the result:

$$h_L = \left(K_1 + K_2 + K_3 + K_4 + \frac{fL}{D} \right) \frac{v^2}{2g} \qquad (6.11)$$

There are four fitting K factors in Eq. (6.11). Each of these factors represents a specific valve or pipe-fitting pressure head loss, fL/D. Notice that this term is not a K term, but rather represents L, the actual straight length of pipe. The reason it is not a K term is that it represents a straight section of pipe. The f factor in Eq. (6.11), including the f factor in each of the K terms, is calculated using Eqs. (6.3), (6.4), or (6.5). Derivation of the K resistance coefficients is reviewed in the next section.

7. The last term, g, is a constant representing the earth's gravitational force, 32.2 ft/s^2. On earth there is always a gravitational force involved, which is derived from the following equation:

$$\text{force} = \text{mass} \times \text{acceleration} = W_m L / T^2 = W_m g$$

In our space calculations, g is removed, and Eq. (6.10) becomes:

$$h_L = \frac{fL}{D} \frac{v^2 \rho F_p \theta^2}{D W_m L C} \qquad (6.12)$$

where W_m = mass, lb
θ = time for acceleration, s
v = velocity of fluid in pipe, ft/s
ρ = fluid density, lb/ft^3
F_p = force of flowing fluid, ft·lb
C = conversion constant

In space, NASA calculates fluid flow by applying equations such as Eq. (6.12). This equation is missing the constant g since Earth's gravity force is missing in space [9].

TABLE 6.1

Nominal pipe size, in	0.5	1	1.5	2	4	6	8	10	12	16	18	30
f_t	0.027	0.023	0.021	0.019	0.017	0.015	0.014	0.014	0.013	0.013	0.012	0.011

Valve and fitting pressure losses

Consider again Eq. (6.11). Each valve and pipe fitting is to receive an independent pipe size resistance coefficient K. A number of works have been published that derive these coefficients for every type and size of pipe fitting and valve. Proposed here is a new variable we shall call f_t. Referring to the *Crane Technical Paper No. 410* [10], these f_t variables are given in Table 6.1.

The following K value equations are listed for various pipe fittings and valves. All use the f_t factor of Table 6.1.

Fitting K value listings.   See Fig. 6.2.

1. Sudden pipe contraction for angle $\theta \leq 45°$:

$$K = \frac{0.8[\sin{(\theta/2)}][1 - (d_1/d_2)^2]}{(d_1/d_2)^4}$$

where d_1 = smaller diameter
d_2 = larger diameter

2. Sudden pipe enlargement for angle $\theta \leq 45°$:

$$K = \frac{2.6[\sin{(\theta/2)}][1 - (d_1/d_2)^2]^2}{(d_1/d_2)^4}$$

3. Pipe entrance:

Inward projection, sharp edge entrance $K = 0.78$
Rounded smooth entrance $K = 0.5$

4. Pipe exit:

$$K = 1.0$$

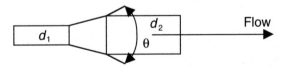

Figure 6.2 Diagram for valve and fitting K calculations.

5. Gate valves for equal inlet and outlet diameters:

$$K = 8 f_t$$

6. Check valves:

Screwed pipe with wye-type body $K = 100 f_t$
Flanged pipe with tee-type body $K = 50 f_t$

7. Ells:

Standard 90° ell $K = 30 f_t$
45° ell $K = 16 f_t$
Short radius 90° ell $K = 42 f_t$

8. Standard tees:

Flow-thru run $K = 20 f_t$;
Flow-thru branch $K = 60 f_t$

9. 180° "U" bend:

$$K = 50 f_t$$

10. Control valves. Valve vendors publish much information about flow-control valves. Much of their information covers what is known as control valve flow coefficients C_v. It is convenient to express the valve capacity and the valve flow characteristics in terms of the flow coefficient C_v. This value is defined as the flow of water at 60°F, in gallons per minute (gpm), at a pressure drop of 1.0 psi across the valve. In a flow valve, therefore, controlling the flow of water at 60°F as the valve opens, keeping a 1.0-psi pressure drop across the valve, the C_v value is the rate of water flow, gpm.

This is a simple way to define control valve characteristics, including their capacity. Using this relationship to the K value derivation herein, the following equations derived by numerous vendors may be applied.

For liquids or gases:

$$C_v = 0.3103 \, d^2 v \left(\frac{\rho}{\Delta P} \right)^{1/2}$$

Please note the C_v value used should be at a midrange of the valve's opening. The vendor's maximum C_v value should never be used for controlling a fluid flow in a piping system. Rather, a midrange C_v valve opening should be used to size the valve. Good design practice is to allow 15 to 25% of the piping run pressure drop to be across the control valve. This percentage is to include inline equipment such as heat exchangers, filters, strainers, and other such equipment in the pipe run. Piping runs should start and terminate at major equipment items, such as pressure flash vessels, fractionation vessels, surge vessels, reactors, and pumps.

11. Gate valves: Wedge-type, disc-type, and plug-type gate valves are common in all process plants, in which each performs with near-equal flow resistance. Therefore each type of gate valve may be represented by:

$$K = 8f_t$$

Conclusions on pipe-fitting equations. Each of the f_t values given in Table 6.1 is the experimental friction factor for that pipe size. This friction factor f_t is not to be mistaken for the friction factor for straight pipe f. They are different discrete values to be applied per Eq. (6.11). K_1, K_2, K_3, and K_4 each represent a different pipe valve or fitting. One could also be a control valve, item 10 in the pipe fitting list. The objective here is to add up all of the fittings expressed as K values and execute Eq. (6.11) to solve the pipe run pressure drop. In summary, this tells how much head h_L the pump must put out in order to do the job.

Conclusions on single-phase flow equations

There are more useful Darcy equation forms than Eq. (6.10). Equation (6.13) is proposed, a more user-friendly equation that calculates the pressure loss ΔP in units of psi.

Factor ΔP with Eqs. (6.10) and (6.7) to get Eq. (6.13):

$$h_L = \frac{fL}{D}\frac{v^2}{2g} \qquad (6.10)$$

$$\Delta P = \frac{\rho h_L}{144} \qquad (6.7)$$

$$\Delta P = 0.0001078\,\frac{fL}{D}\,\rho v^2 \qquad (6.13)$$

The 0.0001078 constant comes from the reciprocal of $1/(144 \times 2g)$. Equation (6.11) now becomes:

$$\Delta P = 0.0001078\left(K_1 + K_2 + K_3 + K_4 + \frac{fL}{D}\right)\rho v^2 \qquad (6.14)$$

What about static leg? It must be added to Eq. (6.14) ΔP. If the flowing fluid elevation rises, then the ΔP of static leg ΔP_s will be a negative number to be subtracted from ΔP. On the other hand, if the flowing fluid has a falling elevation sum, pipe start to pipe termination, ΔP_s will be a positive addition to ΔP. Equation (6.7) is to be applied again, but here with slightly different nomenclature, making Eq. (6.15):

$$\Delta P_s = \frac{\rho h_s}{144} \qquad (6.15)$$

where ΔP_s = static leg ΔP, + for fall, – for rise, psi
h_s = static leg summary rise or fall, ft

The sum of velocity head valves and fittings, and the frictional pipe run loss with the static head gain or loss, make up the entire pipe pressure loss:

$$\Delta P_T = \Delta P + \Delta P_s \qquad (6.16)$$

where ΔP_T = total pipe run pressure loss, psi

Equations (6.1) to (6.16) summarize all single-phase flow solutions for sizing piping with fitting and valve selections. The next challenge is practical application of the equations. Example 6.1 lists 9 steps for solving any single-phase flow problem. This is perhaps the only and best way to resolve the correct methodology for solving most any single-phase fluid flow problem. Each of the following steps flags certain restraints of fluid flow that may apply.

Example 6.1   A designer is to determine the normal and good-practice piping run sizing in a liquified petroleum gas (LPG) recovery plant. A liquids pump discharge will flow the pipe run. A control valve C_v size is also to be determined. The designer has made a field survey showing the following data requirements:

- 35,000 lb/h of liquid is to be pumped to a stabilizer shell/tube heater and then to a fractionation stabilizer's tray with a 16-inlet feed and a 3-in nozzle. A control valve downstream of the shell/tube exchanger is to be installed.
- The liquid density is 42 lb/ft^3 with 1.5 cP viscosity in the exchanger upstream pipe run, and 38 lb/ft^3 with 0.5 cP viscosity downstream.
- The line length from the pump discharge to the exchanger must have eight 90° ells, one check valve, one branched tee, three gate valves, and 150 ft of straight pipe sections.
- After running an exchanger program, FFLW.exe (supplied with this book), the designer has determined that the inline exchanger has an 8-psi pressure loss (the tube side) for the 35,000 lb/h flow rate.

- The pipe run downstream of the heat exchanger will have six 90° ells, one gate valve, and 125 ft of straight pipe sections.
- The pipe run downstream of the exchanger will also have a control valve to throttle the line for good control, allowing any variances of ±20% mass flow rate.
- The pipe run is to terminate with 145 psig pressure at the fractionator inlet nozzle. Here the inlet nozzle is 45 ft above the pump discharge nozzle.

solution   This problem is a common one and one in which most of the equations emphasized will be put into practice.

1. First, divide the total pipe run into two parts, one upstream of the heat exchanger and one downstream. Take an overview of both pipe runs and summarize a general system pressure profile.
     Static leg:

$$\Delta P_s = \frac{\rho h_s}{144} = \frac{42 \times 45}{144} = 13.1 \text{ psi} \qquad (6.15)$$

Exchanger ΔP	8.0 psi
Fitting and line loss (a guess)	3.0 psi
Control valve at 20% total line loss	$24.1 \times 0.2 = 5.0$ psi
Total	29 psi

Please first note that these calculated answers have been rounded off conservatively to the higher values. The reason is that greater acuracy is not justified. The control valve pressure drop selected should be within a tolerance of ±10% of the total system pressure loss the control valve governs.

Second, please note that the control valve's required pressure loss is to be governed by a 20% factor of the overall two-part pipe run pressure drop. This is good and accepted practice. The total shown is our first guess of the pump discharge pressure. Of course, the positive remaining pump suction pressure must also be added to get the pump discharge pressure in psig.

2. Now, to get a good feel for what size pipe is required, run a calculation of the velocity equation, Eq. 6.8:

$$v = 0.0509 \frac{W}{\rho d^2} \qquad (6.8)$$

$$= 2.6 \text{ ft/s} \quad \text{for 4-in standard pipe}$$

$$= 9.33 \text{ ft/s} \quad \text{for 2-in standard pipe}$$

A 2-in size is selected, since good design practice is to *allow up to 12 ft/s before erosion limits and above 3 ft/s for economics.*

3. Sum the K values. Using Table 6.1, find the f_t value of 2-in pipe, which is 0.019. *Note that Table 6.1 f_t values are intended for all pipe fittings and valves.*

First-section pipe run K for ells $\quad\quad\quad K\,30\,f_t = 8 \times 30 \times 0.019 = 4.56$
First-section pipe run K for check valve $\quad\quad 50\,f_t = 0.95$
First-section pipe run K for tee $\quad\quad\quad\quad 60\,f_t = 1.14$
First-section pipe run K for gate valves $\quad\quad K\,8\,f_t = 3 \times 8 \times 0.019 = 0.46$
Pipe entrance loss K value for heat exchanger $\quad 1.0 = 1.00$

$$K \text{ sum} = 8.11$$

4. Calculate the friction factor f for the 2-in pipe run from Eqs. (6.2) and (6.3).

$$\text{Re} = 123.9\,\frac{dv\rho}{\mu} \tag{6.2}$$

$$= 71{,}440$$

where d = pipe diameter 2.06 in
$\quad\quad v$ = velocity, 9.33 ft/s (from Step 2)
$\quad\quad \rho$ = density, 45 lb/ft^3
$\quad\quad \mu$ = viscosity, 1.5 cP

$$\frac{1}{f^{1/2}} = -2\,\log\left(\frac{\varepsilon}{3.7D} + \frac{2.51}{\text{Re}\,f^{1/2}}\right) \tag{6.3}$$

where D = pipe internal diameter, (2.06 in)/(12 in/ft) = 0.1717 ft
$\quad\quad \varepsilon$ = pipe absolute roughness, 0.00015 ft
$\quad\quad \text{Re}$ = Reynolds number, 71,440

Please note that the second part of Eq. (6.3), after the plus sign, is a relatively small value and may be assumed to be 0 to get a quick first estimate of the f value:

$$f = \frac{1}{-2\,\log\,(\varepsilon/3.7D)^2} = 0.019$$

Now add the second part to Eq. (6.3), substituting f as 0.019 in the second half:

$$\frac{1}{f^{1/2}} = -2\,\log\left(\frac{\varepsilon}{3.7D} + \frac{2.51}{\text{Re}\,f^{1/2}}\right) = 7.2538 + 0.0003$$

$$f = \left(\frac{1}{7.254}\right)^2 = 0.019$$

Thus, the first estimate was the final f value calculated, a very good first estimate!

5. The total first section of our pipe run pressure loss problem may now be solved, applying Eq. (6.14):

$$\Delta P = 0.0001078\left(K_1 + K_2 + K_3 + K_4 + \frac{fL}{D}\right)\rho v^2 \tag{6.14}$$

$$= 0.0001078 \left(8.11 + \frac{0.019 \times 150}{0.1717}\right) [45 \times (9.33)^2] = 10.43 \text{ psi}$$

6. Begin the second part of this pipe run from the exchanger to the stabilizer fractionator column.

$$v = 0.0509 \frac{W}{\rho d^2}$$

$$= 11.05 \text{ ft/s} \quad \text{for 2-in pipe} \qquad (6.8)$$

$$\text{Re} = 123.9 \frac{dv\rho}{\mu} \qquad (6.2)$$

$$= 214{,}300 \text{ (last two places rounded off)}$$

where d = pipe diameter, 2.06 in
  v = velocity, 11.05 ft/s
  ρ = density, 38 lb/ft^3
  μ = viscosity, 0.5 cP

$$\frac{1}{f^{1/2}} = -2 \log\left(\frac{\varepsilon}{3.7D} + \frac{2.51}{\text{Re } f^{1/2}}\right) \qquad (6.3)$$

$$= -2 \log \frac{\varepsilon}{3.7D}$$

$$f = 0.019 \quad \text{[second part of Eq. (6.3) not significant]}$$

where D = pipe internal diameter, (2.06 in)/(12 in/ft) = 0.1717 ft
  ε = pipe absolute roughness, 0.00015 ft
  Re = Reynolds number, 214,300

7. The next step is to calculate the pressure loss to the control valve entrance. The designer has located the flow control valve as close as practical to the stabilizer feed entrance. This is a good design location for the control valve. Why? The pressure drop across the control valve results in a two-phase flow, vapor and liquid flowing into the stabilizer feed tray. Thus, the downstream flow is a two-phase flow, and this problem will be finished later, in the section on two-phase flow.

In this step, however, the pressure loss to the control valve is calculated, using Eq. (6.14). For six 90° ells and one gate valve, $K = 3.57$

$$\Delta P = 0.0001078 \left(3.57 + \frac{0.019 \times 125}{0.1717}\right) [38 \times (11.05)^2] = 8.7 \text{ psi}$$

8. Add the static leg pressure drop, Eq. (6.15)

$$\Delta P_s = \frac{\rho h_s}{144} \tag{6.15}$$

$$= \frac{38 \times 45}{144} = 11.88 \text{ psi}$$

where ΔP_s = static leg ΔP, + for fall, – for rise, psi
h_s = static leg summary rise or fall, ft

9. The last step in solving this single-phase part of the problem is the control valve pressure loss. As stated earlier, the pressure loss across the valve will produce two-phase flow in the pipe run from the flow control valve to the stabilizer inlet nozzle. The valve itself, however, will be sized on a single-phase-flow basis. Why? The reason is that the fluid will flash *downstream* of the valve, in however many microseconds. Thus, the design for a single-phase flow control valve will be in accordance with the K equation for the C_v value. The objective here is to determine the correct valve size.

$$C_v = 0.3103 \, d^2 v \, \frac{\rho}{\Delta P}$$

Most every flow control valve is sized 1 to 2 pipe sizes less than the pipe run the valve serves, because the valve restriction opening is considerably smaller than the connected pipe size. We size flow control on the basis of C_v (see Table 6.2). Most every valve vendor publishes C_v values for each valve they make. For good control tolerance choose a valve with a maximum C_v value at least 1.3 times greater than the calculated maximum flow C_v value. This means the designer should try 3 or more C_v values before setting a recommended size for the control valve.

In selecting the control valve size required, first determine the K value required from Eq. (6.14). Since only one K value is involved here, and there is no straight pipe section and we want a 5-psi ΔP, this equation becomes:

$$K = \frac{\Delta P}{\rho v^2} = \frac{5}{38 v^2}$$

There is one more factor, v, necessary to calculate K. Reviewing the C_v equation, and the velocity equation following, reveals that C_v is directly proportional to the velocity v.

For liquid flow:

$$C_v = 0.3103 d^2 v \left(\frac{\rho}{\Delta P} \right)^{1/2}$$

$$v = \frac{0.0509 W}{\rho d^2} \tag{6.8}$$

$$= 11.05 \text{ ft/s} \quad \text{for 2-in pipe}$$

$$= 18.09 \text{ ft/s} \quad \text{for 1.5-in pipe}$$

First try a 1.5-in valve and take a 30% velocity increase to get the C_v value in a good control point (30% less than the maximum C_v value of the valve).

$$C_v = 0.3103 \ d^2 v \left(\frac{\rho}{\Delta P} \right)^{1/2} = 52.2 \qquad \text{(too high for 1.5-in valve size)}$$

where
$$\begin{aligned} v &= 1.3 \times 18.09 = 23.52 \text{ ft/s} \\ d &= 1.61 \text{ in} \\ \rho &= 38 \text{ lb/ft}^3 \\ \Delta P &= 5 \text{ psi} \end{aligned}$$

Next try a 2-in valve size with $1.3 \times 11.05 \text{ ft/s} = 14.37 \text{ ft/s}$. Also increase the valve-required ΔP to 25 psi. This derives a more reasonable 2-in valve size C_v.

Trial 2 for C_v at $v = 14.36$ fps, $d = 2.06$:

$$C_v = 0.3103 \ d^2 v \left(\frac{\rho}{\Delta P} \right)^{1/2} = 23.3 \qquad \text{(too small for 2-in valve size)}$$

Trial 3 for C_v at $v = 23.52 \text{ ft/s}$, $d = 1.5 \text{ in}$, $\Delta P = 25 \text{ psi}$, $\rho = 38 \text{ lb/ft}^3$:

$$C_v = 0.3103 \ d^2 v \left(\frac{\rho}{\Delta P} \right)^{1/2} = 23.3 \qquad \text{(okay for 1.5-in valve size)}$$

For C_v at $v = 18.09$, calculate K for 1.5-in pipe, $\Delta P = 25$ psi:

$$K = \frac{\Delta P}{\rho v^2} = \frac{25}{38 v^2} = 0.0020$$

Check the C_v required for the 2-in valve for a 5-psi ΔP and all else the same:

$$C_v = 0.3103 \ d^2 v \left(\frac{\rho}{\Delta P} \right)^{1/2} = 52.5 \quad \text{(okay for 2-in valve size, but too large for 1.5-in)}$$

TABLE 6.2 Generalized C_v Values

Control valve data	Max C_v (100% open)
1 in valve, full trim	10.0
1.5-in valve, 40% trim size	15.0
1.5-in valve, 100% trim size	34.0
2-in valve, 40% trim size	24.0
2-in valve, 100% trim size	60.0
3-in valve, 40% trim size	48.0
3-in valve, 100% trim size	120.0
4-in valve, 40% trim size	80.0
4-in valve, 100% trim size	200.0
6-in valve, 40% trim size	160.0
6-in valve, 100% trim size	400.0

NOTE: The user is encouraged to use a particular vendor's C_v values.

At this point the valve is sized to make a 25-psi pressure loss for the design flow. Thus, the pump discharge pressure in Step 1 must allow a greater ΔP than the 5 psi originally assumed. Using this pressure loss, the designer determines that a two-phase flow occurs in the remaining run of the pipe to the stabilizer inlet. Thus, a two-phase flow solution is required, which brings us to the next section of this chapter.

Two-Phase Flow

Introduction

An introduction to this section is in order because there is so much conflict as to how and with what to establish a good design and rating of two-phase flow in piping. One of the most outstanding questions is how much liquid holdup occurs in any one specific case. Much has been written about liquid holdup, and many theories have come and gone concerning it. The truth is that at the time of this writing, we don't really know how to mathematically derive a true equation or algorithm to correctly calculate two-phase flow. However, I shall say there is one exception. The following methodology is offered to resolve the two-phase flow problem.

First, consider the fact that there are two principal phase-flow regimes:

- Slug flow regime, in which slugs of liquid and gas of various sizes occur
- Homogeneous flow of gas and liquid, in which liquid and gas phases are thoroughly mixed, representing a homogeneous third phase

Baker presented five flow regimes; the three additions to the preceding were annular flow, plug flow, and wave flow [11]. A quick review of these add-ons reveals that all three resemble each other, each being variations of flow regime in low-Reynolds-number areas.

The myth

Several notable scholars have taken to heart Baker's work on the five regimes of two-phase flow. These scholars made numerous equations for predicting flows, pressure drop, pipe fittings, rises, falls, and so forth. Baker also presented equations for such in his work of 1954. In some applications these equations have proven good and acceptable. In many applications, however, these equations have failed to reveal correct results, even requiring complete equipment replacement. How can these errors be stopped? Looking at Baker's work again, he pointed out one flow regime, froth or bubble flow, to be a homogeneous flow regime. In this regime there are no liquid slugs and no gas flow variations. This regime holds the key to our success.

Baker made an excellent contribution to the technology of two-phase flow. Despite his good work, however, many notable college scholars have taken his work and extended it, introducing much error in the attempt to predict the unpredictable: two-phase slug flow, stratified flow, and the like. Why all this confusion? A simple answer is because when you try to equate these four flow regimes—slug, annular, plug, and wave flows—the result is error in the majority of cases. Why can't it be done? Or why are so many equations in error? The answer is the simple concept with which this chapter began, *experiment*. At the time of this writing, we are lacking the true experimental work needed to correctly establish these flow regime equations. Why hasn't someone or some institution or company laboratory established these experimental needs? The need has not justified the expense. This undertaking has been accomplished to some degree, however, in the case of pipelines, for which very complex equation algorithms have been produced. These are offered in expensive commercial pipeline flow simulation programs. Even results here, however, have received only limited success, with some major pipeline flow errors found. Is there any answer then to this myth of two-phase flow equation solutions? Yes.

A solution for all two-phase flow problems

Consider again Baker's froth-zone flow regime. To have a froth or homogeneous flowing gas-liquid mixture, a high Reynolds number is required. How high should it be? Surprisingly, 200,000. Most every case of refinery, oil and gas, and chemical plant piping involves higher Reynolds numbers for economic pipe sizing. Even pipelines are sized for higher Reynolds numbers. Considering the 200,000 minimum, even a pipe flowing at 3 ft/s would qualify! Then, with this evidence, why not develop equations such as the Darcy equation as has been done for single-phase flow? The answer is that it has been done, and they are presented herein.

In the late 1960s, the American Gas Association, in joint venture with the American Petroleum Institute, awarded Professor A. E. Dukler of the University of Houston the task of determining experimentally the resolution of two-phase flow problems. In 1970, Dukler did indeed accomplish this task, publishing his extensive investigation of two-phase flow equations [12–14]. This work has resolved two-phase flow piping sizing and configuration problems. The key to the success of his work is simply the Reynolds number. If it is maintained above 200,000, success is at hand. In most every pipe flow case, this is easily accomplished simply by making the pipe size small enough. The remainder of this chapter presents a summary of Dukler's work.

Two-Phase Flow Equations

Dukler used three separate equations to resolve all two-phase flow problems. These were:

- Two-phase flow pressure loss due to friction
- Two-phase flow pressure loss or gain due to elevation rise or fall
- Two-phase flow pressure loss due to pipe-fitting acceleration

Each of these pressure losses is given derived equations, and each is a separate entry to be added as an algebraic sum for the total pressure loss. This of course is similar to Eqs. (6.14) and (6.15).

In order to establish these equations, a step-by-step presentation is made. Each step may be used to formulate the solution to any two-phase flow problem. Simply use these steps to calculate any one particular problem of a line segment or segments you may choose.

Step 1. For pressure loss due to friction, first determine the homogeneous flow liquid ratio λ, volume of liquid per volume of mixed fluid flow.

$$\lambda = \frac{Q_{\text{LPL}}}{Q_{\text{GPL}} + Q_{\text{LPL}}} \tag{6.17}$$

where Q_{LPL} = volume of liquid flow W_L/ρ_l, ft³/h
$\quad Q_{\text{GPL}}$ = volume of gas flow W_G/ρ_g, ft³/h
$\quad W_L$ = liquid flow, lb/h
$\quad W_G$ = gas flow, lb/h
$\quad \rho_l$ = liquid density at flow pressure and temperature, lb/ft³
$\quad \rho_g$ = gas density at flow pressure and temperature, lb/ft³

The λ value calculated is valid only over a range in which the pressure loss in the pipe does not exceed 15% of inlet value. This is a considerable handicap if the case involves a large pressure loss. Here the objective is to divide the pipe run into several segments. Each segment will have a different pressure inlet and probably a different temperature due to the gas-flashing cooling effect. This makes the solution longer, but consider the fact that this is a two-phase flow, not subject to easy or short solution in any case.

Step 2. The ratio of two-phase friction factor to the gas-phase friction factor in the pipeline is determined here.

In this step Dukler made numerous experiments, finding a relation between λ and the ratio of f_{TP}/f_o. The following equations are a curve-fit of his curve, relating these two factors:

$$S = 1.281 + 0.478(\log \lambda) + 0.444(\log \lambda)^2$$
$$+ 0.09399999(\log \lambda)^3 + 0.008430001(\log \lambda)^4 \quad (6.18)$$

$$\frac{f_{TP}}{f_o} = 1 - \frac{\log \lambda}{S} \quad (6.19)$$

where f_{TP} = two-phase flow friction factor in the pipe run
f_o = gas-phase friction factor in the pipe run
$\log$ = natural logarithm of base e, 2.7183

Step 3. The Reynolds number Re is calculated in this step. Dukler developed experimental data determining liquid holdup in two-phase flow systems. Re values above 200,000 are free of liquid slugs and holdup. If Re is greater than 200,000, then the flow is in the froth zone, or it is simply homogeneous flow as a mixture. For homogeneous flow, the average density of the two-phase flow fluid mixture is:

$$\rho_m = \rho_l \lambda + \rho_g (1 - \lambda) \quad (6.20)$$

In a similar equation, calculate the average viscosity μ_m, lb/ft·s:

$$\mu_m = \frac{\mu_l}{1488} \lambda + \frac{\mu_g}{1488} (1 - \lambda) \quad (6.21)$$

where μ_l = liquid viscosity, cP
μ_g = gas viscosity, cP

Calculate the mixture flowing velocity v_m, ft/s:

$$v_m = \frac{Q_{GPL} + Q_{LPL}}{\pi D^2\ 900} \quad (6.22)$$

where D = pipe internal diameter, ft

Please note that the 900 factor is 3600 s/h divided by 4, 4 being the square of 2 for reducing D to the pipe radius in feet. Now calculate the two-phase flow Reynolds number:

$$Re = \frac{D v_m \rho_m}{\mu_m} \quad (6.23)$$

Step 4. Calculate the two-phase-flow friction factor f_{TP}. First it is necessary to define f_{TP} from f_o. In charting f_{TP}/f_o against λ, Dukler applied the following published equation for f_o:

$$f_o = 0.0014 + \frac{0.125}{\mathrm{Re}^{0.32}} \qquad (6.24)$$

$$f_{\mathrm{TP}} = \frac{f_{\mathrm{TP}}}{f_o} f_o \qquad (6.25)$$

Knowing Re and the ratio of f_{TP}/f_o, f_{TP} is calculated from Eqs. (6.24) and (6.25). *Remember that Re must be 200,000 or greater for these equation applications.*

Step 5. The pipe friction pressure loss ΔP_f, psi, is calculated in this step:

$$\Delta P_f = \frac{2 f_{\mathrm{TP}} L v_m^2 \rho m}{144 g D} \qquad (6.26)$$

where L = equivalent pipe length, ft
g = acceleration of gravity, 32.2 ft/sec^2

Please note the resemblance of Eq. (6.26) to the earlier Darcy equation. Why is the factor 2 installed? The answer is that Eq. (6.6) is based on the Moody friction factor f and Eq. (6.26) is based on the Fanning friction factor f. These two friction factors are identical, except that the Moody f is exactly four times larger than the Fanning f. Without a long story as to how these two research scientists came up with the same answer, with the exception of one being four times larger in every case, simply accept the fact that one perhaps wanted to be recognized as an independent developer. Nonetheless, the bottom line is that both Eqs. (6.6) and (6.26) are indeed the same equation.

The first of three parts in solving two-phase flow equations is exhibited in Eq. (6.26), which is applicable to any two-phase flow problem.

Step 6. Calculate pressure drop due to elevation changes ΔP_E, psi. The Dukler method requires that the superficial gas velocity v_{sg} be calculated:

$$v_{sg} = \frac{Q_{\mathrm{GPL}}}{\pi D^2 900} \qquad (6.27)$$

This v_{sg} is the velocity of the gas alone in the full cross-section area of the pipe, ft/s. Note that the 900 factor is 3600 s/h divided by 4, 4 being the square of 2 for reducing D to the pipe radius in feet.

Flanigan [15] related a factor he called ϕ to the two-phase gas velocity v_{sg}. As the gas velocity decreases, the ϕ value increases. Equation (6.28) is a reasonable curve-fit of ϕ vs. v_{sg} in Flanigan's resulting curve.

$\phi = 0.76844 - 0.085389v_{sg} + 0.0041264v_{sg}^2$
$$- 0.000087165v_{sg}^3 + 0.00000066422v_{sg}^4 \quad (6.28)$$

This Flanigan curve determines the ϕ value, which is the correction to the static leg rise or fall for the gas phase. As the gas velocity approaches 0, ϕ approaches unity, 1.0. Equation (6.28) should have a range limit that is given as follows:

If $v_{sg} > 50$, then $\phi = 0.04$

If $v_{sg} < 0.5$, then $\phi = 0.85$

$$\Delta P_E = \frac{\phi \rho_l H_T}{144} \quad (6.29)$$

where H_T = height of static leg, − for rise and + for fall, ft

Please note that only the liquid-phase density ρ_l is used in Eq. (6.29). The reason that the average density is not used is that ϕ corrects for the gas-phase static leg ΔP. Equation (6.29) shows the second part of the three pressure losses occurring in two-phase flow. The final pressure loss calculation, acceleration, is covered next.

Step 7. Calculate the pressure drop due to acceleration or pipe fittings ΔP_A, psi. This is usually relatively small compared to the other two, ΔP_f and ΔP_E. In some cases, however, where numerous pipe fittings are involved, it can be very significant. The following equations provide the calculations required for the two-phase flow pipe fittings and valve ΔP_A losses.

Consider first that the K value relations of all fittings can be applied in the same directly proportional relationship as for single-phase fluid flow. Most folks reading this will ask what this is driving at or what does "directly proportional" mean and to what does it apply? What we shall do is take a standard 90° ell and calculate its acceleration pressure loss factor ΔP_{ell}, psi. Then we multiply each fitting in proportion to the single-phase K value factors used: ΔP_A = pipe fitting factor × ΔP_{ell}. The following equations will walk you through this exercise.

The 90° standard ell ΔP_{ell} is calculated as follows:

$$\Delta P_A = \frac{\rho_g Q_{GPL}^2}{1 - \lambda} + \frac{\rho_l Q_{LPL}^2}{\lambda} \quad (6.30)$$

$$\Delta P_{ell} = \frac{\Delta P_A}{3.707e + 10(D/12)^4} \quad (6.31)$$

Note that all the factors required for Eqs. (6.30) and (6.31) have been established in previous steps of this section. Thus, for any 90° standard ell using these two equations, we have the pressure loss established for two-phase flow acceleration. Simply multiply the calculated ΔP_{ell} value by the number of 90° ells in the pipe segment. Remember that if a 15% pressure loss in a pipe segment results, a new pipe segment is required.

Tee angle $\Delta P_A = 3.0\ \Delta P_{ell}$
Tee straight $\Delta P_A = 1.0\ \Delta P_{ell}$
Check valve $\Delta P_A = 2.5\ \Delta P_{ell}$

Using references such as the *Crane Technical Paper No. 410,* other pipe-fitting values can be proportionally factored. For two-phase flow pipe entrance and exit, however, new, more accurate equations are presented in this book, as shown here:

Pipe sharp-edge entrance:

$$\Delta P_A = 2f_{TP}[6.469(\log D) + 24]\ \frac{\rho_m}{144}\ \frac{v_m^2}{32.2} \tag{6.32}$$

Pipe sharp-edge exit:

$$\Delta P_A = 2f_{TP}\ [14.403(\log D) + 42]\ \frac{\rho_m}{144}\ \frac{v_m^2}{32.2} \tag{6.33}$$

These new equations, Eqs. (6.32) and (6.33), have now been applied to numerous field trials and tests, finding good results. Please note the resemblance to Eq. (6.6). The log D factor is a plot of pipe diameter vs. pressure loss on a log-log scale.

Step 8.   Summary of total two-phase pressure loss ΔP_T, psi:

$$\Delta P_T = \Delta P_f + \Delta P_E + \Delta P_A$$

A final note for applying this two-phase flow procedure is important here. *All preceding 8 steps are made on the assumption that Re is 200,000 or greater. This is important. Otherwise, welcome to the myth.*

Two-phase flow program TPF

The code shown in Fig. 6.3 is the key code listing from the program TPF supplied in the CD with this book. A brief review of Fig. 6.3 will show the same equations listed in this figure. The part of the code not shown is the input of data code and the output data answer code. These two code sections (general templates) are covered in Chap. 9.

```
250 QLPL = WL / DENL: QGPL = WG / DENG: AL = QLPL /
    (QLPL + QGPL)
251 S = 1.281 + .478 * (Log(AL)) + .444 * ((Log(AL)) ^ 2)
    + .09399999 * ((Log(AL)) ^ 3) + .008430001 * ((Log(AL))
    ^ 4)
252 FTPO = 1 - ((Log(AL)) / S)
260 VM = (QLPL + QGPL) / (2827.43 * (D / 12) ^ 2)
261 RL = AL: DENTP = (DENL * ((AL) ^ 2) / RL) + DENG *
    ((1 - AL) ^ 2) / (1 - RL)
262 UTP = (UL / 1488) * AL + (UG / 1488) * (1 - AL)
263 RETP = (D / 12) * VM * DENTP / UTP
270 FO = .0014 + (.125 / (RETP ^ .32)): FTP = FTPO * FO
280 DPF = 2 * FTP * (L * 12 / D) * (DENTP / 144)
    * ((VM ^ 2) / 32.2)
290 VSG = QGPL / (2827.43 * (D / 12) ^ 2)
291 THETA = .76844 - .085389 * VSG + .0041264 * (VSG ^ 2)
    - .000087165 * (VSG ^ 3) + .00000066422 * (VSG ^ 4)
295 If VSG > 50 Then THETA = .04
296 If VSG < .5 Then THETA = .85
297 DPHT = THETA * DENL * HT / 144
300 Rem PIPE FITTING DELTA P , DPEL, CALC
310 DPA1 = ((DENG * (QGPL ^ 2)) / (1 - RL))
    + ((DENL * (QLPL ^ 2)) / RL)
311 DPEL = DPA1 / (37070000000# * ((D / 12) ^ 4))
320 DPAT = ELL * DPEL + TEES * DPEL + TEEA * 3 * DPEL + VLV
    * (8 / 20) * DPEL + CHK * (5 / 2) * DPEL + ENT *
    (2 * FTP * (6.469 * (Log(D)) + 24) * (DENTP / 144)
    * ((VM ^ 2) / 32.2)) + EXT * (2 * FTP * (14.403 *
    (Log(D)) + 42) * (DENTP / 144) * ((VM ^ 2) / 32.2))
390 Rem CALC OF DELTA P, PSI
400 TOTDP = DPF + DPHT + DPAT
```

Figure 6.3   Two-phase flow program TPF key code listing.

Note that the variable nomenclature in the code listing is different from that in the book text. For example, ρ_l is liquid density as given in the text, and DENL is the liquid density in the TPF program code.

Calculation examples

Example 6.1, Conclusion   In this example, the liquid density was 38 lb/ft^3, with a viscosity of 0.5. Upon running a flash for 30 mol% propane and 70% butane (54 MW liquid), the designer finds that 20% of the molar flow flashes,

TABLE 6.3   Two-Phase Flow Pipe Run Segment

	Feed, mol/h	Flash, liquid, mol/h	Flash, liquid, lb/h	Flash, gas, mol/h	Flash, gas, lb/h
Propane	194	77	.	117	
Butane	454	441		13	
Total	648	518	29,090	130	5910
MW		56.2		45.4	
Density, lb/ft³	32		0.89		
Viscosity, cP	0.5		0.012		

forming 90% propane and 10% butane in the gas phase with a 50 MW gas. Since we initially had 35,000 lb/h or 35,000/54 = 648 mol/h flow, the molar balance is now as shown in Table 6.3.

Using Table 6.3, the two-phase flow part of Example 6.1 will now be worked. The procedure includes 8 steps, applying all of the equations shown.

solution   Two-phase flow after the control valve pressure drop:

Step 1   Calculate λ:

$$\lambda = \frac{Q_{\mathrm{LPL}}}{Q_{\mathrm{GPL}} + Q_{\mathrm{LPL}}} = 0.1204 \qquad (6.17)$$

where Q_{LPL} = volume of liquid flow W_L/ρ_l = 29,090/32 = 909.1 ft³/h
  Q_{GPL} = volume of gas flow W_G/ρ_g = 5910/0.89 = 6640 ft³/h
  W_L = liquid flow, 29,090 lb/h
  W_G = gas flow, 5910 lb/h
  ρ_l = liquid density at flow pressure and temperature, 32 lb/ft³
  ρ_g = gas density at flow pressure and temperature, 0.89 lb/ft³

Step 2   Calculate f_{TP}/f_o:

$$S = 1.281 + 0.478(\log \lambda) + 0.444(\log \lambda)^2$$
$$+ 0.09399999(\log \lambda)^3 + 0.008430001(\log \lambda)^4 = 1.5364 \quad (6.18)$$

where $\lambda = 0.1204$

$$\frac{f_{\mathrm{TP}}}{f_o} = 1 - \frac{\log \lambda}{S} = 2.3779 \qquad (6.19)$$

where f_{TP} = two-phase flow friction factor in the pipe run
  f_o = gas-phase friction factor in the pipe run
  log = natural logarithm of base e, 2.7183

Step 3   Calculate the Reynold's number Re of the mixture density ρ_m, lb/ft³:

$$\rho_m = \rho_l \lambda + \rho_g (1 - \lambda) = 4.64 \text{ lb/ft}^3 \qquad (6.20)$$

$$\mu_m = \frac{\mu_l}{1488} \lambda + \frac{\mu_g}{1488} (1 - \lambda) = 4.772\text{e-}05 \text{ lb/ft·s} \qquad (6.21)$$

where μ_l = liquid viscosity, cP
μ_g = gas viscosity, cP

Calculate the mixture flowing velocity v_m, ft/s:

$$v_m = \frac{Q_{GPL} + Q_{LPL}}{\pi D^2 \, 900} = 41.06 \text{ ft/s} \qquad (6.22)$$

where D = pipe internal diameter, (3.06 in)/(12 in/ft) (3-in pipe assumed)

Trial 1, 3-in pipe:

$$\text{Re} = \frac{D v_m \rho_m}{\mu_m} = 1.02e + 06 \qquad (6.23)$$

for comparison, try 1.5-in pipe.

Trial 2, 1.5-in pipe, again using Eqs. (6.22) and (6.23):

$$v_m = 148.3 \text{ ft/s}$$

$$\text{Re} = 1.93e + 06$$

where $D = (1.61 \text{ in})/(12 \text{ in/ft})$

This 1½-in pipe size is also okay.
   Trial 1, the 3-in pipe size, appears the better choice regarding the velocity. Notice that either the 1.5-in or the 3-in pipe size would have been an okay choice.

Step 4   Calculate the friction factor f_{TP}:

$$f_o = 0.0014 + \frac{0.125}{\text{Re}^{0.32}} = 0.0029 \qquad (6.24)$$

From Step 2, $f_{TP}/f_o = 2.3772$.

$$f_{TP} = \frac{f_{TP}}{f_o} f_o = 0.0069 \qquad (6.25)$$

This f_{TP} value, 0.0069, is the Fanning friction factor, whch is also exactly one-fourth the Moody friction factor calculated for single-phase flow. The Moody friction factor was used earlier in this chapter for single-phase flow.

Step 5   Calculate pipe straight segment friction pressure loss ΔP_f:

$$\Delta P_f = \frac{2 f_{TP} L v_m^2 \rho_m}{144 g D} = 0.55 \text{ psi} \qquad (6.26)$$

where L = equivalent pipe length, 6 ft
g = acceleration of gravity, 32.2 ft/s^2

Note that the flow control valve is placed as close to the column as feasible, 6 ft. This two-phase pipe segment begins at the flow control valve and ends at the fractionation column inlet nozzle.

Step 6 Calculate elevation change for two-phase flow (no elevation changes occur).

Step 7 Calculate the pressure loss ΔP_A of pipe fittings. There is one fitting loss, the pipe exit into the column. To calculate the exit loss, use Eq. (6.33) for pipe sharp-edge exit:

$$\Delta P_A = 2f_{TP}[14.403(\log D) + 42]\ \frac{\rho_m}{144}\ \frac{v_m^2}{32.2} \qquad (6.33)$$

$$= 1.353 \text{ psi}$$

where $f_{TP} = 0.0069$
$D = 3.06$ in
$\rho_m = 4.64$ lb/ft^3
$v_m = 41.06$ ft/s

Step 8 Summary of total two-phase pressure loss ΔP_T, psi:

$$\Delta P_T = \Delta P_f + \Delta P_E + \Delta P_A = 0.55 + 0 + 1.353 = 1.90 \text{ psi}$$

Computer program examples

To check these hand calculations, run the CD program TPF. The input screen is displayed in Fig. 6.4 and the output answers in Fig. 6.5. Please note that all input data in Fig. 6.4 is the same as that used in the eight two-phase steps of Example 6.1. All answers in Fig. 6.5 agree with the eight-step two-phase flow hand calculations. The small differences are negligible and are due to the hand-calculation round-offs.

Example 6.2 Figure 6.6 exhibits this two-phase flow problem, a thermosyphon reboiler. The solution to the problem is locating the vertical reboiler far enough below the liquid level shown in Fig. 6.6 to produce the flow desired through the reboiler, C to E. The fluid flow from C to D is the reboiler tube side. In this flow zone, 8 to 40% of the circulated fluid is vaporized. This percentage of vaporization is normal in all thermosyphon reboilers. In our example here, 25% by weight will be selected as the vaporized percentage. How do we know this will happen? Consider the liquid level height L_F. If L_F is set to the right measure, this 25% vaporization will happen, as the freewheeling liquid flows around the wheel of points A, B, C, D, and E. Also note, however, that segments between points C, D, and E will be two-phase flow.

In setting the vaporization percentage, 25% here, we do not actually set the amount of vapor to be vaporized; rather, we set the amount of liquid that spins around from point A to point E, and back to A again. The amount of vapor to be vaporized is fixed by the overall fractionation column, the amount of vapor boilup to be sent up the fractionation column. The vertical-

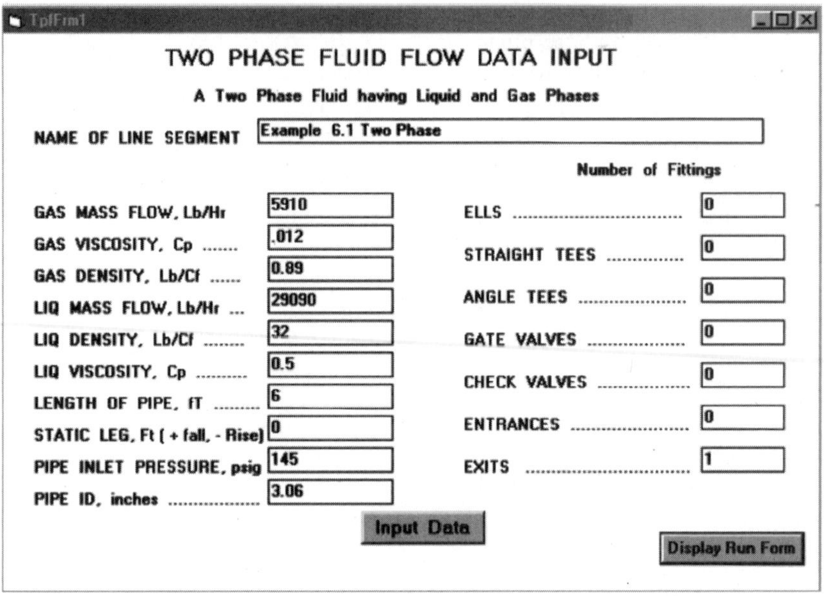

TpfFrm1

TWO PHASE FLUID FLOW DATA INPUT

A Two Phase Fluid having Liquid and Gas Phases

NAME OF LINE SEGMENT	Example 6.1 Two Phase
GAS MASS FLOW, Lb/Hr	5910
GAS VISCOSITY, Cp	.012
GAS DENSITY, Lb/Cf	0.89
LIQ MASS FLOW, Lb/Hr ...	29090
LIQ DENSITY, Lb/Cf	32
LIQ VISCOSITY, Cp	0.5
LENGTH OF PIPE, fT	6
STATIC LEG, Ft (+ fall, - Rise)	0
PIPE INLET PRESSURE, psig	145
PIPE ID, inches	3.06

Number of Fittings

ELLS	0
STRAIGHT TEES	0
ANGLE TEES	0
GATE VALVES	0
CHECK VALVES	0
ENTRANCES	0
EXITS	1

Input Data

Display Run Form

Figure 6.4 TPF input screen.

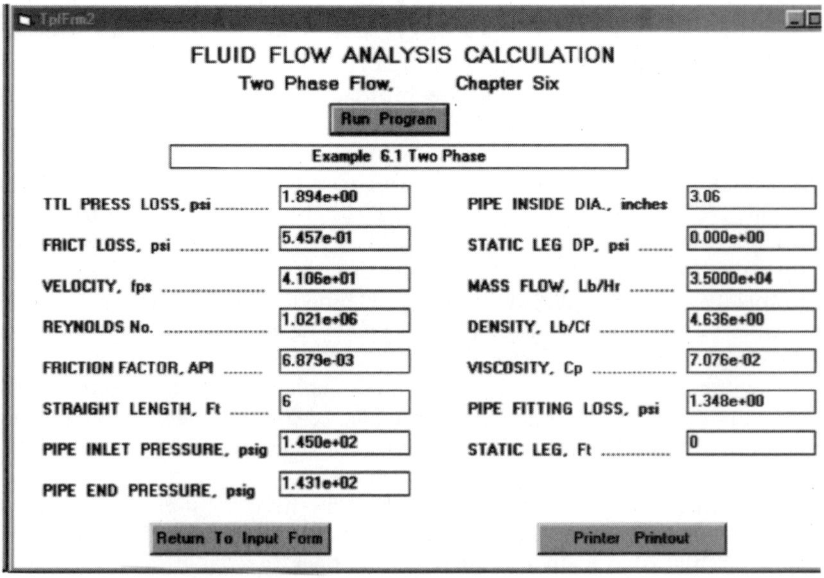

TpfFrm2

FLUID FLOW ANALYSIS CALCULATION

Two Phase Flow, Chapter Six

Run Program

Example 6.1 Two Phase

TTL PRESS LOSS, psi	1.894e+00	PIPE INSIDE DIA., inches	3.06
FRICT LOSS, psi	5.457e-01	STATIC LEG DP, psi	0.000e+00
VELOCITY, fps	4.106e+01	MASS FLOW, Lb/Hr	3.5000e+04
REYNOLDS No.	1.021e+06	DENSITY, Lb/Cf	4.636e+00
FRICTION FACTOR, API	6.879e-03	VISCOSITY, Cp	7.076e-02
STRAIGHT LENGTH, Ft	6	PIPE FITTING LOSS, psi	1.348e+00
PIPE INLET PRESSURE, psig	1.450e+02	STATIC LEG, Ft	0
PIPE END PRESSURE, psig	1.431e+02		

Return To Input Form

Printer Printout

Figure 6.5 TPF calculation screen.

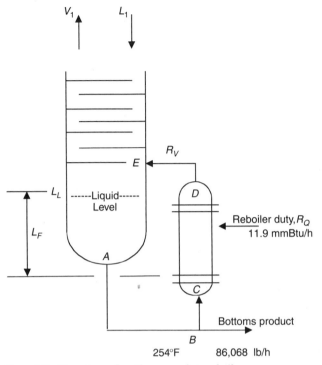

Figure 6.6   Two-phase flow thermosyphon reboiler.

type thermosyphon reboiler has a big liquid type of flywheel, off which a bottoms liquid stream spins. The bottoms stream here is the 86,068-lb/h liquid stream shown in Fig. 6.6. In this problem, we shall again use a material balance table, Table 6.4.

TABLE 6.4   Thermosyphon Reboiler Pipe Run

	Liquid leaving column, lb/h	Liquid to reboiler, lb/h	Flash vapor, lb/h	Liquid through reboiler, lb/h	Bottoms product stream, lb/h
Vapor			103,346		
Liquid	499,452	413,384		310,038	86,068
MW			72.57		
Density, lb/ft³	31.67	31.67	1.25	31.67	31.67
Viscosity, cP	0.103	0.103	0.0095	0.103	0.103

The first part of the problem is assuming a liquid level height L_F. The problem then of course becomes trial-and-error, since the assumed L_F must be corrected and the run repeated. In most cases, however, the assumption of a slight excessive height is sufficient to solve the problem conservatively, since it is always feasible to run the column with a slightly lower bottoms level than it is designed for. It is the higher level that will require another trial run.

For the case here, the computer program has calculated the reboiler sizing:
Reboiler Size Required

- 920 tubes, inside diameter 0.57 in
- Tube length 8 ft
- One tube pass

Since the tubes are 8 ft long, and allowing 7 ft for the exchanger head, pipe outlet riser, and one ell, a good first estimate for L_F is

$$L_F = 8 \text{ ft} + (7 \text{ ft} - 4 \text{ ft}) = 11 \text{ ft}$$

Please note that a 4-ft vertical vapor space is assumed for the column bottoms section.

At this point, the pipe pressure profile should be run from point E to point B. Starting at point E, it is good to assume any reasonable column bottoms operating pressure. In this example, a column bottoms pressure of 128 psia is assumed, as this is the value calculated in the simulation run. It is important to note here that the assumed pressure value of point E must equal or exceed its calculated pressure value. Put another way, the pressure profile of loop points E-A-B-C-E will be calculated, and this final calculated value of E is to equal or exceed the assumed E value. Of course, if the calculated E value excessively exceeds the assumed value, then a smaller L_F value should be assumed and another trial-and-error run made.

For our purpose here it is necessary to consider only the liquid height L_F measure to the face of the bottom tube bundle, point C. The reason is that the vertical heights of B to C and B to A tend to cancel each other. The L_F height is therefore 11 ft. Consider L_F the driving force that will make this big liquid flywheel spin.

Now the program FFLW will be run. The input and output screens are displayed in Figs. 6.7 and 6.8.

Several pipe sizes were run with the FFLW pipe simulator shown in Figs. 6.7 and 6.8. The preferred velocity in thermosyphon suction flow piping is 2 to 5 ft/s. As seen in the screen in Fig. 6.8, 3.5 ft/s was achieved, using 16-in pipe. One size smaller pipe will be tried in the pipe run from B to C, since over 86,000 lb/h is shaved off as the bottoms product.

Please observe that point B pressure has increased by approximately 2 psi above the starting point E pressure, 128 psia. Why? The liquid static leg L_F, coupled with a low piping friction loss, has caused this pressure increase. Please note, also, that point B actually extends several feet below point C. This is a canceling static leg effect since the elevation distance of A to B is equal to the elevation distance of B to C. Point C is to be taken as the point at the tube bundle face entrance. Another factor that should be realized is that the tee-fitting pressure loss is included in the segment run from point A to point B. Thus, no tee pressure losses will be counted in the run from B to C.

The next procedure is calculation of the single-phase flow pressure loss from point B to point C. This flow will be less than the previous one, since 83,086 lb/h is split off as product, leaving a 413,384 lb/h feed flow to the reboiler. FFLW will next be run for segment B to C. The input is shown in Fig. 6.9 and the output in Fig. 6.10.

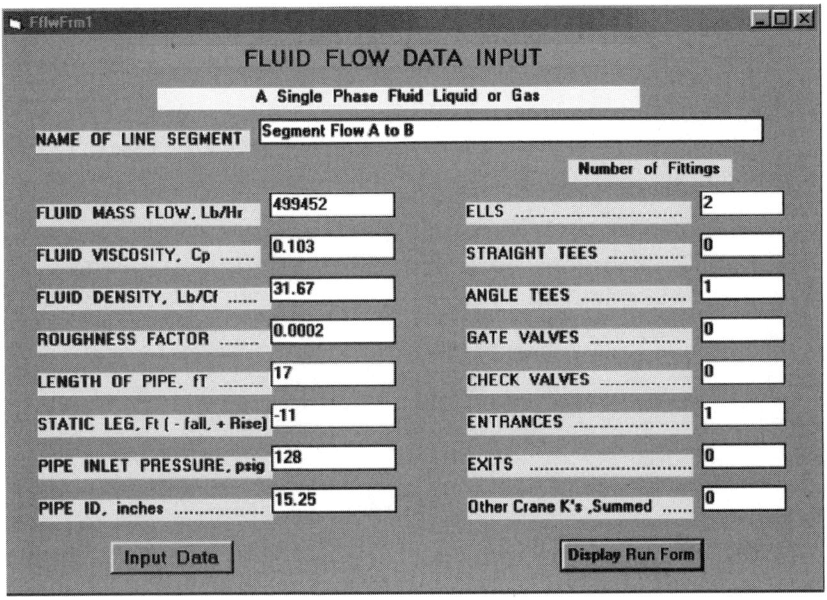

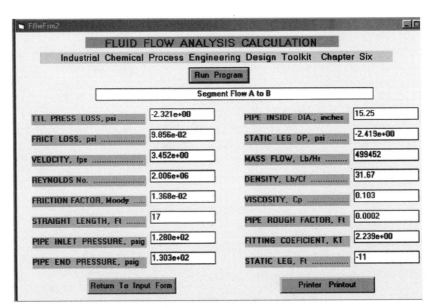

Figure 6.7   Single-phase fluid flow, points E to B.

Figure 6.8   Single-phase fluid flow, points E to B.

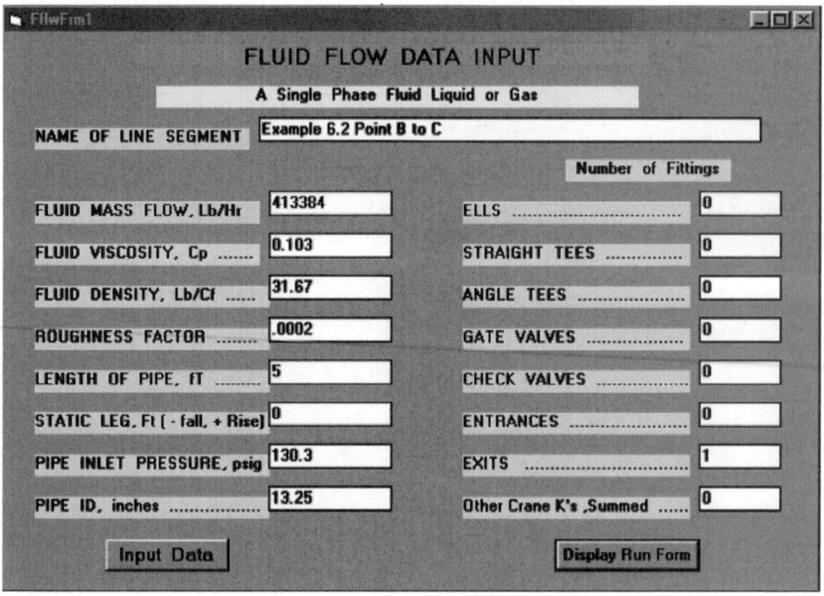

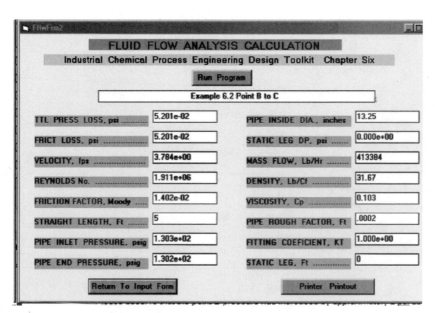

Figure 6.9 FFLW input screen.

Figure 6.10 FFLW output answers.

Please note in Fig. 6.9 that only one pipe fitting was input, a pressure loss for the fitting into the exchanger bottom shell head. A 5-ft straight section of pipe was also input, allowing sufficient pipe riser distance.

Figure 6.10 shows very little pressure loss, approximately 0.1 psi for the 14-in pipe size. This is good, since the desired velocity of 2 to 5 ft/s is maintained and the majority of the pressure drop is allowed for the two-phase flow tube and pipe section, C to E segment. Note, also, that no accounting has been made for static leg pressure loss, since this is canceled per previous discussion.

The next pipe segment, the exchanger tube segment, is the point C to point D section. This is a vertical 920-tube bundle, one pass, and it will have both single-phase and two-phase flow. A conservative approach to determining the correct pressure loss is to assume that the first third of the vertical tube bundle has fully single-phase flow and the remaining length has fully two-phase flow. This means that all vapor that will vaporize will be assumed to be in full flow in this final two-thirds of the tube section. This is an important concept, presented here as a methodology for thermosyphon reboiler calculations. The author has applied this assumption to numerous vertical thermosyphon reboilers over the past 30 years, finding both ratings and design to give excellent results. The answers have been found to be slightly conservative, within a +15% pressure profile tolerance of accuracy. This 15% tolerance is based on the total pressure loss of the looped circuit, E-A-B-C-D-E. This is indeed a good report on a new method, a method you are encouraged to use on your next vertical thermosyphon reboiler rating or design task.

There is a condition on this rule of thirds—it is to be limited to areas of latent heat and not to the tube area dedicated to sensible heat. Most reboilers of this type comply to the requirement for very little sensible heat because the large flywheel of circulating liquid is near or exactly equal to the fractionator column bottoms temperature. Therefore, only latent heat applies, and this rule of thirds also applies.

Figures 6.11 and 6.12 show the input screen and the output answers for the tube section pressure loss. Please note that the pressure loss for only one tube is analyzed, since all 920 tubes are assumed to have equal flow and vaporization. Thus, the pressure profile made for one tube will be the same for the full tube bundle.

Flow for one tube:

$$\text{Liquid first-third section} = \frac{413{,}384}{920} = 449.3 \text{ lb/h}$$

Referring to Fig. 6.11, please observe that one-third of the tube length, 2.67 ft, was input. A 2.67 elevation rise was also input. Note in Fig. 6.12 how the major part of the tube pressure drop is due to this elevation rise.

The next part of the tube bundle pressure drop calculation is the two-phase flow (Fig. 6.13).

$$\text{Tube two-phase flow vapor rate} = \frac{103{,}346}{920} = 112.3 \text{ lb/h}$$

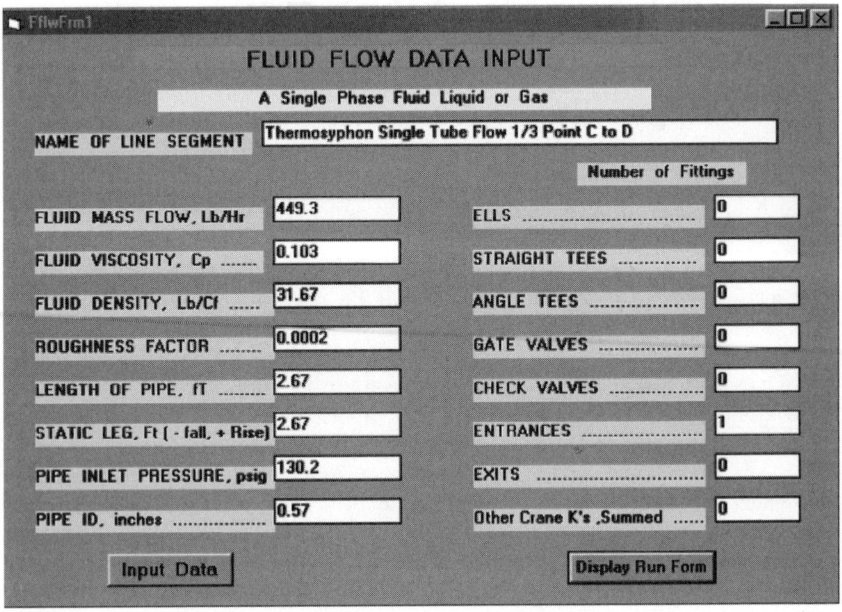

Figure 6.11   FFLW input screen for tube section pressure loss.

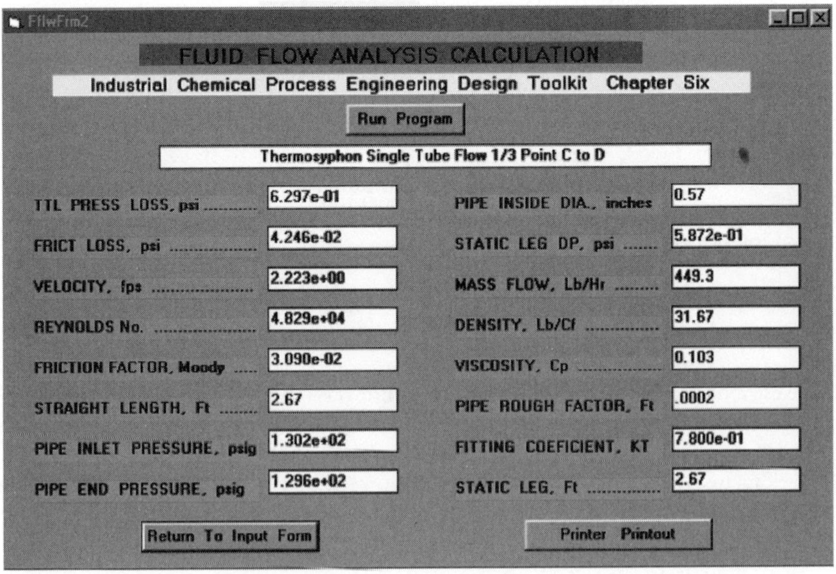

Figure 6.12   FFLW output for tube section pressure loss.

$$\text{Tube two-phase flow liquid rate} = \frac{310{,}038}{920} = 337.0 \text{ lb/h}$$

The input pressure for the two-phase flow portion of the tube will be 129.6 psia, which is the ending pressure of Fig. 6.12, the single-phase portion.

Noting that the two-phase flow vertical rise is a large part of the pressure drop through the tube bundle, it is good here to review Step 6 for ΔP_E, the change in pressure gain or loss due to elevation. Here the ΔP_E value will be a loss, since the flow is going up. The v_{sg} value must first be calculated.

$$v_{sg} = \frac{Q_{\text{GPL}}}{\pi D^2 900} = 14.08 \text{ ft/s} \tag{6.27}$$

$$Q_{\text{GPL}} = \frac{W_g}{\rho_g} = \frac{112.3}{1.25} = 89.84 \text{ ft}^3/\text{h}$$

Flanigan [15] related the factor ϕ to the two-phase gas velocity, v_{sg}. As the gas velocity decreases, the ϕ value increases. Equation (6.28) is a reasonable curve-fit of ϕ vs. v_{sg}.

$$\phi = 0.76844 - 0.085389 v_{sg} + 0.0041264 v_{sg}^2$$
$$- 0.000087165 v_{sg}^3 + 0.00000066422 v_{sg}^4 = 0.167 \tag{6.28}$$

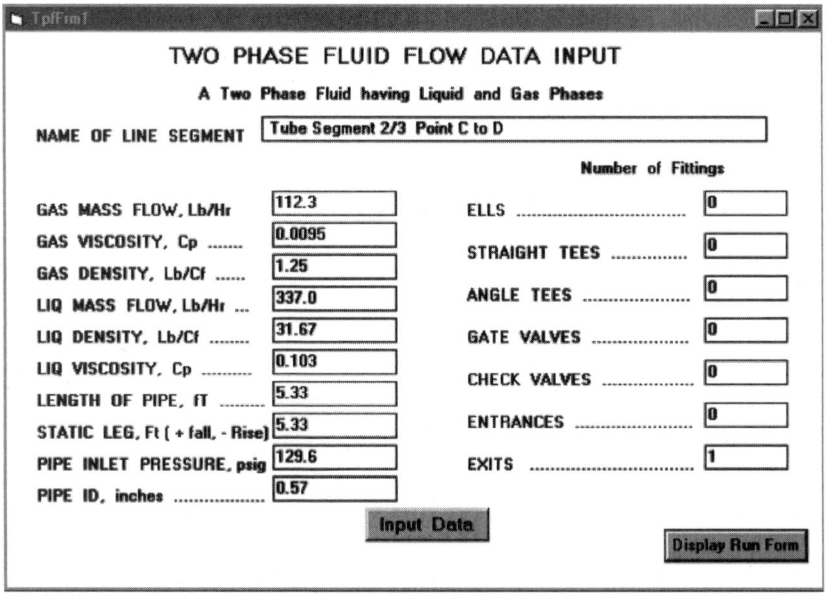

TWO PHASE FLUID FLOW DATA INPUT

A Two Phase Fluid having Liquid and Gas Phases

NAME OF LINE SEGMENT   Tube Segment 2/3 Point C to D

Number of Fittings

GAS MASS FLOW, Lb/Hr	112.3	ELLS 0
GAS VISCOSITY, Cp	0.0095	STRAIGHT TEES 0
GAS DENSITY, Lb/Cf	1.25	
LIQ MASS FLOW, Lb/Hr ...	337.0	ANGLE TEES 0
LIQ DENSITY, Lb/Cf	31.67	GATE VALVES 0
LIQ VISCOSITY, Cp	0.103	CHECK VALVES 0
LENGTH OF PIPE, fT	5.33	
STATIC LEG, Ft (+ fall, - Rise)	5.33	ENTRANCES 0
PIPE INLET PRESSURE, psig	129.6	EXITS 1
PIPE ID, inches	0.57	

Input Data

Display Run Form

Figure 6.13   TPF input for tube section pressure loss.

This curve determines ϕ, which is the correction to the rise or fall static leg for the gas phase. As the gas velocity approaches 0, the ϕ value approaches unity, 1.0. Equation (6.28) should have the following range limit:

If $v_{sg} > 50$, then $\phi = 0.04$

If $v_{sg} < 0.5$ then $\phi = 0.85$

Here v_{sg} is less than 50 and greater than 0.5; therefore, $\phi = 0.167$.

$$\Delta P_E = \frac{\phi \rho_l H_T}{144} = \frac{0.167 \times 31.67 \times 5.33}{144} = 0.196 \text{ psi} \qquad (6.29)$$

where H_T = height of static leg, 5.33 ft

Please note that only the liquid-phase density ρ_l is used in Eq. (6.29). The reason that the average density is not used is that ϕ corrects for the gas-phase static leg ΔP.

This ΔP_E value agrees perfectly with the static leg pressure drop calculated by the TPF program, 0.196 psi, as shown in Fig. 6.14.

Please note that the Reynolds number has always been above the 200,000 minimum limiting value. Should any application of this methodology reveal an Re value that falls below 200,000, a smaller pipe size should be considered. If this condition applies to existing piping, then be assured of possible slug flow, and welcome to the myth.

The final leg of this series is the two-phase flow of the pipe riser from the reboiler top head to the fractionation column bottoms vapor section, point D to point E. This is exhibited in Figs. 6.15 and 6.16.

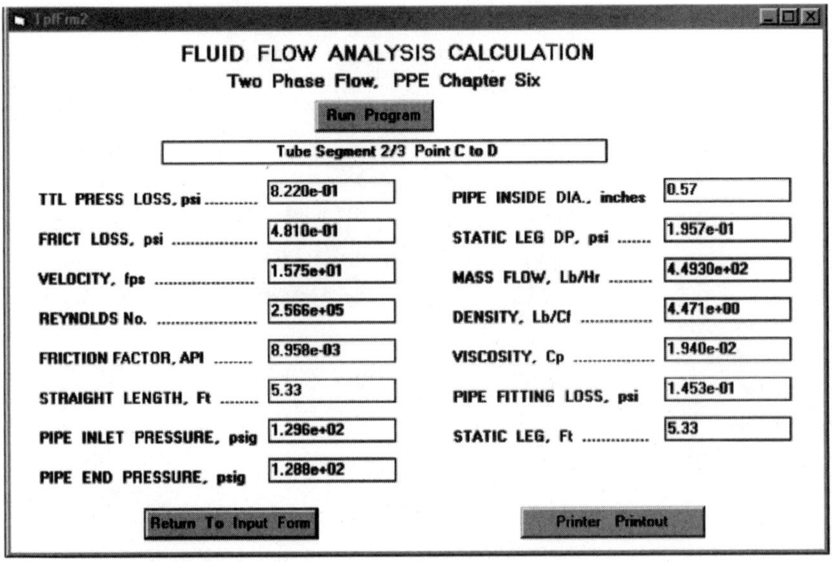

FLUID FLOW ANALYSIS CALCULATION
Two Phase Flow, PPE Chapter Six

Run Program

Tube Segment 2/3 Point C to D

TTL PRESS LOSS, psi	8.220e-01	PIPE INSIDE DIA., inches	0.57
FRICT LOSS, psi	4.810e-01	STATIC LEG DP, psi	1.957e-01
VELOCITY, fps	1.575e+01	MASS FLOW, Lb/Hr	4.4930e+02
REYNOLDS No.	2.566e+05	DENSITY, Lb/Cf	4.471e+00
FRICTION FACTOR, API	8.958e-03	VISCOSITY, Cp	1.940e-02
STRAIGHT LENGTH, Ft	5.33	PIPE FITTING LOSS, psi	1.453e-01
PIPE INLET PRESSURE, psig	1.296e+02	STATIC LEG, Ft	5.33
PIPE END PRESSURE, psig	1.288e+02		

Return To Input Form   Printer Printout

Figure 6.14 TPF output for tube section pressure loss.

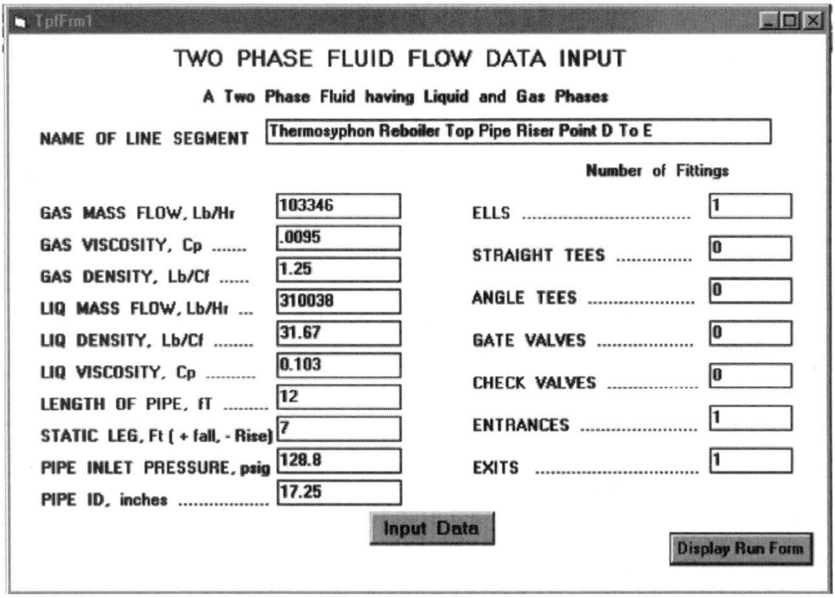

Figure 6.15   TPF input for riser pressure loss.

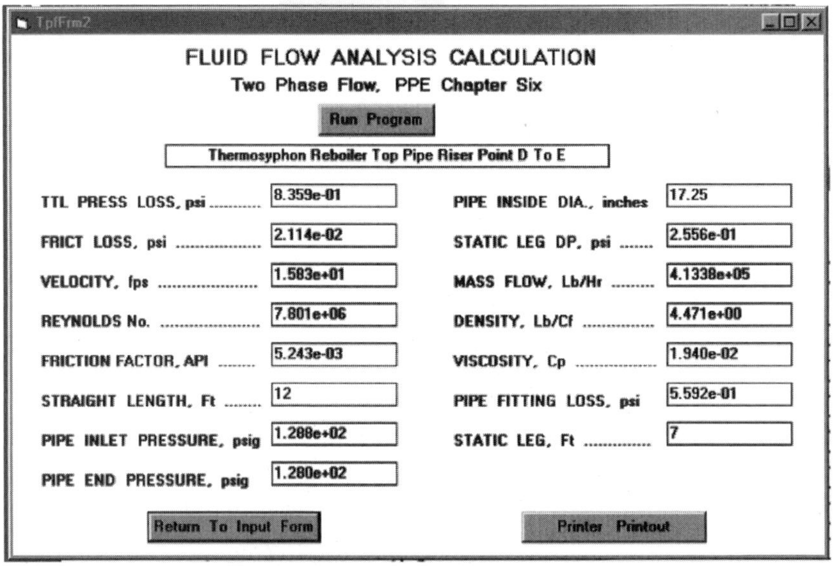

Figure 6.16   TPF output for riser pressure loss.

Please note that the calculated E value is exactly the starting assumed E pressure value, 128.0 psi. Perhaps the reader or student believes the author cheated and ran several attempts to get this perfect match? Believe it or not, the author actually made this one and only run. How? Look again at the initial assumption of L_F.

Initial assumption: Since the tubes are 8-ft long, and allowing 7 ft for the exchanger head, pipe outlet riser, and one ell, a good first estimate for L_F is

$$L_F = 8 \text{ ft} + (7 \text{ ft} - 4 \text{ ft}) = 11 \text{ ft}$$

Note that a 4-ft vertical vapor space is assumed for the column bottoms section.

Having made this initial assumption of a value of 11 ft, L_F how can the author be so lucky? Remember, the driving force is this static leg that pushes the liquid flywheel around the A-B-C-D-E points. A good rule for first L_F estimates is to take the tube length times 1.4, which should roughly equal the minimum static leg driving force required. Here, in Example 6.2, we found that 11 ft was a good first estimate. However, be advised that this calculated L_F is a minimum value. Therefore, add at least 1 ft more in vertical height to ensure adequate driving force. See Fig. 6.17, which contains all pipe size determinations.

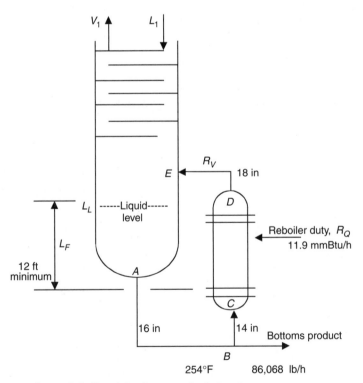

Figure 6.17 Reboiler piping loop, required pipe sizes.

Overview

The goals of Chap. 6 have been twofold: first, to give the reader the necessary tools to understand the methodology for resolving any fluid flow problem, and second, to open doors of understanding where few technical books have ventured. These goals have been accomplished if you followed through the simplified steps, which may be used as templates for any piping fluid flow rating and design problem.

In closing, one note of accomplishment needs to be mentioned, and that is regarding thermosyphon reboiler design and rating. The author has failed to find any technical book over years of practice that clearly demonstrates the methodology of complete problem solving in this area. It is trusted that users will find the methodologies herein of benefit in filling this void.

References

1. Daugherty, R. L., and Ingersoll, A. C., *Fluid Mechanics,* McGraw-Hill, 1954, Equations 5.8–5.10, pp. 71–77.
2. Daugherty and Ingersoll, "Hydraulic Gradient," pp. 81–84.
3. Daugherty and Ingersoll, "Reynolds Numbers," pp. 161–163.
4. Colebrook, C. F., "Turbulent Flow in Pipes," *Journal of the Institute of Civil Engineers,* London, February 1939.
5. Moody, Lewis F., "Friction Factors for Pipe Flow," *ASME* 66:671, 1944.
6. Chen, His-Jen, "Extract Solution to the Colebrook Equation," *Chem. Eng.,* February 16, 1987.
7. Daugherty and Ingersoll, Equation 8.19, p. 167.
8. Daugherty and Ingersoll, "Pipe-Friction Equation," p. 202.
9. Kern, Donald Q., *Process Heat Transfer,* McGraw-Hill, New York, 1950, p. 35.
10. The Crane Company, "Flow of Fluids Through Valves, Fittings, and Pipe," Pipe Friction Data Table, *Crane Technical Paper No. 410,* King of Prussia, PA, 1988, p. A-26.
11. Baker, O., "Two Phase Flow," *Oil and Gas Journal,* July 26, 1954, p. 185.
12. Dukler, A. E., "Gas-Liquid Flow in Pipelines, Research Results," University of Houston, Houston, TX; American Gas Association/American Petroleum Institute, New York, May 1970.
13. Dukler, A. E., "American Gas Association Project NX-28," American Gas Association, New York, 1969.
14. Dukler, A. E., "Gas-Liquid Flow in Pipelines," *AICHE Journal,* 1964, p. 38.
15. Flanigan, O., "Two Phase Flow Gas Velocity Variable," *Oil and Gas Journal,* March 10, 1958, p. 132.

Liquid-Liquid Extraction

Nomenclature

Δ_ρ	density difference between liquid phases, lb/ft^3
a	packing area, ft^2/ft^3
D_C	continuous-phase density, lb/ft^3
D_D	dispersed-phase density, lb/ft^3
DIA_1	extractor internal diameter, column section 1, ft
DIA_{1R}	required extractor column internal diameter for raffinate phase, ft
DIA_{1S}	required extractor column internal diameter for solvent phase, ft
F_e	packing factor
$FLOOD_{R1}$	extractor loading flood percentage, raffinate flow rate basis
$FLOOD_{S1}$	extractor loading flood percentage, solvent flow rate basis
H	Henry's law constant, atm/mole fraction of solute in liquid phase
H_f	Henry's law constant for solute in feed liquid phase
H_s	Henry's law constant for solute in solvent liquid phase
K_D	equilibrium constant; distribution coefficient
MW_f	molecular weight of feed without solute
MW_s	molecular weight of solvent without solute
O_j	interfacial surface tension from Fig. 7.12, dyn/cm
p	partial pressure of solute, atm
$RDEN_1$	raffinate density, column section 1, lb/ft^3
S	entering solvent, lb

S_1	solvent rate, column section 1 (solute included), lb/h
SDEN$_1$	solvent density, column section 1, lb/ft^3
S_{e1}	lb solute per S entering extractor
S_{e2}	lb solute per S exiting extractor
U_C	viscosity of continuous phase, cP
V_C	continuous-phase velocity, ft/h
VCVD	Ratio of V_C to V_D, ft^3/h or ft/h
V_D	dispersed-phase velocity, ft/h
V_{S1}	actual velocity of solvent phase, ft/h
V_{W1}	actual velocity of raffinate phase, ft/h
W	exiting raffinate, lb
W_1	raffinate rate, column section 1 (solute included), lb/h
W_C	actual continuous-phase rate, lb/h
W_D	actual dispersed-phase rate, lb/h
W_{e1}	lb solute per W exiting extractor
W_{e2}	lb solute per W entering extractor
X	Figure 7.11 x-axis
XAB	mole fraction of raffinate dissolved in solvent
XBA	mole fraction of solvent dissolved in raffinate
XCA	mole fraction of solute dissolved in raffinate
XCB	mole fraction of solute dissolved in solvent
x_m	molar concentration of solute in liquid phase
X_r	concentration of solute in the raffinate
Y	Figure 7.11 y-axis
Y_e	concentration of solute in the extract
y_m	mole fraction of solute in solvent

The introduction to this book discussed a historic event involving a liquid-liquid extraction problem. This chapter will address this problem and all other liquid-liquid extraction problems, extractor rating, and extractor column design. For our needs of separating components, some components form *azeotropes,* where the bubble point and dew point lines cross on a plot of temperature vs. pressure. This is a good application for liquid-liquid extraction, whereby a solvent can be mixed with the feed, extracting the desired component or components. These extracted components are later fractionated by distillation where the distilled component separations have large boiling point differences and no azeotropes. This is a good application of liquid-liquid extraction. There are numerous others, some of which are shown in Table 7.1.

TABLE 7.1   Liquid-Liquid Extraction Processes

Feed	Solute	Solvent
Water	Acetic acid	Benzene
Lubricating motor oil	Naphthenes	Furfural
Water	Phenol	Chlorobenzene
LPG	H$_2$S	MEA

MEA, monoethylnolamine.

General Definitions

There are three major components in extraction. These are:

- The *solute,* which is the component or components to be removed from the feed by a contacting solvent.

- The *feed,* which is a stream containing a solvent and the solute to be removed in the extraction process. The feed is called the raffinate stream after it begins to lose the solute.

- The *solvent,* which is the liquid that absorbs the solute. The solvent is called the extract stream when it leaves the contactor.

The feed stream (raffinate stream) and the solvent (extract stream) must be two immiscible streams. More simply, these two liquid streams must be insoluble or exhibit very low mutual solubility.

Liquid-liquid extraction involves contacting the feed stream (raffinate stream) with an immiscible liquid solvent in which one or more of the feed stream components is soluble. The component or components that are absorbed by this solvent are called the solute. Thus two different liquid phases are formed after addition or mixing of the solvent with the feed. The component that is more soluble in the solvent than in the feed will transfer to the solvent, a process known as *absorption.* The transferred component or components is called the solute.

Thus in liquid-liquid extraction there are two immiscible liquids of which at least one single component is soluble in both the feed and the contacting solvent. This component—the solute—transfers from one liquid phase to the other. This extraction transfer process falls into three categories:

1. The solute transfer occurs due to the solute being more soluble in the solvent liquid phase than in the feed (raffinate) liquid phase.

2. The solute transfer occurs due to the solvent phase having a greater mass flow or the solvent having a much smaller quantity of solute. Either or both of these situations may occur.

3. The solute transfer occurs because of a chemical reaction between the solute and the solvent, such as MEA or H_2S. H_2S reacts with MEA, forming a salt that is subsequently regenerated, freeing the MEA for recycling.

The most common of these three transfer processes is the first. This leads to the question of how soluble a solute is in any particular solvent. This is addressed in the following text. As for the solvent having a much greater flow rate than the feed rate, this leads to the question of how much contact time is necessary for solute transfer to occur. This questions of time, contact efficiency, and contactor size will also be addressed in later text. Finally, when there is a chemical reaction such as between H_2S and MEA, the chemical bonding of reaction does not occur until the H_2S has dissolved into the MEA aqueous phase (MEA-water). Thus, even for a chemical reaction, the mass transfer depends to a degree on the solubility of the solute.

Liquid Extraction Equilibria

The preceding definitions all point out that there are three components involved in liquid-liquid extraction. This is true even if each of these three components is itself a group of multiple components. The concept here is that each group within the multicomponent system plays the role of solute, feed-raffinate, or solvent-extract. The objective is to identify each as a discrete component, even if it is a component mixture, and apply the methodology as given herein.

Having identified discrete components as our criterion, the next problem is how to define the equilibrium of three components. In Chap. 2, for example, the vapor-liquid equilibrium of a flash—a single component transfer between two phases—was derived. A certain quantity of a component, called the y fraction, is vaporized into the vapor phase as an equal amount of the same component, called the x fraction, is dissolved in the liquid phase. The K value (the equilibrium constant of $K = y/x$) is used to determine the component distribution results. This same logic may be used here. Although we cannot use these same K values, as they do not apply, we can apply another database.

Treybal, in his book *Liquid Extraction* [1], works equilibrium material balances with triangular coordinates. The most unique and simple way to show three-phase equilibrium is a triangular diagram (Fig. 7.1), which is used for extraction unit operation in cumene synthesis plants [2]. In this process benzene liquid is used as the solvent to extract acetic acid (the solute) from the liquid water phase (the feed-raffinate). The curve D,S,P,F,M is the equilibrium curve. Note that every point inside the triangle has some amount of each of the three components. Points A,

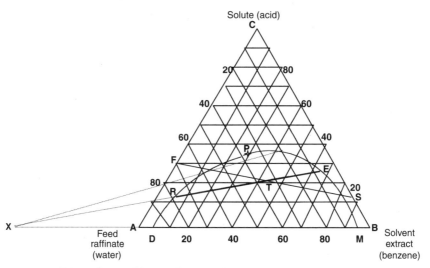

Figure 7.1   Triangular equilibrium diagram of water, benzene, and acetic acid.

B, and C each represent 100% by weight of their respective components on the triangular plot.

Unlike the flash equilibrium of Chap. 2, the extract and raffinate phases (compared to vapor-liquid) are miscible to some degree, as seen in the triangle diagram. Point D, for example, shows how much of the solvent (benzene) can dissolve into the feed (water) at equilibrium of the two liquid phases. Similarly, point M shows how much feed water will dissolve into benzene at equilibrium.

For a better understanding of liquid-liquid equilibrium, a brief study of a three-component triangle diagram is recommended. Using Fig. 7.1 as a reference, the following observations should be realized for any such diagram:

- All points to the left of curve segment D,R,P are considered outside the three-component, two-liquid-phase envelope and behave as one liquid phase only.

- All points to the right of curve segment D,R,P and inside the curve area are considered within the three-component, two-liquid-phase region. Here, at any point inside this curve D,R,P,E,M, two immiscible liquid phases exist at any point on the curve.

- Curve D,R,P,E,M is referenced to one temperature only (80°F). Curve variance occurs at other temperatures.

- All points to the right, to the left, and above curve D,R,P,E,M are a single liquid phase.

- Point F is the water (the feed-raffinate) with an amount of acetic acid to be extracted.

- Point S is the entering benzene (the solvent-extract) with an amount of unstripped acetic acid remaining.

- Drawing a line between S and F and making a point T represents a mixture of all the components inside the equilibrium curve.

- Point T, however, will separate into two liquid phases. If the extractor consists of one true equilibrium stage, this mixture (point T) separates into two phases represented by points R and E.

- Line R-E is called a *tie-line*. Numerous tie-lines can be made that approach point P. The R-E tie-line is an equilibrium line along which any three component mixture combinations will result in a fixed two-liquid-phase equilibrium at points R and E. One theoretical equilibrium stage is, of course, assumed.

- Point P is called the *plait point*. Only a single phase exists as the plait point is approached. The plait point may be considered akin to the vapor-liquid–phase envelope critical point from Chap. 2, where the vapor-liquid phases become one phase.

- Dotted lines beginning at the right side of the equilibrium curve and extending to a convex focal point X, called the *operating point* [1], are each equilibrium tie-lines like R-E. Thus any number of equilibrium tie-lines can be made, beginning at any point on the curve side M,E,P and extending to the convex focal point or operating point X. With the exception of the R-E line segment, all of these lines are shown as dotted lines in Fig. 7.1.

- Only one equilibrium tie-line, such as R-E, is required to fix the convex point X, which is then determined by making the dotted line shown through points B and A to the intersect of the line through points R and E. Any other equilibrium tie-line can then be made simply by drawing a line from X through the curves as done for the tie-line R-E.

- The equilibrium tie-line intersect on curve segment D,R,P represents a feed-raffinate three-component liquid composition (wt%) in equilibrium with the same tie-line intersect point on curve segment P,E,M. This P,E,M curve point is the three-component composition solvent-extract liquid phase.

- Beginning with the first equilibrium stage (assuming more than one theoretical stage is applied), and as the tie-lines progress upward on the D,R,P,E,M envelope curve, the solute concentration C increases in both liquid phases. The concentration of solvent B in the raffinate phase A, and the concentration of raffinate A in the solvent phase B, also both increase. This raffinate-solvent opposite-phase concentration increase is due to the increased transfer of the solute, which is totally soluble in both liquid phases.

- The solute equilibrium concentration C is normally always greater in the solvent liquid phase than in the feed-raffinate phase on the same tie-line. Note the slope of the tie-lines (the dotted lines).

- Concentrations have been referenced in Fig. 7.1 as weight percentages. Molar percentages can as easily be applied. The choice depends on the laboratory data available.

- All points on curve segments D,R,P (the feed-raffinate liquid phase) and P,E,M (the solvent-extract liquid phase) are discrete, separate, and immiscible liquid phases.

At this point you are probably feeling somewhat buffaloed. You have waded through numbers so intense that you are now exhausted and are wondering if you can really make any sense out of this liquid-liquid equilibrium method. Well, please hold on, as we will now turn to a simplified method that requires very limited equilibrium data. In fact, only one point of equilibrium data is sufficient in many applications for the method about to be introduced.

The Rectangular X-Y Equilibrium Diagram

The equilibrium diagram in Fig. 7.2 is considerably different from the triangular diagram in Fig. 7.1. It is the same extraction liquid-liquid equilibrium, with water the feed-raffinate, benzene the solvent-extract, and acetic acid the solute. The complete solution has been performed in Microsoft Excel, but other commercial software spreadsheets may also be used. Please notice first, as shown in the boxed area in columns A and B, that the equilibrium curve data have been entered. Next we will see how such a curve is made.

The question now is, how can a two-dimensional x-y rectangular plot represent a three-dimensional triangular plot? In answer, the first criterion to explain is the equilibrium distribution coefficient:

$$K_D = \frac{Y_e}{X_r} \qquad (7.1)$$

where K_D = distribution or equilibrium coefficient
X_r = concentration of solute in the raffinate
Y_e = concentration of solute in the extract

K_D is reasonably constant for the first 20 to 25% of solute in both liquid phases shown in Fig. 7.1. Above this area, K_D begins to change, usually increasing. However, most extraction operations are main-

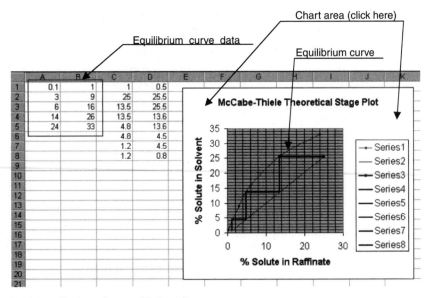

Figure 7.2 Rectangular equilibrium diagram.

tained in this 20%-or-less solute concentration range. For this reason, many liquid-liquid extraction equilibrium curves on a rectangular diagram can be a straight line drawn from the 0,0 point through only one K_D data point (K_D having been derived for solute concentrations of 25% or less in both liquid phases). In most liquid-liquid extraction operations, the solute concentration stays in this 25%-or-less range in both liquid phases. In fact, most selected solvent rates are large enough to keep the solute equilibrium in this concentration range. Please note that even though the feed may contain 40 to 50% solute or more, the solvent can have a flow rate great enough to achieve this solute concentration of 25% or less in the first equilibrium stage of the extractor. This 25%-or-less solute concentration will be the first stage in both liquid phases.

This phenomenon has been applied to many an operating plant extractor, even after the plant has been designed for other criteria. Indeed, some plants have added parallel extractors to achieve this lower solute concentration operating practice. One major reason so many operators pursue this lower solute concentration is that as the solute concentration increases above 25 or 30%, the equilibrium of both the solvent in the feed-raffinate and the feed-raffinate in the solvent increases. This increased switching of the feed and solvent means the liquid phases become less immiscible and therefore a very poor extractor separation is made. Remember that as the plait point P is approached, only one liquid phase exists. It is therefore recommended that first-stage extraction solute equilibrium concentrations be kept relatively low—say, not above 20% and preferably 10 to 15%. The second, third, fourth, and further theoretical equilibrium stages should then easily do the finishing job.

The Henry's law constant in gas-liquid absorption is very similar to the K_D in Eq. (7.1). The Henry's law constant is:

$$p = H * x_m \qquad (7.2)$$

where H = Henry's law constant, atm/mole fraction of solute in liquid phase
p = partial pressure of solute, atm
x_m = molar concentration of solute in liquid phase

The mol liquid in the H constant may be that of either the feed-raffinate or the solvent. x_m is the solute expressed as mol solute per mol of either feed or solvent in the liquid state. The atm in the H constant is the total pressure, which here is the vapor pressure of the solute at the system temperature.

A unique database is available for any extraction system for which component Henry's law constants are available. Please refer to the following presentation:

1. At equilibrium for any liquid-phase temperature, the vapor pressure of the solute in the feed-raffinate phase is equal to the vapor pressure of the solute in the solvent phase.

2. The derivation of the K_D distribution equation is:

$$p = y_m * H_s = x_m * H_f$$

and

$$\frac{y_m}{x_m} = \frac{H_f}{H_s}$$

$$K_D = \frac{y_m}{x_m} * \frac{MW_s}{MW_f}$$

and for a final usable K_D equation:

$$K_D = \frac{H_f}{H_s} * \frac{MW_s}{MW_f} \qquad (7.3)$$

where H_f = Henry's law constant for solute in feed liquid phase
H_s = Henry's law constant for solute in solvent liquid phase
K_D = equilibrium constant; distribution coefficient

MW_f = molecular weight of feed without solute
MW_s = molecular weight of solvent without solute
y_m = mole fraction of solute in solvent

A good resource for H_f and H_s, is Lange [3]. Although this nearly exhaustive reference source does not list Henry's law constants as such, the book does provide huge resources on gas solubility in various solvents, from which Henry's law constants can be calculated using the following equation (see App. C):

$$H_f = \frac{p}{x_m}$$

The Henry's law constant H is intended for relatively dilute concentrations of the solute in the solvent or the feed. Above about 20% by weight, these constants become inaccurate. This is true especially above 30%, where inaccuracies are too large for any validation or reliability. At these higher solute percentages, exacting laboratory data readings are strongly advised. For solute concentrations of 20% or less, however, the Henry's law constants given here are reasonably accurate and may be applied.

Notice that in Eq. (7.1), Y_e and X_r are both expressed as weight percentages of the solvent and feed-raffinate liquid phases, respectively. This means mathematically that the solute increments (solute gain or loss) of the two liquid phases must be added to the discrete solvent or feed to get the true weight percentage at any point of the extractor, including the inlets and outlets of the extractor. The following equation derivations may be applied to derive the proper rectangular curve generation and operating line:

$$W(W_{e2} - W_{e1}) = S(S_{e2} - S_{e1})$$

and

$$S = W \frac{W_{e1} - W_{e2}}{S_{e2} - S_{e1}} \qquad (7.4)$$

where S = entering solvent, lb
S_{e1} = lb solute per S entering extractor
S_{e2} = lb solute per S exiting extractor
W = exiting raffinate, lb
W_{e1} = lb solute per W exiting extractor
W_{e2} = lb solute per W entering extractor

To convert from these W_e and S_e values to X and Y values, or vice versa, it is necessary to execute the following equations:

$$Y_1 = \frac{S_{e1}}{1 + S_{e1}} \qquad (7.5)$$

and

$$S_{e1} = \frac{Y_1}{1 - Y_1} \qquad (7.5a)$$

$$Y_2 = \frac{S_{e2}}{1 + S_{e2}} \qquad (7.6)$$

$$S_{e2} = \frac{Y_2}{1 - Y_2} \qquad (7.6a)$$

$$X_1 = \frac{W_{e1}}{1 + W_{e1}} \qquad (7.7)$$

$$W_{e1} = \frac{X_1}{1 - X_1} \qquad (7.7a)$$

$$X_2 = \frac{W_{e2}}{1 + W_{e2}} \qquad (7.8)$$

$$W_{e2} = \frac{X_2}{1 - X_2} \qquad (7.8a)$$

Any point on the X-Y equilibrium curve may be found using the following equation:

$$K_D = \frac{Y_1}{X_1} \qquad \text{or} \qquad X_1 * K_D = Y_1 \qquad (7.9)$$

and combining Eqs. (7.1), (7.5), and (7.7):

$$\frac{W_{e1}}{1 + W_{e1}} * K_D = \frac{S_{e1}}{1 + S_{e1}} \qquad (7.10)$$

The subscript 1 and 2 denote the inlet and outlet of the liquid-phase streams. Take note that these subscripts denote the equilibrium curve points (points 1 and 5 in Fig. 7.2). Throughout this book the extractor is assumed to be countercurrent flowing raffinate and solvent streams. Y_1 is therefore the weight percentage of solute in the inlet solvent stream that is in equilibrium with the X_1 stream. X_1 denotes the weight percentage of the solute in the raffinate stream leaving the extractor. The

Y_1 stream denotes the weight percentage of the solute in the solvent entering the extractor column.

Now S_{e1} may be determined by converting Eq. (7.10) to:

$$S_{e1} = \frac{W_{e1}}{K_D + W_{e1} * K_D + W_{e1}} \tag{7.11}$$

An equation for W_{e1} may be derived similarly and given as:

$$W_{e1} = \frac{K_D * S_{e1}}{1 + S_{e1} + (K_D * S_{e1})} \tag{7.12}$$

Equations (7.11) and (7.12) may be used to determine the pounds of solute at any point on the respective equilibrium curve, including the entrance and exit points of the extractor. K_D may also be calculated from:

$$K_D = \frac{S_{e1}/(1 + S_{e1})}{W_{e1}/(1 + W_{e1})} \tag{7.13}$$

Equation (7.13) is one of the more useful equations in field operation data gathering. For any actual extraction process over time, numerous K_D values can be calculated. These are excellent and reliable data to use for plant expansions or new extractor units, or for rating any similar existing extractor. Another way in which K_D equilibrium data can be obtained is through actual operating data.

Many times, I have been asked, "How do I get the correct liquid-liquid equilibrium data?" I trust that the preceding text clearly answers this question. However, I request that users of this field gathering technique share findings with me so that others may equally benefit. Referring to our present-day laws regarding nondisclosure of proprietary data, we know company nondisclosure policies prevent the majority of these findings from being made public. Perhaps this is one reason the Japanese have advanced ahead of the good old USA in many technical areas. Some of the data can, however, be distributed by those who see this need and are open to sharing. A hearty thank you to those companies and individuals.

We have looked at four K_D equilibrium resources: laboratory data, triangular equilibrium data from publications, Henry's law constants, and operating field data. All four are reliable data sources if applied with limitations as described. All can be applied to the rectangular equilibrium curve (the distribution coefficient equilibrium curve) as shown in Fig. 7.2, which includes a listing of laboratory data for the extraction process shown. Please notice the data table in the boxed area of columns

A and B. These tabled data are the shown equilibrium curve points on the curve.

Column A in Fig. 7.2 is the weight percentage of solute in the feed-raffinate liquid phase. Point 1 in column A is the concentration of the exiting solute in the raffinate, and point 5 is the solute concentration in the feed-raffinate liquid phase entering the extractor column. Column B is the weight percentage of solute in the solvent liquid phase. Point 1 in column B is the concentration of the solute in the entering solvent liquid phase, and point 5 is the concentration of the solute in the exiting solvent liquid phase in the extractor column.

Each of the line numbers 1 through 5 on the left side of Fig. 7.2 corresponds with a point on the equilibrium curve. Now observe Fig. 7.3, which shows the source data for this equilibrium curve. To find this screen on your computer, simply bring up Excel and open the program called MTPlot72 on the CD supplied with this book. Then click on the

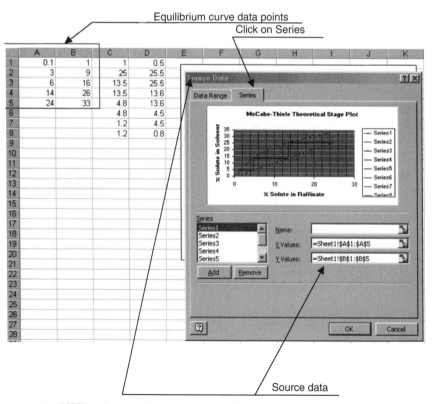

Figure 7.3 MTPlot72 equilibrium curve source data.

chart area of the display. (The chart area referenced is shown in Fig. 7.2.) Next click on Charts at the top of the computer display that appeared as soon as you clicked on the chart area of Fig. 7.2. A charts menu is now displayed. Click on Source Data in this menu window. The displayed source data window, as shown in Fig. 7.3, should now be seen. Next click on the Series tab. The complete Fig. 7.3 should now be seen on your computer screen.

If you have any trouble getting Fig. 7.3 to come up on your computer, please read the previous paragraph again and carefully follow the steps. Considerable effort has been expended here to lead you through this exercise of making a plot using Excel. This is an excellent tool and one the process engineer should find of unequaled value in countless similar applications. You do not need much familiarity with Excel to use this wonderful tool of our trade. One more important note here is that the purpose of this exercise is to give you a useful template that you may apply to many other similar applications. Please understand that this is not intended to be a complete guide to how to use and manipulate Excel or any other spreadsheet program. More simply, use the CD program MTPlot72 as a template for application to your particular needs in liquid-liquid extraction. Please refer to App. B for another display of MTPlot72.

The following text will give you a walkthrough of the use of this spreadsheet and the McCabe-Thiele equilibrium curve. MTPlot72 is applicable as a template in solving any liquid-liquid problem.

In Fig. 7.3 the equilibrium curve data are referenced to the spreadsheet columns A and B and line points 1 through 5. Look at how this notation is made in Excel:

X values = Sheet1!\$A\$1:\$A\$5

Y values = Sheet1!\$B\$1:\$B\$5

The Series notations designate the curve, the operating line, and each horizontal or vertical line making up the equilibrium stage steps. Each one of these listed diagram parts is denoted a Series. Delimiters are the dollar sign (\$) and the colon (:). You can change the equilibrium curve to any curve desired simply by increasing or decreasing the number 5 to any other number desired and by inputting the corresponding curve data points in columns A and B.

Example 7.1   The next part of the McCabe-Thiele diagram is the operating line. To generate the operating line, the problem for Fig. 7.2 must be stated. The following conditions are given. Find the required solvent rate and number of theoretical equilibrium stages to meet the treated water specification of 1.0% acid remainder. Benzene will be the solvent. Also determine what solvent inlet purity is required to meet this water specification. A step-by-step methodology will next be applied that may also be used as a template for solving any extractor problem, whether it involves design or rating:

solution

Step 1. Determine the feed-raffinate material balance. The feed—water plus acetic acid—enters the extractor at 1000 lb/h and carries 334 lb of dissolved acid. W_{e1} and W_{e2} are calculated from the given data and the following equations:

$$W_{e1} = \frac{X_1}{1 - X_1} = 0.3333 \qquad (7.7a)$$

$$W_{e2} = \frac{X_2}{1 - X_2} = \frac{0.01}{1 - 0.01} = 0.01 \qquad (7.8a)$$

where $W = 1000$ lb/h
$X_1 = 334/(1000 + 334) = 0.25$ or 25%

Step 2. Select solvent inlet purity. The solvent (benzene) enters the extractor at a point that must be reasonably true and obtainable for the equilibrium curve. Please note that we cannot cross the equilibrium curve and that the McCabe-Thiele steps will be to the right of the curve. The left side would be for points of miscible liquids, not two-phase. Since the water is to contain 1.0% solute at exit, a good estimate for the solvent entering is 0.5 wt % acid. Reviewing the Fig. 7.2 equilibrium curve, in order to meet this treated water 1.0% acid specification, the concentration of acid in the inlet counterflow solvent must be below 1.0%. We may assume the solvent has been nearly stripped free of acetic acid. Thus, 0.5% acid in the inlet solvent (benzene) is a reasonable assumption. Plotting the McCabe-Thiele equilibrium steps will show that this is a good and reasonable estimate to achieve the treated water specification of 1.0% acid remainder. Then,

$$S_{e1} = \frac{Y_1}{1 - Y_1} = \frac{0.005}{1 - 0.005} = 0.005 \qquad (7.5a)$$

Step 3. Select Y_2. Reviewing the equilibrium curve for the solvent outlet loading, make a reasonable estimate for the solvent liquid-phase loading Y_2. Here we have two degrees of freedom or, more simply, two unknowns, which are Y_2 (or S_{e2}) and S. Assuming either will set the other by solving Eq. (7.4), it is prudent here to consider the limitations of both Y_2 and S. First, consider Y_2, which is the amount of acid (solute) loading pick-up allowable in the solvent. Review Fig. 7.1 and notice that as the solute concentration in the solvent approaches the plait point P, the two liquid phases approach each other, becoming only one phase that will fail the extractor operation. Thus roughly 25% should be considered the upper limit for Y_2. In determining limitations of S, review the equilibrium curve in Fig. 7.2. Notice that if we increase the value of the solvent rate, with all other variables held constant, solvent loading decreases and the operating line tends to shorten. This means fewer theoretical stages are required to do the same extraction treatment job. Thus the extractor column will be shorter, but will also have a larger diameter requirement.

For this run, choose a Y_2 value and calculate the corresponding solvent rate required. An S value will be chosen in another example problem. Since the upper limit of Y_2 here is roughly 25%, select 25.5%.

$$S_{e2} = \frac{Y_2}{1 - Y_2} = \frac{0.255}{1 - 0.255} = 0.3423 \qquad (7.6a)$$

Step 4. Calculate S. Using the preceding information, the amount of required solvent (benzene) can now be calculated with Eq. (7.4):

$$S = W \frac{W_{e1} - W_{e2}}{S_{e2} - S_{e1}} \qquad (7.4)$$

$$S = 1000 \frac{0.3333 - 0.01}{0.3423 - 0.005} = 959 \text{ lb/h}$$

Thus the extractor column raffinate outlet rate and the solvent inlet rate are approximately equal. This is indeed the minimum solvent rate allowed, since a lower rate will overload the solvent, referencing the plait point. This rate will also set the required minimum extractor column diameter. For some refinery-type extraction operations, such as lube oil extractors, where relatively much larger solvent-raffinate rates apply, this method for determining minimum solvent rate is very economical and desirable.

Step 5. Install the operating line in the McCabe-Thiele diagram. Refer to Fig. 7.4 and notice where the operating line data input are. We have assumed a straight operating line since both liquid phases—the solvent and raffinate—lose essentially no mass to each other. Put another way, the solvent (benzene) starts with 959 lb/h at the bottom of the extractor column, and benzene is withdrawn at the top section of the column at a rate of 959 lb/h. Also, the raffinate (water) is fed to the column top section at a rate of 1000 lb/h and withdrawn at the bottom section at a rate of 1000 lb/h. Of course, the solute is added to these rates per equilibrium transfer from one phase to the other. Thus a straight line is installed from the top of the column operating points, 1 and 0.5, to the bottom operating points, 25 and 25.5, as shown in Fig. 7.4. The boxed area in columns C and D contains the operating line data points. Column C contains the raffinate (x-axis) points. Column D contains the solvent (y-axis) points.

Step 6. Perform the McCabe-Thiele equilibrium stage steps. Figure 7.5 displays the data points inputted for the first equilibrium step. As in all the other steps, here too the user may change any of the data in the boxed area of columns C and D before bringing up the Source Data window.

Please note here that the starting point for the first stage step was the bottom of the column step that was the last data point used for the operating line in Fig. 7.4. Here we are taking steps up the column, starting with the column bottom at the point shown in Fig. 7.5. This is somewhat confusing because the curve shown in these figures is actually inverted, with the curve

Operating line data points

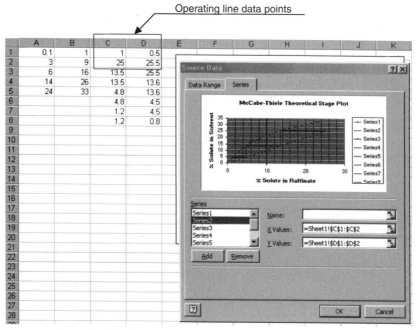

Figure 7.4   Operating line data in McCabe-Thiele diagram.

First equilibrium stage step from column bottom

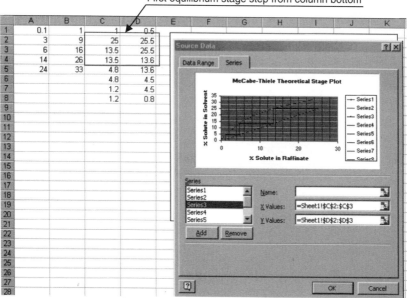

Figure 7.5   Data points inputted for the first McCabe-Thiele equilibrium stage step.

top being the column bottom. The rectangle diagram could have been drawn with the 0,0-X-Y point at the upper right corner. As this greatly deviates from conventional rectangular diagrams, much more confusion would have been created. The Fig. 7.4 and 7.5 curves should be the easier to conceive if you keep in mind the actual locations of the column tops and bottoms.

To do the first step as shown in Fig. 7.6, start at the operating line point, $Y = 25.5, X = 25$. Then, for the equilibrium point intersect, you may use trial and error until you get a good matching point on the curve. Please notice that when you try any point the line is immediately drawn on the Excel diagram. Note that this first curve intersect point is $Y = 25.5, X = 13.5$. The y-axis is the solvent solute weight percentage and the x-axis is the raffinate solute weight percentage. Note also that these numbers are inputted in columns C and D as shown in Fig. 7.5. Column C is the x-axis (solute percentage in the raffinate phase) and column D is the y-axis (solute percentage in the solvent phase).

The second part of the first step as shown in Fig. 7.6 is inputted similarly, starting on the previous curve intersect point, $Y = 25.5, X = 13.5$. Then, on line 4 of Fig. 7.5, the operating line point intersect is found again by trial and error. It should be pointed out here that this trial and error is very rapidly achieved, since only one number is found by trial and error, which here is Y. As shown in Fig. 7.5, Y is found to be 13.6. For more convenience and accuracy, Fig. 7.6 can be enlarged. For most applications, however, the $\pm 0.5\%$ accuracy shown here is most acceptable and adequate.

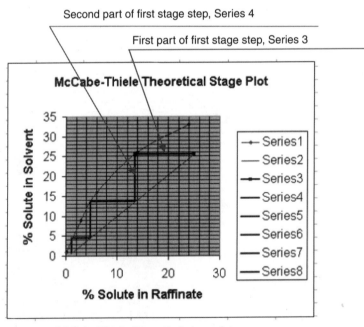

Figure 7.6 McCabe-Thiele theoretical stage plot.

The remaining three equilibrium steps are accomplished using the same procedure. Please note that the last step falls just slightly over the required raffinate (water) specification—1.0% solute (acetic acid) and 1.2% remainder. This is, of course, well within our ±0.5% accuracy, and for most applications this would be acceptable. For the present case, however, we will assume that this specification is critical and that 1.0% or less solute remainder must be achieved.

Now notice how small an equilibrium stage step was made for the third step in Fig. 7.6. It is very obvious that the next stage step will be even smaller, yet will require as much of the extractor column physical size as the first equilibrium stage step required. Thus, for a critical specification of 1.0% or less as stated here, a fourth stage is required. This fourth stage ensures that the specification will be met.

In the preceding example, the material balance equation, Eq. (7.4), was solved by assuming a maximum loading limit for the solvent. Also as stated previously, a smaller solvent loading could be applied if the extractor column diameter is adequately designed. Next, therefore, it is good to fix a much lower solvent loading, which requires a larger solvent rate as seen in Example 7.2. Fewer equilibrium stages will be required compared to Example 7.1. This would, of course, make the 1.0% solute remainder in the exiting raffinate easier to achieve.

Example 7.2 In this example, assume all the same items stated in Example 7.1. Here, however, we will start with step 3, in which a solvent loading Y_2 is fixed:

Step 3. Select Y_2. In Example 7.1, a maximum solvent loading of 25.5% solute was chosen. In this example, choose a lower loading of 12%, about half the previous value. Calculate S_{e2}:

$$S_{e2} = \frac{Y_2}{1 - Y_2} = \frac{0.12}{1 - 0.12} = 0.1364 \qquad (7.6a)$$

Step 4. Calculate S. With the S_{e2} value calculated in step 3, the solvent rate S can now be calculated from Eq. (7.4):

$$S = W \frac{W_{e1} - W_{e2}}{S_{e2} - S_{e1}} \qquad (7.4)$$

$$S = 1000 \frac{0.3333 - 0.01}{0.1364 - 0.005} = 2460 \text{ lb/h}$$

Step 5. Install the operating line in the McCabe-Thiele diagram. Figure 7.7 displays the completed diagram with the replaced operating line, using Fig. 7.6 as a template. Refer to Fig. 7.4 again and note that the following change was made to the series 2 data points:

column D, line 2: replace 25.5 with 12

When you make this change to the operating line and click on the spreadsheet, you see an immediate change in Fig. 7.6. The first stage step really

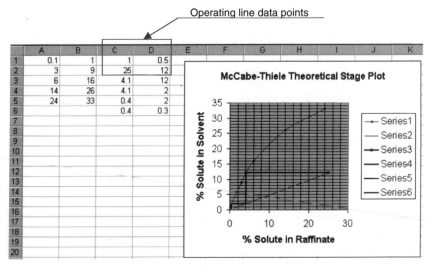

Figure 7.7   Replaced operating line in McCabe-Thiele diagram.

looked bad. Keep this operating line, however, and go to step 6, where we will run the equilibrium stage steps.

Step 6. McCabe-Thiele equilibrium stage steps. Refer again to Fig. 7.3, noting how to find this Source Data window and the included Series data. It is best to start without any equilibrium stage steps on the diagram. Therefore, click on each of the lines that make up the Example 7.1 stage steps, and then press the Delete key. This will clear your diagram, leaving the only equilibrium curve and the operating line. Now input the new values of Y (column D) and X (column C) as shown in lines 2 and 3 of Fig. 7.8.

Referring to Fig. 7.8, there are two inputs necessary for each Series line: the data point numbers and the X and Y values as shown for the Series3 data. The following code lines tell the program where to draw the lines.

X values = Sheet1!\$C\$2:\$C\$3

Y values = Sheet1!\$D\$2:\$D\$3

These X-Y values are for the first line of the first stage, named Series3. The second part of the first stage line is named Series4 and is denoted:

X values = Sheet1!\$C\$3:\$C\$4

Y values = Sheet1!\$D\$3:\$D\$4

There are two more such series lines to input with their respective data points. Again, use trial and error to determine the one data point for each step, using the McCabe-Thiele plot as in Example 7.1. After installing the last series line, your McCabe-Thiele plot should be as shown in Fig. 7.9.

The increased solvent rate has reduced the number of required equilibrium stages to only two. Please observe that in Fig. 7.8 our last step well exceeded

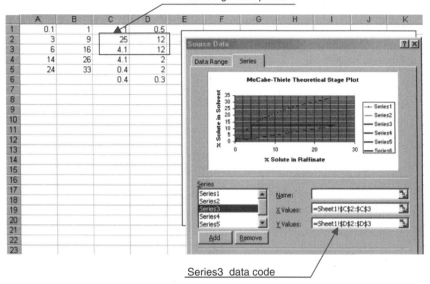

Figure 7.8   Series3 data in McCabe-Thiele diagram.

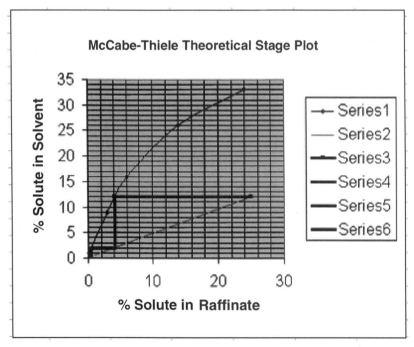

Figure 7.9   McCabe-Thiele theoretical stage plot with series data inputted.

the minimum 1.0% solute remainder in the raffinate (water). A final value of 0.4% solute in the raffinate is shown in Fig. 7.8, line 5, column C. Figure 7.9 also reveals this low 0.4% number. The obvious conclusion here is that it is good practice to check out higher solvent rates if the equipment design allows.

Extractor Operation

Figure 7.1 shows the true equilibrium of the two immiscible phases; however, this true equilibrium is never achieved in actual extractor column operation. Some deviation from the ideal equilibrium is always present. Given enough equilibrium stages, however, nearly true equilibrium can be achieved. As shown in Figs. 7.7 through 7.9, the operating line of the extractor column simply approaches equilibrium. Notice how close the operating line comes to the equilibrium curve. The closer this operating line is to the equilibrium curve, the closer the system is to true equilibrium. To actually achieve equilibrium would require too many stages. Generally, therefore, one to eight theoretical stages are sufficient for almost every extraction operation.

An extractor column is generally a tall, vertical packed tower that has two or more bed sections. Each packed bed section is typically limited to no more than 8 ft tall, making the overall tower height about 40 to 80 ft. Tower diameter depends fully upon liquid rates, but is usually in the range of 2 to 6 ft. Liquid-liquid extractors may also have tray-type column internals, usually composed of sieve-type trays without downcomers. These tray-type columns are similar to duoflow-type vapor-liquid separation, but here serve as contact surface area for two separate liquid phases. The packed-type internals are more common by far and are the type of extractor medium considered the standard. Any deviation from packed-type columns is compared to packing.

Packing vendors offer specialized packing for liquid-liquid extraction. Some of these offerings are:

Koch Engineering Company	SMV structured packing, metal or plastic
Sulzer Chemtech Company	Gauze structured packing, type CY
Norton	IMPT random metal packing

These packing materials come in various shapes and sizes, and each has preferred sizes for different column diameters.

Dispersed phase

A good question here is which liquid phase should be the dispersed phase and which should be the continuous phase. There are at least four conditions upon which to make this choice:

1. *Minimize liquid-film resistance.* Since the organic substance has a greater film resistance compared to an aqueous substance, it is advantageous to disperse the organic into the aqueous phase. Dispersing the organic phase generally reveals lower resistance to mass transfer, and therefore offers a higher packing efficiency by virtue of being the dispersed as opposed to the aqueous phase.

2. *Maximize the interfacial area.* Consider and identify which liquid phase has the greater volumetric flow. If the phase with the greater flow rate is dispersed, the resulting droplet interfacial area will be greater. Of course the phase with the greater flow rate should always be the dispersed phase, as explained later in this chapter. If the interfacial area is greater, the extraction efficiency is greater, requiring less extractor volume for the job. One may ask the question here, "Why or how is the interfacial area comparably increased?" Think about it. If you increase the flow rate, given the same droplet size for any flow rate, the interfacial area increases simply because the number of droplets increases.

3. *Maximize column capacity.* Consider which phase has the higher viscosity. Make the higher-viscosity phase the dispersed phase. If the lower-viscosity phase were the dispersed phase, the dispersed-phase droplet rise time would be greater. Thus column capacity is increased if the higher-viscosity phase is the dispersed phase. The droplet rise time is decreased accordingly, and the column capacity is therefore increased.

4. *Minimize packed extractor bed coalescence.* Consider the packing material to match your choice of continuous phase. The function of the continuous phase is to wet the packing. Two research scientists, Ballard and Piret [4], found that when the dispersed phase wetted the packing, it flowed as a film over the packing. This result is undesirable, as it considerably reduces mass transfer. Thus the continuous phase should in all cases be the packing wetting medium. How do you achieve this? First, ensure in your operations that the packing is submerged in the continuous phase before the dispersed phase is started. Second, select the proper material for the continuous phase using the generalized choices in Table 7.2 [5].

TABLE 7.2   Packing Material Choices in Regard to Continuous Phase

Continuous-phase wetted packing	Preferential continuous-phase type
Ceramic packing	Aqueous solvents
Plastic packing	Organic solvents
Metal packing	Aqueous or organic solvents

Table 7.2 is a general guide to which type of solvent-packing combination should be used for the continuous phase to avoid coalescence of the two phases. Water will form a stabilized emulsion, called *coalescence,* with many organic-type compounds. These stabilized emulsions are very difficult to break. Other solvent coalescence combinations also occur. They are discussed in considerable detail in Chap. 4. Settling time, added emulsion-breaking chemicals, temperature increase, and electrostatic field coagulation are the four principal process methods used to break these emulsions. Therefore, whenever a new extraction process is planned, it is mandatory to have in hand proven process compatibilities between the solvent phases—both feed and solvent— and the solute transferred. If actual operating experience is not available, then well-explored laboratory data are mandatory.

General Extraction Operation

When choosing a dispersed phase, it is good to consider each of the four previously listed guidelines. When a particular solvent fulfils the criteria of three of the four listings, it is best to go with such an advisory. The important thing to do here is to review both liquid phases, categorizing them for each of these four preferences.

A last note about the continuous phase is the fact that it must completely immerse the packing section where the mixing of the two phases takes place. The inner phase between the two liquid phases is therefore to be near the extractor column's dispersed-phase outlet. The extract stream, having gained the transferred solute, exits the column at the opposite end from where the raffinate stream exits. The raffinate stream is the inlet feed stream containing the extracted solute.

The objective of extraction is to selectively remove a solute from one liquid phase into a second liquid phase. The end result is that the solute is much more easily fractionated out of the second liquid solvent than from the first. There are a number of typical liquid-liquid extraction operations, some of which are listed in Table 7.3.

TABLE 7.3   Typical Liquid-Liquid Extraction Operations

Feed-raffinate	Solute	Solvent-extract
Refinery lubricating oil stock	Naphthenes	Furfural
LPG	H_2S	20% MEA
Gasoline	Mercaptans (R-S-H)	NaOH solution
Butane-butylene	Isobutylene	98% sulfuric acid
Reformate	Aromatics	Diethylene glycol
Water	Phenol	Chlorobenzene
Water	Acetic acid	Benzene
Water	Methyl-ethyl-ketone	Toluene

Equilibrium Stages Overview

To this point, we have covered how to determine the number of equilibrium stages required for any type of extractor. Three components, any of which may be a mixed group as shown in Table 7.3, make up this equilibrium stage assembly. Each equilibrium stage provides inner-liquid-phase contact between the two liquid phases, transferring the solute between the two liquid phases. A rectangular diagram curve provides a discrete, unique, and accurate method of applying the well-known McCabe-Thiele equilibrium step diagram. With this diagram, a material balance derivation of a step-by-step method has been presented, which may also be applied to any extraction process. Thus a means of equilibrium stage determination has been presented, offering a complete material balance for any extraction unit operation. The next part of this chapter deals with choosing an extractor contactor type, choosing the hydraulic mixing media, and sizing the extractor contactor.

Choosing the Contactor

Choosing the type of contactor depends heavily on how many theoretical stages are needed. In the past, if only one stage was required, a single mechanical mixer was used. Also, for only one stage, a simple inline static mixer may be used with a downstream settling drum for separation of the two liquid phases. In some applications rotating equipment is used to mix the two immiscible liquid phases and separate them downstream in a settling drum or tank.

In most cases, although items of mechanical rotating equipment appear to be good applications and are excellent mixers, an undesirable result is coalescence. In fact, coalescence is so severe that in some cases a stable emulsion occurs between the two liquid phases that is very difficult to break. These stabilized emulsions that result from mechanical mixing occur due to the shearing action of the mixing blades. When the dispersed-phase droplets are formed, this shearing action tends to promote emulsion-type bonding between these dispersed droplets and the two inner phases of liquid. This emulsion is a stabilized single phase composed of the two liquid phases.

Such inner-phase bonding may indeed grow out of control. The end result is frequent shutdowns. For this reason alone, most operating companies have chosen the packed column–type extractor, even if only one stage is required. The packed column–type extractor works by exploiting the gravity flow difference between the two liquid-phase densities only. Shearing problems are therefore nonexistent. Operators have found excellent results with a number of packing types, as shown in Tables 7.2 and 7.4.

TABLE 7.4   HETS, Packed Bed Depths Required for Modern Packings

Theoretical stages per bed	Packing size		
	1 in	1.5 in	2 in
1.5	4.3 ft	5.2 ft	6.2 ft
2.0	7.1 ft	8.5 ft	10.1 ft
2.5	9.8 ft	11.8 ft	14.0 ft

HETS, relative height equivalent to a theoretical stage.
SOURCE: R.F. Strigle, Jr., *Random Packings and Packed Towers*, Gulf Publishing, Houston, TX, 1987. Reprinted with permission.

Sieve plate columns may also be considered as an alternative to packing in extractor columns. A sieve tray is simply a stainless steel plate, placed in the column, having perforated holes. Like a fractionation tray, a sieve plate may also have a downcomer. The downcomer, although not absolutely required, is used for the continuous liquid-phase flow (down or up) through the mixing section of the column. Otherwise, without downcomers, the continuous phase flows through the perforated holes in the opposite direction of the dispersed phase. Usually three to six sieve tray–type plates make up one theoretical stage of the extractor column. There are limited data for determining the number of sieve plates that make one theoretical stage. Such data have not been deemed reliable enough to disclose herein. Perhaps one of my readers will accept this challenge and succeed.

Packing-type media in extractor columns are by far the preferred extractor type internals. In fact, Strigle [6] reveals the bed depth typically required per theoretical stage using current packing sizes. Such packing sizes are referenced to Norton chemical process products, metal Norton IMTP packing, Norton plastic pall rings, and Norton ceramic Intalox saddles. Through the courtesy of Gulf Publishing, this table is reprinted here as Table 7.4.

In keeping with the industry's worldwide choice, packed column–type liquid-liquid extraction column sizing will be addressed throughout the remainder of this chapter.

Stage Efficiency

Many authorities have claimed to be able to predict stage efficiency. For example, Treybal [7] gives the height of transfer units vs. flow velocity ratios as well as the height of transfer units vs. tower bed packing height. Many others have since made equal contributions, but none, prior to the 1980s, has resolved all problems in extractor efficiency determination. A comprehensive review of the literature to date has revealed a method, used here, that complies with most of the present-day notable authorities. First, tests performed by Nemunatis [8] showed that packed-column extraction efficiency is independent of continuous-phase velocity at 80 ft/h or greater. However, the dispersed-phase velocity produced considerable difference. Eckert [9] showed that the rate of mass transfer decreased rapidly as the residence time of the dispersed-phase droplets in the continuous phase increased. This increase in droplet residence time is an indicator that the droplet particles are impacting less, that is, they are having fewer collisions with the countercurrent continuous phase. Thus the reproduction of dispersed-phase droplets as often as is feasible tends to improve efficiency. This limits the packed-bed height, since reproduction of dispersed-phase droplets requires packed-bed dispersed-phase collection and redistribution to another packed-bed section. The conclusions that should be made here are:

1. Liquid-liquid extractor packing-bed height should be in the range of 6 to 10 ft.

2. When packing heights of more than 10 ft are required, redistribution trays or spray rings should be used to keep reasonable efficiency.

Eckert [9] showed that a relative height equivalent to a theoretical stage (HETS) vs. the dispersed-phase velocity revealed the packed-column efficiency, or simply the required height, to make one theoretical stage. (See Fig. 7.10). Eckert and others [6, 8] have shown that normally the theoretical packed-column stage requires 2.5 ft of column packed height. All this of course refers strictly to liquid-liquid extraction processing. Also, the continuous-phase velocity V_C (ft/h) and the dispersed-phase velocity V_D (ft/h) are referenced to the liquid-phase

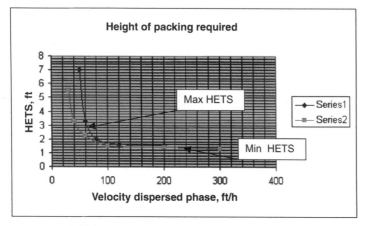

Figure 7.10 HETS vs. dispersed-phase velocity reveals packing height needed for one theoretical stage.

velocity through the full cross-sectional area of the extractor column. Accordingly, notice in Fig. 7.10 how the two curves tend to approach a packing height of 2 ft as the dispersed-phase velocity increases. Each curve of Fig. 7.10 represents one dispersed-phase velocity. Although more curves could have been made, these two curves may be interpolated within a V_D range of 10 to 150 ft/h with good results. If higher rates than this are used, the V_C scale in Fig. 7.10, will most likely exceed flooding velocity limits. More simply, outside the ranges in Fig. 7.10, the system is most likely to flood due to high velocities.

Thus, given here are two sources to determine the packed-bed height required per theoretical stage—Table 7.4 and Fig. 7.10. The recommendation is therefore that you review both referenced methods and select the one requiring the greater HETS, which is the more conservative answer. For rating an existing extractor in service, previous operations data should govern how much packing height makes up one theoretical stage. A third curve could therefore be added as part of Fig. 7.10 for any particular established operating case or cases. The importance of acquiring actual field operating data cannot be overstressed. Get as much information as possible and assemble it in a format such as those shown in Table 7.4 and Fig. 7.10.

Seibert et al. [10] found that liquid-liquid extraction HETS varied appreciably with variance of the dispersed-phase velocity. This effect of V_D is shown in Fig. 7.10. You are advised to compare this HETS with the HETS found in Table 7.4 and use the greater HETS for your answer.

Our industry is starved for data such as that shown in Fig. 7.10 and Table 7.4. Perhaps books and technical journal articles will eventually supply much missing information on these liquid-liquid extraction process areas. Nonetheless, these data as presented herein have been used successfully in many extractor design and rating solutions, and will no doubt be used in the same applications in the future. These tried and proven data are the best choice if actual field operating data with laboratory backup are not available.

Column Capacity

Determination of theoretical stage number and stage efficiency, or more simply HETS, has been established for any extraction process. The next item of order is the packed extractor column flooding limit. Just as fractionation columns must be sized for vapor and liquid, liquid-liquid extraction columns must be sized for flood limits.

Flooding in liquid-liquid extraction is similar to vapor-liquid flooding in that the controlling liquid-phase flow rate will cause blockage of the opposite liquid-phase flow if the diameter is too small. However, liquid-phase channeling can happen if the column diameter is too large. Keeping the packed-column diameter flood numbers at 50% or greater and keeping the upper flood limit at 90% or less is therefore advised. Having set these goals of a 50% minimum and a 90% maximum extractor packed-column flood limit, some brief research on current-day extraction flooding criteria and causes is in order.

Once theoretical stage efficiency and required packing height per stage have been determined, the next objective is to calculate the extractor column diameter requirement. This, of course, is the flood condition of the extractor. As explained earlier, the objective here is to achieve a 50 to 90% flood operating range for the extractor loading. Here it is prudent to first discuss flooding—what it is and how it occurs.

Flooding is an event in which one of the two countercurrent phases ceases to flow in the direction in which it should by gravity forces. Thus, in a flood condition, the flow of the flooded phase exits the column at the exit of the other liquid-phase. This necessitates a unit shutdown, which is a costly operation that entails production loss.

Flooding is principally caused by liquid phases coalescing due to high velocity of the dispersed phase or high velocity of the continuous phase. In the case of dispersed-phase high velocity, coalescing occurs when the wetted packing surfaces become dominantly captured by the dispersed phase. This happens when the dispersed-phase velocity becomes so great that it begins to disperse the continuous-phase wetted surface areas. The dispersed phase begins to coalesce, causing a third-phase interfacial surface layer. Eventually, as the dispersed-phase velocity increases, the coalescence becomes so great that the continuous-phase flow reverses, flowing out the exit with the dispersed phase. As for continuous-phase high velocity, the continuous-phase velocity can only reach a certain level before it tends to reverse the flow of the dispersed phase. A third interfacial-phase layer will form due to the collisions of the dispersed-phase droplets, eventually exiting with the continuous-phase liquid.

$$X = \frac{U_C}{\Delta_\rho} \left(\frac{O_i}{D_C}\right)^{0.2} F_e^{1.5} \tag{7.14}$$

$$Y = (V_D^{1/2} + V_C^{1/2}) \frac{D_C}{a} U_C \tag{7.15}$$

where Δ_ρ = density difference between liquid phases, lb/ft^3
   a = packing area, ft^2/ft^3
   D_C = continuous phase density, lb/ft^3
   F_e = packing factor
   O_i = interfacial surface tension, dyn/cm
   U_C = viscosity of continuous phase, cP
   V_C = continuous phase velocity, ft/h

V_D = dispersed phase velocity, ft/h
X = Figure 7.11 x-axis
Y = Figure 7.11 y-axis

Figure 7.11 shows modified Crawford-Wilke [11] correlation plot curves. Note that the y-axis is a type of Reynolds number, as discussed in Chap. 6. This y-axis number is similar to the Reynolds number, having density (D_C), viscosity (U_C), and velocity (V_C and V_D). If you review Chap. 6 and the Reynolds number, the same dimensional analysis is seen in the order given in Fig. 7.11 on the y-axis. The x-axis relates to viscosity (U_C), surface tension (O_i), density (D_C), and packing size factor (F_e). Originally the square root of the x-ordinate was used in the Crawford-Wilke correlation plotted against such a Reynolds number. Also, only one curve was made in this original work, the top curve labeled Crawford-Wilke in Fig. 7.11. This top curve represents the point at which the continuous phase is saturated with solute, in equilibrium condition. Eckert [9] reported that when V_C is increased, beginning at $V_C = 0$, the system floods before V_C reaches this saturation Crawford-Wilke curve.

Upon this discovery, Eckert began to research the Norton company's data files [9]. His findings revealed that the flood points were accurate at the family of curves shown in Fig. 7.11. Please note that these curves

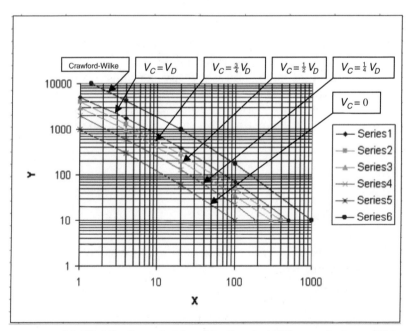

Figure 7.11   Modified flooding velocities of the Crawford-Wilke correlation and Eckert velocity ratios [9, 11].

depict the ratio of V_C to V_D at which the system floods. Eckert varied these ratios from $V_C = 0$ to $V_C = V_D$. Each curve in Fig. 7.11 represents a flood line of the extractor column. Note that V_C can never be greater than V_D without being in the flood condition.

Nemunatis [8] observed the same flooding conditions below the saturation top curve (the Crawford-Wilke curve). In that instance, flood was noted at 20% of this saturation curve. Strigle [6] suggests that design not exceed a 12% value of the shown family of 100% flood curves. Note again that, whatever the ratio of V_C to V_D, the curve for that specific ratio in Fig. 7.11 is indeed the 100% flood value for that particular extractor column condition. Therefore it is wise to place a loading 12% below flood levels before the extractor column is flooded.

These findings contribute to the work of Eckert [9], in which V_C values are held constant for each curve generated in the Fig. 7.11 plot. Thus Fig. 7.11 should be used to determine the required extractor column diameter and/or rate an existing column size for new loadings. Eckert's work agrees with the earlier Craford-Wilke single-curve finding, but Eckert also provided a way to establish reasonable factoring by use of the V_C to V_D ratio.

If you took note of Eq. (7.14), you noted the variable O_i, the interfacial surface tension constant. We know how to get surface tension constants for various materials from sources such as Lange [3], but what about interfacial surface tension? Here interfacial surface tension refers to the surface between the two liquid phases in the extractor. So how do we obtain or calculate this interfacial surface tension? For the answer, please see Fig. 7.12.

Taking account that the two liquid phases are immiscible, Trebal [7] derived a unique way to determine the interfacial surface tension without basing the data on the surface tensions of individual components. (See Fig. 7.12.) Please note that only the mole fractions of the three principals (solvent, raffinate, and solute) are required as inputs. In reviewing the x-axis of Fig. 7.12, Eq. (7.14), note that if the solvent dissolved in the raffinate and the raffinate dissolved in the solvent are both near zero, then the first part of the x-axis equation may be dropped, using only:

$$X = \frac{XCA + XCB}{2} \tag{7.16}$$

After an X value is found, the interfacial surface tension is easily read from Fig. 7.12.

The original Crawford-Wilke correlation [11] was based on eight different binary liquid systems, the packing of which is listed in Table 7.4. The void fractions for packings had a maximum value of 0.74. The liquid properties ranged as follows:

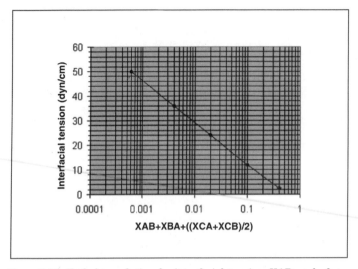

Figure 7.12 Trebal correlation for interfacial tension. XAB, mole fraction of raffinate dissolved in solvent; XBA, mole fraction of solvent dissolved in raffinate; XCA, mole fraction of solute dissolved in raffinate; XCB, mole fraction of solute dissolved in solvent.

Liquid viscosity	0.58 to 7.8 cP
Liquid density difference	9.4 to 37.2 lb/ft^3
Interfacial surface tension	8.9 to 44.8 dyn/cm

The modified correlation of Fig. 7.11 is based on data from the Crawford-Wilke work from six different tower packings having efficiency values up to 94%. All of these flooding data were based on hydraulic flow data with no mass transfer of solute. The extrapolation of Fig. 7.11 or its use outside the preceding physical property data ranges is not advised. Laboratory and/or pilot plant work is in order if data outside these values are needed.

The LiqX program on the accompanying CD uses a curve-fit equation to represent Fig. 7.11. In doing so, LiqX solves for the V_C to V_D ratio value in Fig. 7.11. Once this ratio is known, then one of the curves in the curve family of Fig. 7.11 is selected. The X value is calculated using Eq. (7.14). The Y value of Fig. 7.11 is then solved in Eq. (7.17) by installing the X value calculated from Eq. (7.14). Equation (7.18) is one of these curve-fit equations.

$$\text{VCVD} = \frac{W_C/D_C}{W_D/D_D} \qquad (7.17)$$

where D_C = continuous phase density, lb/ft^3
D_D = dispersed phase density, lb/ft^3

VCVD = ratio of V_C to V_D, ft^3/h or ft/h
W_C = actual continuous phase rate, lb/h
W_D = actual dispersed phase rate, lb/h

Please note that V/column area = ft/h. In Eq. (7.17), however, the column cross-sectional area would cause the values in the numerator and the denominator to be the same. An area value is therefore not shown in this equation.

$$Y_{25} = \exp\,[7.6087 - 0.8955 * \log X + 0.037543$$
$$* (\log X)^2 - 0.018665 * (\log X)^3 + 0.0013862 * (\log X)^4] \quad (7.18)$$

In Eq. (7.18), Y_{25} represents the y-axis value of the curve-fit equation for $V_C = 0.25\,V_D$. Other curves in Fig. 7.11 will also have Y value curve-solving equations similar to Eq. (7.18). Use any of the Fig. 7.11 curves, or simply interpolate between the curves for any V_C to V_D ratio less than 1.0.

Finally, the solved Y value for the extractor is used in Eq. (7.19) to solve for VCVD$_2$ and subsequently the V_C value for the extractor, Eq. (7.20). Note that this V_C value is the 100% flood rate (ft/h) for the continuous phase in the extractor column. Please also note that the V_D value is the 100% flood rate (ft/h) for the dispersed phase.

$$\text{VCVD}_2 = \left(Y * \frac{a * U_C}{D_C}\right)^{0.5} \quad \text{solution of } V_C^{0.5} + V_D^{0.5} \quad (7.19)$$

$$V_C = \left(\frac{\text{VCVD}_2}{1 + (1/\text{VCVD})^{0.5}}\right)^2 \quad (7.20)$$

Before applying Eqs. (7.21) through (7.27), the continuous and dispersed phases must be identified as being the solvent or raffinate phases.

$$V_{S1} = \frac{S_1}{3.1416 * \text{SDEN}_1 * (\text{DIA}_1/2)^2} \quad (7.21)$$

where V_{S1} = actual velocity of solvent phase, ft/h

$$VW_1 = \frac{W_1}{3.1416 * \text{RDEN}_1 * (\text{DIA}_1/2)^2} \quad (7.22)$$

where DIA_1 = extractor internal diameter, column section 1, ft
RDEN_1 = raffinate density, column section 1, lb/ft^3
S_1 = solvent rate, column section 1 (solute included), lb/h
SDEN_1 = solvent density, column section 1, lb/ft^3

V_{W1} = actual velocity of raffinate phase, ft/h
W_1 = raffinate rate, column section 1 (solute included), lb/h

Equations (7.21) and (7.22) refer to a specific column size, i.e., the internal diameter DIA_1 (ft). The number 1 in DIA_1 refers to the top part of the column. DIA_2 would, of course, refer to the bottom part of the column. You may ask, "How do I know the correct column diameter to use in these equations?" The answer is, you don't, as this is indeed a trial-and-error solution method. In many cases you may have an existing column size for which you wish to find the limiting loads both for solvent and for raffinate.

Equations (7.21) and (7.22), together with Eqs. (7.24) and (7.25), determine the flood percentage of the extractor in reference to the particular column section for which the flow rates S_1 and W_1 are taken. The flood loading range sought has a lower limit of 50% and an upper limit of 80%. A flood percentage equal to or anywhere between these two limits is the objective for a feasible, operable extractor. Thus a flood value of 50 to 80% in Eq. (7.24) or (7.25) is acceptable. For maximum design solvent and raffinate rates, an 80% flood loading from these equations is more desirable because the normal operation rates should be lower than this but still within these flood loading limits.

Once you calculate V_C in Eq. (7.20), you can easily find the dispersed-phase velocity using Eq. (7.23). Note that you now have the 100% flood values determined. This means you can now calculate the actual liquid-phase velocities from Eqs. (7.21) and (7.22) and ratio them to these 100% flood velocities. Thus this ratio times 100 is the flood condition of your extractor column. Please see Eqs. (7.24) and (7.25), where this is done.

$$V_D = \frac{V_C}{\text{VCVD}} \tag{7.23}$$

Assuming the solvent phase has been designated the dispersed phase, Eqs. (7.24) through (7.27) apply:

$$\text{FLOOD}_{S1} = \frac{V_{S1}}{V_D} * 100 \tag{7.24}$$

$$\text{FLOOD}_{R1} = \frac{V_{W1}}{V_C} * 100 \tag{7.25}$$

where FLOOD_{R1} = extractor loading flood percentage, raffinate flow basis
FLOOD_{S1} = extractor loading flood percentage, solvent flow basis

At this point the extractor flood number is calculated for the continuous and dispersed liquid phases. See Eqs. (7.24) and (7.25). The higher flood percentage governs. It is common for the dispersed-phase flood rate to be the higher flood percentage and therefore the governing liquid phase. Nonetheless, either may be the higher flood percentage. One more step may now be made for design purposes. Many cases, such as installation of a new extractor, require calculation of the diameter for the extractor column. This is a simple calculation now because V_D and V_C have been calculated in Eqs. (7.20) and (7.23).

Earlier in this section, the desired flood range of 50 to 90% was presented. These limits are again emphasized here. The lower limit of 50% is fixed because a lower flood number introduces channeling of liquid phases. Here, little contact occurs between the two phases, and thus solute transfer is very poor. For this reason a 50% or higher flood is necessary; in fact, 80% flood is a good target, since it ensures ample capacity loading for the maximum rates and thereby the lowest equipment investment cost. Loadings of V_C and V_D between 80% and 90% are feasible, but this removes all capacity, which could prove insufficient for those times when a little more loading is required. Some loading surges should be designed with an allowance—say 10%—for surges and for recycled off-spec startup product and the like. In summary, design a new extractor for 80% maximum rate loadings. Since we already know the floading velocities V_D and V_C for any extractor from Eqs. (7.20) and (7.23), we can now use these flood velocities to calculate the extractor diameter required for an 80% flood. These calculations are given in Eqs. (7.26) and (7.27).

$$\text{DIA}_{1S} = \left(\frac{W_D}{3.1416 * \text{SDEN}_1 * 0.8 * V_D} \right)^{0.5} * 2 \qquad (7.26)$$

$$\text{DIA}_{1R} = \left(\frac{W_C}{3.1416 * \text{RDEN}_1 * 0.8 * V_C} \right)^{0.5} * 2 \qquad (7.27)$$

where DIA_{1R} = required extractor column internal diameter for raffinate phase, ft
DIA_{1S} = required extractor column internal diameter for solvent phase, ft

Equations (7.26) and (7.27) are based on 80% extractor column flood. (Note the 0.8 factor in both equations.) Please note that V_C and V_D are the superficial full-column internal diameter velocities. *Superficial* simply means that you calculate these velocities as though the respective liquid phase were the only material in the full-column cross-sectional area moving at the calculated velocity rate. More simply, divide the flow rate (lb/h) by the column internal area times respective density ($ft^2 \times lb/ft^3$) to get the column's liquid-phase velocities. This logic is applied in these two equations deriving diameters required for 80% flood. Please note that both the dispersed phase and the continuous liquid phase are checked for the required diameter. The larger calculated diameter governs.

Example 7.3   Example 7.3 will be an extension of Example 7.2, where the extraction of acetic acid is accomplished using the solvent benzene. The only difference will be the rates of solvent (benzene) and raffinate (water). The same ratios of feed, solute, and solvent will be maintained, only all rates will be multiplied by 10 in order to present a full-scale operation. The theoretical stage number will be maintained as calculated in Example 7.2 (2.0 stages). The following seven steps will be performed in a step-by-step procedure, applying the previous equation logic. Here we will solve for the column diameter required, the type of packing required, and depth of packing required. The following data are given in accordance with Example 7.2:

Feed raffinate, water rate	10,000 lb/h, 62 lb/ft³, 18 MW
Solute (acid) rate	3,340 lb/h, 48 lb/ft³, 64 MW
Solvent (benzene) rate	24,600 lb/h, 52 lb/ft³, 78.1 MW

solution

Step 1. First, solve the VCVD ratio, applying Eq. (7.17). Before solving this equation, you must first choose a dispersed liquid phase. Here the solvent having the greatest flow is chosen. You must also determine where the solvent enters the column and which column section you will check.

It is good practice to check both the top and bottom sections of the extractor column for diameter sizing, as either can be the determining loading size. Since the solvent is the much larger load factor by virtue of its mass flow, select the top section of the column for sizing. Note also that the solvent (benzene) will have nearly all of the solute loading as it leaves the top section. We must therefore include the solute in the top section loading for the diameter determination. Now Eq. (7.17) may be solved:

$$\text{VCVD} = \frac{W_C/D_C}{W_D/D_D} \tag{7.17}$$

$$\text{VCVD} = \frac{10,000/62}{27,940/51.5} = 0.2973$$

Notice that the mass flow of the solvent phase in Eq. (7.17) includes the solvent and solute summed and the solvent liquid-phase density calculated on the mass weight percentage factored for both the solvent and the solute.

Step 2. Calculate O_i. In order to find O_i it is first necessary to solve Eq. (7.14). Here it will be assumed that little or no raffinate dissolves into the solvent and that little or no solvent dissolves into the raffinate. XAB and XBA values are therefore zero in Eq. (7.16). Also, in our example, nearly all the extract dissolves in the solvent, and therefore Eq. (7.16) becomes the mole fraction of solute in the solvent divided by 2:

$$XAB + XBA + \frac{XCA + XCB}{2} = \frac{52/(315 + 52)}{2} = 0.071 \qquad (7.28)$$

From Fig. 7.12, O_i = 14 dyn/cm.

Step 3. Having determined a VCVD ratio and O_i values, the next step is solving Eq. (7.14) and then finding the flood values from Fig. 7.11.

$$X = \frac{U_C}{\Delta_p}\left(\frac{O_i}{D_C}\right)^{00.2}(F_e)^{1.5} = 1.58 \qquad (7.14)$$

From Fig. 7.11, Y = 1600.

Step 4. The objective in this step is to use the Y value determined in Fig. 7.11 to calculate V_C. Equation (7.19) is used to calculate $VCVD_2$. Substituting the VCVD ratio, a V_C value is calculated using Eq. (7.20). This is an ethical and valid procedure using two equations and two unknowns.

$$VCVD_2 = \left(Y * \frac{a * U_C}{D_C}\right)^{0.5} \qquad \text{solution of } V_C^{0.5} + V_D^{0.5} \qquad (7.19)$$

$$VCVD_2 = \left(1600 * 63 * \frac{0.25}{62}\right)^{0.5} = 20.16$$

$$V_C = \left(\frac{VCVD_2}{1 + (1/VCVD^{0.5})}\right)^2 \qquad (7.20)$$

$$V_C = \left(\frac{20.16}{1 + (1/0.297^{0.5})}\right)^2 = 50.6 \text{ ft/h}$$

Step 5. Calculate the dispersed solvent-phase velocity next, using Eq. (7.23):

$$V_D = \frac{V_C}{VCVD} \qquad (7.23)$$

$$V_D = \frac{50.6}{0.297} = 170.4 \text{ ft/h}$$

Step 6. Calculate the dispersed- and continuous-phase flood values using the V_C and V_D values from Eqs. (7.24) and (7.25). At this point you need to assume a column diameter. A good guess for a first trial is a 3-ft internal diameter. A second diameter trial may be necessary, but you need to start at this point again only for another diameter assumption.

$$V_{S1} = \frac{S_1}{3.1416 * SDEN_1 * (DIA_1/2)^2} \qquad (7.21)$$

$$V_{S1} = \frac{27,940}{3.1416 * 51.5 * (3.0/2)^2} = 76.75 \text{ ft/h}$$

$$V_{W1} = \frac{W_1}{3.1416 * \text{RDEN}_1 * (\text{DIA}_1/2)^2} \tag{7.22}$$

$$V_{W1} = \frac{10,000}{3.1416 * 62 * (3/2)^2} = 22.82 \text{ ft/h}$$

Now calculate the flood percentage:

$$\text{FLOOD}_{S1} = \frac{V_{S1}}{V_D} * 100 \tag{7.24}$$

$$\text{FLOOD}_{S1} = \frac{76.75}{170.4} * 100 = 45\%$$

$$\text{FLOOD}_{R1} = \frac{V_{W1}}{V_C} * 100 \tag{7.25}$$

$$\text{FLOOD}_{R1} = \frac{22.82}{50.6} = 45\%$$

Since both flood numbers are equal, no selection is necessary. Normally you should select the greater flood loading for the controlling number.

Having calculated a flood value for the assumed diameter discussed earlier, it is now time to find the preferred diameter for the extractor loadings. This preferred diameter is, of course, an 80% extractor column diameter loading.

$$\text{DIA}_{1S} = \left(\frac{W_D}{3.1416 * \text{SDEN}_1 * 0.8 * V_D} \right)^{0.5} * 2 = 2.24 \text{ ft for 80\% flood} \tag{7.26}$$

$$\text{DIA}_{1R} = \left(\frac{W_C}{3.1416 * \text{RDEN}_1 * 0.8 * V_C} \right)^{0.5} * 2 = 2.25 \text{ ft for 80\% flood} \tag{7.27}$$

where $\text{RDEN}_1 = 62 \text{ lb/ft}^3$
$\text{SDEN}_1 = 52 \text{ lb/ft}^3$
$V_C = 50.6 \text{ ft/h}$
$V_D = 170.4 \text{ ft/h}$
$W_C = 10,000 \text{ lb/h}$
$W_D = 27,940 \text{ lb/h}$

Both are close loadings with the raffinate phase slightly larger. This is mainly due to the fact that the continuous phase floods at a lower V_C. The flood numbers are truly too close to call for any significant difference. Thus a diameter of 2.25 ft is required for the desired 80% flood loading.

Step 7. In this step the height of a theoretical stage (HETS) in the column packing is determined. Once the HETS is determined, the total bed depth (or height, as it may be called) is determined. Please review Table 7.4 and Fig. 7.10. Using these references, select the larger HETS for your column packing height. Please note that 1.5-in structured packing has been selected and that the previously calculated V_D is 76.75 ft/h. Two theoretical stages are required.

From Fig. 7.10, the HETS is found to be 2 ft. For two stages, a total bed depth of 4 ft is required, as calculated in Example 7.1. Table 7.4 shows that a total bed depth of 8.5 ft is required. Since 8.5 ft is less than the 10-ft limit on packing height per bed, only one bed is required.

Running the LiqX Program

Next run the executable program on the CD supplied with this book. Duplicate Fig. 7.13 by clicking on the Run Program button. Please note that all the answers you have calculated are displayed as the answers in Fig. 7.13. The difference in the answers is due to the equation accuracy used for curve-fitting and the inaccuracies we have made in rounding off and reading the data charts.

The program allows the user to input the X and Y values. You may directly calculate the X value and read the Y value directly from Fig.

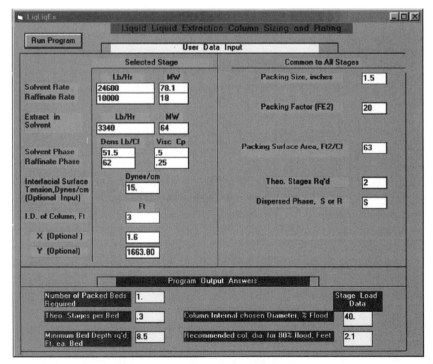

Figure 7.13 LiqX screen capture, user-inputted diameter of 3 ft.

7.11. Whether to read Fig. 7.11 or allow the program to calculate the Y value with algorithms is up to you. Both answers serve as a check in that they should be nearly equal. If they are not, you may be out of range of the program, as well as Table 7.4 and Fig. 7.13. Almost all problems should, however, be in these ranges.

Some discussion of the extractor loading flood sensitivity to the column diameter is merited here. Please notice the difference in the flood numbers in Figs. 7.13 and 7.14 (40% vs. 44.3%). The only difference between these two computer runs is a slight change in the user-inputted diameter. You may also input direct calculations and chart readings for the X and Y values. Why not try several different inputs to see what happens?

The LiqX algorithms achieved answers close to those found with hand calculations. Some judgment must be used to choose the better answer. It is good to go with the more conservative answer, which here is the one found by hand calculation. These hand calculations have derived a 2.5-ft internal diameter for 80% flood, whereas LiqX came up with a value of 2.1 ft. For other problems, LiqX's answers could just as easily be the more conservative ones. The 80% flood is a good design

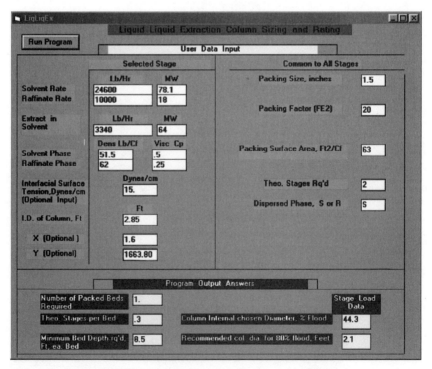

Figure 7.14   LiqX screen capture, user-inputted diameter of 2.85 ft.

criterion and is strongly recommended for new units. You may add 10% to the program's answer for new equipment good design practice.

This completes our review of the current state of the art in liquid-liquid extraction equipment design and rating. I only wish I had had this technique in hand 45 years ago when my supervisor came to me with a very disturbed voice and the adjectives to go with it, telling me my cold acid unit was really messed up and being shut down. I can only now pass the challenge to the upcoming generation in heartfelt hope that you will indeed overcome this deficiency with the techniques I have shown you in this book. The challenge is now yours. Good luck and deepest wishes for your successful service!

References

1. Treybal, Robert E., *Liquid Extraction,* McGraw-Hill, New York, 1951, pp. 156–160.
2. Treybal, Fig. 6.43, p. 173.
3. Dean, John A., and Lange, Norbert Albert, "Thermodynamic Properties," *Lange's Handbook of Chemistry & Physics,* 15th ed., McGraw-Hill, New York, 1998, Sec. 6.
4. Ballard, J. H., and Piret, E. L., *Ind. Eng. Chem.* 42:1088, 1950.
5. Strigle, Ralph F. Jr., *Random Packings and Packed Towers,* Gulf Publishing, Houston, TX, 1987, Table 11-1, p. 241.
6. Strigle, Table 11-8, p. 260.
7. Treybal, Robert E., *Liquid Extraction,* 2d ed., McGraw-Hill, New York, 1963, p. 132.
8. Nemunatis, R. R., *Chem. Eng. Prog.* 67(11):60, 1971.
9. Eckert, J. S., *Hydrocarbon Processing* 55(3):117, 1976.
10. Seibert, A. F., Humphrey, J. L., and Fair, J. R., *Evaluation of Packings for Use in Liquid-Liquid Extraction Processes,* University of Texas, 1985.
11. Crawford, J. W., and Wilke, C. R., *Chem. Eng. Prog.* 47(8):423, 1951.

Chapter 8

Process Equipment Cost Determination

Nomenclature

ACF	annual cash flow, $
BHP	brake horsepower, hp
CC	total construction cost, $
C_p	specific heat,
C_p/C_v	$(MW \times C_p)/[(MW \times C_p) - 1.986]$
eff	compressor efficiency, %
F_a	auxiliary equipment factor, $
F_m	material type factor, $
F_p	pressure rating factor, $
F_{ps}	shell-side pressure rating factor, $
F_{pt}	tube-side pressure rating factor, $
F_s	tray spacing factor, $
F_t	tray type factor, $
GHP	gas horsepower, hp
H_{ad}	adiabatic horsepower, ft · lb/lb
i	annual interest rate, %
k	gas specific heat ratio, C_p/C_v
LC	land cost, $
MW	molecular weight of gas stream
P_1	inlet pressure, psia
P_2	outlet pressure, psia
P_o	plant worth at startup time, $

R universal gas constant, 1545/MW

SV   salvage value, $

T construction or payment schedule, yr

T_i inlet temperature, °R

WC   working capital, $

w_i gas rate, lb/min

Z average gas compressibility factor of compressor inlet and outlet gas streams

The process engineer is always challenged to determine and design refinery, chemical plant, and oil and gas production equipment. Equally challenging is the determination of discrete equipment cost for the equipment the engineer has determined and designed. In many situations he or she is also assigned the cost analysis project. How do we as process engineers find or calculate equipment cost ethically, reasonably, and accurately. Not only is this a challenge for the major equipment fabrication shop cost of free onboard (FOB shipping point), but it is our duty to determine its complete turnkey field installation cost, including the cost of all necessary major equipment supporting items. This chapter offers a methodology to meet this challenge with what is commonly called the *factor cost analysis method*. These major equipment items become what will be called in this chapter *equipment modules*. Equipment module cost shall be factored herein to include the following:

- Piping
- Concrete
- Steel
- Instruments
- Electrical
- Insulation
- Paint

Additional costs are as follows:

- Labor
- Indirect cost

The labor is both shop fabrication cost and the field erection cost. The indirect cost cannot be neglected, as it also is a major factored cost, like labor.

Factored cost methods each have a ±25% error accuracy and in some cases a ±20% error accuracy, which is more likely if only one or at most three major items of equipment are involved. For greater equipment numbers, the 25% error accuracy is more probable. This difference is largely due to indirect cost uncertainty. In nearly all budgetary cost determinations by the process engineer, this method is most acceptable.

This chapter offers cost curves of equipment types updated to year 2000. Much data is referenced to actual vendor cost of delivered equipment, FOB shop. Greater accuracy requires direct vendor quotations for the particular equipment.

More accurate cost estimating is called *definitive cost estimating*. This is the next step up from the factored method, and it involves determining the quantity of each item together with detailed sizing. For example, here a reasonably close estimate is made for all piping runs, length, and size of each pipe. The definitive cost estimate yields a ±10 to 15% error accuracy. The extent of detail includes vendor guarantee quotations and preliminary piping and instrumentation drawing (P&ID) equipment, instrumentation, piping, and electrical layout with quantity discounts.

A third refinement is the detailed design cost estimate. This cost estimate has a ±0 to 5% error accuracy. It requires a complete engineering design detailing and can be executed only after at least half of the engineering design package is completed for construction. Most refinery, chemical plant, and oil/gas production projects execute this type of cost estimating halfway through the detailed design phase of engineering.

The process engineer will, during the course of design, have to make many decisions on alternative processing routes. The key element in the decision must be cost-effectiveness.

The technical literature is rife with detailed methods of cost-estimating process equipment. For the purpose of this chapter, the costs assembled in Table 8.1 will orient the novice designer to the approximate price of common types and sizes of plant equipment. The values shown are the purchased, as opposed to the installed, cost of equipment. A few items worthy of note from Table 8.1 are as follows:

- Carbon steel costs are typically half those of stainless steel construction.

- Centrifugal, motor-driven pumps are the only type of process equipment that is relatively cheap.

- The prices shown for distillation columns are misleadingly low. Most of the cost of a conventional distillation column is associated with the reboiler, condenser, pumps, reflux drum, and internal trays or packing. Equipment costs can be ratioed by the 0.6 power with size to obtain rough prices for larger or smaller sizes.

The total cost of building a process plant comes as quite a shock to the entry-level process engineer. Most surprising are the tremendous

TABLE 8.1   A Sample of Equipment Costs for Common Process Application
(2001 costs)

Equipment	Costs, $
Shell and tube heat exchanger, all carbon steel, 16-foot tubes, low-pressure service, 1500 ft² heat exchanger surface	19,200
Same as previous item, except all 304 stainless steel construction, 1500 ft² exchanger surface	35,200
A centrifugal pump putting up 100-psi differential head, and pumping 200 gpm	4,000
A centrifugal, motor-driver compressor rated for 2000 brake hp	880,000
A reciprocating motor-driven compressor rated for 2000 brake hp	1,120,000
A 25-tray distillation column, carbon steel, rated for 300 psig and 8-ft diameter. Does not include reboiler, condensers, trays, pumps, and so on	208,000
Same tower as in previous item, 50 trays	304,000
25 316 stainless steel valve trays for an 8-ft-diameter tower	64,000
25 carbon steel valve trays for an 8-ft-diameter tower	32,000
2000 ft³ of stainless steel, perforated ring tower packing	80,000
Air-cooled exchanger, carbon steel, 100,000 ft² extended surface, including fan and motor	224,000
A water cooling tower, 20,000 gpm	640,000
A box or cylindrical fired heater, 500-psig tube pressure rating, carbon steel tubes, 100 mmBtu/h heat duty absorbed, no air preheater	1,920,000
Vertical drum, 10,000-gal capacity, low-pressure service, 304 stainless steel	72,000

costs associated with erecting, instruments, and piping up a relatively inexpensive pump or drum.

Installation factors for several major classes of process equipment are summarized in Table 8.2. The factors are typical but not necessarily accurate for a specific case. They will vary with the total cost of each equipment category; that is, the cost for installing 10 identical heat exchangers is less per exchanger than the cost of installing a single exchanger. Also, plant location, labor efficiency, and general economic conditions can markedly affect installation costs. However, the factor shown should be sufficiently accurate to permit selecting between alternatives.

The installation costs are based on the individual pieces of equipment that form part of a major process unit. As such, the cost of the entire process unit is defrayed among the individual drums, compressors, pumps, towers, and so on. Hence, the installation costs cover such items as the following:

- Foundations
- Process piping and insulation

TABLE 8.2   Installation Factors*

Equipment category	Factor
Fired heaters	3.0
Finfan air coolers	2.8
Shell and tube heat exchangers	3.3
Shop-fabricated drums	6.6
Shop-fabricated reactors	5.3
Shop-fabricated distillation towers	5.9
Field-fabricated vessels	7.0
Pumps and drivers	4.3
Compressors	2.8
Skid-mounted, preassembled units, such as package dryers or desalters	3.0
Alloy fractionation trays	2.4

* Multiply equipment quotes received from suppliers by these factors to obtain approximate installation costs. Based on lump-sum contract labor and materials.

- Instrument loops
- Control valves
- Utility systems inside the unit
- Control room
- Engineering
- Firefighting equipment
- Startup manuals
- Sewers

The installation factors do not cover costs such as steam and power-generation facilities needed to support the process unit's operation or feed and product tankage.

Installation Cost for Individual Process Items

If a $100,000 heat exchanger is to be added to an existing crude unit, the cost of installing it will normally be far less than the $230,000 calculated from Table 8.2. This is because the sewers, control room, and instrument loops are already in place. The foundation, piping, and insulation of the exchanger might cost $80,000 to $160,000. It is simply not possible to estimate an installed cost of one or two items of process equipment without considering each case on an individual basis.

Almost any category of process equipment can be purchased on the used-equipment market. Some plants have been almost entirely constructed from such equipment. The following recommendations will help to avoid trouble:

- Purchase of an entire plant, intact, from the original owner is usually okay.
- Purchasing individual pieces of equipment is fraught with hazard.
- Used shell-and-tube heat exchangers should be disassembled for inspection prior to purchase. Check that the tube-and-shell baffles coincide with the as-built drawing.
- Plan to retray used distillation towers.
- The installation costs for used equipment are frequently higher than for custom-built equipment.

Other than saving procurement time for long-lead-time items, it is not often cost-effective to purchase used equipment. The fundamental reason is that the major part of the cost of a project is the installation cost. Also, one is never sure about the actual condition of used process equipment.

Major Plant Costs

Table 8.3 summarizes the actual or estimated prices to build a variety of chemical and refinery process plants. The stated costs do not include associated tankage, utilities, effluent treatment, service roads, general-purpose buildings, spare parts, or all the other components required to complete a major project. These additional offsite facilities are typically considered to add 50% onto the cost of a project. [1, 2]

Cost Estimation

Although detailed discussions of various capital cost–estimating methods are not part of the intended scope of this work, some comments are pertinent.

All capital cost estimates of industrial process plants can be classified as one of four types:

1. Rule-of-thumb estimates
2. Cost-curve estimates
3. Major equipment factor estimates
4. Definitive estimates

The capital cost data presented in this work are of the second type, cost-curve estimates.

TABLE 8.3   Typical Costs of Some Common Chemical and Refinery
Process Units (based on 2001 costs)

Unit	Cost, $ million
Sour water stripper, 500 gpm	3
Aromatic extraction and fractionation, 20,000 barrels per stream day (BSD)	29
Depentanizer, 50,000 BSD	5
Hydrogen plant based on natural gas feed, 40 million standard cubic feet per day (mmscfd)	48
Propylene concentration unit, 5000 BSD	12
Coal gasification, 50,000 mmBtu/d	224
Stack gas scrubber, 100,000 lb/h of fuel burned	24
Gas/oil hydrocracker, 30,000 BSD	152
Three-way gasoline fractionator, 16,000 BSD total feed	10
Delayed coker, 600 tons per day (tpd)	64
Light resid desulfurizer, 50,000 BSD	112
Claus sulfur recovery unit, 200 long tons per day (ltpd)	11
Amine stripper, 200 ltpd	10
Sulfur plant tail-gas unit, 200 ltpd	6
Naphtha reformer, 40,000 BSD	88
Crude unit with both atmospheric and vacuum distillation, 200,000 BSD	96
Fluid catalytic cracking unit for gas/oil, 25,000 BSD	77
Butane splitter (ISO and normal) 10,000 BSD	8
Sulfuric acid alkylation unit, 15,000 BSD	50
Diesel oil or #2 oil hydrosulfurizer, 40,000 BSD	45
Caustic regenerator for dissolved mercaptans, 2000 BSD	6
Gasoline sweetening, 30,000 BSD	3
LPG treater for HD-5, 10,000 BSD	5
Mercaptan extraction for alkylation unit feed stock, 18,000 BSD	3

Rule-of-thumb estimates.   The rule-of-thumb estimates are, in most cases, only an approximation of the magnitude of cost. These estimates are simply a fixed cost per unit of feed or product. Some examples are as follows:

Complete coal-fired electric power plant	$693/kW
Complete synthetic ammonia plant	$57,750/tpd
Complete petroleum refinery	$5,775/bpd
Alkylation unit	$2,310/bpd

These rule-of-thumb factors are useful for quick, ballpark costs. Many assumptions are implicit in these values, and the average deviation from actual practice can often be more than 50%.

Cost-curve estimates.   There is a power exponent–ratio method of estimating corrections for the major deficiency in Table 8.3, previously illustrated by reflecting the significant effect of size or capacity on cost. These exponent-ratios indicate that costs of similar process units or plants are related to capacity by an equation of the following form:

$$\frac{\text{Plant A cost}}{\text{Plant B cost}} = \frac{(\text{plant A capacity})^x}{\text{plant B capacity}}$$

This relationship was reported by Lang [3], who suggested an average value of 0.6 for the exponent x. Other authors have further described this function.

Cost curves of this type have been presented for petroleum refinery costs in the past. The curves presented herein have been adjusted to eliminate certain costs such as utilities, storage, offsite facilities, and location cost differentials. Separate data, included in indirect cost, provide the cost of these items. The facilities included have been defined in an attempt to improve accuracy.

It is important to note that most of the cost plots have an exponent that differs somewhat from the 0.6 value. Some of the plots actually show a curvature in the log-log slope, which indicates that the cost exponent for these process units varies with capacity. Variations in the log-log slope (cost exponent) range from about 0.5 for small-capacity units up to almost 1.0 for very large units. This curvature, which is not indicated in the previously published cost curves, is due to paralleling equipment in large units and to disproportionately higher costs of very large equipment in large units, such as vessels, valves, and pumps. The curvature in the log-log slope of cost plots has been recently described by Chase [4].

The cost-curve method of estimating, if carefully used and properly adjusted for local construction conditions, can predict actual costs within 15%. Except in unusual circumstances, errors will probably not exceed 25%.

Major equipment factor estimates.   Major equipment factor estimates are made by applying multipliers to the costs of all major equipment required for the plant or process facility. Different factors are applicable to different types of equipment, such as pumps, heat exchangers, and pressure vessels. Equipment size also has an effect on the factors.

It is obvious that prices of major equipment must first be developed to use this method. This requires that heat and material balances be completed in order to develop the size and basic specifications for the major equipment.

This method of estimating, if carefully followed, can predict actual costs within 10%.

A shortcut modification of this method uses a single factor for all equipment. A commonly used factor for petroleum refining facilities is 3.0. The accuracy of this shortcut is of course less than when using individual factors.

Definitive estimates. Definitive cost estimates are the most time-consuming and difficult to prepare, but they are also the most accurate. These estimates require preparation of plot plans, detailed flow sheets, and preliminary construction drawings. Scale models are sometimes used. All material and equipment is listed and priced. The number of labor hours for each construction activity is estimated. Indirect field costs, such as crane rentals, costs of tools, and supervision, are also estimated. This type of estimate usually results in an accuracy of ±5%.

Summary form for cost estimates

The items to be considered when estimating investment from cost curves are as follows:

Process units

Heaters, furnaces, heat exchangers, pressure vessels, pumps, compressors, and so forth

Storage facilities

Tanks, pressure vessels

Steam systems

Cooling water systems

Offsites

Special costs

Used equipment credit

Contingencies

Total

Storage facilities. Storage facilities represent a significant investment cost in most refineries. Storage capacity for crude oil and products varies widely at different refineries. In order to properly determine storage requirements, the following must be considered: the number and type of products, method of marketing, source of crude oil, and location and size of refinery.

Installed costs for "tank farms" vary from $18 to $28 per barrel of storage capacity (the higher cost is for smaller capacity). This includes tank, piping, transfer pumps, dikes, fire protection equipment, and tank-car or truck-loading facilities. The value is applicable to low-vapor-pressure products such as gasoline and heavier liquids. Installed cost for butane storage ranges from $46 to $104 per barrel, depending on size. Costs for propane storage range from $58 to $150 per barrel. (These costs are as of 2001.)

Land and storage requirements. Each refinery has it own land and storage requirements, depending on location, with respect to markets and crude supply, methods of transportation of the crude and products, and number and size of processing units.

Availability of storage tanks for short-term leasing is also a factor, as the maximum amount of storage required is usually based on shutdown of processing units for turnaround at 18- to 24-month intervals, rather than on day-to-day processing requirements. If sufficient rental tankage is available to cover turnaround periods, the total storage and land requirements can be materially reduced, as the land area required for storage tanks is a major portion of refinery land requirements.

Three types of tankage are required: crude, intermediate, and product. For a typical refinery that receives the majority of its crude by pipeline and distributes its products in the same manner, about 13 days of crude storage and 25 days of product storage should be provided. The 25 days of product storage is based on a 3-week shutdown of a major process unit. This generally occurs only every 18 months or 2 years, but sufficient storage is necessary to provide products to customers over this extended period. A rule-of-thumb figure for total tankage, including intermediate storage, is approximately 50 barrels of storage per barrel per day (bpd) crude oil processed.

Steam systems. An investment cost of $35 per lb/h of total steam generation capacity is recommended for preliminary estimates (costs as of 2001). This represents the total installed costs for gas- or oil-fired, forced-draft boilers operating at 250 to 300 psig, and all appurtenant items such as water treating, deaerating, feed pumps, yard piping for steam, and condensate.

Total fuel requirements for steam generation can be assumed to be 1200 Btu, lower heating value (LHV), per pound of steam. A contingency of 20% should be applied to preliminary estimates of steam requirements. Water makeup to the boilers is usually 5 to 10% of the steam produced.

Cooling water systems. An investment cost of $58 per gpm of total water circulation is recommended for preliminary estimates. This represents the total installed costs for a conventional induced-draft cooling tower, water pumps, water-treating equipment, and water piping. Special costs for water supply and blowdown disposal are not included.

The daily power requirements, kWh/day, for cooling water pumps and fans is estimated by multiplying the circulation rate in gpm by 0.6. This power requirement is usually a significant item in total plant power load and should not be ignored.

The cooling tower makeup water is about 5% of the circulation. This is also a significant item and should not be overlooked. An omission factor, or contingency of 15%, should be applied to the cooling water circulation requirements.

Other utility systems. Other utility systems required in a refinery are electric power distribution, instrument air, drinking water, fire, water, sewers, waste collection, and so forth. Since these are difficult to estimate without detailed drawings, the cost is normally included in the offsite facilities.

Offsites. Offsites are the facilities required in a refinery that are not included in the costs of major facilities. A typical list of offsites follows. Obviously, the offsite requirements vary widely between different refineries. The values shown here can be considered typical for grassroots refineries when estimated as outlined in this text.

Crude oil feed, BPSD	Offsite costs, % of total major facilities costs*
Less than 30,000	20
30,000 to 100,000	15
More than 100,000	10

* Major facilities as defined herein included process units, storage facilities, cooling water systems, and steam systems.

Offsite costs for the addition of individual process units in an existing refinery can be assumed to be about 15 to 20% of the process unit costs. Items considered to be offsites in this work are as follows:

Electric power distribution

Fuel oil and fuel gas facilities

Water supply and disposal

Plant air systems

Fire protection systems

Flare, drain, and waste disposal systems

Plant communication systems

Roads and walks

Railroads

Fencing

Buildings

Vehicles

Ethyl blending plants

Special costs. Special costs include land, spare parts, inspection, project management, chemicals, miscellaneous supplies, and office and laboratory furniture. For preliminary estimates, these costs can be estimated as 4% of the cost of the process units, storage, steam systems, cooling water systems, and offsites. Engineering costs and contractor fees are included in the various individual cost items.

Contingencies. Most professional cost estimators recommend that a contingency of at least 15% be applied to the final, total cost determined by cost-curve estimates of the type presented herein.

The term *contingencies* covers many loopholes in cost estimation of process plants. The major loopholes include cost data inaccuracies, when applied to specific cases, and lack of complete definition of facilities required.

Escalation. All cost data presented in this book are based on U.S. Gulf Coast construction averages for 2001. This statement applies to the process unit cost curves, as well as values given for items such as cooling water systems, steam plant systems, storage facilities, and catalyst costs. Therefore, in any attempt to use the data for current estimates, some form of escalation or inflation factor must be applied. Many cost index numbers are available from the federal government and from other published sources. Of these, the *Chemical Engineering* Plant Cost Index is the most readily available and probably the most commonly used by estimators and engineers in the U.S. refining industry. A comparison of years is shown here.

Year	*Chemical Engineering* Plant Cost Index
1970	125.7
1980	261.2
1985	325.4
1995	381.1
2000	398

If these indices are compared by converting them to a 1970 base, the following values are obtained:

Year	*Chemical Engineering* Plant Cost Index
1970	100.0
1980	209
1985	260.3
1995	304.9
2000	318.4

The use of these indices is subject to errors inherent in any generalized estimating procedure, but some such factor must obviously be incorporated in projecting costs from a previous time basis to a current period.

Escalation or inflation of refinery investment costs is influenced by items that tend to increase costs, as well as by items that tend to decrease costs. Items that increase costs include obvious major factors, such as the following:

1. Increased cost of steel, concrete, and other basic materials on a per-ton basis

2. Increased cost of construction labor and engineering on a per-hour basis

3. Increased costs for higher safety standards and better pollution control

Major equipment modular material cost method. The modular cost method was first established by Guthrie [2]. This same methodology is applied here, but with updated and added cost factoring.

The base cost of a major equipment module is herein defined as the specific equipment item fabrication cost, such as that of a fractionation column, a field-erected furnace, a shell/tube heat exchanger, or process pump. Each of these items will be given a fabrication cost. A figure cost curve will be presented for each major equipment item, and will have cost of the item vs. a key process variable. These key process variables could be the furnace absorbed-heat duty, the shell/tube exchanger tube surface area, and the pump discharge pressure/gpm product number.

Determining the value of these key variables will fix a fabrication cost for the particular major equipment item. Simply refer to the particular major equipment item type in this chapter and read the cost curve at the value of the key variable.

One more step is necessary to fix a major equipment item modular material cost. This involves the particular selection of the item materials, the operating pressure, and the equipment type. Adding all three of these additional supplemental costs to the base cost from the given curve gives us the total material cost. Please note that adding these three items is more simply done by multiplying the curve chart base cost by factors. See Tables 8.1 and 8.2. These factors are determined again by the same key variable that was used to determine the base from the cost curve. Some equipment items offer a selection of the varied types, such as a canned-type pump vs. an inline type. The materials, pressure, and the equipment type are each given factors to add to the principal major equipment base fabrication cost.

Labor cost. Once the total material cost is determined, labor cost is factored from this total major item material cost. Labor cost factors are based on new equipment installation.

Indirect cost. Indirect cost includes a large list of varied supporting equipment, utilities, and land for the modular major equipment cost method proposed herein. These items are commonly called offsites, which include cost of such items as roadways, land, buildings, warehouses, spare parts, maintenance shops, and electric power and water utilities. In addition to offsites, the indirect cost also includes the field-erection equipment, such as erection cranes, temporary construction buildings, welding supplies, and trucks. All of these items are necessities for the major equipment modules.

Indirect cost normally ranges from 20 to 35% of material plus labor of the modular major equipment cost. In most cost analysis of complete processing units, the indirect cost is very significant. Indirect cost factors should therefore always be a part of the summary sheet's total cost analysis and never overlooked. For most refinery, chemical plant, and oil- and gas-treating facilities, a 25% indirect cost factor is a well-accepted number in the industry, worldwide. The 25% factor of the material plus labor is therefore advised and is used in this chapter.

Process furnace cost. Process furnaces are used for the larger-sized process heating unit operations, more specifically for heating duties of 30 mmBtu/h and larger, up to 500 mmBtu/hr. (The abbreviation mmBtu/h simply means millions of British thermal units per hour.)

The structural profiles are usually box-type or A-frame with multitube banks and integral stacks.

Figure 8.1 is based on field-erected cost for furnaces designed to elevate hydrocarbon streams to 700°F at 500 psi (maximum) with absorbed heat duties in excess of 10 mmBtu/h. All tube banks are carbon steel, except pyrolysis or reformer furnaces, which have stainless tubes. Process furnaces generally have overall fuel efficiency of 75% (lower fuel heating value conversion to actual absorbed heat). Although most furnace vendors claim much higher efficiencies, this 75% efficiency value is

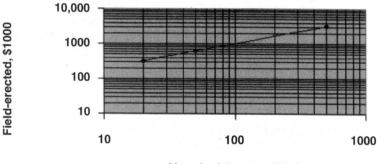

Figure 8.1   Furnace base cost.

found true in most furnaces after a year or more of operation. This efficiency factor also holds true for oil-fired and gas-fired systems.

In Table 8.4, add each applicable factor, then multiply by Fig. 8.1 base cost. You may interpolate between values shown.

Process furnace total base cost = Base cost (Fig. 8.1) × $(F_t + F_m + F_p)$ × escalation index

The escalation index is simply the yearly increase of the particular major equipment item, referred to in *Chemical Engineering*'s published index [5].

The next cost factoring is applying all the supporting equipment costs that make up the entire furnace module. These supporting items are listed in Table 8.5.

The total modular factor, 218.7%, multiplied by the base cost, is equal to the particular equipment module field-erected and ready to commission.

TABLE 8.4   Process Furnace Cost Factors

	Base cost, $1000		
Material and type	40	400	4000
Pyrolysis, F_t	1.12	1.10	1.09
Reformer, F_t	1.40	1.35	1.34
Chromium/molybdenum, F_m	0.27	0.25	0.24
Stainless, F_m	0.38	0.35	0.34
	Tube design pressure, psi		
Type	100	1000	1200
Pyrolysis, F_p	0	0.15	0.50
Reformer, F_p	0	0.25	0.70

TABLE 8.5   Furnace Module Associated Cost Factor Additions, in Percent

Piping	17.7
Concrete	8.1
Steel	—
Instruments	3.9
Electrical	2.0
Insulation	2.0
Paint	—
Subtotal	33.7
Labor	32
Total direct materials and labor	65.7
Base cost factor	100
Total direct cost	165.7
Indirect cost factor (× direct cost)	1.32
Total modular factor	218.7

Direct-fired furnace. Fired heaters are designed to increase the process temperature of oil and gas streams. This increase of temperature in most every case does not change molecular structure. Thus, temperatures up to 500°F maximum with 400°F design are very common. Designs are usually cylindrical, with vertical radiant tube banks fired by oil/gas combination burners.

Figure 8.2 is based on shop fabrication up to 15 mmBtu/h heat absorption units and field-erected 20 mmBtu/h heat absorption units and larger. Direct-fired heaters generally are 30 mmBtu/h duty or less, although field erections have been much greater, 50 mmBtu/h or greater.

Figure 8.2 is based on direct-fired heaters, both shop and field-erected, with carbon steel tubes, and includes all direct materials, shop and field labor, and subcontractor overhead and profit.

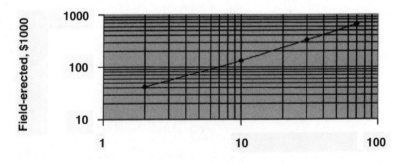

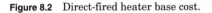

Absorbed duty, mmBtu/h

Figure 8.2   Direct-fired heater base cost.

TABLE 8.6 Direct-Fired Heater Cost Factors

Material and type	Base cost, $1000		
	4	40	400
Heating media, F_t	1.35	1.33	1.32
Chromium/molybdenum, F_m	0.47	0.45	0.44
Stainless, F_m	0.77	0.75	0.73
Type	Tube design pressure, psi		
	100	1000	1100
Cylindrical, F_p	0	0.15	0.50
Heating media, F_p	0	0.15	0.30

In Table 8.6, add each applicable factor, then multiply by Fig. 8.2 base cost. You may interpolate between values shown. Add dollars for foundations, external piping, and other factors as provided in Table 8.7.

The total modular factor, 227.5% for the direct-fired heater, multiplied by the base cost, is equal to this particular equipment module field-erected and ready to commission.

Shell/tube exchanger cost. The most common major equipment item in most any process is the shell/tube exchanger. Classification is defined by the construction characteristics, such as floating head, U-tube, fixed-tube sheet, or kettle-type reboiler. These type designations are TEMA designs. In fixing a cost analysis of all TEMA designs, a floating-head-type shell/tube design has been chosen because it is slightly more expensive compared to the other choices and therefore slightly more conservative.

TABLE 8.7 Direct-Fired Heater Module Associated Cost Factor Additions, in Percent

Piping	16.5
Concrete	9.0
Steel	—
Instruments	4.8
Electrical	3.0
Insulation	4.0
Paint	—
Subtotal	37.3
Labor	30
Total direct material and labor	67.3
Base cost factor	100
Total direct cost	167.3
Indirect cost factor (× direct cost)	1.36
Total modular factor	227.5

Alloys and high pressure are usually involved on the tube side. However, both shell-side and tube-side metal alloy additions are accounted for here.

Figure 8.3's cost curve is based on floating-head shell/tube exchanger construction with carbon steel shell and tubes. The cost figures may reasonably be extrapolated up to 10,000 ft^2 of bare outside tube surface area. The subsequent tables, Tables 8.8 to 8.12, include factors for design type, materials of construction, and design pressures up to 1000 psi. Cost factors for foundations, field materials, field labor, and indirect cost may be obtained from these tables. Add each applicable factor, then multiply by Fig. 8.3 base cost. You may interpolate between values shown.

The shell/tube exchanger base cost is as follows:

$$\text{Total base cost} = \text{base cost} \times (F_t + F_m + F_{pt} + F_{ps}) \times \text{escalation index} \quad (8.1)$$

The escalation index is simply the yearly increase of the particular major equipment item, referred to *Chemical Engineering*'s published index. The equipment cost index is 438 as of January 2001. If you are calculating an equipment cost for, say, 2007, and this same referenced cost index for that year is 450, then the cost index is 450/438, or 1.027. These yearly equipment cost index numbers, 438 and 450, come from *Chemical Engineering*'s monthly equipment cost index. Of course, 450 is an assumed index number, since the 2007 index has not been published as of this writing.

Table 8.12 reveals the supplemental equipment and labor costs that make up the completed module material and labor cost.

$$\text{S/T exchanger } M + L \text{ total direct cost, } \$ = 2.313 \times \text{total base cost, } \$ \quad (8.2)$$

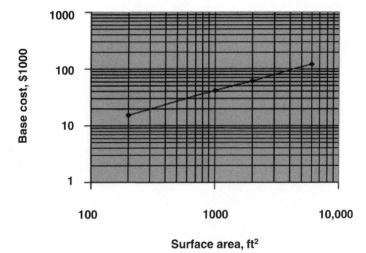

Figure 8.3 Shell and tube exchanger base cost.

TABLE 8.8   Shell/Tube Heat Exchanger Cost Factors

Design type	Base cost, $1000		
	4	40	400
Kettle reboiler, F_t	1.37	1.35	1.34
U-tube, F_t	0.87	0.85	0.84
Fixed-tube sheet, F_t	0.81	0.79	0.78

TABLE 8.9   Shell/Tube Heat Exchanger Cost Factors

Design type	Tube bundle design pressure, psi		
	100	1000	2000
Kettle reboiler, F_{pt}	0.05	0.07	0.11
U-tube, F_{pt}	0.05	0.07	0.12
Floating head, F_{pt}	0.04	0.06	0.10

TABLE 8.10   Shell/Tube Heat Exchanger Cost Factors

Type	Shell-side design pressure, psig		
	100	1000	1100
Kettle reboiler, F_{ps}	0.15	0.25	0.50
U-tube, F_{ps}	0.10	0.18	0.40
Floating head, F_{ps}	0.10	0.18	0.40

TABLE 8.11   Shell/Tube Heat Exchanger Cost Factors

Material type, shell/tube	Tube OD surface area, ft^2		
	100	1000	10,000
Admiralty/admiralty, F_m	1.10	1.25	1.40
Molybdenum/molybdenum, F_m	1.80	1.90	2.10
Carbon steel/molybdenum, F_m	1.40	1.65	1.90
Stainless steel/stainless steel, F_m	3.0	3.25	4.0
Carbon steel/stainless steel, F_m	2.0	2.5	3.25
Monel/monel, F_m	3.5	4.0	5.0
Carbon steel/monel, F_m	2.25	3.0	4.25
Titanium/titanium, F_m	11.0	12.5	15.0
Carbon steel/Titanium, F_m	5.0	8.0	10.0

The total $M + L$, material and labor cost, is shown in Table 8.12 as a direct cost factor, $231.3. This is a percentage-type cost factor, a percentage of the total base cost. All materials of the fabrication shop and field materials are included. Labor in both the fabrication shop and field are equally included. Thus, this single figure in Table 8.12 is a factor for deriving the total direct cost of a new shell/tube heat exchanger fabricated and field-installed, ready to commission.

TABLE 8.12   Shell/Tube Exchanger Module Associated Cost Factor Additions, in Percent

Piping	45.0
Concrete	5.0
Steel	3.0
Instruments	11.0
Electrical	2.0
Insulation	4.8
Paint	0.5
Subtotal	71.3
Labor	61
Total direct material and labor	131.3
Base cost factor	100
Total direct cost	231.3
Indirect cost factor (× direct cost)	1.38
Total modular factor	319.2

Some discussion is due here regarding the makeup of an equipment module. What is an equipment module per se? It is the essential supplemental materials and equipment that make the equipment item a functional part of the overall processing plant or facility for which it is designed. Further, a module of equipment herein includes all the piping, instrumentation, concrete, steel, and so forth that interconnect to other modules of like completeness. Together these modules make up the entire plant or facility, which then receives raw materials and produces the finished products. An oil and gas field-production treating facility, a gas-treating plant, a refinery, or a chemical plant may be made up of these modules, establishing the total investment cost of a complete packaged plant installed and ready for commissioning.

Indirect cost.   Indirect was reviewed earlier in this chapter. In Table 8.12 an indirect cost factor of 1.38 is given for shell/tube type heat exchangers. Much of this cost factor is weighted with the field-erection cost of cranes and other heavy construction equipment items, such as welding trucks and heavy equipment transport logistics. The total shell/tube module cost is then:

Total shell tube module cost factor = direct cost × indirect cost factor

or

Total shell tube module cost factor = 231.3 × 1.38 = 319.2

For a quick estimate of a shell/tube exchanger module package installed cost, multiply the vendor shop quotation of the unit times 319.2. If you need a quick ±25% vendor shop FOB cost, then use Fig.

8.3 and the appropriate associated table's F factors, as well as the yearly escalation index ratio.

Module cost. Table 8.12 and the associated Fig. 8.3 and table's F factors thus make up the entire shell/tube modular cost. All other types of major equipment items are to be factored for cost calculations, applying similar figures and tables throughout this chapter.

Air cooler cost. Heat rejection from a process plant is a common process operation. Using the environmental air for this heat rejection cooling is also a common means to provide this needed heat rejection by means of an air cooler. Air coolers always come with a fan to flow air over and through the tubes. See Chap. 6 for a good review of the air finfan cooler integral parts. Air cooler design always includes the extended fins on the tube's outside walls. The common means of determining the size of the air cooler is the total bare tube surface area in units of ft^2. Here the fin area added is called the *extended surface area*. See Chap. 6 again for extended surface area sizes. The term *bare tube surface area* refers to the OD surface area of the tubes without the fin area added. In this chapter, the bare tube OD surface area will be used to relate the air cooler size to cost.

Figure 8.4 displays the air finfan unit cost in carbon steel materials. This figure includes the cost of the tube bundle, fan, fan electric motor, bundle casing, inlet/outlet casing manifolds, steel structure, stairways, and ladders. Tube pressure rating, material, and labor for both fabrication shop and field assembly are added factors per the associated tables. (See Tables 8.13 to 8.16.)

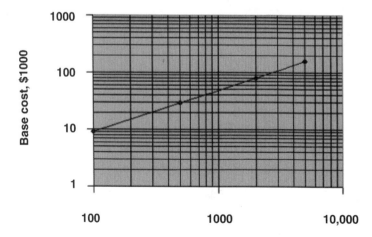

Figure 8.4 Air finfan cooler base cost, \$, Fig. 8.4 cost × $(F_t + F_p + F_m)$ × escalation index.

TABLE 8.13   Tube Pressure Rating Cost Factors

Tube pressure rating	Factor, F_p
150 psig	1.00
250 psig	1.05
500 psig	1.10
1000 psig	1.15

TABLE 8.14   Tube Length Cost Factors

Tube length	Factor, F_t
16 ft	0
20 ft	0.05
24 ft	0.10
30 ft	0.15

TABLE 8.15   Tube Material Cost Factors

Tube material	Factor, F_m
Carbon steel	0
Aluminum	0.50
SS 316	2.00

TABLE 8.16   Air Finfan Cooler Module Associated Cost Factor Additions, in Percent

Piping	13.9
Concrete	1.4
Steel	—
Instruments	4.0
Electrical	10.2
Insulation	—
Paint	0.6
Subtotal	30.1
Labor	32
Total direct material and labor	62.1
Base cost factor	100
Total direct cost	162.1
Indirect cost factor (× direct cost)	1.39
Total modular factor	220.5

Note for Tables 8.14 and 8.15: If these tables are applied without Table 8.13, then add 1.00 to the F factors.

Process pressure vessel cost. Process pressure vessels are always designed in accordance with the current ASME code. These major equipment items are always cylindrical metal shells capped with two elliptical heads, one on each end. Installation can be either vertical or horizontal. Vertical is generally a fractionation-type column with internal trays or packing, although the smaller-height vertical vessels (less than 15 ft) are mostly two-phase scrubber separators. The horizontal vessel is generally a two- or three-phase separation vessel.

Two charts, Figs. 8.5 and 8.6, have been produced to cover the pressure vessel cost range. Figure 8.5 addresses vertical vessels only, and Fig. 8.6 horizontal vessels. These cost figures are as of 2001. The height notations in Fig. 8.5 and the length notations in Fig. 8.6 are the seam-to-seam dimensional measure of the vessel. Both figures have been extrapolated beyond 10-ft diameters. The diameters 10 ft or less refer to actual vendor quotations, as of 2001. As always, however, all vendor quotations are not the same, with some vendors having large percentage in cost difference, such as 30% less than other fabrication vendors. This difference is largely due to fabrication location and labor. Our

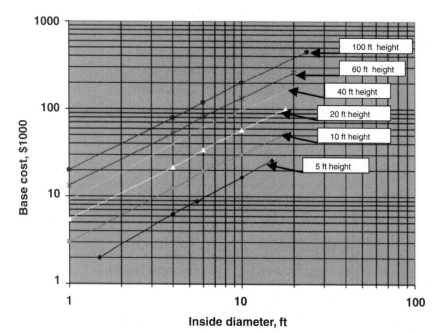

Figure 8.5 Vertical pressure vessel base cost, $, Fig. 8.5 cost $\times$ ($F_p + F_m$) $\times$ escalation index.

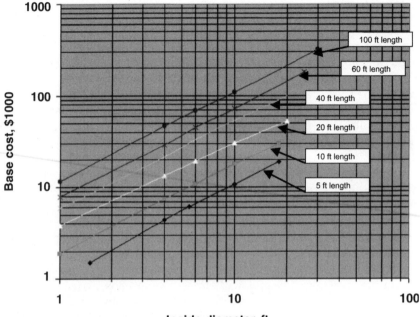

Figure 8.6   Horizontal pressure vessel base cost, $, Fig. 8.6 cost $\times (F_p + F_m) \times$ escalation index.

current-day worldwide quotation offerings are very competitive and for the most part equal in quality. We should, however, be cautious of such low vendor quotations as this 30% figure deviation. In many cases this is a red flag for low-quality work, even perhaps below ASME code specifications. For later years you need to include an escalation index cost factor as previously explained.

The base cost given in Figs. 8.5 and 8.6 represents pressure vessels fabricated in carbon steel for a 50-psig internal pressure in accordance with the ASME code. This base cost includes average allowance for vessel nozzles, manways, and vessel supports. Factors are used to adjust pressures up to 3000 psi and higher. At 1500-psig design pressures and higher, the vessel walls become very thick, >3 in. Such a heavy wall requires a special fabrication cost that has been included in the pressure factors, F_p. These pressure factors are given in Table 8.17. Table 8.18 provides the material types, other than carbon steel, that are the F_m factors.

Figure 8.7 provides fractionation column internal trays that are to be added to the overall pressure vessel cost. A packed-column internal (random packing) cost is also given as cost per cubic foot of column-packed volume.

TABLE 8.17   Process Vessel Shell Cost

Pressure, psi	Pressure factor F_p
100	1.25
200	1.55
300	2.0
400	2.4
500	2.8
600	3.0
700	3.25
800	3.8
900	4.0
1000	4.2
2000	6.5
3000	8.75
5000	13.75

The height and the length numbers given in Figs. 8.5 and 8.6 are weld-seam-to-weld-seam measures for the full length of the cylindrical part of the vessel. This measure is also the measure between the weld seams of the two elliptical heads.

The most common vertical vessel internals are trays or packing. Tray cost increases considerably with increase in column diameter. Packing likewise increases with column diameter increase, due to column volume increase. (See Tables 8.19 to 8.21.)

Now apply Figs. 8.5 and 8.6 in one of the following equations for the vessel base cost.

Vertical vessel base cost, $ = Fig. 8.5 cost $\times$ $(F_p + F_m)$ $\times$ escalation index

Horizontal vessel base cost, $ = Fig. 8.6 cost $\times$ $(F_p + F_m)$ $\times$ escalation index

The next steps for deriving the pressure vessel's total modular cost is now given in Tables 8.22 and 8.23.

Process pump modular cost. Figure 8.8 reveals centrifugal pump cost, including both pump and electric motor driver. This cost figure is for 2001. These are general-purpose process pumps of API standard speci-

TABLE 8.18   Process Vessel Shell Cost

Shell material	F_m factor, clad	F_m factor, solid
Carbon steel	1.0	1.00
Stainless 304	2.30	3.50
Stainless 316	2.60	4.25
Monel	4.60	9.75
Titanium	4.89	10.6

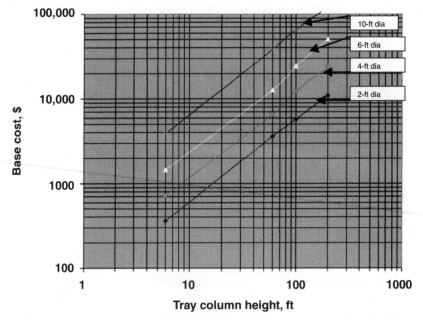

Figure 8.7 Fractionation tray cost (shop-installed), \$, base tray cost $\times (F_t + F_s + F_m) \times$ escalation index. Add this tray cost to the vertical vessel's overall module cost.

fication. Figure 8.8 displays the x axis as gpm times the psi differential head pressure output of the pump. The x axis applies equally to a 100-gpm pump at 1000 psi and to a 5000-gpm pump at 20 psi of head differential. Use Fig. 8.8 for either extreme and all cases in between. Extending Fig. 8.8 beyond the 100,000 x axis value is deemed within reason of large pump base cost for x axis values up to \$100,000.

TABLE 8.19   Fractionation Column
Tray Cost, F_s Factor

Tray spacing, in	F_s factor
24″	1.0
18″	1.3
12″	2.0

TABLE 8.20   Fractionation Column
Tray Cost, F_t Factor

Tray type	F_t factor
Sieve plate	0
Valve tray	0.3
Bubble cap	1.6

TABLE 8.21   Fractionation Column
Tray Cost, F_m Factor

Tray material	F_m factor
Carbon steel	0
Stainless steel	1.5
Monel	8.5

TABLE 8.22   Vertical Pressure Vessel
Module Associated Cost Factor
Additions, in Percent (total module
cost = base cost × total modular
factor + tray cost)

Piping	62
Concrete	11
Steel	8.2
Instruments	12.5
Electrical	4.2
Insulation	7.5
Paint	1.5
Subtotal	106.9
Labor	102
Total direct material and labor	208.9
Base cost factor	100
Total direct cost	308.9
Indirect cost factor (× direct cost)	1.37
Total modular factor	423.2

TABLE 8.23   Horizontal Pressure
Vessel Module Associated Cost Factor
Additions, in Percent (total module
cost = base cost × total modular factor)

Piping	42
Concrete	8
Steel	—
Instruments	7.0
Electrical	6.0
Insulation	6.0
Paint	0.7
Subtotal	69.7
Labor	65
Total direct material and labor	134.7
Base cost factor	100
Total direct cost	234.7
Indirect cost factor (× direct cost)	1.34
Total modular factor	314.5

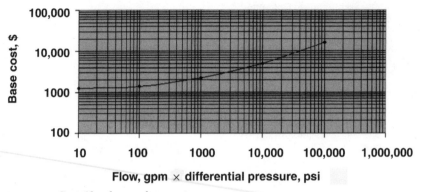

Figure 8.8   Centrifugal pump base cost.

Figure 8.8 costs include the pumping unit, coupling and gear, mechanical seal, and the base plate. All are FOB fabrication shop costs for 2001. The basis of material is cast iron. Use Table 8.24 for F_m. See Table 8.25 for associated cost factors.

Suction pressure of most API pump services is 150 psig or less. For greater suction pressures of 200 psig up to 1000 psig, use an F_p of 1.50.

Gas compressor packaged system cost.   Compressing gas is a very common unit operation in every type of process plant known, such as oil and gas production facilities, gas-treating plants, refineries, and petrochemical plants. Process gases are required in a wide range of capacities, pressures, and temperatures. In designing compressors, a complex calculation setup is required for both a centrifugal and a reciprocating-type gas compressor. These two types of compressors, reciprocating and centrifugal, are commonly used in most every gas compression case worldwide, probably 99.5% of all services. Other types of gas compressors are rotary screw, plunger-type, and the gear-type. These are seldom used, except for instrument air compression and small chemical-type gas injection, such as chlorine gas injection into municipal water for water purification. In view of these statistics, only the centrifugal and the reciprocating com-

**TABLE 8.24   Pumping Unit Base Cost,
$ = Fig. 8.8 Cost × (F_p + F_m) × Escalation Index**

Pumping unit material	F_m
Cast iron	1.00
Cast steel	1.85
Stainless	2.45

Note: If pump suction pressure is greater than 150 psig, then $F_p = 1.50$; otherwise, $F_p = 0$.

TABLE 8.25 Pumping Module Associated Cost Factor Additions, in Percent (Total module cost = base cost × total modular factor.)

Piping	32
Concrete	5.0
Steel	—
Instruments	3.0
Electrical	32.0
Insulation	3.2
Paint	0.8
Subtotal	76.0
Labor	65
Total direct material and labor	141.0
Base cost factor	100
Total direct cost	241.0
Indirect cost factor (× direct cost)	1.30
Total modular factor	313.3

pressor types are cost-analyzed in this book. These two compressor types make up nearly all of the gas compressor costs in every application of any significant cost. In this book, our quest is major equipment cost, with minor equipment additions as supplemental.

The centrifugal compressor is generally a higher-capacity machine than the reciprocating compressor. The centrifugal compressor ranges from 400 to over 100,000 actual cubic feet per minute (acfm). Centrifugal compressor suction pressures are higher than those of the reciprocating compressor. Centrifugal compressor suction pressures are seldom designed for less than 50 psig. The centrifugal also generally has a comparably lower discharge pressure, ranging from 150 psig to approximately 4000 psig.

The reciprocating gas compressor, compared to the centrifugal, has a much wider range between design suction and design discharge pressures, less than 1.0 psig suction to over 10,000 psig. The gas capacity of reciprocating compressors, however, is considerably less than that of the centrifugal compressor, being limited to approximately 25,000 acfm.

These comparisons between the centrifugal and the reciprocating compressor types have assumed multistages of both compressor types and a single inline series flow for both types. It is obvious that both types of these compressors could be arranged in parallel flow for most any gas capacity. The more feasible and lower-cost compressor will, however, be capable for the full gas capacity load as a single series multistage compressor train.

Figure 8.9 displays the cost of both the centrifugal and the reciprocating types of gas compressors. Please note that the x axis is scaled for the

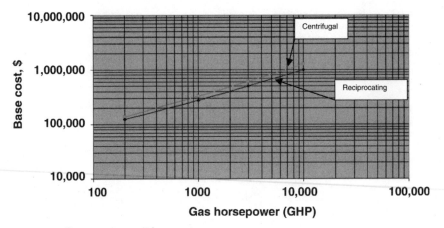

Figure 8.9 Compression unit base cost.

compressor horsepower, hp. These curves may be straight-line extended to over 5000 hp with reasonably accurate results, ±20% of shop FOB cost, as of 2001.

Figure 8.9 costs include the basic compressor unit, ranging from 200 hp to over 5000 hp. The driver cost is a separate unit cost to be read from Fig. 8.10.

For centrifugal compressor supplemental equipment, use Table 8.26. Here supplemental equipment includes such items as inlet suction gas filter-scrubber pressure vessels, suction coolers, intercoolers, and aftercoolers. Air finfan cooling heat exchangers are assumed for the supplemental cost in Tables 8.26 and 8.27. Compressor unit base cost = (base cost from Fig. 8.9 and Table 8.25 or 8.26 F_a) × escalation index.

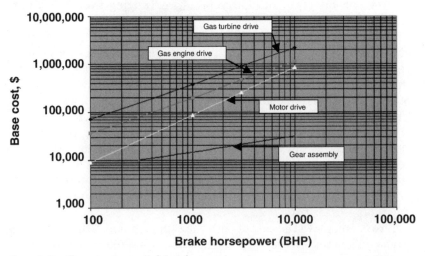

Figure 8.10 Compression unit driver base cost.

TABLE 8.26   Centrifugal Compressor Auxiliary Equipment Cost Factor F_a

Gas hp	F_a, $
100	12,240
500	25,920
1,000	38,000
5,000	79,500
10,000	110,000

The key to reading Fig. 8.9 and these accompanying tables for the F_a cost in dollars is of course finding the horsepower of the compressor. It is therefore necessary to determine the required horsepower from the common given variables, which normally are as follows:

P_1 = inlet pressure, psia

P_2 = outlet pressure, psia

k = gas specific heat ratio, C_p/C_v

C_p/C_v = (MW ∗ specific heat C_p)/[(MW ∗ specific heat C_p) − 1.986]

MW = molecular weight of gas stream

Z = average gas compressibility factor of compressor inlet and outlet gas streams

R = universal gas constant, 1545/MW

T_i = inlet temperature, °R = °F + 460

$$H_{ad} = \frac{ZRT_i}{(k-1)/k} \left[\left(\frac{P_2}{P_1} \right)^{(k-1)/k} - 1 \right] \tag{8.3}$$

where H_{ad} = adiabatic horsepower, ft · lb/lb

Most compressor manufacturers use the adiabatic horsepower calculation, Eq. (8.3), which is therefore recommended and applied here.

$$\text{GHP} = \frac{w_i H_{ad}}{\text{eff } 33,000} \tag{8.4}$$

TABLE 8.27   Reciprocating Compressor Auxiliary Equipment Cost Factor F_a

Gas hp	F_a, $
100	27,800
500	66,700
1,000	86,500
5,000	159,000
10,000	270,000

where GHP = consumed gas horsepower

w_i = gas rate, lb/min

eff = compressor efficiency: for reciprocating, usually 73% or 0.73; for centrifugal, usually 78% or 0.78

$$BHP = GHP + 50 \qquad (8.5)$$

Equation (8.5) adds the horsepower consumed by the compressor bearings, seal losses, and gearbox losses. The 50 hp is an average for compressors running from 500 to 1500 hp. Larger-hp cases should allow more of this deficiency loss, although most vendors claim this 50 hp is never realized in their machines, even at 3000 hp. This is really hard to accept, as case histories of most centrifugal and reciprocating compressors reveal these losses of bearings, gearboxes and seals to be between 50 and 100 hp. This higher loss, 100 hp, is found more prevalent in reciprocating-type compressors, especially between 1000 and 2000 hp. Therefore, a good rule here is as follows:

1. Add 50 hp to the preceding GHP value, in Eq. (8.4), to derive the BHP of centrifugal.

2. Add 100 hp to Eq. (8.4)'s GHP to derive reciprocating-type BHP + VE hp.

Equation (8.5) does not account for the volumetric efficiency (VE) loss of reciprocating-type compressors. Volumetric efficiency is simply the clearance allowed in the compressor cylinder head in which this compressed gas volume is allowed to mix with the inlet gas to be compressed in the next compression stroke. Thus this compressed gas in the cylinder clearance is recycling, which makes the output of the compressor less efficient, the larger this clearance volume. Volumetric efficiencies usually range fron 6 to 20%. An average reciprocating compressor VE should be 10%. For conservative cost calculating, use 15% of overall GHP for the VE efficiency.

For example, if you have calculated 1000 hp for a reciprocating compressor, using Eqs. (8.3) and (8.4), then you should equate the following:

BHP = 1000 hp for GHP + 100 hp for gears, seal, and bearings + 150 hp for VE

BHP = 1250 hp required for the driver hp delivery and for Fig. 8.9

The centrifugal-type compressor is advantaged here, since it doesn't have a VE efficiency (cylinder clearance) to account for. There are no cylinders in a centrifugal-type compressor. More and more major oil/gas producers are therefore installing centrifugal compressors in their newly sited field facilities, rather than the reciprocating type. Another reason major gas/oil-producing companies are choosing centrifugal over reciprocating compressors is maintenance cost. Centrifugal is a little more expensive in initial investment; however, in the first two years with maintenance cost added, the centrifugal is the lower investment by at least 10 to 20%.

Remember, however, that for high head discharge pressures and for low gas rates, say below 10 Mscfd, reciprocating-type compressors are still the preferred choice because of their investment cost and capabilities of high heads, above 2000 psi.

Regardless of the particular compressor case for the hp you finally derive, Fig. 8.9 is applicable to all cases for both reciprocating and centrifugal-type compressors. See Table 8.28 for compressor module associated cost factors.

Every compressor has a driver. The centrifugal compressor may be driven by a gas-fired turbine, a steam turbine, or an electric motor. Centrifugal compressors larger than 300 hp are normally driven by a gas-fired turbine. Reciprocating compressors are normally driven by an internal combustion (cylinder-type) gas engine. This engine is similar to the gasoline engine you have in your automobile. However, the reciprocating gas compressor–driven engine is fueled by natural gas or methane-rich gas rather than by gasoline. Again using the derived total BHP required, Fig. 8.10 may be used to find the vendor shop FOB cost of these drivers. This cost (the driver) should be added to the cost summary page of the project cost, as later shown.

TABLE 8.28 Compressor Module Associated Cost Factor Additions, in Percent (total module cost = base cost × total modular factor)

Piping	16.0
Concrete	5.0
Steel	—
Instruments	12.0
Electrical	7.0
Insulation	1.5
Paint	0.5
Subtotal	42.0
Labor	32
Total direct material and labor	64.0
Base cost factor	100
Total direct cost	164.0
Indirect cost factor (× direct cost)	1.35
Total modular factor	221.4

The next need for the compressor package is the driver. Most driver types applied to <500-hp loads are usually electric motors at 440-V, 60~ standard. Greater loads generally call for gas-fired or steam turbines. Please see Figs. 8.10 and 8.11. These figures display the cost of adding the driver of choice.

Figure 8.10 covers gas-fired turbine and cylinder-type engines, and Fig. 8.11 covers steam turbine drivers. Motor drivers and gear assembly cost is also shown in Fig. 8.10. Having a required compressor BHP, the driver cost of choice may be read from one of these two figures. The gear assembly cost is applicable to any of the drivers given in Figs. 8.10 and 8.11.

Figures 8.10 and 8.11 are based on the driver FOB fabrication shop cost only. For the completed driver package, additional items, such as a surface condenser for the condensing steam turbine or a starting motor or air piston starter for the gas engine, are required. Tables 8.29 to 8.32 are each dedicated to providing factoring for these specific costs. Such costs will again be based on the compressor horsepower, BHP.

The first of these four tables is Table 8.29. Table 8.29 lists a factor F_a that includes costs for a shell/tube steam surface condenser, steam vacuum jets with piping manifold, condensate pumps, and a condensate accumulator vessel. Please note here that plant-inhibited closed-loop cooling water is assumed to be available.

The next table, Table 8.30, is similar to the previous, but without any condenser. A steam condensate accumulator with condensate pumps is required; however, many installations simply return the exhaust steam in an existing LP header directly, without any supplemental equipment other than piping added to the noncondensing turbine driver. Piping is

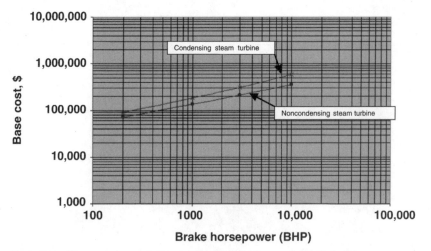

Figure 8.11 Compression unit driver base cost (condenser not included).

TABLE 8.29   Condensing Turbine Driver Supplemental Items F_a, $

Brake horsepower (BHP)	F_a, $
100	18,000
500	28,800
1,000	61,400
10,000	126,400

TABLE 8.30   Noncondensing Turbine Driver Supplemental Items F_a, $

Brake horsepower (BHP)	F_a, $
100	7,500
500	12,300
1,000	14,800
10,000	29,500

TABLE 8.31   Gas Engine Driver Supplemental Items F_a, $

Brake horsepower (BHP)	F_a, $
100	61,200
500	97,200
1,000	111,600
10,000	223,200

TABLE 8.32   Gas-Fired Turbine Driver Supplemental Items F_a, $

Brake horsepower (BHP)	F_a, $
100	22,600
500	71,600
1,000	165,000
10,000	440,000

approximately one-half of the cost factor in Table 8.30. If your case requires no added steam condensate accumulator, then take one-half of the dollar values shown for F_a.

Table 8.31 provides the added cost for a gas engine driver. These supplemental equipment items are the starter (electric motor or air piston), air tanks, air filters, cooling water pumps, and local piping. An air-cooled radiator is included in Fig. 8.10, the engine cost curve. The radiator of course requires water circulation pumps. For 500 BHP or less, the assumption is made that cooling water of the recirculated inhibited type is available from outside circulated cooling water plant utility distributions. In the larger gas engines, above 500 BHP, an air finfan cooling exchanger is added to Table 8.31 costs. Plant utility cooling water at the rate of approximately 8 gpm per 100 BHP is required without this added air finfan cooler above the 500 BHP gas engine driver capacity.

The last supplemental equipment cost factor table for compressors is the gas turbine driver supplemental equipment table, Table 8.32. Here there must also be a starting electric motor or pressurized air. In place of having a captured air compressor, instrument air may be used, which would reduce the F_a value here by approximately one-fourth. Since gas-fired turbines have a very high-level effluent temperature, 1200°F plus, it is common to have a waste heat recovery exchanger, which reduces this high-level temperature to a lower range of 900 to 600°F. This recovered heat may be used for any number of heating needs in a gas- and oil-production facility, gas plant, or refinery. A heat recovery exchanger of this order is therefore included in Table 8.32, F_a factor.

Since the labor and indirect cost is already factored in Table 8.28, another similar table for these cost additions is not required. Simply apply the following equation to calculate the driver cost, and add this cost to the summary cost page in the equipment cost section. The summary cost page will be reviewed and discussed later in this chapter.

Driver cost, $ = (base cost of Figs. 8.10 and 8.11 + F_a) × escalation index

This cost should be accounted as part of the direct material and labor cost of the compressor equipment module.

Plant utility steam generators. Steam generation throughout the industry is a common means of energy distribution for process heating and for driving steam turbines. Steam is used for driving vacuum jets, producing near absolute vacuum in countless vacuum distillation processes. Steam is also used as a stripping medium in most every refinery, in crude columns and in crude column sidestream strippers. The need for steam generation is so common that it is an integral part of every refinery worldwide. So it is important to provide the process engineer with a good understanding of costs for steam-generating equipment.

Steam-generating equipment is similar to a gas-fired furnace in that fuel gas is most always the heating source of the steam generator. The furnace is generally an A-type tube-shaped frame inside a rectangular box in which the steam is on the tube side.

Steam generator cost. Figure 8.12 reveals the cost of almost every steam generator as of 2001. The figure may be extrapolated to 2000 psig. The cost shown is for a complete steam unit, installed and ready to commission. A surge high-pressure steam drum and receiving condensate sys-

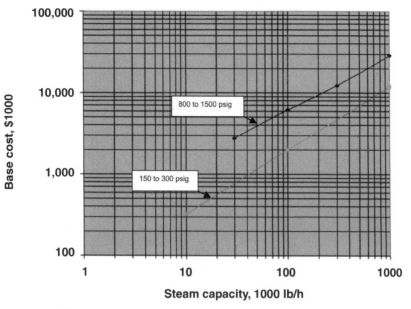

Figure 8.12 Steam generator cost.

tem and pumps are included, as well as a deaerator unit cost with pumps and 10% of total steam load condenser for low-pressure steam. Fuel gas scrubber and fuel gas manifold are costs included in Fig. 8.12 with complete instrument control setup in the central control room. Water treatment is not included, nor is provision for water makeup sources.

Cooling water tower cost. With Fig. 8.13, you may estimate the complete cost of a cooling water system. The cooling water tower base cost curve shown includes the complete tower of current-day materials, such as fiberglass-reinforced plastics (FRPs). Figure 8.13 is the complete packaged system cost, which includes the concrete foundation with water-retention basin, the main circulation pumps, all modular piping, all instrument controls, electrical switchgear, electrical, fans, and chemical treatment injection equipment. Simply add the cost of distribution piping to derive the cooling system cost of a complete plant. If you cost-analyze the complete plant by this modular cost method, you simply need to add the cost of a supply header and a return header. Water distribution headers generally range from $50/ft for 3-in pipe to $150/ft for 12-in pipe, installed cost.

Refrigeration package system cost. The cost curve in Fig. 8.14 is based on a −20°F refrigerant outlet temperature, which is the temperature of the majority of process plant refrigerant levels. Higher-temperature

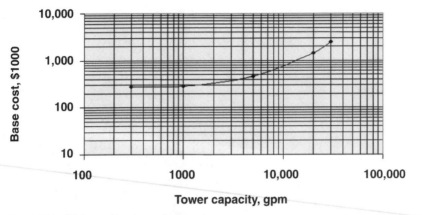

Figure 8.13   Water cooling tower base cost.

refrigerant levels result in fewer installed costs, such as 10% less for
0°F. Using Fig. 8.14 for all higher-temperature levels is advised, however, since this cost difference is conservative and also a marginal difference. The Fig. 8.14 curve is the installed cost for materials and
labor, including centrifugal compressors, drivers, condensers, evapora-
tors, piping, insulation, electrical, steel, and foundations.

API cone roof storage tank cost. Every refinery, chemical plant, and
oil/gas-processing plant or field-production facility has API-type cone
roof tanks. These tanks are made of steel-plate materials and are field-
assembled, welded, and erected. Figure 8.15 displays a full installed
cost range of these tanks from 20- to 50-ft heights and from 20- to 200-
ft diameters. You may extrapolate these curves to 5 ft height and to
over 400 ft diameter. Tank heights greater than 45 ft are generally not

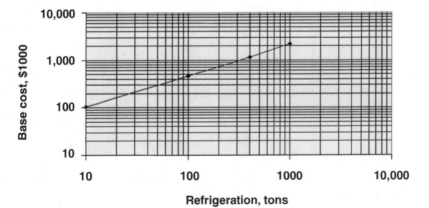

Figure 8.14   Installed refrigeration package cost.

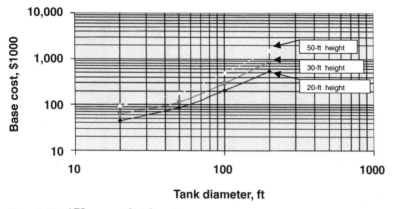

Figure 8.15   API cone roof tank cost.

made due to the tank-wall thickness requirement exceeding the economic value of the tank cost, as compared to the cost of installing
another tank. These costs include all labor, foundation, dike containment, instrumentation, piping, stairwells, paint, and insulation. Spe-
cialty equipment items, such as bayonet heaters and tank internal
baffling, are not included.

Ethane storage is commonly accomplished using API-type cone roof
tanks with an inner tank placed inside a larger tank, and a vacuum
maintained in the annular space between the outside tank and inner
tank wall. For these tank costs, multiply the Fig. 8.15 cost by 1.9. The
outside tank is well insulated. Heat loss does occur, however, and
refrigeration is required to condense the vapors released for recycling.
A 10- to 20-ton refrigeration package is generally adequate for a 2- to
4-tank system, each tank 50 ft in diameter by 30 ft high. See Fig. 8.14
for adding refrigeration cost.

Spherical LPG storage tank cost.   Liquid petroleum gas (LPG) storage
pertains mainly to propane, butane, and pentane liquids. These liquids approach the ambient temperature as stored liquids, ranging
from 60 to 90°F and from 130- to 160-psig pressures. Spherical tanks
may be economically designed for these temperatures and pressures
for up to over 30,000 bbl. Ethane and ethylene are generally stored in
refrigerated API cone roof tanks, as explained in the previous paragraph.

Figure 8.16 displays spherical tank cost for liquid storage ranging
from 300 to 20,000 bbl. This cost curve may be extrapolated to over
30,000 bbl. Fabrication economics show cost minimized by installing
multiple spherical tanks rather than going over 35,000 bbl per tank.
Figure 8.16 gives the curve based on the maximum liquid volume con-

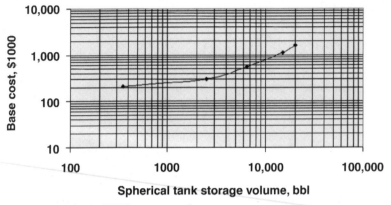

Figure 8.16   Spherical LPG storage tank cost.

tainment permissible in the tank. Approximately 15% of the total tank volume must be maintained in the spherical tank's top dome as vapor space. One major reason for this vapor space is emergency safety pressure relief. During an emergency pressure release, such as for fire, the liquid level is not to spill out through the relief valve. Ample vapor space above the liquid level is therefore maintained to allow liquid expansion during a safety valve release due to fire. Another reason for this vapor space is assurance of avoiding overfill and subsequent liquids release. A pressure control valve maintains proper tank pressure. However, in the event of liquids overfill, downstream safety valve release will occur, as well as with the tank safety valve. Why should this liquid release happen? Remember that transfer pump discharge shutoff head? This shutoff pump head most likely is over 200 psi, which far exceeds the safety valve release pressure, even considering the static pressure head loss.

These spherical tanks are not insulated, since the stored liquids approach the ambient temperature in most countries. Also, insulation problems result from water moisture condensing on the metal walls under the insulation.

Most companies do without refrigeration for vented gas recovery, since vented gases are marginally small. However, some producers, especially in tropical locations, find good economics in installing refrigeration packages to recover these vented LPG gases. Allow 350 Btu/ft² for one-half of the spherical tank outside surface area to determine the refrigeration duty required.

Figure 8.16 is the complete modular LPG storage spherical tank cost, including piping, stairwells, platforms, instruments, controls, steel, foundation, and dike containment. Both material and labor are included in the Fig. 8.16 cost curve, installed and ready to commission.

Modular equipment cost spreadsheet. Coverage of most every major equipment item has now been given a base cost curve reference from which a specific base cost can be obtained. This base cost is simply the major equipment item cost FOB vendor shop without any supplemental equipment, such as piping, steel, foundation, or installation labor added. At the beginning of this chapter the modular equipment–type cost factor was introduced. Several of these cost factors, labeled *materials factors,* are applied in a spreadsheet form for each major equipment item. Please see Table 8.33 and observe these materials factors, such as 62.0 given for the piping. This 62.0 factor is to be regarded as a percentage factor to be multiplied by the base cost of $1,712,000 to get a product showing piping material cost. The spreadsheet here automatically does this multiplication for the piping and each of the other associated items—concrete, steel, instruments, and so forth. The labor factor is also a multiplier of the shown base cost to get the total installed associated labor cost. These associated costs, labor and asso-

TABLE 8.33   Modular Equipment Spreadsheet Cost Analysis

PROJECT: CC5 recovery, 28-Jul-01

EQUIPMENT TYPE: Fractionation columns

Item name	New equipment cost
IC5 Fractionator, top sect., 10′ ID × 100′	$ 427,000
IC5 Fractionator, btm. sect., 10′ ID × 100′	427,000
NC5 Fractionator, top sect., 8′-6″ ID × 93′	319,000
NC5 Fractionator, btm. sect., 8′-6″ ID × 97′	328,000
CC5 Fractionator, top sect., 4′-0″ ID × 74′	110,000
CC5 Fractionator, btm. sect., 4′-0″ ID × 74′	110,000
Total	$1,721,000
Used equipment credit	$ 0

ASSOCIATED COSTS:

Materials	Factor	Cost
Piping	62.0	$1,067,020
Concrete	11.0	189,310
Steel	8.2	141,122
Instruments	12.5	215,125
Electrical	4.2	72,282
Insulation	7.5	129,075
Paint	1.5	25,815
Total associated costs, material		$1,839,749
Labor	102.0	$1,755,420
Total associated costs		$3,595,169
Total direct costs		$5,316,169
Indirect costs	37.0	$1,966,983

ciated materials, are then added to show the total associated cost, $3,595,169. The total direct cost, $5,316,169, is calculated simply by adding the major equipment item base cost, $1,721,000 + $3,595,169 = $5,316,169.

The last item of the spreadsheet in Fig. 8.33 is indirect costs. These costs pertain to numerous items such as cranes, construction scaffolding, welding trucks, freight, land cost, civil earth work, roads, and construction storage. Indirect costs may range from 20 to 35% of major equipment modular cost material + labor that was given near the beginning of this chapter.

The spreadsheets from Tables 8.33 to 8.38 are given in the CD as the EquipData.Xls applications program. You can use this spreadsheet as a template for any plant cost application desired. You may alter the associated material cost factors to any values you find correct. The values given in these tables are found to be good and sound to this date by the author. More specific associated factors for each case, however, may require variance as actual equipment cost data is obtained. Remember, these costs are accurate to the degree of ±25% error. Lower error requires, first, a definitive engineering cost estimate involving thousands of engineering labor hours for these spreadsheets. Second, accurate vendor quotations are required for every equipment item. Here, however, in this chapter methodology, equipment costs are curve-factored referencing certain vendor cost FOB fabrication shops, 2001.

Example   Isopentane (IC5), normal pentane (NC5), and cyclopentane (CC5) are to be separated by means of distillation. A 5000-bpd rich feed with these components is received. The mixed feed containing these and many other components—including ethane, propane, butane, through benzene—is received as a liquid. A three-column distillation train, in series, will be installed to produce IC5, NC5, and CC5 spec product liquid streams. Methane, ethane, propane, and butane have been removed in an upstream stabilizer column. Only trace ethane and propane are remaining in the feed stream feeding the first column, IC5.

A 100-trayed column will fractionate the incoming mixed IC5, NC5, and CC5 stream, first separating the IC5 in a 100-trayed column. The IC5 overhead product stream will be liquid and will be 99% mol or greater IC5 purity, which exceeds the 98.5% mol specification. The remaining bottoms stream, NC5, and CC5 are fed to the second fractionator, downstream. This second fractionator is the NC5 column (also a 100-trayed column), where the NC5 overhead stream meets the NC5 specification of over 99% purity NC5. The overhead product is also a liquid product stream at 180°F and 60 psig pressure.

The third and last column of this three-column series is the CC5 distillation column where the column overhead specification meets or exceeds 81.5% CC5 purity in the overhead liquid product. This is an 80-trayed column with 18-in tray spacing.

TABLE 8.34   Modular Cost Analysis—Coolers and Condensers

PROJECT: CC5 recovery, 28-Jul-01	
EQUIPMENT TYPE: Air finfan coolers	
Item name	New equipment cost
IC5 Column overhead condenser, 13,572 ft^2	$310,000
NC5 Column overhead condenser, 4,838 ft^2	180,000
NC5 Product cooler, 880 ft^2	44,000
CC5 Column overhead cooler, 2149 ft^2	90,000
CC5 Product cooler, 280 ft^2	18,000
Total	$642,000
Used equipment credit	$ 0

ASSOCIATED COSTS:		
Materials	Factor	Cost
Piping	13.9	$ 89,238
Concrete	1.4	8,988
Steel	0.0	0
Instruments	4.0	25,680
Electrical	10.2	65,484
Insulation	0.0	0
Paint	0.6	3,852
Total associated costs, material		$ 193,242
Labor	32.0	$ 205,440
Total associated costs		$ 398,682
Total direct costs		$1,040,682
Indirect costs	39.0	$ 405,866

The IC5 and NC5 columns have 100 trays each, which requires both to be over 200 ft tall, allowing 18-in tray spacing. This 18-in tray spacing is a good design for this close boiling point–type fractionation. Whenever a vertical column reaches 100 ft in design, the wind and foundation cost factors become very significant. Therefore, columns taller than 120 ft are rarely built because they are cost-prohibitive.

It is more economical to install two towers in series than to erect one column taller than 200 ft. Two 100-ft-tall columns are therefore used in this application for the IC5 fractionator and two 100-ft-tall columns each for the NC5 fractionator. Thus four columns, each 100 ft tall, are required to make the IC5 and NC5 overhead product streams. A bottoms transfer pump between each column pair is of course required to transfer the bottoms liquid to the downstream top column tray. The vapors flow in a countercurrent direction from the bottom tower top to the bottoms section of the upper column. The transfer pumps placed between these two columns transfer 1000 gpm of liquid from the bottom tray of one to the top tray of the second column, which is the bottom column of the two.

A process simulation of this three-column distillation train has been made, finding and establishing the number of actual trays, column diame-

TABLE 8.35   Modular Cost Analysis—Heat Exchangers

PROJECT: CC5 recovery, 28-Jul-01	
EQUIPMENT TYPE: Shell/tube heat exchangers	
Item name	New equipment cost
IC5 Column reboiler, 2 ea., 1220 ft^2	$140,000
NC5 Column reboiler, 3 ea., 983 ft^2	186,000
CC5 Column reboiler, 1 ea., 240 ft^2	31,000
Total	$357,000
Used equipment credit	

ASSOCIATED COSTS:		
Materials	Factor	Cost
Piping	45.0	$160,650
Concrete	5.0	17,850
Steel	3.0	10,710
Instruments	11.0	39,270
Electrical	2.0	7,140
Insulation	4.8	17,136
Paint	0.5	1,785
Total associated costs, material		$254,541
Labor	61.0	$217,770
Total associated costs		$472,311
Total direct costs		$829,311
Indirect costs	38.0	$315,138

ters, overhead air finfan cooler duties, reboiler duties, and pumping capacity heads required. The previous paragraphs (under "Modular Equipment Cost Spreadsheet") explained the various cost assembly on an equipment-type modular summary sheet. Table 8.33, the fractionation column module cost spreadsheet, gave the cost summary for this type of major equipment. Please observe that each particular major type of equipment is grouped into one spreadsheet for summarizing that particular modular type.

Note that each equipment type listed is to be considered as discrete and separate from the other equipment items of the same type. Why? For the convenience of grouping, all associated equipment factors being the same, items such as the six fractionator columns in Table 8.33 can be summed as one table, covering all six module costs.

In the manner of Table 8.33, Tables 8.34 to 8.37 each pertain to a particular type of major equipment item. The base cost of Table 8.33 fractionators was obtained from Figs. 8.5 and 8.7 and the associated tables accompanying these two fractionator column figures. The particular major equipment cost from previous parts of this chapter was also used to find the base cost of major equipment items in Tables 8.34 to 8.37. In simpler terms, please use the numerous base cost tables given in this chapter to find the particular equipment items you may require for any project. Note again that the base cost here means the major equipment item cost FOB the vendor shop. Asso-

TABLE 8.36   Modular Cost Analysis—Horizontal Pressure Vessels

PROJECT: CC5 recovery, 28-Jul-01

EQUIPMENT TYPE: Horizontal pressure vessels

Item name	New equipment cost
IC5 Reflux accum., horiz., 12'-0" ID × 28'-0"	$ 62,500
NC5 Reflux accum. horiz., 10'-0" ID × 25'-0"	57,000
CC5 Reflux accum. horiz., 6'-0" ID × 12'-0"	21,000
Total	$140,500
Used equipment credit	$ 0

ASSOCIATED COSTS:

Materials	Factor	Cost
Piping	42.0	$ 59,010
Concrete	8.0	11,240
Steel	0.0	0
Instruments	7.0	9,835
Electrical	6.0	8,430
Insulation	6.0	8,430
Paint	0.7	984
Total associated costs, material		$ 97,929
Labor	65.0	$ 91,325
Total associated costs		$189,254
Total direct costs		$329,754
Indirect costs	34.0	$112,116

ciated equipment items such as piping, steel, and instrumentation, must be added (factors) as in Tables 8.33 to 8.37.

Vertical and horizontal base cost method.   Pressure vessels, both horizontal and vertical, are cost-evaluated by diameters and ends welded seam-to-seam. Pressure and material type has considerable cost variance, but may be simply a multiplier factor. See Figs. 8.5 to 8.7 and the associated cost factor tables. It is important to note that Table 8.22 gives the associated cost factors of any vertical vessel, and Table 8.33 contains the same associated equipment factors. The reason is simply the fact that these same factors pertaining to vertical vessels also apply to fractionation columns. The complete module cost is given in Table 8.33 for the total fractionation column modules.

Air finfan overhead condensers.   Table 8.34 gives the modular cost of the overhead condensers and product coolers. Since this plant is to be installed in Saudi Arabia, cooling tower water is not available. Also, cooling water is not necessary since the products will be stored in spherical tanks requiring 150 to 180°F cooling. This is easy to obtain, even in Saudi Arabia, with air finfan coolers. The square feet of surface area shown in Table 8.34 is the bare tube surface area of each cooler. Please note this is not the extended surface of the outside tube area, but rather the outside bare tube surface area in square feet. Figure 8.4 was used to find the base cost of these air coolers.

TABLE 8.37   Modular Cost Analysis—Pumps

PROJECT: CC5 recovery,	28-Jul-01
EQUIPMENT TYPE: Pumps	

Item name	New equipment cost
IC5 Bottoms pump, 120 gpm, 100 psi head, 2 rq'd	$ 16,500
IC5 Tower pump, 1300 gpm, 100 psi head, 2 rq'd	76,000
IC5 Column reflux, 1300 gpm, 100 psi head, 2 rq'd	76,000
NC5 Bottoms pump, 30 gpm, 128 psi head, 2 rq'd	37,000
NC5 Tower pump, 1000 gpm, 130 psi head, 2 rq'd	76,000
NC5 Column reflux, 1000 gpm, 130 psi head, 2 rq'd	76,000
CC5 Bottoms pump, 250 gpm, 130 psi head, 2 rq'd	33,400
CC5 Tower pump, 250 gpm, 130 psi head, 2 rq'd	33,400
CC5 Column reflux, 250 gpm, 132 psi head, 2 rq'd	35,000
Total	$459,300

Used equipment and assoc. cost credit

ASSOCIATED COSTS:

Materials	Factor	Cost
Piping	32.0	$ 146,976
Concrete	5.0	22,965
Steel	0.0	0
Instruments	3.0	13,779
Electrical	32.0	146,976
Insulation	3.2	14,698
Paint	0.8	3,674
Total associated costs, material		$ 349,068
Labor	65.0	$ 298,545
Total associated costs		$ 647,613
Total direct costs		$1,106,913
Indirect costs	30.0	$ 332,074

Reboilers.   Figure 8.3 was used to find the reboiler base cost. Please observe that each fractionation unit (three units required) has a reboiler section. These are thermosyphon-type reboilers, shell/tube units, heated by hot circulated oil.

Pumps.   Pump base costs were found in Fig. 8.8. Here pump product of capacity, gpm times the respective head, psi, gives a factor that is used to fix each particular pump base cost. Again, as in all the other equipment module spreadsheets, the associated equipment table factors—including piping, electrical, steel, concrete, and instrumentation—are multipliers to derive the full pump module cost. Of course labor and indirect costs are factored as well in Table 8.25.

Table 8.38 is the summary table of problem cost analysis. Please observe in Table 8.38 that each of the major equipment items are shown as a single entry, such as fractionation columns, air finfan coolers, and pumps. Next observe all of the line-item associated costs are also summed as single-item costs that include piping, concrete, steel, instruments, and electrical. Each of

TABLE 8.38   Modular Cost Analysis Summary

PROJECT: CC5 recovery, 28-Jul-01

PROJECT SUMMARY:	
Equipment type	New equipment cost
Fractionation columns	$1,721,000
Pumps	459,300
Shell/tube heat exchangers	357,000
Horizontal pressure vessels	140,500
Air finfan coolers	642,000
Miscellaneous equipment	
Total new equipment cost	$3,319,800
Used equipment credit	$ 0

ASSOCIATED COSTS:	
Materials	
Piping	$ 1,522,894
Concrete	250,353
Steel	151,832
Instruments	303,689
Electrical	300,312
Insulation	169,339
Paint	36,110
Total associated costs, material	$ 2,734,529
Total material costs	$ 6,054,329
Labor	$ 2,568,500
Labor items estimated separately	$ 0
Total direct costs	$ 8,622,829
Total indirect costs	$ 3,132,177
Sales tax and freight, 10.0% of material	$ 605,433
Client supervision, 5.0% of total	$ 846,605
Outside engineering, 12.0% of total	$ 2,031,853
Contingency, 10.0% of total	$ 1,693,211
Total project installed cost	$16,932,107

these associated cost items is totaled from its corresponding associated item in Tables 8.33 to 8.37. Following this, notice that all labor and indirect costs are totaled in Table 8.38 from these five previous tables, in like manner. The final figure, $16,932,107, in Table 8.38 is the bottom line of all investment direct and indirect cost, including sales tax, freight, client supervision, engineering fees, and contingency.

The last four cost figures before the final totaled cost in Table 8.38—sales tax, freight, client supervision, engineering fees, and contingency—are shown as factors. These factors in the same identified order are 10% of materials, 5% of total, 12% of total, and 10% of total. Notice that the 10% for sales tax and freight is exactly 10% of the total materials cost in Table 8.38. Likewise, the other percentages are exactly that percent of the total project's bottom-line cost in Table 8.38.

The grandeur of Tables 8.33 to 8.38 is the fact that you may use the spreadsheet in the CD supplied with this book and change any number desired. These are the identical tables given in the Excel spreadsheet EquipData. You may apply this spreadsheet to countless cost problems simply by using it as a template. Any of the tables may also be converted to any type of major equipment item, such as a furnace. Notice that you can change the associated cost factors in any of these five tables. In the accompanying CD spreadsheet program, change any of these factors and notice all of the affected cost values change, including the summary spreadsheet, Table 8.38.

Economic Evaluation

Economic evaluations are generally carried out to determine whether a proposed investment meets the profitability criteria of the company or to evaluate alternatives. There are a number of methods of evaluation, and a good summary of the advantages and disadvantages of each is given in Perry [6]. Most companies do not rely on one method alone, but utilize several to obtain a more objective viewpoint.

As this chapter is primarily concerned with cost-estimation procedures, there will be no attempt to go into the theory of economics, but equations are presented for the economic evaluation calculations. A certain amount of basic information is needed to undertake the calculations for an economic evaluation of a project.

Definitions

Depreciation. Depreciation arises from two causes: deterioration and outdatedness (economically). These two causes do not necessarily operate at the same rate, and the one having the faster rate determines the economic life of the project. Depreciation is an expense, and there are several permissible ways of allocating it. For engineering purposes, depreciation is usually calculated by the straight-line method for the economic life of the project. Frequently, economic lives of 10 years or less are assumed for projects of less than $250,000.

Working capital. The working capital (WC) consists of feed and product inventories, cash for wages and materials, accounts receivable, and spare parts. A reasonable figure is the sum of these items for a 30-day period.

Annual cash flow. The annual cash flow (ACF) is the sum of the earnings after taxes and the depreciation for a one-year period.

Sensitivity analysis. Uncertainties in the costs of equipment, labor, operation, and raw materials—as well as in future prices received for products—can have a major effect on the evaluation of investments. It is important in appraising the risks involved to know how the outcome would be affected by errors in estimation, and a sensitivity analysis is made to show the changes in the rate of return due to errors of estimation of investment costs, raw material, and product prices. These will be affected by the type of cost analysis performed (rough-estimate or detailed analysis), stability of the raw material and product markets, and the economic life of the project. Each company will have its own bases for sensitivity analyses. However, when investment costs are derived from the installed-cost figures in this book, the following values are reasonable:

	Decrease by	Increase by
Investment cost	15%	20%
Raw material costs	3%	5%
Product prices	5%	5%
Operating volumes	2%	2%

Product and raw material cost estimation. It is very important that price estimation and projections for raw materials and products be as realistic as possible. Current posted prices may not be representative of future conditions, or even the present value to be received from an addition to the quantities available on the market. A more realistic method is to use the average of the published low over the past several months.

Return on original investment. This method is also known as the engineer's method, the du Pont method, or the capitalized earning rate. It does not take into account the time value of money; because of this, it offers a more realistic comparison of returns during the latter years of the investment. The return on original investment is defined as follows:

$$\text{ROI} = \frac{\text{average yearly profit}}{\text{original fixed investment} + \text{working capital}} \times 100$$

The return on investment should be reported to two significant figures.

Payout time. The payout time is also referred to as the *cash recovery period* and *years to payout*. It is calculated by the following formula and is expressed to the nearest one-tenth year:

$$\text{Payout time} = \frac{\text{original depreciable fixed investment}}{\text{annual cashflow}}$$

If the annual cashflow varies, the payout time can be calculated by adding the cash income after income taxes for consecutive years, until the sum is equal to the total investment. The results can be reported to a fractional year by indicating at what point during the year the cashflow will completely offset the depreciable investment.

Discounted cashflow rate of return (DCFRR). This method is called the *investors' return on investment, internal rate of return, profitability index, interest rate of return,* or *discounted cashflow.* A trial-and-error solution is necessary to calculate the average rate of interest earned on the company's outstanding investment in the project. It can also be considered the maximum interest rate at which funds could be borrowed for investment in the project, with the project breaking even at the end of its expected life.

The discounted cashflow is basically the ratio of the average annual profit during construction and earning life to the sum of the average fixed investment, working capital, and the interest charged on the fixed and working capital that reflects the time value of money. This ratio is expressed as a percentage rather than a fraction. Discounted cashflow is discussed here.

In order to compare investments having different lives or with variations in return during their operating lives, it is necessary to convert rates of return to a common time basis for comparison. Although any time may be used for the comparison, the plant startup time is usually the most satisfactory. Expenditures prior to startup, and income and expenditures after startup, are converted to their worth at startup. The discussion that follows is based on the predicted startup time as the basis of calculation.

Expenditures prior to startup. The expenditures prior to startup can be placed in two categories: (1) those that occur uniformly over the period of time before startup and (2) lump-sum payments that occur at some point before the startup time.

Construction costs are generally assumed to be disbursed uniformly between the start of construction and the startup time, although equivalent results can be obtained if they are considered to be a lump-sum disbursement taking place halfway between the start of construction and startup. The present worth of construction costs that are assumed to occur uniformly over a period of years T prior to startup can be calculated using either continuous interest compounding or discrete (annual) interest compounding.

Continuous interest compounding

$$P_o = \frac{CC}{T} \frac{e^{iT} - 1}{i}$$

Discrete (annual) interest compounding

$$P_o = \frac{CC}{T} \frac{\ln [1/(1 + i)]}{[1/(1 + i)]^T - [1/(1 + i)]^{T-1}}$$

where P_o = worth at startup time
        CC = total construction cost

T = length of construction period before startup, yr
i = annual interest rate

The cost of the land is a lump-sum disbursement and, in the equation given, it is assumed that the land payment coincides with the start of construction. If the disbursement is made at some other time, then the proper value should be substituted for T.

Continuous interest compounding

$$P_o = LC\ e^{iT}$$

Discrete (annual) interest compounding

$$P_o = LC(1 + i)^T$$

where LC = land cost
T = years before startup time that payment was made
i = annual interest rate

Expenditures at startup. Any costs that arise at startup time do not have to be factored, but have a present worth equal to their cost. The major investment at this time is the working capital, but there also may be some costs involved with the startup of the plant that would be invested at this time.

Income after startup. The business income is normally spread throughout the year, and a realistic interpretation is that 1/365 of the annual earnings is being received at the end of each day. The present-worth factors for this type of incremental income are essentially equal to the continuous-income present-worth factors. Even though the present worth of the income should be computed on a continuous-income basis, it is a matter of individual policy as to whether continuous or discrete compounding of interest is used. The income for each year can be converted to the reference point by the appropriate equation.

Continuous-income, continuous-interest

$$P_o = ACF\ \frac{e^i - 1}{i}\ e^{-in}$$

Continuous-income, with interest compounded annually:

$$P_o = ACF\ \frac{[1/(1 + i)]^n - [1/(1 + i)]^n - 1}{\ln{(1 + i)}}$$

where ACF = annual cash flow for year N
n = years after startup
i = annual interest rate

For the special case where the income occurs uniformly over the life of the project after startup, the calculations can be simplified.

Uniform continuous-income, continuous-interest

$$P_o = \text{ACF} \frac{e^i - 1}{i} \frac{e^{in} - 1}{ie^{in}}$$

Uniform continuous-income, with interest compounded annually

$$P_o = \text{ACF} \frac{[1/(1 + i)]^n - 1}{\ln [1/(1 + i)]}$$

There are certain costs that are assumed not to depreciate and that are recoverable at the end of the normal service life of the project. Among these are the cost of land, working capital, and salvage value of equipment. These recoverable values must be corrected to their present worth at the startup time.

Continuous interest

$$P_o = (\text{SV} + \text{LC} + \text{WC})e^{-in}$$

Interest compounded annually

$$P_o = (\text{SV} + \text{LC} + \text{WC}) \left(\frac{1}{1+i} \right)^n$$

where   SV = salvage value, \$
        LC = land cost, \$
      WC = working capital, \$
        i = annual interest rate, decimal/yr
       n = economic life, yr

For many studies, the salvage value is assumed equal to dismantling costs and is not included in the recoverable value of the project.

It is necessary to use a trial-and-error solution to calculate the discounted cashflow rate of return, because the interest rate must be determined that will make the present value at the startup time of all earnings equal to that of all investments. An example of a typical balance for continuous-income and continuous-interest with uniform annual net income follows.

Case Study Problem: Economic Evaluation

$$\text{LC } e^{iT} + \text{CC} \frac{e^{iT} - 1}{i} \frac{1}{T} + \text{WC}$$

$$- \text{ACF} \frac{[(e^i - 1)/i] (e^{in} - 1)}{ie^{in}} - (\text{SV} + \text{LC} + \text{WC}) + \frac{1}{e^{in}} = 0$$

TABLE 8.39 Investment Costs and Utilities

Item	BPCD	% onstream[a]	BPSD	$(×10³) 2001	Power, MWh/day	Water, gpm[b] Cooling	Water, gpm[b] Process	Stm., Mlb/h	Fuel, MBtu/h
Desalter	150,000	96.90	154,800	2,250	18	416	131	37.5	438
Atm crude still	150,000	96.90	154,800	25,500	75	626		31.5	138
Vac. pipe still	84,090	96.90	86,800	12,750	17	3,504		39.0	237
Coker	35,640	96.10	37,080	76,500	75	2,475		6.8	113
Hydrotreater	27,060	96.80	27,960	8,250	54	5,637		31.2	468
Cat reformer[c]	37,395	96.80	38,625	34,650	113	15,581			203
Cat cracker	48,450	95.70	50,625	58,500	291	16,800			485
Alkylation unit	11,190	97.20	11,505	21,000	42	28,800		5.1	291
Hydrocracker[c]	25,395	97.10	26,160	74,048	368	9,930	141	99.5	380
H2 unit, mmscfd	38.4	97.10	39.6	15,900	30	9,150			390
Gas pit., mmscfd	54.3	96.90	57.2	20,250	186	46,380			51
Amine treater, gpm	349	96.90	876	6,600	12	3,735			
Claus sulfur, long ton/day	131	96.90	135	4,650	3	135	74	(−34.5)	
Stretford unit, long ton/day	10	96.90	15	4,650	32				
				365,498	1316	143,169	346	231.7	3194
CW system (+15%)				12,375	99	165,000	8250		
Steam system (+20%)				13,875		8623	27	277.5	333
Subtotal				391,748	1415				3527
Storage				229,980					
Subtotal				621,728					
Offsites (15%)				93,258					
Subtotal				714,986					
Location (1.1)									
Special costs (1.04)									
(0.3156)(714986) =				311,162					
Contingency (1.15)									
Total (2001)				1,026,148					

[a] 365 days per year onstream.

[b] Process water and makeup water consumption.

[c] Catalyst cost for initial charge included in unit construction cost; $7,725,000 for reformer and $3,924,000 for hydrocracker.

All of the values are known except i, the effective interest rate. An interest rate is assumed and, if the results give a positive value, the trial rate of return is too high, and the calculations should be repeated using a lower value for i. If the calculated value is negative, the trial rate is too low, and a higher rate of return should be tested. Continue the trial calculations until the rate of return is found that gives a value close to zero. The return on investment should be reported only to two significant figures.

Case Study Problem: Economic Evaluation   The estimated 2001 construction costs of the refinery process units and their utility requirements are listed in Table 8.39.

TABLE 8.40   Summary of Operating Costs

Item	$/yr ($\times 10^3$)
Royalties	3,924.00
Chemicals and catalysts	12,955.00
Water makeup	1,087.00
Power	21,756.00
Fuel[a,b]	522,086.00
Insurance	8,208.00
Local taxes	16,419.00
Maintenance	73,882.00
Miscellaneous supplies	2,462.00
Plant staff and operation	11,676.00
Total[c]:	677,455.00

Case-Study Problem: Economic Solution

Storage costs (based on 50 bbl of storage per BPCD crude oil processed. Assume 21 days storage provided for n-butane; n-butane: 7890 bpd $\times$ 21 days = 165,690):

Total storage = 7,500,000 bbl
Spheroid = 165,690
General = 7,334,310

Cost:	$ ($\times 10^3$)
General @ $30/bbl (7,334,310) ($30) =	220,035
Spheroid @ $60/bbl (165,690) ($60) =	9,945
Total storage costs =	229,980

[a] Fuel quantity is deducted from refinery fuel gas, propane, heavy fuel oil, and some distillate fuel.

[b] Fuel cost (alternative value basis):

Fuel gas	1707 mmbtu/hr @ $4.40/mmBtu	180,259.00
Propane	4460 BPCD @ $36.96/bbl	164,822.00
No. 2 FO	1260 BPCD @ $60.48	76,205.00
Hvy FO	2520 BPCD @ $40.00/bbl	100,800.00
		552,086.00

(552,086) (365) = 190,561,536/yr

[c] Additional items, such as corporate overhead, research, and development vary among refineries and companies. These items are omitted here.

TABLE 8.41   Refinery Annual Summary

	BPCD	($/bbl)	$/yr (×10³)
Inputs:		Costs	
North Slope crude	150,000	24.35	1,333,163
i-Butane	3,510	25.40	32,541
n-Butane	3,690	23.10	31,112
Methane, mmscfd	3,990	4.00[a]	12,534
			1,409,350
Products:		Realizations	
Gasoline	98,460	40.00	1,437,516
Jet fuel	19,305	40.00	281,853
Distillate fuel (No. 2)	22,080	37.80	304,638
Coke, ton/day	1,889	20.00[b]	13,787
Sulfur, long ton/day	144	45.50[c]	5,781
			2,043,575

[a] $/Mscf
[b] $/ton
[c] $/long ton

The cooling water, steam systems, and water makeup requirements were calculated according to the guidelines of historic cases.

The estimated plant construction is expected to start in August 2001, and the refinery process is expected to begin in August 2003. Inflation rates of 10% per year are used to bring the costs to their values in 2003. These dates are integrally calculated on an annual compounded-interest basis.

The working capital is assumed to be equal to 10% of the construction costs. A review of the refinery staffing requirements indicates that approximately 139 people will be required to operate the refinery, exclusive of maintenance personnel. The maintenance personnel are included in the 4.5% annual maintenance costs. An average annual salary of $40,000, including fringe benefits, is estimated.

TABLE 8.42   Costs and Revenues

		$/yr (×10³)
Gross income		2,043,575
Production costs:		
Raw materials	1,406,843	
Operating costs	423,410	
Depreciation	68,283	
		1,904,535
Income before tax		145,040
Less federal and state income tax		74,898
Net income		70,142
Cashflow		138,425

TABLE 8.43 Total Investment

Construction costs	$1,365,801,000
Land cost	27,316,500
Working capital	136,579,500
Total:	$1,529,697,500
Return on original investment	4.6%
Payout time	9.9 years
Discounted cashflow rate of return	5.9%

Investment Cost

2001 investment cost = $1,026,147.
Inflation rate estimated at 10% per year.
Completion date scheduled for August 2003.
Estimated completed cost = $(1,026,147)(1.10)^3 = \$1,365,801$.

Royalty Costs

Unit	BPCD	$/bbl	$/CD
Hydrorefiner	27,060	0.015	405
Cat reformer	37,395	0.045	1682
FCC	48,450	0.04	1938
Alkylation	11,190	0.15	1679
Hydrocracker	25,395	0.04	1016
(6722)(365) = $2,453,250			

TABLE 8.44 Sensitivity Analysis
(In Percent; DCFRR on Original Basis = 5.9%)

	Investment cost		Raw materials cost		Product prices		Operating volume	
	+20	−15	+5	−3	+5	−5	+2	−2
DCFRR	3.9	6.9	2.9	6.9	8.9	1.9	5.9	4.9
Change in DCFRR	−2.0	+1.0	−3.0	+1.0	+3.0	−4.0	0	−1.0

Chemical and Catalyst Costs $/CD

Desalter: 300 lb/day × $1.04/lb	312
Hydrorefiner: 27,060 bpd × 0.04	1,082
Catalytic reformer: 37,395 bpd × 0.16	5,983
FCC: 48,300 bpd × 0.24	11,592
Alkylation: 3360 lb HF/day × $1.99/lb	6,686
2235 lb NaOH/day × $0.31	693
Hydrocracker: 25,395 bpd × 0.32	8,126
Hydrogen unit	1,030
Amine treater: 81 lb/day × $1.52/lb	123
Stretford unit	77
	35,704

(35,704)(365) = $13,031,960/yr

Note: In the above analysis, the annual costs for insurance, local taxes, and maintenance were not changed with varying investment costs. Royalties and catalyst replacement costs were not changed with operating volumes.

A 20-year life will be assumed for the refinery, with a dismantling cost equal to salvage value. Straight-line depreciation will be used. The federal tax rate is 48% and the state rate is 7%.

Investment costs and utility requirements are summarized in Table 8.39, operating costs in Table 8.40.

A summary of overall costs and realizations is given in Table 8.41, annual costs and revenues in Table 8.42, and total investment costs in Table 8.43, together with payout time and rates of return on investment. The effects of changes in investment and raw material cost, product prices, and operating volume are shown in Table 8.44.

References

1. Hall, S. R., et al., "Process Equipment Costs," *Chem. Eng.,* April 5, 1982.
2. Guthrie, K. M., "Capital Cost Estimating," *Chem. Eng.* 76(6):114–142, 1969.
3. Lang, H. J., *Chem. Eng.* 55(6):112, 1948.
4. Chase, J. D., *Chem. Eng.* 77(7):113–118, 1970.
5. Fouhy, Ken (ed.), "Economic Cost Indicator," *Chem. Eng.* 107(8):162, 2000.
6. Perry, R. H., and Chilton, C. H., *Perry's Chemical Engineer's Handbook*, 5th ed., McGraw-Hill, New York, 1973, pp. 25-11–25-23.

9

Visual Basic Software Practical Application

Introduction

This chapter is dedicated to all of you who are committed and dedicated to good old General Working (GW) Basic. Like me, you don't want to change to this complex Windows Visual Basic (VB)*. I have written numerous programs in the old GW Basic and have produced very user-friendly input-output screens using special, commercially sold software packages. Unfortunately, this is all history and we now apply only VB in the Microsoft Windows environment.

For those of you who are not familiar with these historic Basic language programs, such as the most recent QuickBasic and the IBM Basic, may I affirm that you, too, have the same difficulty of applying the inputs and outputs of VB.

What can we do? Well, I first researched a number of Visual Basic books, finding few if any clear explanations of how you make the switch from the simple Basic of old to the new VB by Microsoft. I found that the books that came with VB 3.0 were the best guides. I then spent considerable time (several months) on many trials, learning and applying this VB. Well then, must we each spend this same time learning? I answer an emphatic *no!*

My recommendation is simply to follow the example I submit here. The real difference between this new VB and the old is simply the input and output screens with their respective input/output code differences. All the motors that run and produce your answers from your previous GW basic inputs are the same, and may remain unchanged in

---

* Visual Basic is copyrighted by Microsoft Corporation.

VB. In fact, you can even install the same code you used from your previous GW basic programs. Wonderful, right?

For those of you who have not experienced the practice of making your own programs, even simple, short ones, I express my sincere encouragement. I give you very detailed instructions in this book. Those who have programming experience in Basic languages of the past may find some of my explanations repetitive. I present them for those who do not have previous experience. Even those who have Basic language background experience should find useful the simplicity of the examples as a guide and template to facilitate the development of many practical programs. One more point I wish to make is that I have applied VB version 3.0 throughout this book. Any VB edition you may have or use will work just as well.

The plan in this book is to guide you through a classic example program named RefFlsh.mak. This program is discussed and defined more fully in Chap. 2. This chapter covers inputs and outputs in VB, using RefFlsh.mak. Clear, precise, step-by-step examples are given, pointing out these new VB input and output screens.

RefFlsh.mak is by no means a short and simple program, but is rather large. I have chosen it because it contains almost every tool you should need to do your own VB programming. Also, you can use the same VB templates I present to duplicate and produce your own VB programming. RefFlsh.mak is included on the CD accompanying this book.

In keeping with my promise that you can use the same GW Basic code, I have kept all the old code I used in the GW Basic code in the RefFlsh.mak program code. This GW Basic is the same as QuickBasic, IBM Basic, and all the other common Basic languages. This code is installed with the " ' " comment line. As you can see, the old code for input-output is more extensive than the code required for VB. I submit, therefore, that VB should be considered an improvement rather than an added burden.

Figures 9.1 to 9.10, presented in the following sections, are sketches of actual VB screen images [1]. Later versions of VB will be found to be similar. You should be able to apply these figures to most any version of VB.

Making the VB Screens and Data Input Code

Form1 and txtN text box

Here's your first step. If you have a copy of VB, bring it up in Windows. You should see a Form1 and a small menu with Form1.frm in it. See Fig. 9.1*a*.

If you follow the steps in the first part of this chapter, you will make a file for VB called RefFlsh.mak. See Fig. 9.1*b*, which shows the file RefFlsh.mak. The default file name Project1.mak is replaced with this name. When you bring up an xxx.mak file in VB active program mode, you will get a window like that shown in Fig. 9.1*b*. Simply follow the steps shown to get the program's active screens. Of course, if you run the executable program file, xxx.exe, accompanying this book, you immediately get an active VB screen ready for your inputs or for running the program's defaulted values.

You may highlight files CMDIALOG.VBX and GRID.VBX and delete them under Files Menu, as these files are not used in this book. You may restore them at any time, as they will be in the VB stored files in your VB directory.

For those of you who do not have a copy of VB, simply bring up your executable copy of RefFlsh.exe on your Windows PC screen. The first screen form presented is the executable form type. Unfortunately, you cannot see the code, so I have copied it here for each of these input/output steps we shall walk through.

Here's the code for text box labeled No. of Selected Components:

```
Sub txtN_Change ()
End Sub
```

This text box is named txtN. The code under the Input Feed button will indeed make use of this name. The number of components you will choose is to be input in this box. When running the program, simply type in your choice of components.

Project1

View Form   View Code

Form1.frm   Form1

CMDIALOG.VBX

GRID.VBX

Figure 9.1*a*   Form1.frm in Visual Basic menu.

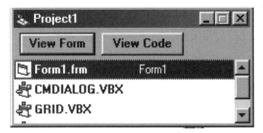

Click second on View Form

REFFLSH.MAK

View Form   View Code

ANSRUN.FRM   AnsRun

DAT.FRM   DEdit

MODULE1.BAS

Click first on DAT.FRM; highlight it

Figure 9.1*b*   RefFlsh.mak file.

The actual picture of what you should see on your computer screen is shown in Fig. 9.2.

12-Step how-to list.   For those of you who have a copy of VB, the following simple 12-step how-to is given. If you are writing this program, and starting with a new form, the following steps are made to put this text box, txtN, input into your program.

1. Click on the text box tool at the left of your VB screen (assuming you have started VB with a new form on your screen). Please see Fig. 9.3.

2. Move your mouse pointer to where you want to place this text box on the form. Notice your mouse pointer is now a crossbar instead of an arrow.

3. Click (left mouse key) at the selected location and hold down your mouse key. Now drag the crossbar shown and notice a box outline

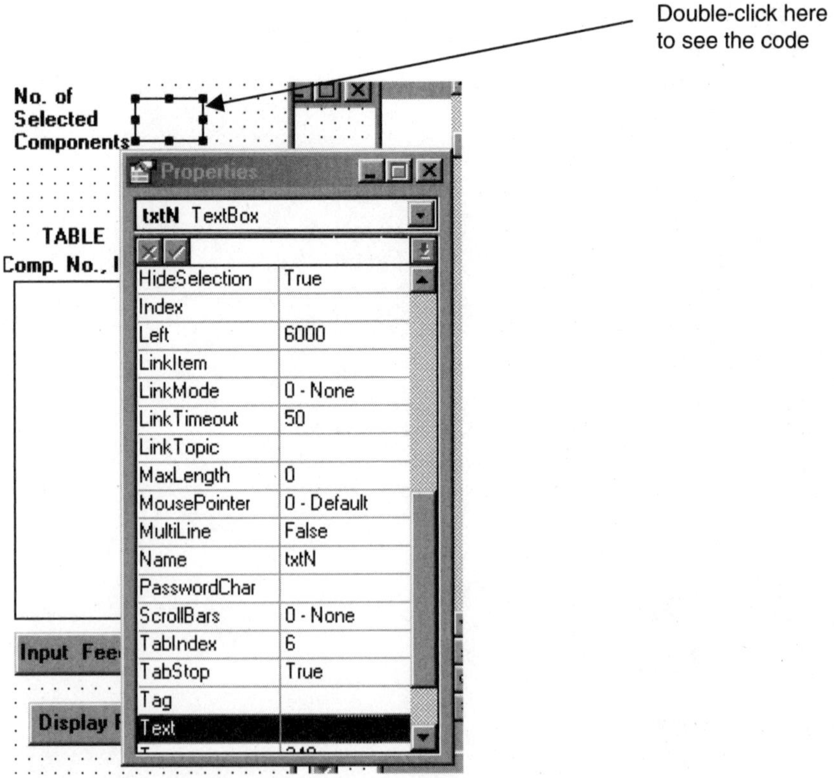

Figure 9.2   Text box txtN.

is formed. Let up on the key when you obtain the size of the text box desired. Again, Fig. 9.3 shows the final form of the text box.

4. Click once on the box you just made.

5. Notice at the bottom of your screen you have a menu bar, and on this menu bar click on Properties. A Properties menu now appears, visible and active. (After item #4, clicking this text box you made, the Properties menu is actually now a window menu smaller than and behind the form you are using. Another way is clicking this background Properties menu; it will appear in front, visible and active.) See Fig. 9.4.

6. Under the Properties menu at Name, replace the Text1 name with your selection, txtN. You can choose any name you wish for a program. This name simply becomes a variant that may represent a numeric variable or other such as a file name, which we will cover later.

7. Again under the Properties menu replace the Text item with blank spaces. Notice this also blanks out the text box you made on the Form1. Now when you run the program, only what you install in the text box will be shown.

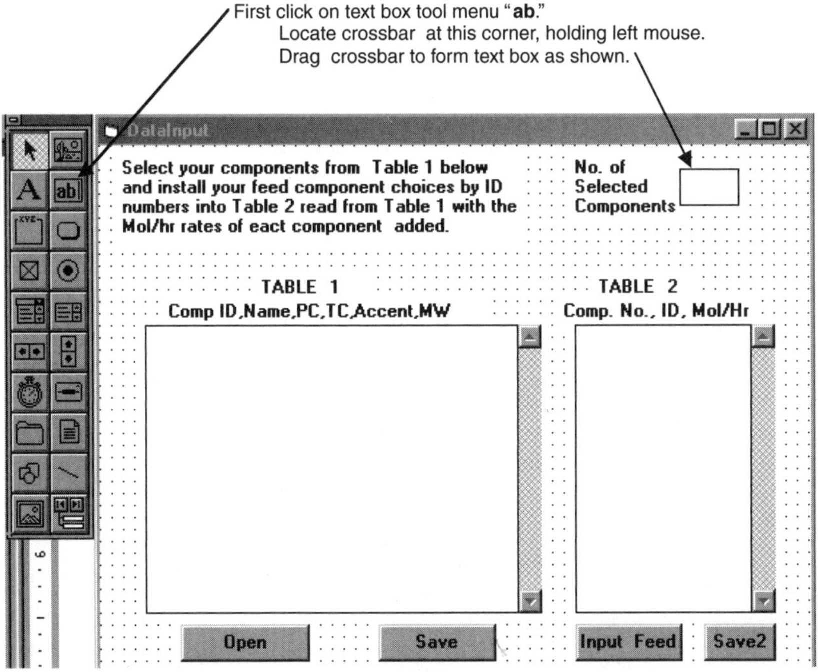

First click on text box tool menu "**ab**."
Locate crossbar at this corner, holding left mouse.
Drag crossbar to form text box as shown.

Figure 9.3 Final form of the text box.

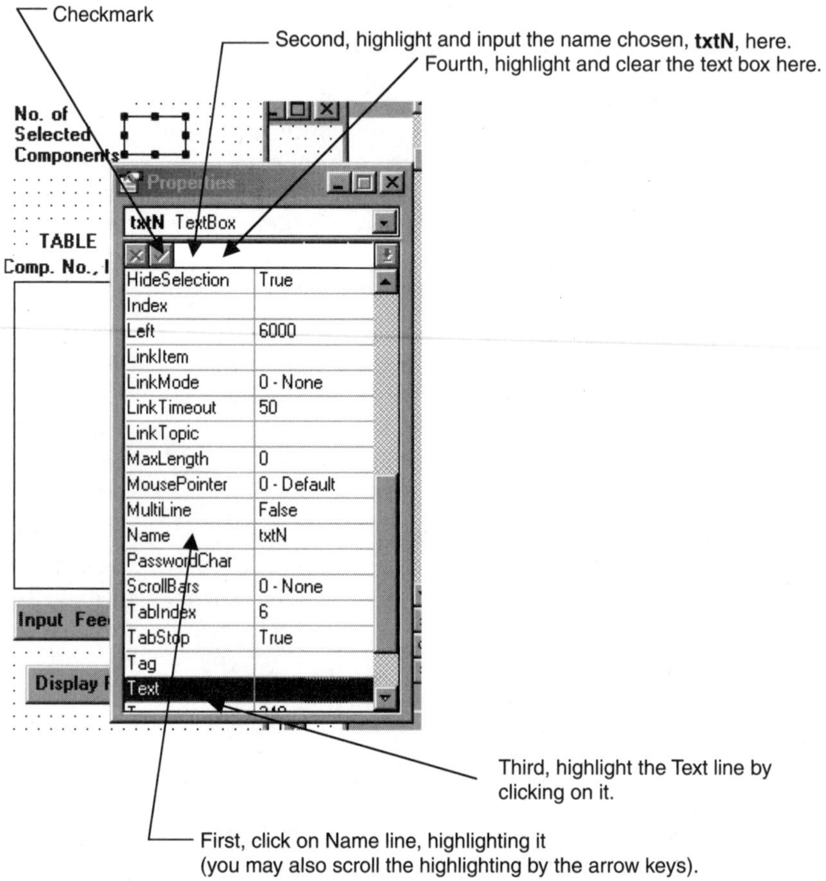

Checkmark

Second, highlight and input the name chosen, **txtN**, here.

Fourth, highlight and clear the text box here.

No. of
Selected
Components

TABLE
Comp. No.,

txtN TextBox	
HideSelection	True
Index	
Left	6000
LinkItem	
LinkMode	0 - None
LinkTimeout	50
LinkTopic	
MaxLength	0
MousePointer	0 - Default
MultiLine	False
Name	txtN
PasswordChar	
ScrollBars	0 - None
TabIndex	6
TabStop	True
Tag	
Text	

Input Fee

Display

Third, highlight the Text line by
clicking on it.

First, click on Name line, highlighting it
(you may also scroll the highlighting by the arrow keys).

Figure 9.4   Properties menu.

8. As a last item on this active Properties menu, click the checkmark at the top of the Properties menu. This inputs your changes into VB.

9. Now click anywhere on Form1. The form with the text box you made reappears as the front item, visible and active. Next click twice on the text box you made. Notice the code window appears with the same code again as previously. The statement Change() is the default statement and the one we commonly keep. This line allows you to input your choice anytime in program execution. Your running program input will now always be the txtN program variant, just as GW Basic says: Let txtN = 12. Note that nothing is required in this code window—no additions and no changes.

10. Next click anywhere on the Form1 area. Click on the background Properties menu again, bringing it in front and active. See Fig. 9.5.

11. At the Properties item Caption, replace the Form1 with your form title choice. In this program I chose Datainput. As the essential data input of the program, this seemed a good choice for a title. By *caption* is simply meant a heading or a title. I don't know why the VB designers used such nomenclature here. I would simply have used the word *Title*. Oh well.

12. Again, at the Properties item Name, replace Form1 with your choice for a form name. I chose DEdit. This stands for data edit. That's what you do on this form, edit data, change it, add some. Please see Fig. 9.6. This name will also be shown in the RefFlsh.mak menu of Fig. 9.1. This is the name that the VB program code recognizes. It is important, as you will see in later steps. In this step, also replace the name Project1.mak under the File menu with the name Ref-Flsh.mak.

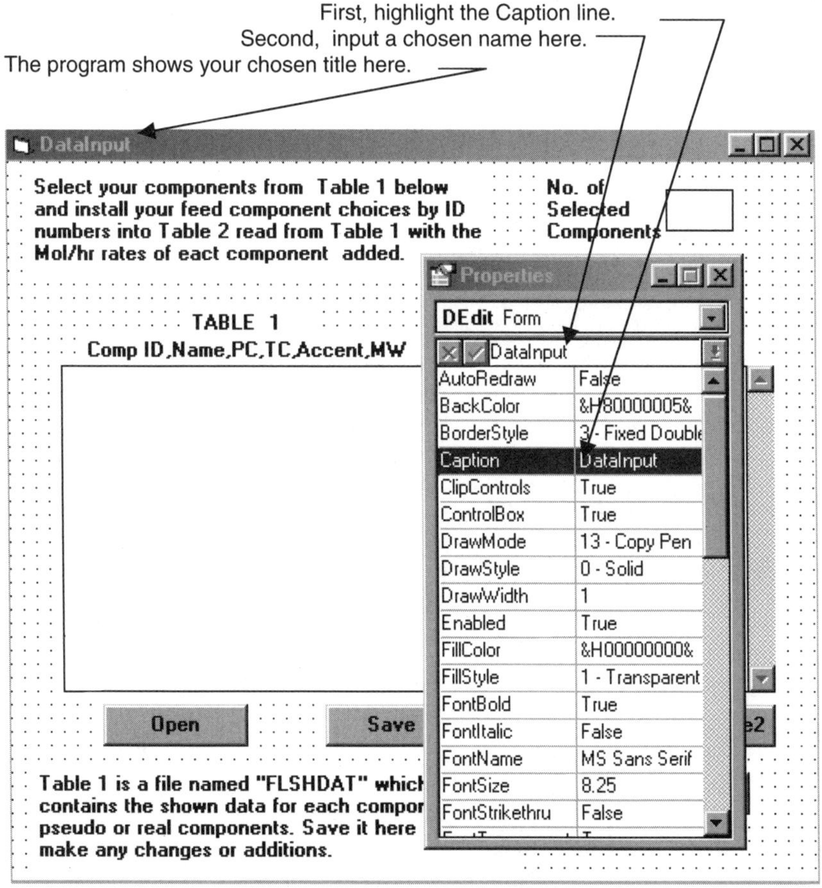

First, highlight the Caption line.
Second, input a chosen name here.
The program shows your chosen title here.

Figure 9.5 Selecting a caption for the form.

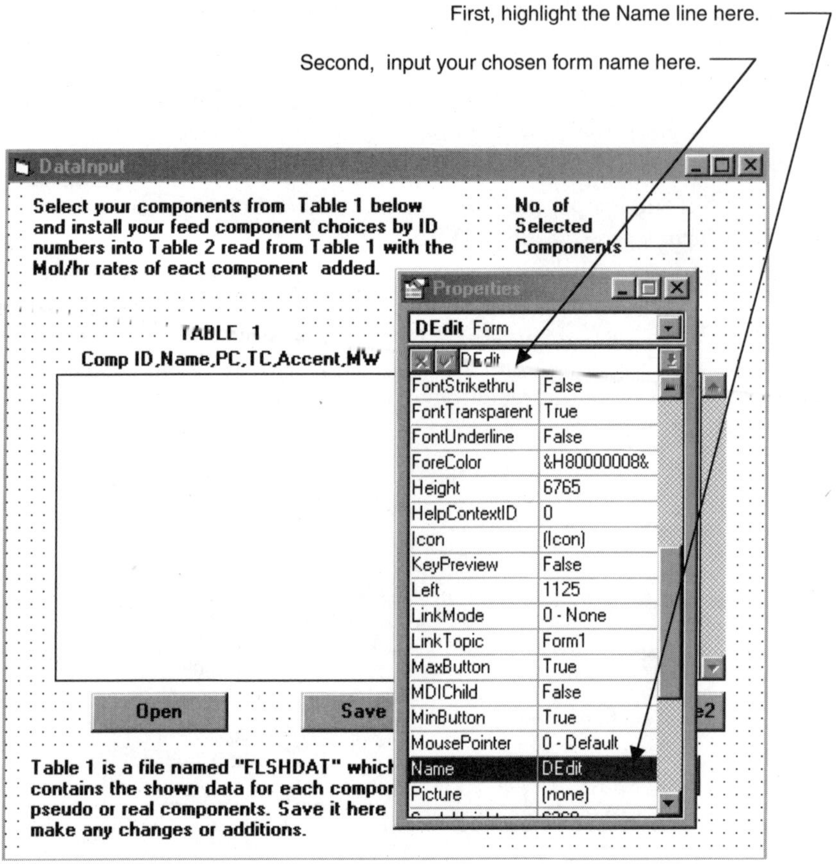

First, highlight the Name line here. ⟍

Second, input your chosen form name here. ⟍

Figure 9.6 Naming the form.

The Properties menu. By this point in our progress we have made a new text box and given it a name. We have also given a title (caption) and a name to the Form1 we used. The following thus summarizes:

Properties menu

Item	Input
Form1:	
Caption	Datainput
Name	DEdit

Properties menu

Item	Input
Text box:	
Name	txtN
Text	(blanked)

Three steps for a label. Let's next input a label near the text box we made. A label is made the same way you made the text box. Simply click on the label icon tool (the big **A**) and locate your mouse crossbar at the point you want. Then do just as you did for the text box in sizing and locating. You can also relocate any label or text box simply by clicking on it, holding the left mouse key down, and dragging it to the desired location.

As for the text box, each label also has a Properties menu. Here's the Properties menu summary for the Label2 installed for the text box in Fig. 9.7.

Properties menu

Item	Input
Label2:	
Caption	No. of Selected Components

Please note this label has the name Label2 since another label on this form was named Label1. Also note you must type in your caption text on the Properties input line, which is next to the checkmark at the top of the menu. See Fig. 9.7. I find leaving the label default name Label1, Label2, Label3, and so on a convenience. Labels are more clearly identified by their label text. I won't cover any further details

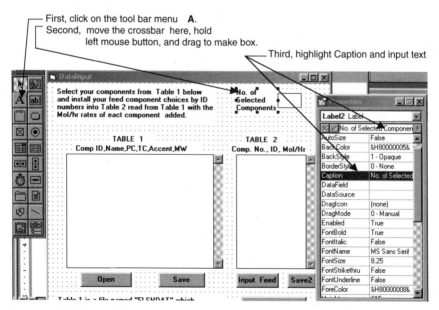

Figure 9.7 Making a label.

about labels; you clearly can see each label on the two forms presented in both the VB editing mode and the execution mode.

txtContent text box with scrollbar

Next make the largest text box figure on this first form, the DataInput form. See Fig. 9.8. Install the following on the textbox Properties menu:

Properties menu

Item	Input
Text box:	
Name	txtContent
Text	(Blank)
Multiline	True
Scrollbar	2 (Vertical Scrollbar)

In this text box, named txtContent, you will install the major portion of data for running the program. Your data input will be a multi-

First click on text box tool menu **ab** which is the text box icon.
Second, locate crossbar as shown and drag, holding mouse button to size box.
Third, click on Properties at bottom of screen, displaying this menu.
Fourth, highlight these lines, one at a time.
Fifth, choose selection or input text as described here.

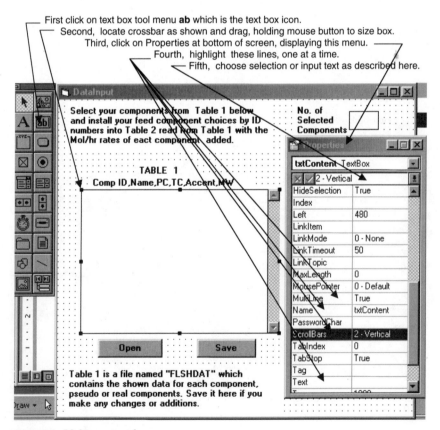

Figure 9.8 Making a text box.

line type and will be scrollable. Hundreds of data lines could be installed here. Figure 9.8 displays the actual screen and how to make these inputs.

Open CommandButton.   Next make a new box called a *command box.* This is the box-type icon right below the text box icon you used. Just as you did for the text box, simply click on the CommandButton icon and then locate a new box, sizing it on the form, DataInput. See Fig. 9.9. Here you will see the caption Command1. Next input the following in this command button's Properties menu:

Properties menu

Item	Input
CommandButton:	
Caption	Open
Name	btnOpen

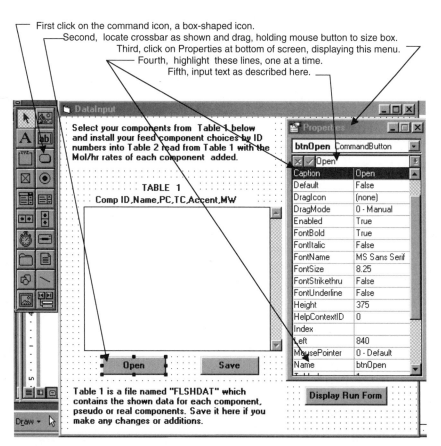

First click on the command icon, a box-shaped icon.
Second, locate crossbar as shown and drag, holding mouse button to size box.
Third, click on Properties at bottom of screen, displaying this menu.
Fourth, highlight these lines, one at a time.
Fifth, input text as described here.

Figure 9.9   Making a command button.

Figure 9.8 is the most complete and comprehensive of all figures shown so far in this chapter and is self-explanatory, as the previous text in this chapter explained the input. When one first tries to understand how to use VB, he or she is often confused about which menu in the window the book text refers to. Figure 9.8 should help you overcome such a difficulty.

A Program Example   The program RefFlsh uses the Open command button to install a data file named flshdat. The following program code is to be stored in the Open command button. If you could view the code contained in this button, you would see the following:

```
Sub btnOpen_Click ()                               'This line is the VB
                                                    default line and
                                                   'needs no changes

Open "FLSHDAT" For Input As #1                      ' This line in VB
FLSHDAT                                             opens the file—make
                                                    sure computer is
                                                    defaulted to the file
                                                   ' which contains most
                                                    of the data to be
                                                   'used.
   For J = 1 To 60                                  ' Limits number of
                                                    data comp to 60
Input #1, J, OO(J), BA(1, J), BA(2, J),            ' J= Component ID
BA(3, J)                                            number,
OO(J) =
                                                   ' the name,
BA(1,J)=PC,BA(2,J)=TC
                                                   'and BA(3,J)=the
                                                    acentric factor
                                                   ' The input #1 reads
                                                    the file

FLSHDAT.
   If OO(J) = "end" Or OO(J) = "END"               ' This line stops
GoTo STARTINPUT                                     file reading
      Next J                                       ' into the program.
                                                    You must include
                                                   ' this data line in
                                                    your file, FLSHDAT,
                                                   ' as the last line
                                                    of data.

STARTINPUT:
   Close #1                                        ' Closes the file
                                                    FLSHDAT
Open "FLSHDAT" For                                 'This line opens the
Input As #1                                         file FLSHDAT
                                                   'again
```

```
txtContent = Input&(LOF(1), #1)        'This line writes
                                       FLSHDAT into the
                                       'textbox named
                                       txtcontent.
     Close #1                          'Closes the file
FLSHDAT
End Sub                                 ' A default line in
                                       VB
                                       'which closes
                                       subroutine
```

At the end of several of the preceding lines you see the " ' " quote mark, which in VB allows comments not affecting the code execution in any way. If you could see a copy of VB I have written in the Open button command, you wouldn't see these comments in the code. I list these comments here for a clear understanding of the whys and hows of the code used. If you wish, you may add these comments to the code lines you write.

We have now brought in a file from outside the program named FLSHDAT. In the program execution mode, when the user clicks on the Open command button, VB writes the FLSHDAT file into the program for executable data. This is the same as writing the subject data into our program line by line. Such data may be called upon for use anywhere in the program. The data is exhibited in the text box named txtContent.

Making the Save command button. Next I will cover the Save command button, which will save any changes or additions you make on the exhibited file in the txtContent textbox. Here the accomplishment will be as follows:

- Save any change you wish to make to the file FLSHDAT.

- Any changes made to the txtContent text box file are be immediately written into the program for updated data and subsequent execution.

To do this, make a command button just like you did for the Open button, and input the following in the respective Properties menu:

Properties menu

Item	Input
Command button:	
Caption	Save
Name	btnSave

Now if you could view the Save button code with my line comments added, you would see the following:

```
Sub btnSave_Click ()
If txtContent.Text = "" GoTo EndSub      ' In the event you click
                                           on this command button
                                         ' before first clicking
                                           the "OPEN" button the
                                         'entire file, FLSHDAT
                                           would be lost. This
                                         ' line is therefore
                                           installed to prevent
                                           such loss.
Open "FLSHDAT" For Output As #1          ' This line opens the
                                           file "FLSHDAT located
                                           with the other RefFlsh
                                           files"
                                         ' for writing
                                         ' the contents of the
                                           textbox, txtContent,
                                           into it.
Print #1, txtContent.Text                ' This line writes the
                                           contents of the textbox
                                         ' txtContent into the
                                           file #1 or FLSHDAT.
EndSub:
Close #1                                 ' Closes the file
                                           FLSHDAT.

End Sub
```

With the Open command button and the Save command button, you may enter any data desired into the file FLSHDAT, remove any data desired, or modify any data, as long as you keep the same format of data input line by line. Of course, if you don't keep this same format, the RefFlsh.exe executable program will fail with an illegal function call. You can indeed make any other format changes you wish in these Open and Save command buttons.

If you have made these command buttons and have added the VB code as shown, activate a VB run and try writing a new line of data on your file in the text box, txtContent. Notice the Enter at the end of a line brings up a new line of data you may install. A Delete key at the beginning of this blank line will also delete the blank line. You can also delete any selection of data simply by highlighting it with your mouse and then pressing Delete. No data changes are made to the file FLSH-DAT, however, until you click the Save command button.

Observe there are no spaces between the FLSHDAT file data, and its string data is input in " " marks. Each data input on the input lines is separated by a comma. There are precisely five data items on each line. To change such a format, you must change the input line "Input #1, J,

OO(J), BA(1,J), BA(2,J), BA(3,J)." The OO(J) data input on each line is always the name of the component. BA(1,J) is always the critical pressure (in psia), and BA(2,J) is always the critical temperature (in °F). BA(3,J) is always the component acentric factor. Each line defines the data input for one component. Note also that the component number identifier, J, becomes the component's identifier in the next data input text box.

Making the txtFeed text box

The user will select component choices from Table 1 (text box txtContent) and install each by ID number in a Table 2 text box that has been named txtFeed. Now you must make the text box, txtFeed, the same way you made the txtContent text box. See Fig. 9.8 again, where the procedure is the same. The Table 2 text box is given the Property menu items shown here. Next click once on the Properties menu bar at the bottom of your screen. Properties menu inputs for txtFeed text box are as follows:

Properties menu

Item	Input
Textbox:	
Name	txtFeed
Text	(Blank)
Multiline	True
Scrollbar	2 (Vertical Scrollbar)

Sizing these scrolling text box figures is a matter of choice. I have found the sizes presented are adequate for all applications I have encountered.

InputFeed using the command button. As with the data input text box, txtContent, this text box, txtFeed, will also be given an Open button and a Save button. Since this is Table 2, I will call this Save button Save2. I might just as well have titled the Open button here Open2. Perhaps those of you who are working with a VB programming copy may wish to do so. In place of calling this an Open2 button, however, I called it InputFeed button, hoping to keep it as simple as possible.

Here are the Properties menu item default changes for the Input-Feed button. The code input for this command button, InputFeed, follows. All code lines in this chapter are from RefFlsh.

```
Sub InputFeed_Click ()
N = txtN                          ' Defines the num-
                                    ber of components
                                    you will select.
```

```
If N = "" Then N = 12                    ' Defaults a value
                                         if left blank.

txtN = N
FT = 0
Open "FeedDat" For Input As #1           ' Opens the file
                                         named FeedDat

For I = 1 To N
Input #1, I, J, F(I)                     ' Reads the file
                                         FeedDat and inputs
                                         the data
Q(I) = OO(J): P(I) = BA(1, J): T(I) =    ' The above line
  BA(2, J): W(I) = BA(3, J): CJ(I) = J   transfers data to
                                         the program
                                         ' data variables.
If J = 99 GoTo start3                    ' Stops data input
                                         if end of file.
                                         Note you should
                                         ' always install
                                         this number at the
                                         end of the file.
FT = FT + F(I)                           ' Calculates the
                                         Mol/Hr sum, FT
Next I
start3:
Close #1                                 ' Closes File
                                         FeedDat.

Open "FeedDat" For Input As #1           ' Opens File
                                         FeedDat
txtFeed = Input$(LOF(1), #1)             ' Writes file to
                                         TextBox named
                                         txtFeed
Close #1                                 ' Closes file
                                         FeedDat
Open "trnsf1" For Output As #1           ' Opens a file named
                                         trnsf1

For I = 1 To N + 1
If I = N + 1 Then Print #1, "Total";     ' Writes the column
Tab(14); FT: GoTo start4                 sum, FT, to ' the
                                         file.
Print #1, I; Tab(7); CJ(I); Tab(14); F(I)  ' Writes the shown
                                         data, comp No.,
                                         Comp ID,
                                         ' and mol/hr compo-
                                         nent flow rate to
                                         the file.
Next I
start4:
```

```
Close #1                              ' Closes the file,
                                        trnsf1.

End Sub
```

Save2 command button. Next make the Save2 command button. Click on it once and bring up the Properties menu. Make the following inputs on this menu:

Properties menu

Item	Input
Command button:	
Caption	Save2
Name	SaveFeed

Next double-click on the command button named InputFeed. If you could view the executable file you have with this book, you would see the following code. I list the code here with my explanatory comments:

```
Sub SaveFeed_Click ()
If txtFeed.Text = "" GoTo EndSub2     ' This code line voids
                                        writing to the file
                                        "FeedDat"
                                      ' if the txtFeed text box
                                        is blank.
Open "FeedDat" For Output As #1       ' Open the file
                                        "FeedDat" for writing to.
Print #1, txtFeed.Text                ' Write contents of the
                                        text box, txtFeed, to
                                      ' the file "FeedDat".
EndSub2:
Close #1                              ' Closes the file
                                        "FeedDat".

End Sub
```

This completes the data input needed for running this program. It is therefore necessary to start a new form, as we have used all available space in this first form.

Running the data input form

It is prudent to review what has been accomplished. Now execute the run mode of the program. The Form1 screen appears with its blank text boxes as made. Click once on the Open button below Table 1. The files FlshDat and FeedDat have each been included in the CD supplied with this book. When you bring up these files under these command buttons, make sure you have established them where you have installed the xxx.mak or xxx.exe files. Here's what you get:

```
HERE'S WHAT AN ENGINEER NEEDS

TABLE 1
1,"H2",188.1,238.4,0
2,"N2",492.9,227.3,0.03468
3,"C02",1069.9,547.7,.225
4,"H?S",1306.5,672.5,.09509
5,"C1",673.1,343.2,.00573
6,"C2",709.8,550.0,.09381
7,"C3",617.4,665.9,.1497
8,"IC5",483.,828.7,.22035
9,"NC5",489.5,845.6,.24914
10,"C8",362.1,1023.4,.40158
11,"C9",332.,1070.,.44592
12,"BP120",320,864,.16846
13,"BP140",379,895,.22841
14,"BP160",389,920,.25831
```

In Table 1 notice a comma separates discrete inputs. Each line input is for a specific component. The first item is the component's ID, or identification number. The second input on each line is the component's name in an abbreviated text. In sequence, critical pressure, critical temperature, and acentric factor are each input, separated by commas. *Do not install any blank spaces.*

In Table 1 you may input any data you wish, make changes, or add data as you wish. Please notice the editing user-friendliness. Delete by highlighting and pressing the Delete key. Insert any add-on lines of component data. You must, however, keep this exact format of line data input. Any deviations, such as leaving out a comma or replacing a comma with a period, will shut down your program run with an error message.

Any changes you make to Table 1 or Table 2 and then save becomes the respective file. Since you must save any desired changes, it is a good practice, and strongly advised, to make a backup copy of FlshDat and keep it in a safe place. In fact you should make a backup copy of all the data files you may make. FlshDat2, FlshDat3, -4, -5, and so on is a good suggestion. Then you can rename any one particular to the FlshDat file name for program usage. Please note also that the program calls for the file in the C:\ppevb directory. For those of you who have the VB program, you may change the input program VB line code to any directory and drive you desire for calling the FlshDat file. You may also rename the file if desired.

Table 2 is similar to Table 1, with the exception of the different file form. Here the user chooses from Table 1 the components desired to run. The first item on each line is the consecutive number of the component, beginning with component number 1. The second item (separated by a comma) on each line is the component's ID taken from Table 1. The third and last item of each line is the respective component's mol/h flow rate to be used in the program. The following is a typical example:

```
TABLE 2
1,5,0.004
2,6,0.05
3,7,0.12
4,9,10
5,11,30
6,14,40
7,15,150
8,18,300
9,24,250
10,28,200
10,99,0
```

You may input any of the components selected from Table 1 into Table 2 simply by inserting into Table 2 the ID number of the component from Table 1. Any number of components up to 60 may be installed. Those of you with a copy of VB may change a few program code lines to any number of components desired.

Notice that the last line of the Table 2 input has the ID number 99. This is required, as it signals VB that this is the end of the file. There is another way, but this is an easy and convenient method to allow the user to signal VB to stop reading the file. You could have much more data after this line that you wish to keep in the file, but have not used.

The same is true for the data file, FlshDat, in which the last line of data reads End for the component name. This line must be installed for the program to run. As for Table 2, the Table 1 file also stops being read by VB. You may have much more data you wish to keep in the file after the "end" line that isn't to be read by the program. Simply install such data after the component named End. Note, you may insert the "end" component line anywhere you wish in the file FlshDat to stop the program data reading.

In conclusion, you now are presented a Table 1 that automatically presents a database, a database for which you may easily pick and choose components, installing your choices into Table 2. These chosen components with their respective molar rates become the data input for running the program.

It is important to realize that both Tables 1 and 2 are files you can make, edit, and save on the Form1 screen in this program. Table 1 is the database required to run the program, and Table 2 is the user's input per choice of components to be run in the program. Every time the program is executed, the previous Table 2 components chosen are presented. Please note, you should click the Save2 command button each time you modify or renew the Table 2 component data.

Calling up a new second form, ProgRun

Click on the File Windows menu bar at the top of your screen. Note, you are presented the option to start a new form on this file menu. Do so. The Properties menu for this new form is next brought up by clicking anywhere on this new form and then clicking on the Properties button on the bottom menu bar of Windows. Make the following changes to this new Properties menu:

Properties menu

Item	Input
Form2:	
Caption	ProgRun
Name	AnsRun

You are ready to start the program answer form, with one exception. The Global dimension statement of all variables and arrays must be declared. The Global dimensioning will ensure that all variables made in Form1 will be applied equally in Form2. Visual Basic uses a code module to do this. Global dimensioning of variables must be made here; otherwise, this program simply will not run. Please note, this code module is separate from all forms, being in a way a form itself.

Code Module form. Just as you did for the new form, again click on the Windows upper bar at File and then click on New Module. See Fig. 9.10. The default name Module1.bas code screen appears. Of course, the program you are viewing already has this Module1.bas code module. You can view this code by clicking on the DAT button on the menu bar at the bottom of your Windows screen. Highlight the Module1.bas menu item and then click View on this DAT.MAK menu. The code I have installed is exhibited next. Since you do not have a copy of this VB code on your CD files, this code follows, with my comments added. (You may also see it in Fig. 9.10.)

```
DefStr O                    ' All variables and arrays
                            beginning with an "O" are
                            ' declared strings.
```

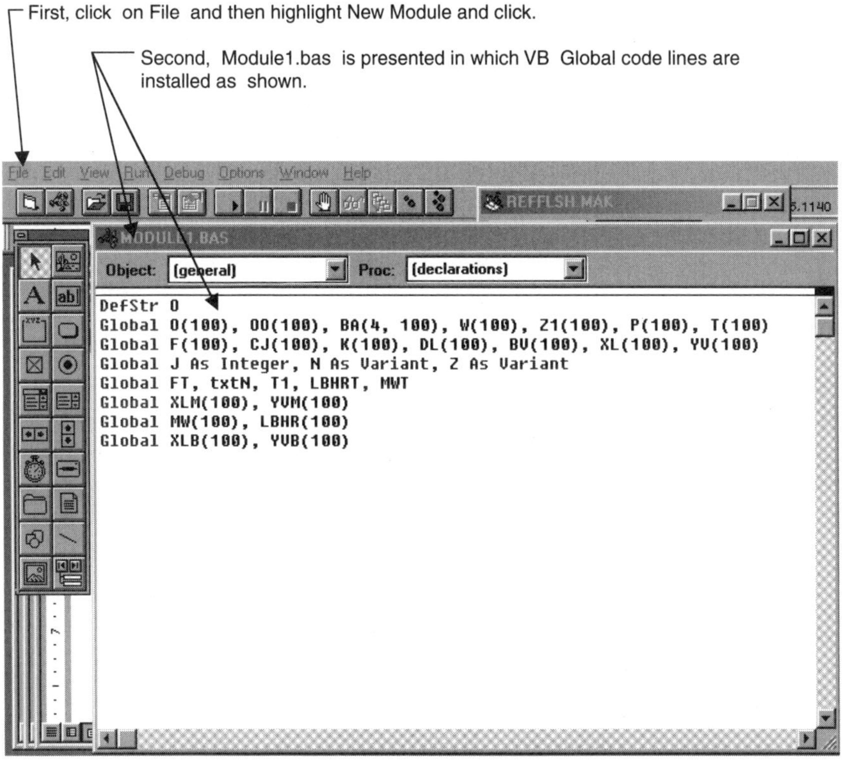

First, click on File and then highlight New Module and click.

Second, Module1.bas is presented in which VB Global code lines are installed as shown.

File Edit View Run Debug Options Window Help

REFFLSH.MAK

MODULE1.BAS

Object: [general]   Proc: [declarations]

```
DefStr O
Global O(100), OO(100), BA(4, 100), W(100), Z1(100), P(100), T(100)
Global F(100), CJ(100), K(100), DL(100), BU(100), XL(100), YU(100)
Global J As Integer, N As Uariant, Z As Uariant
Global FT, txtN, T1, LBHRT, MWT
Global XLM(100), YUM(100)
Global MW(100), LBHR(100)
Global XLB(100), YUB(100)
```

Figure 9.10   Entering the Global dimension code.

```
                           'Following lines declare
                           all program variables and
                           arrays
Global O(100), OO(100), BA(3, 100), W(100), Z1(100), P(100),
  T(100)
Global F(100), CJ(100), K(100), DL(100), BV(100), XL(100),
  YV(100)
Global J As Integer, N As Variant, Z As Variant
Global FT, txtN
Global XLM(100), YVM(100)
```

Close when you have finished reading. By the way, you should always close these code view screens to minimize the clutter on your Windows screen. You may also add to or subtract or modify this code module per any program changes you may desire.

When finished, your DataInput form should look like Fig. 9.11. Note that it contains the defaulted values shown when you run the executable RefFlsh program. Click on Open first and then Run Prog. This inputs the data shown into the program. Make sure you have a backup copied of the current file of Table 1 or Table 2 if you wish to save either without the changes you make. If you make changes to these tables and click on Save or Save2, the files as shown with your changes will replace the previous respective file. It is also important to note you can run this program only with the route directory defaulted to the directory that contains these two program files, FlshDat and FeedDat. You may read these files at any time by activating your executable copy of RefFlsh or simply by opening either file in the DOS Editor and prompting where the file is located, such as C:\Ppevb, print edit FlshDat. The file contents will then be shown in the DOS Editor program as they are in Fig. 9.11.

Display Run Form command button.   Now go back to Form1, as there is yet one more small command button to install on it. This command button will be one that brings up Form2. See Fig. 9.11, the lower right-hand corner. Make another command button with the following Properties menu changes for it. Remember the command icon is that square box as shown in Fig. 9.9.

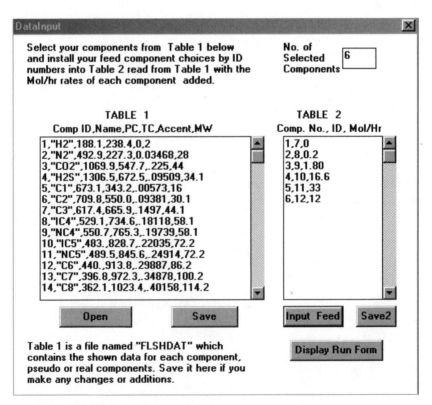

Figure 9.11 The finished DataInput form.

Properties menu

Item	Input
Command button:	
Caption	Display Run Form
Name	RunForm

Since you do not have the VB code, I have listed the Display Run Form code, with my comments added:

```
Sub RunForm_Click ()
Load AnsRun                  ' Loads the form2, named "AnsRun"
                             for program
                             ' execution.

AnsRun.Show                  ' Displays the AnsRun form2 on
                             your screen.
End Sub
```

txtP1 and txtT1 text box

Now we shall start the text box development on the second display screen. Please refer to Fig. 9.12. Make two text boxes as shown on your Form2 screen. These are the pressure and temperature text boxes. Bring up their respective Properties menus for the following, the txtP1 being nearest the left of the form:

Properties menu

Item	Input
Text box:	
Name	txtP1
Text	(Blank)

Properties menu

Item	Input
Text box:	
Name	txtT1
Text	(Blank)

Here's what the second form should look like when finished. The program RefFlsh default values are shown. Simply click on Run Prog and View Feed command buttons to get this screen on your executable program. Please note, however, you first must input the first form you made by clicking on the Open and Input Feed command buttons. The first form, named DataInput, must be executed before the second form, ProgRun, will run.

The preceding two text box default codes are used. Double-click on the respective text box to bring up the VB code, and input the following code lines for this text box named txtP1. The codes are (assuming you

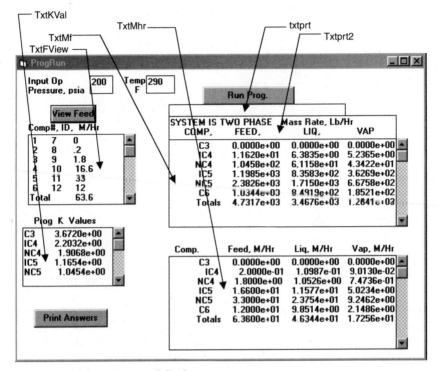

Figure 9.12   Making the second display screen.

have made these text boxes on your VB Properties menu screen, named txtP1 and txtT1):

```
Sub txtP1_Change ()          ' allows the user to input any
                             desired pressure, psia.
                             ' Required input.
End Sub
```

The same applies to the txtT1 code. Temperature, °F, is required input by the user. Please notice the labels are added separately. See Figure 9.3 for adding labels.

txtFView text box

Make another text box on Form2 for viewing the feed input you made. See Fig. 9.12. Sizing this box as shown is suggested. This box should scroll and allow multiline viewing for each component chosen; one line is dedicated for each component. Component number, component ID number, and the component mol/h flow rate will each be exhibited. If desired, the user may return to Form1 and make changes to the feed input as many times as desired. This is a unique enhancement of VB: if answers are not within expectations, the user may rapidly modify the feed until the desired results are delivered. The text boxes for P1 and T1 may be changed as well. This is especially unique for determining dew point and bubble points at varied temperatures and pressures. Note that any user changes in this box are not read into the program. The InputFeed and Save2 command buttons for Table 2 are the means by which any changes are to be made. See Form1.

Make the following inputs for the txtFView text box:

Properties menu

Item	Input
Text box:	
Name	txtFView
Text	(Blank)
Multiline	True
Scrollbar	2 (Vertical Scrollbar)

No changes are required for the txtFView code that is defaulted in VB. Note that the keyword Change is defaulted in this code.

View Feed command button. Make another command button with the following Properties menu inputs:

Properties menu

Item	Input
Command button	
Caption	View Feed
Name	ViewFd

This command button may be clicked at any time to bring up a file for exhibit named trnsf1. Please recall that this is the file in which the input feed choices were stored when the InputFeed command button was clicked. This file is therefore to be written in the txtFView text box. The following code is for this command button, View Feed:

```
Sub ViewFd_Click ()

Open "trnsf1" For Input As #1    ' Opens the file named trnsf1
txtFView = Input$(LOF(1), #1)    ' Writes the file trnsf1 to
                                   the textbox, txtFView


Close #1


End Sub
```

Making the VB Screens and Output Answer Code

Run Program command button. This command button, Run Prog., is the heart of the program, the program's motor. Here under this button, most all of the program code will be stored for execution. Previously I mainly covered inputs on Form1, and P1, T1 data inputs here on Form2. Now it is time to make this Run Program command button that will generate output answers in five text boxes as shown on Form2. For reasons of copyright protection, however, the publisher retains the right not to reveal the program code.

For those of you who are walking through the writing of this VB program with me, next make a command button with the following inputs on its Properties menu:

Properties menu

Item	Input
Command button:	
Caption	Run Prog.
Name	Command1

I left the defaulted name, Command1, on the Properties menu since this is indeed the program's chief button that must be clicked each time the program is run. Whenever you make changes to the input files, Table 1 or 2, or make P1 or T1 text box changes, this Run Prog. command button must be clicked to get the answers per changes.

Now for the code under this Run Prog. command button. The copyrighted main body of code is not to be revealed and is of limited interest here. I therefore will list only the code specific to exhibiting the answers in their respective text boxes. After all, our objective is to reveal how to write inputs and outputs in Visual Basic rather than studying algorithms of a particular VB program.

Notice I have used text boxes for both the input and the output answers. I found this the simpler and more user-friendly way to work VB. Yet, I have taken nothing from the available enhancements of VB.

Each of the following and remaining text boxes will be given VB code reference under this command button, Run Prog. Putting it another way, whenever the program calls for printing out an answer, I will simply show you how to print out the answers in VB. This command button, Run Prog., has enough code written to make all program calculations. A limited amount of its code will be covered here to adequately provide you with the input-output functions.

"txtPrt" message text box

The first text box immediately under Run Prog. is named txtPrt. See Fig. 9.12. This is a very unique answer box, as it displays a special message if you don't choose a temperature and pressure within the flash zone. If the temperature and pressure are not in the flash zone, the message "System is all liquid" or "System is all vapor" will appear in this text box. Otherwise, no message will appear in this box.

Make a new text box as shown and make the following inputs to its Properties menu:

Properties menu

Item	Input
Text box	
Name	txtPrt
Text	(Blank)

The following code is input in VB under the command button for running the program, Run Prog.:

Program code for text message screen display, txtPrt text box

```
658 SK = 0
660 If DPTS < 1! GoTo 680
670 GoTo 690
680 txtprt = "SYSTEM IS ALL VAPOR":     ' Prints message
                                        GoTo 2000 shown in
                                        txtprt
                                        ' text box.

690 If BPTS < 1! GoTo 710
700 GoTo 720
710 txtprt = "SYSTEM IS ALL LIQUID":    ' Prints message
GoTo 2000                               shown in txtprt
                                        ' text box.

720 If DPTS > 1! And BPTS > 1!
GoTo 740
```

DPTS represents the sum value of component mole fractions divided by their respective K values. The K value is the component's equilibrium value, y/x. BPTS represents the sum value of the component mole fraction times its respective K value. If the sum DPTS is less than 1.0, it means that the respective K values and temperature are far too great for the system to be within the flash zone where both vapor and liquid exist. BPTS is the same, except the K values are not large enough for any vapor to exist. The code lines that make this calculation are as follows:

```
620 DPTS = (Z1(I) / K(I)) + DPTS
630 BPTS = (Z1(I) * K(I)) + BPTS
```

where $K(I) = I$ component equilibrium constant at system temperature and pressure

$Z1(I) = I$ component mole fraction in feed

$DPTS$ = sum of each component $Z(I)/K(I)$

$BPTS$ = sum of each component $Z(I) * K(I)$

Note that the original GW Basic or QuickBasic line identification numbers remain in the program. This program is shown in its original form before I installed any VB insertions. The input of txtprt on lines 680 and 710 simply replaced the GW Basic code word Print. Thus the VB text box named txtPrt receives one of the preceding messages if the value of either DPTS or BPTS is less than 1.0. I confirm that the conversion to VB from GW Basic is really simple.

txtPrt2 message text box

Similar to text box txtPrt, txtPrt2 presents another message if the test condition on line 720 is true. See line 720 from the following program listing under the command button, Run Prog. Please see Fig. 9.12 again, which points out this text box.

For those of you who are going step-by-step through this program development, now make a new text box for a double line right below the text box named txtPrt. Make the following inputs to the Properties menu of this new text box:

Properties menu

Item	Input
Text box	
Name	txtPrt2
Text	(Blank)

Code for a message of text.   The code written for this text box, txtPrt2, is copied here from the code installed under the Run Prog. command button.

```
720 If DPTS > 1! And BPTS > 1! GoTo 740
730 txtprt2 = "SYSTEM IS NOT TWO PHASE": GoTo 2000
740 txtprt2 = "SYSTEM IS TWO PHASE MassRate, Lb/Hr   COMP,
    FEED,   LIQ,   VAP"
```

Note that the text box name txtPrt2 replaces the code word Print in GW Basic. When the system test is true for two-phase, line 720, then the message shown on line 740 is written in text box txtPrt2. COMP, FEED, LIQ, and VAP are the header labels for the flash data to be displayed in the next text box.

txtMF answer text box

Make a new text box, right below the text box you just made. Size it for the same length but make it for multilines of text.

This text box contains the printed data under the line headers of COMP, FEED, LIQ, and VAP. This header input came from the preceding text box named txtPrt2. See code line 740.

The pounds-per-hour rate of each component is written to this multiline text box from the file trnsf2. This text box is named txtMF and has the following Properties menu inputs:

Properties menu

Item	Input
Text box	
Name	txtMF
Text	(Blank)
Multiline	True
Scrollbar	2 (Vertical Scrollbar)
Alignment	1—Right justify

Multiple lines of component data are written by the program in this text box named txtMF. One line for each component is written. This is an answer text box showing the multicomponent flash. The respective code for writing these lines of answer data is as follows:

Program Code for trnsf2 File

```
Open "trnsf2" For Output As #1        ' Opens the file named
                                        trnsf2 for writing to.

For I = 1 To N
Print #1, Format(O(I), "00000000"); Tab(11); Format(Z1(I),
  "0.0000"); Tab(25); Format(XL(I),
  "0.0000"); Tab(38); Format(YV(I),   ' Writes shown data to
  "0.0000")                             file, transf2.
                                      ' Each "Print #1" starts
                                        a new data line.
Next I
Close #1                              ' Closes file trnsf2
Open "trnsf2" For Input As #1         ' Opens the file named
                                        trnsf2 for reading
                                        from.

txtMF = Input$(LOF(1), #1)            ' Writes the file trnsf2
```

```
                                    data to the text box
                                  ' named txtMF.
Close #1
```

This code is listed under the Run Prog command button. It is found following the nonactive line number 1140 in the program listing. Each time the user clicks this command button, the program iterates fully, producing new answers if any data input changes were made.

The VB keyword Format is used here for the first time. This VB code word sets numeric data answers in decimal places or scientific or other notation as may be desired. Note the necessary syntax. This is a unique way in which you may exhibit your answers conveniently. The program writes all the text installed into the file named trnsf2. Such data are the answer results of the program written in the format as shown, following the Print #1 VB command keyword. Note that each repeated Print #1 starts a new line in the file trnsf2. Please observe also that after the trnsf2 file is made, it is then opened again and copied to the txtMF text box. This activity is done in the last three lines of the preceding program code.

"txtMHR" answer text box

The next text box to be made is named txtMHR. Please refer again to Fig. 9.12. This name stands for component mol/h flow rates summary. This text box will be a near duplicate of the previous one, replacing the mole fractions with molar rates. Therefore, make a similar text box, sizing it wider as shown on your screen. Make the following inputs to the respective text box Properties menu:

Properties menu

Item	Input
Text box	
Name	txtMHR
Text	(Blank)
Multiline	True
Scrollbar	2 (Vertical Scrollbar)
Alignment	1—Right justify

Code for txtMHR text box.   The code for txtMHR is listed under the Run Prog command button. This code is very similar to the code for txtMF. It may be found the same as for the preceding code under the unused basic program line number 1140. The code with comments added is listed following:

Program Code for RefFlsh, txtMHR
```
Open "trnsf3" For Output As #1
For I = 1 To N + 1
```

```
If I = N + 1 Then Print #1, "Totals"; Tab(11); Format(FT,
   "0.0000e+00"); Tab(25); Format(XLMS,
"0.0000e+00"); Tab(38); Format(YVMS, "0.0000e+00"):
  GoTo Start7                        ' Prints sums of the shown
                                     ' columns at column btm.

Print #1, Format(O(I), "00000000"); Tab(11); Format(F(I),
   "0.0000e+00"); Tab(25);
Format(XLM(I), "0.0000e+00"); Tab(38); Format(YVM(I),
   "0.0000e+00")
                                     ' Writes shown data to file
                                     trnsf3.
                                     ' Each Print #1 starts a new
                                     line.
Next I
Start7:
Close #1
Open "trnsf3" For Input As #1        'Opens the file named trnsf3
                                     for reading from.
txtMHR = Input$(LOF(1), #1)          'Writes the file trnsf3 data
                                     to the text box
                                     ' named txtMHR

Close #1
```

All said for the previous text box txtMF applies equally here to this text box code, txtMHR. The Totals label with column sums line has been added. Also note how the format has been changed to scientific notation rather than fixed-decimal printing.

The contents of files trnsf1, trnsf2, and trnsf3 have no variance upon the program, as each time you run the program these files are renewed per the program automatic data inputs.

In opposition to this, the contents of files FlshDat and FeedDat do make considerable variance. These files are Table 1 and Table 2, respectively; therefore, care should be taken that they are backed up and stored in a recoverable diskette or CD. Table 2, the FeedDat file, can, however, be remade each time the program is run. After all, Table 2 is to be fixed per the user's choices of the components with their flow rate inputs. Table 2 therefore retains the same component choices with molar flow rates per the last running of the program. I recommend that you, the user, take care to back up your Table 2 file the same as you did with Table 1. You may have a long component listing with inputs of exacting, tedious molar rates of each component. Saving this FeedDat file, Table 2, is prudent and wise for any future runs. Remember, you should back up these files as FeedDat2 and FlshDat2.

txtKVal text box

The last text box to be described on Form2 is the text box named "txtK-Val." Please refer again to Fig. 9.12. As the name indicates, this is the text box showing the derived K equilibrium values of $K = y/x$. The y is vapor phase mole fraction of a component. The x is the mole fraction of the same component in the liquid phase. This is a handy exhibit to determine if the program-derived K values compare favorably with other known K values of different origins. If you attempt to make K value changes in this text box, these changes will not be used by the program. Only those program-derived K values will be executed in this program. This is a good improvement you may wish to make on this program, that is, install code to replace the program derived K values with those installed in the text box txtKVal.

In fact, further carrying the thought of making your own program, you can install your own algorithms under the command button Run Program. You could choose any K value generator you wish, using the VB screens herein as templates to input and output your answers.

This being the last text box to be made by those of you who are going through this program redevelopment (I suggest you again review the 12 steps I developed for you in the first part of this chapter), I will briefly repeat the procedure here, which is repeatable for any desired text box. In brief, you simply click once on the text box icon tool menu at the left side of your screen. That's the one with the letters ABC enclosed in a box. Then move your mouse pointer (now a crossbar) to a desired location. The location shown on the program Form2 is suggested for making this box, to be named txtKVal. Locate this crossbar at the upper left-hand corner of this desired text box. Now press down on your left mouse button, not letting up on it until you have the desired size of the text box txtKVal. VB shows an outline form of the text box before you let up on the mouse button. See Figs. 9.8 and 9.12.

Now that you have made this text box, you must name it and set certain other properties necessary for this application. Therefore, click once on this new text box. Next bring up your Properties menu, which is probably behind the Form2 window you are using. Click on this Properties menu window to bring it up, or you may click on the bottom menu bar on the PROP selection. Here are the inputs required for the new text box properties:

Properties menu

Item	Input
Text box:	
Name	txtKVal
Text	(Blank)
Multiline	True
Scrollbar	2 (Vertical Scrollbar)

Code now must be installed to input the multicomponent K value listing. The name of each component also is to be listed in this box. As shown it will be a vertical, scrollable box. As in the other text boxes having the calculated answer outputs from this program, this code will be listed under the Run Prog. command button. This code will be found immediately following line number 657 in the program listing. This code with comments added is listed here:

Program Code for FlshRef, trnsf2

```
657:
Open "trnsf2" For Output As #1      'Open trnsf2 file for writing
                                     data to.
For I = 1 To N
Print #1, O(I); Tab(10);            ' Write this data to file
Format(K(I), "0.0000e+00")          trnsf2.
                                     ' Each Print #1 starts a new
                                     line of data.
                                     ' One line of data is written
                                     for each
                                     ' component.
Next I
Close #1                            ' Closes trnsf2 file.
Open "trnsf2" For Input As #1       ' Opens trnsf2 file for
                                     reading data from
txtKVAL = Input$(LOF(1), #1)        ' Writes read data to text
                                     box txtKVAL
Close #1                            ' Closes trnsf2 file.
```

This print code is located in the program listing after the code line loop of 183–650. This loop calculates each of the component K values and could be bypassed if one wishes to input new K values. The user would thereby input new K values in the program text box txtKval.

One last comment about the Run Prog. command button is that this entire program listing is essentially listed under this button. The logic here is you want as little activity as necessary to input and run your programs. A one-button execution is therefore most feasible.

Printing the Output

Print Answers command button

The program execution is now completed, and your answers are on your computer's screen forms 1 and 2. The next and last step is the option of printing the desired results and making a hard copy. The last icon item we install on form2 is therefore a command button called Print Answers. When this button is clicked, your printer will make a hard copy of the selected text boxes. The text boxes the program chooses are the form2 output answer box and the Input Feed text box under the View Feed command button.

Make a last command button named Command2, its default name, and give it a caption called Print Answers. See Fig. 9.12. Make this command button the same as you did the preceding text box. Next click once on it to bring up its Properties menu. Make the following inputs:

Properties menu

Item	Input
Command button:	
Caption	Print Answers
Name	Command2

Visual Basic line code for driving a printer

The printer code is the next to be written under this command button. This code listing follows with comments added:

```
Sub Command2_Click ()
Printer.Print                    ' Prints a blank line for making
                                 a line space on the printed
                                 page.

Printer.Print " Component        ' This actual printing is
Mol Fractions        "           printed as shown.
Printer.Print "COMP    FEED
LIQ      VAP"
Printer.Print txtMF              ' Prints the entire contents of
                                 the text box
                                 ' named txtMF

Printer.Print
Printer.Print
Printer.Print Label5             ' Prints the contents of Label5
                                 at this
                                 ' location.

Printer.Print txtMHR
Printer.NewPage                  ' Starts a new page.
Printer.Print
Printer.Print
Printer.Print Label4
Printer.Print txtFView
Printer.Print
Printer.Print Label3
Printer.Print txtKVAL
Printer.EndDoc                   ' Sends print contents to printer
                                 for printing.

End Sub
```

This listing is in a simple form and is indeed easy to understand and apply. As of this writing, I have not yet found any VB book explaining how to print out your results.

Books and technical journals seem to bypass the easy way to send printouts to a printer. Have our knowledgeable technical code writers missed this need? It appears they indeed have. Therefore the preceding printing code I present is justified without any further error catchers other than what you have in your computer.

The printer code I present under the command button Print Answers is unique in that you simply write a code line, Printer.Print txtMHR. The entire data contents of the text box txtMHR is thereby output by your default printer in Windows 98 or other Windows software you may be using. Each line of data is printed in the same form as presented on your PC screen. This is indeed a unique way to obtain your printouts as well as the PC screen data.

Conclusion

I trust you have gained a sound basic understanding of VB programming. I encourage you to apply it in all categories of engineering and sciences—and most of all, to filling in those gaps we have all encountered in the chemical engineering profession.

References

1. Microsoft Corporation, *Programmer's Guide for Microsoft Visual Basic,* Programming System for Microsoft Windows Version 3.0, Microsoft, Redmond, WA, 1993.

10

Advanced Fluid Flow Concepts

Nomenclature

α	angle between V and u in rotating machinery, measured between their positive directions; angle of bed slope to a horizontal datum line
β	angle between v and u in rotating machinery, measured between $+v$ and $-u$; Mach angle
ε	height of surface roughness
θ	any angle
μ	absolute or dynamic viscosity, cP
ν	kinematic viscosity (μ/ρ), ft²/s
ρ	density (w/g)
σ	surface tension, lb/ft
τ	shear stress, lb/ft²
A	any area, ft²
A	cross-sectional area of a stream normal to the velocity, ft²
a	linear acceleration, ft/s²
B	any width, ft
B	length of weir crest, ft
B	width of turbine runner or pump impeller at periphery
B	bed width of open channel, ft
C	any coefficient
C	Chézy's coefficient
C_c	coefficient of contraction for orifices, tubes, and nozzles
C_d	coefficient of discharge for orifices, tubes, and nozzles
C_v	coefficient of velocity for orifices, tubes, and nozzles

C_f	average skin coefficient for total surface
c	wave or acoustic velocity (celerity), ft/s
c	specific heat, Btu/lb · °F
c_f	local skin-friction coefficient
c_p	specific heat at constant pressure, Btu/lb · °F
D	diameter of pipe, turbine runner, or pump impeller, ft
E	enthalpy $(I + pv/J)$, Btu/lb*
E	specific energy in open channels $(y + V^2/2g)$
E	linear modulus of elasticity, lb/ft²
E_V	volume modulus of elasticity, lb/ft²
F	any force, lb
f	friction factor for pipe flow
G	total weight, lb
g	acceleration of gravity, 32.174 ft/s²
H	total head $(p/w + z + V^2/2g)$
h	any head, ft
h_a	atmospheric pressure in equivalent height of fluid
h_f	head lost in friction
I	plane moment of inertia, ft⁴
J	778 ft · lb/Btu
K	any constant
K	equivalent volume modulus for fluid in elastic pipe
k	any coefficient of loss
K	c_p/c_v for gases
L	length, ft
M	constant for venturi or orifice meter
m	mass, (G/g), slugs
N_F	Froude number $(V/(gL)^{1/2})$
N_M	Mach number (V/c)
N_R	Reynolds number $(LV\rho/\mu = LV/v)$
n	Kutter or Manning coefficient of roughness
p	fluid pressure, lb/ft²
p_a	atmospheric pressure
p_v	vapor pressure
Q	rate of flow, ft³/s

---

* In thermodynamics, this is usually written as $h = u + pv/J$.

q	volume rate of flow per unit width of rectangular channel
q	heat flow in fluid, Btu/lb
R	hydraulic radius (A/P), ft
R	gas constant
r	any radius
S	slope of energy gradient (h_f/L)
S_0	slope of channel bed
S_w	slope of stream surface
s	specific gravity of a fluid
T	absolute temperature ($°F + 460°$)
T	period of time for travel of a pressure wave, s
t	time, s
U	uniform velocity of fluid, ft/s
u	local velocity of fluid, ft/s
V	mean velocity of fluid, ft/s
V_r	radial component of velocity ($V \sin \alpha = v \sin \beta$)
V_u	tangential component of velocity ($V \cos \alpha = u - v \cos \beta$)
V_L	total volume, ft^3
v	relative velocity
v	specific volume $(1/w)$, ft^3/lb
u, v, w	components of velocity in general x, y, z directions
W	rate of discharge (wQ), lb/s
w	specific weight $(1/v)$, lb/ft^3
x	a distance, usually parallel to flow
y	a distance along a plane in hydrostatics, ft
y	total depth of open stream, ft
y_c	critical depth of open stream, ft
y_0	depth for uniform flow in open stream, ft
Z	height of weir crest above bed of flume, ft
z	elevation above any arbitrary datum plane, ft

Dynamics of Fluid Flow

Kinetic energy of a flowing fluid

A body of mass m and velocity V possesses kinetic energy $= mV^2/2$, or $V^2/2$ per unit of mass. As there are g units of weight per unit of mass, the kinetic energy per unit of weight is $V^2/2g$.

In the flow of a real fluid, the velocities of different particles will usually not be the same, so it is necessary to integrate the kinetic energies of all portions of the stream to obtain the total value. Consider the case where the axial components of velocity vary across a section. If u is the local axial velocity component at a point, the mass flow per unit time through an elementary area dA is $\rho\, dQ = \rho u\, dA = (w/g)\, u\, dA$. Thus the flow of kinetic energy per unit time across the area dA is $(\rho u\, dA)u^2/2 = (w/2g)u^3\, dA$, so the total kinetic energy possessed by the fluid and physically carried across the entire section per unit time is:

$$\frac{\rho}{2} \int_A u^3\, dA = w/2g \int_A u^3\, dA \tag{10.1}$$

If the exact velocity profile is known, the true kinetic energy of a flowing stream can be determined by integration of this equation.

It is convenient to use the mean velocity V and a factor a such that for the entire section the true average value is:

$$\text{Kinetic energy per unit weight} = \frac{\alpha\,(V^2)}{2g} \tag{10.2}$$

The flow across the entire section is $W = g\rho Q = wQ = wAV$, so the total kinetic energy transmitted is:

$$W\alpha\,\frac{V^2}{2g} = wAV\acute{\alpha}\,\frac{V^2}{2g} = \acute{\alpha}\,\frac{w}{2g}\,AV^3 \tag{10.3}$$

Equating Eq. (10.3) to Eq. (10.1), we obtain:

$$\acute{\alpha} = \frac{1}{AV^3} \int_A u^3\, dA \tag{10.4}$$

As the average of cubes is greater than the cube of the average, the value of a will always be more than 1. The greater the variation in velocity across the section, the larger the value of a. For laminar flow in a circular pipe, $\acute{\alpha} = 2$; for turbulent flow in pipes, a ranges from 1.01 to 1.15. In some instances it is very desirable to use the proper value of a, but in many cases the error caused by neglecting it is negligible. As precise values of a are seldom known, it is customary to disregard it and assume that the kinetic energy is $V^2/2g$ per unit weight of fluid.

Potential energy

The potential energy of a particle of fluid is measured by its elevation above any arbitrary datum plane. We are usually interested only in differences of elevation, and therefore the location of the datum plane is determined solely by convenience.

Internal energy

Internal energy is more fully presented in texts on thermodynamics, as it is thermal energy; but in brief it is the energy due to molecular activity, which is a function of temperature. Internal potential energy is due to the forces of attraction between molecules. A perfect gas is one for which these forces are assumed to be zero, and hence by Joule's law the internal energy of a perfect gas is a function of its temperature only and is independent of pressure and volume. For gases at usual temperatures and pressures, the distance between molecules is so great that forces are negligible, and hence in most cases real gases can be treated as ideal gases.

The zero of internal energy may be taken at any arbitrary temperature, as we are usually concerned only with differences. For a unit weight $I = c_v T$, where c_v is specific heat at constant volume, in this text taken as Btu/lb · °F. Thus I is in Btu/lb. The definition given is for sensible internal energy and applies to liquids as well as to gases. For a change of state, as when water is evaporated into steam, there may be a great change in volume. For example, a pint of water, when transformed into steam, may occupy a volume of several cubic feet and even as much as 1.000 ft³ or more, depending upon the pressure. The change in potential energy at constant temperature due to the separation of the molecules against the forces of attraction between them is called the *internal latent heat*.

General equation for steady flow of any fluid

Figure 10.1 shows a stream of fluid for which an equation will be derived by using the fundamental principle of mechanics that the external work done upon any system is equal to the change of energy of

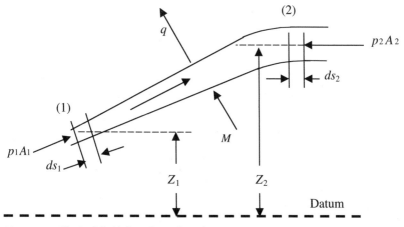

Figure 10.1 Typical fluid flow through a pipe.

the system. At section (1) of Fig. 10.1 the total force is $p_1 A_1$. Assume that in a brief time dt this section moves a short distance ds_1; then the force will do work $(p_1 A_1) ds_1$ on the body of fluid between sections (1) and (2). In similar manner, at (2) the external work due to the pressure and the motion is $-(p_2 A_2) ds_2$, the minus sign indicating that the force and the displacement are in opposite directions.

Also, between (1) and (2) there might be a machine that works on the fluid and puts mechanical energy into it, which will be denoted by M (ft · lb per lb of fluid flowing). If the machine were a turbine, the fluid would do work upon it and M would be negative. Heat q (Btu per lb of fluid flowing) may be transferred from the fluid to the surroundings, but if the heat flow is into the fluid, the value of q is negative. During the time dt the weight of fluid entering at (1) is $w_1 A_1 ds_1$, and for steady flow an equal weight must leave at (2) during the same time interval. Hence the energy that enters at (1) during the time dt is $w_1 A_1 ds_1 (z_1 + \acute{\alpha}_1 V_1^2/2g + JI_1)$, while that which leaves at (2) is represented by a similar expression. (In the foot-pound-second system, $J = 778$ ft · lb/Btu.)

Equating the work done and the transfer of thermal energy to the change in the energy of the fluid and at the same time factoring out $w_1 A_1 ds_1 = w_2 ds_2$:

$$\frac{p_1}{w_1} - \frac{p_2}{w_2} + M - Jq = (z_2 - z_1) + \left(\acute{\alpha}_2 \frac{V_2^2}{2g} - \acute{\alpha}_1 \frac{V_1^2}{2g}\right) + J(I_2 - I_1) \quad (10.5)$$

This equation applies to liquids, gases, and vapors and to ideal fluids as well as to real fluids with friction. The only restriction is that it is for steady flow.

In turbulent flow there are other forms of kinetic energy besides that of translation, described earlier. These are the rotational kinetic energy of eddies initiated by fluid friction and the kinetic energy of the turbulent fluctuations of velocity. These are not represented by any specific terms in Eq. (10.5) because their effect appears indirectly. The kinetic energy of translation can be converted into increases in p/w or z, but these other forms of kinetic energy can never be transformed into anything but thermal energy. Thus they appear as an increase in the numerical value of I_2 over the value it would have if there were no friction or else produce an equivalent change in the numerical values of some other terms.

The general energy equation Eq. (10.5) and the continuity equation are two important keys to the solution of many problems in fluid mechanics. For compressible fluids it is necessary to have a third equation, which is the equation of state, i.e., the relation between pressure, temperature, and specific volume (or specific weight or density).

In many cases Eq. (10.5) is greatly shortened owing to the fact that certain quantities are equal and thus cancel each other or are zero. Thus, if two points are at the same elevation, $z_1 - x_2 = 0$. If the conduit is well insulated or if the temperature of the fluid and that of its surroundings are practically the same, q may be taken as zero. On the other hand, q may be very large, as in the case of flow of water through a boiler tube. If there is no machine between sections (1) and (2) of Fig. 10.1, then the term M drops out. If there is a machine present, the work done by or upon it may be determined by solving Eq. (10.5) for M.

Flow work

The term p/w appearing in Eq. (10.5), or its equivalent pw, is called *flow work*, because it represents external work done during flow and is not present if the flow is zero. Unlike potential, kinetic, and thermal energies, this does not represent energy possessed by the fluid; it is merely energy that is transmitted across a section by pressure and motion, just as energy may be transmitted by a belt although the belt does not possess energy.

In a nonflow process the energy stored in a fluid by compression is represented by $\int_2^1 p\, dv$. Each small increment of gas compressed dv, multiplied by the pressure p, and added to all other small increments equals the energy stored in the gas. If the fluid is incompressible, such as a liquid, dv is zero. For any real liquid, the compressibility is small, and the stored "pressure energy" is negligible. For a gas or a vapor, this store of energy may be very appreciable.

For a flow process, as in a compressor, it is necessary not only to compress the fluid but also to draw it into the machine at a low pressure and then discharge it against a higher pressure. Thus a flow process involves a complete cycle, represented by $\int_2^1 v\, dp$. For an incompressible fluid, this is the area of a rectangle, the value of which is $v^2(p_1 - p_2)$. For a compressible fluid, it is necessary to have a relation between p and w. For example, if pw^k is a constant, the work of a flow process is k times the value given.

Energy equation for liquids

For liquids—and even for gases and vapors when the change in pressure is small—the fluid may be considered as incompressible for all practical purposes, and thus we may take $w_1 = w_2 = w$ as a constant. In turbulent flow the value of a is only a little more than unity, and, as a simplifying assumption, it will now be omitted. Then Eq. (10.5) becomes:

$$\left(\frac{p_1}{w} + z_1 + \frac{V_1^2}{2_g}\right) + M - Jq = \left(\frac{p_2}{w} + z_2 + \frac{V_2^2}{2_g}\right) + J\,(I_2 - I_1)$$

Fluid friction produces eddies and turbulence, and these forms of kinetic energy are eventually transformed into thermal energy. If there is no heat transfer, the effect of friction is to produce an increase in temperature so that I_2 becomes greater than I_1. In most cases, with liquids this increase in temperature is of no worth and is considered a loss of useful energy, although the total energy of the liquid remains constant.

Suppose there is a loss of heat q at such a rate as to maintain the temperature constant so that $I_2 = I_1$. In this event there is an actual loss of energy from the system that is equal to the mechanical energy that has been converted into thermal energy by friction.

The transfer of heat may be of any value and may be in either direction, i.e., it may be either positive or negative. But whatever the value of q may be, the fluid friction is the same. Thus the loss of useful energy per pound of liquid caused by friction may be represented by:

$$h_f = J\,(I_2 - I_1) + Jq = Jc\,(T_2 - T_1) + Jq \tag{10.6}$$

where c is specific heat. If the loss of heat is greater than h_f, then T_2 will be less than T_1. If there is any absorption of heat, T_2 will be greater than the value that would be produced by friction but, as q is then negative, the value of h_f is unaffected. Because of the magnitude of J, which is 778 ft · lb/Btu, a larger value of h_f produces only a very small rise in temperature if there is no heat transfer, or only a very few Btu of heat transfer if the flow is isothermal.

If there is no machine between sections (1) and (2) of Fig. 10.1, the energy equation becomes:

$$\frac{p_1}{w} + z_1 + \frac{V_1^2}{2g} = \frac{p_2}{w} + z_2 + \frac{V_2^2}{2g} + h_f \tag{10.7}$$

In some cases the value of h_f may be very large, and, although for any real fluid it can never be zero, there are cases where it is so small that it may be neglected with small error. In such special cases

$$\frac{p_1}{w} + z_1 + \frac{V_1^2}{2g} = \frac{p_2}{w} + z_2 + \frac{V_2^2}{2g} \tag{10.8}$$

and from this it follows that

$$\frac{p}{w} + z + \frac{V^2}{2g} = \text{constant} \tag{10.9}$$

The equation in either of these last two forms is known as *Bernoulli's theorem,* in honor of Daniel Bernoulli, who presented it in 1738. Note that Bernoulli's theorem is for a frictionless, incompressible fluid only.

The equation of continuity and the Bernoulli theorem together show, for a stream of incompressible fluid, that (a) where the cross-sectional area is large and the streamlines are widely spaced, the velocity is low and the pressure is high, and (b) where the cross-sectional area is small and the streamlines are crowded together, the velocity is high and the pressure is low. Hence a flow net gives a picture not only of the velocity field but also of the pressure field.

Energy equation for gases and vapors

If sections (1) and (2) of Fig. 10.1 are chosen so that there is no machine between them and if a is assumed to be unity for simplicity, as in the preceding section, Eq. (10.5) becomes:

$$\left(\frac{p_1}{w_1} + JI_1 + z_1 + \frac{V_1^2}{2g}\right) - Jq = \left(\frac{p_2}{w_2} + JI_2 + z_2 + \frac{V_2^2}{2g}\right) \quad (10.10)$$

For a gas or a vapor the quantity p/w is usually very large compared with $z_1 - z_2$ because of the small value of w, and therefore the z terms are usually omitted. But $z_1 - z_2$ should not be ignored unless it is known to be negligible compared with the other quantities.

As the p/w ($= pw$) and the I terms are usually associated for gases and vapors, it is customary to replace them by a single term called *enthalpy*, indicated by a single symbol such as $E = I + pw/J$, where E is in Btu/lb. (In thermodynamics h is used instead of E, but the letter h has so many uses in fluid mechanics that it seems desirable to use a different letter here. Values of E for vapors commonly used in engineering, such as steam, ammonia, sulfur dioxide, freon, and others, may be obtained by the use of vapor tables or charts. For a perfect gas and practically for real gases $E = c_p T$, where c_p is specific heat at constant pressure. For air at usual temperatures the value of c_p is 0.24 Btu/lb · °F, but it increases with increasing temperature. Values for this and for other gases may be found in thermodynamics texts and in handbooks. With these changes, Eq. (10.10) becomes:

$$JE_1 + \frac{V_1^2}{2g} - Jq = JE_2 + \frac{V_2^2}{2g} \quad (10.11)$$

This equation may be used for any gas or vapor and for any process. Some knowledge of thermodynamics is required in order to evaluate the E terms, and in the case of vapors it is necessary to use vapor tables or charts, because their properties cannot be expressed by any simple equations.

Isothermal flow of a gas.   For constant temperature $I_1 = I_2$ and pw is a constant or $p_1/w_1 = p_2/w_2$. Consequently $E_1 = E_2$, and hence for this case Eq. (10.11) becomes:

$$\frac{V_2^2 - V_1^2}{2g} = -Jq \tag{10.12}$$

which shows that an isothermal flow is accompanied by an absorption of heat when there is an increase in kinetic energy. This equation applies either with or without friction. If there is friction, less heat is absorbed from the surroundings. Since q is numerically less, there is less increase in kinetic energy.

Adiabatic flow. If heat transfer q is zero and the same assumptions regarding the z and a terms are made as before, the energy equation for any fluid becomes:

$$JE_1 + \frac{V_1^2}{2g} = JE_2 + \frac{V_2^2}{2g} \tag{10.13}$$

and since $778 \times 2g$ is practically 50,000, this may be written:

$$V_2^2 - V_1^2 = 50,000\,(E_1 - E_2) \tag{10.14}$$

For a $E = c_p T$, and therefore for gases:

$$V_2^2 - V_1^2 = 2gJc_p\,(T_1 - T_2) = 50,000c_p\,(T_1 - T_2) \tag{10.15}$$

For vapors, Eq. (10.14) is convenient, but for gases a different expression from Eq. (10.15) is often useful. From thermodynamics:

$$C_p = \frac{kR}{(k - I)J}$$

And for a perfect gas, $pw = RT$. Therefore, for a perfect gas:

$$\frac{V_2^2 - V_1^2}{2g} = \frac{k}{k - I}\,(p_1v_1 - p_2v_2) \tag{10.16}$$

The preceding equations are valid for flow either with or without friction. For ideal flow without friction, the values of E_2, T_2, and v_2 are all determined as the result of a frictionless (or isentropic) expansion from p_1 to p_2. Fluid friction increases the numerical value of all these quantities, and it is sometimes desirable to designate these larger values as E_2', T_2', and v_2'.

Sometimes it is feasible to calculate these terminal values with friction, but more often it is convenient to use the ideal frictionless values in the equations and then multiply the ideal result, such as V_2, by an experimental factor to obtain the correct value with friction.

Adiabatic flow of a perfect gas without friction. An ideal frictionless process is represented where a gas is compressed, the area indicating the flow work, which is equal to the change in kinetic energy if no mechanical work in a machine is involved. Thus:

$$\frac{V_2^2 - V_1^2}{2g} = -\int_1^2 v \, dp \tag{10.17}$$

The integral may be evaluated since $pw^k = p_1 v_1^k = p_2 v_2^k = $ constant for a frictionless adiabatic. The result is Eq. (10.16), which is valid for any process if the proper numerical values are inserted.

As v_2 is usually an unknown quantity, it may be replaced in the case of a frictionless flow of gas by means of the preceding relations, and then Eq. (10.16) becomes:

$$\frac{V_2^2 - V_1^2}{2g} = \frac{k}{k-1} p_1 v_1 \left[1 - \left(\frac{p_2}{p_1} \right)^{(k-1)/K} \right] \tag{10.18}$$

This equation may also be transformed into other equivalent expressions by means of the perfect gas laws. One other form is:

$$\frac{V_2^2 - V_1^2}{2g} = \frac{k}{k-1} RT_1 \left(1 - \frac{T_2}{T_1} \right) \tag{10.19}$$

General case. In practice many flow problems approach either the isothermal or the adiabatic process, but in the event that neither applies it is possible to use Eq. (10.5) or Eq. (10.10). The same procedure may be used for vapors when they depart too widely from the perfect gas laws. If the pressure ratio p_2/p_1 is of the order of 0.95 or more, the error will be negligible if the fluid is treated as incompressible and the equations for liquids are used.

Momentum equation

The energy equations for compressible fluid in the previous section contain no specific term for friction. This is because fluid friction transforms the kinetic energy of eddies into thermal energy and is therefore represented by changes in the numerical values of other terms. In order to obtain a term for friction, it is necessary to use the principle that force equals rate of change in momentum and thereby avoid any thermal terms.

For simplicity, consider a horizontal pipe of uniform diameter. If τ_0 is the shear force at the wall per unit area, the resultant force is $-dp \, \pi \, D^2/4 - \tau_0 \pi \, D \, dx$. As dp is inherently negative, the net result is a force directed toward the right. The rate of increase in momentum is $d(pV^2) \pi \, D^2/4$, but pV is constant for steady flow, and so this may be expressed as $pV \, dV \, \pi \, D^2/4$. Equating force to rate of change in momentum, factoring out $\pi \, D^2/4$, and rearranging, the result is:

$$\frac{dp}{p} + V \, dV = -\frac{4}{D} \frac{\tau_0}{\rho} \, dx \tag{10.20}$$

If p is replaced by w/g, the equation may also be written

$$\frac{dp}{w} + \frac{V \, dV}{g} = -\frac{4\tau_0}{Dw} \, dx = -dh_f \tag{10.21}$$

If the fluid were incompressible, this equation could be integrated to give $(p_1 - p_2)/w + (V_1^2 - V_2^2)/2g = h_f$, which is the same as Eq. (10.7) for horizontal pipe and is not even restricted to uniform diameter.

Although the momentum equations and the energy equations are identical for an incompressible fluid, they do not coincide for a compressible fluid because the integration of dp/p or dp/w will not give the same result as the p terms in Eq. (10.10) for a compressible fluid. A few illustrations will be presented.

Isothermal flow of a gas. In Eq. (10.21) dp/w is replaced by its equivalent $v \, dp$, which is more customary with gases. For isothermal flow $v = p_1v_1/p$, and consequently

$$\int_1^2 v \, dp = p_1v_1 \log_e \frac{p_2}{p_1}$$

As $\int_1^2 V \, dV/g = (V_2^2 - V_1^2) \, 2g$ and, by Eq. (10.12), this in turn equals $-Jq$, the fluid friction in isothermal flow of a gas expressed in ft · lb/lb is:

$$h_f = p_1v_1 \log_e \frac{p_1}{p_2} + \frac{V_1^2 - V_2^2}{2g} = p_1v_1 \log_e \frac{p_1}{p_2} + Jq \tag{10.22}$$

Example 10.1. Assume isothermal flow of air in a long pipe such that in a certain distance there is a decrease in pressure from 90 to 15 psia owing to friction. If the initial velocity is 100 ft/s, then at a pressure of 15 psia the specific volume will be six times as great. If the pipe has a uniform diameter, then $V_2 = 6V_1 = 600$ ft/s. The heat transferred into the air in the pipe from the surroundings is $-Jq = (600^2 - 100^2)/2g = 5436$ ft · lb per lb of air, or 7 Btu per lb of air. If the air temperature is 80°F or $T = 540°$, then:

$$p_1v_1 \log_e \frac{90}{15} = RT \log_e 6 = 53.3 \times 540° \times 1.795 = 51,770 \text{ ft · lb per lb of air}$$

Inasmuch as $Jq = -5436$ ft · lb per lb of air, the value of the friction is $h_f = 51,770 - 5,436 = 46,334$ ft · lb per lb. An interesting observation is that because the air receives heat in its flow, the total energy at section (2) of Fig. 10.1 is greater than at section (1). Thus friction merely means a degradation of mechanical energy but not a loss of energy.

Adiabatic flow of a gas. For a frictionless flow $h_f = 0$; but, as an illustration, take Eq. (10.21), replace dp/w with its equivalent $v\, dp$, and then use the relation that $pw^k = p_1 v_1{}^k$. Then:

$$\int_1^2 v\, dp \left[\frac{k}{k-1}\right] (p_2 v_2 - p_1 v_1) \tag{10.23}$$

From Eq. (10.16):

$$\int_1^2 \frac{V\, dV}{g} - \left[\frac{k}{k-1}\right] (p_1 v_1 - p_2 v_2)$$

These two are equal in value but opposite in sign, thereby checking the initial assumption that $h_f = 0$.

The preceding is a special case, but for an adiabatic flow with friction the relation that pv^k is a constant does not apply, and instead the relation between p and v is more complicated. As an illustration, consider flow in a pipe of area A at the rate of W lb/s, and assume that at some point the values p_1, v_1, and T_1 are known, while at some other point p_2, v_2, and T_2 are replaced by any values p, v, and T. As $V = Wv/A$, it follows from Eq. (10.16) that:

$$\frac{W^2 v^2}{2gA^2} + \frac{k}{k-1} pv = \frac{W^2 v_1^2}{2gA_1^2} + \frac{k}{k-1} p_1 v_1 = X \tag{10.24}$$

where X is a constant evaluated from known conditions at section (1) of Fig. 10.1. Thus in general:

$$pv = \frac{k-1}{k}\left[X - \frac{W^2}{2gA^2} v^2\right] \tag{10.25}$$

It does not appear to be practical to use the relation between p and v given by Eq. (10.25) to integrate Eq. (10.21). The better procedure would be to assume a series of values of v, compute the corresponding values of p, and then evaluate $fv\, dp$ arithmetically. However, there is little necessity for this, and Eq. (10.25) is presented here merely to show the possible complexity of the relation between p and v for adiabatic flow with friction.

Example 10.2. Assume air at a pressure of 150 psia and 140°F or 600° abs in a tank where $V_1 = 0$ and air flows out through a long pipe at the rate of 135.5 lb/s · ft² of pipe area. At a distant point where the specific volume is assumed to be 5.18 ft³/lb, Eq. (10.25) gives $p = 5750$ psfa or 40 psia. From $pw = RT$, the temperature is 559° abs, or 99°F, showing a temperature drop of 41°F. However, for a frictionless adiabatic expansion from 150 to 40 psia, the temperature would be 600°/(150/40) $(k - 1)k = 411$° abs, or 49°F, giving a temperature drop of 189°F. The continuity equation gives the velocity at this section as 702 ft/s.

Additional equations are necessary in order to determine the length of pipe in which this pressure drop occurs. However, it is evident that the greater the rate of flow, the shorter the length needed for the same pressure drop.

Head

In Eq. (10.7) every term represents a linear quantity. Thus p/w, called *pressure head*, is an equivalent height of liquid representing a pressure p in a fluid of specific weight w; z is elevation head; and $V^2/2g$ is velocity head. It is obvious that h_f must be a linear quantity also; it is called *friction head* or *lost head*. The sum of the first three is called *total head* and is denoted by H, where:

$$H = \frac{p}{w} + z + \frac{V^2}{2g} \tag{10.26}$$

For a frictionless, incompressible fluid, $H_1 = H^2$, but, for any real liquid:

$$H_1 = H_2 + h_f \tag{10.27}$$

Which is merely a brief way of writing Eq. (10.7). It is obvious that, if there is no input of work M by a machine between sections (1) and (2) of Fig. 10.1, the total head must decrease in the direction of flow. If there is a pump between sections (1) and (2), then:

$$M = H_2 - H_1 + h_f \tag{10.28}$$

And in general H_2 is greater than H_1. But if there is a turbine between sections (1) and (2), then:

$$M = H_1 - H_2 - h_f \tag{10.29}$$

And H_1 must always be greater than $H_{12} + h_f$. In Eqs. (10.28) and (10.29), the conventions of plus and minus signs, as indicated in Fig. 10.1, have been ignored, but it is understood that for the pump M is work done on the fluid, while for the turbine M is work done by the fluid. In deriving Eq. (10.5) the term $w\,A\,ds$ representing a weight of fluid was factored out. Thus every term in Eq. (10.5) and subsequent equations, except Eqs. (10.14), (10.15), and (10.20), represents not only a linear quantity but also work or energy in ft · lb/lb. If any such linear value of head is multiplied by rate of weight flow, the product is power, as the dimensions are ft × lb/s. Hence the linear quantity head also represents power per pound of fluid per second. Thus the expression for power is:

$$P = WH \qquad (10.30)$$

In the application of this expression H may be any head for which the corresponding power is desired.

As an illustration, suppose the surface of water in a reservoir is 500 ft above the level of the site of a power plant and that there is available a flow of 200 ft³/s. In this case, $H = z = 500$ ft and $P = 62.4 \times 200 \times 500 = 6{,}240{,}000$ ft · lb/s, or 11,340 hp, which is the total power available. If in the pipeline the friction loss is $h_f = 25$ ft, then the power lost by friction is Wh_f and is seen to be 567 hp. Again, suppose a nozzle discharges 50 lb of water per second in a jet with a velocity of 120 ft/s $H = V^2/2g = 224$ ft. Then the available power in the jet is $50 \times 224/550 = 20.3$ hp. The expression *power equals force applied times velocity of the point of application of the force* cannot be used in the preceding case, because it has no physical significance, as there is no force applied to anything, nor is there any point of application. In the case of a jet, the force it might exert would depend upon what happened when it struck an object, and the power produced would depend upon the velocity of the object. But the available power of the jet is a definite quantity, no matter what it acts upon or whether it ever acts upon anything.

Pressure

Pressure over a section. Strictly speaking, the equations that have been derived apply to flow along a single streamline. They may, however, be used for streams of large cross-sectional area by taking average values, as has been explained in the case of velocity. The case of average pressure over a section will now be considered.

Consider a small vertical prism of a stream that is flowing horizontally in a pipe. The vertical forces are those on the top and bottom of the prism and the force of gravity, which is the weight of the prism. Velocities and acceleration are horizontal and do not affect the vertical forces. An equilibrium equation in a vertical direction is $p_1A + wAy = p_2A$, where y is the height of the prism and A is its cross-sectional area. Thus $p_2 - p_1 = wy$, which is similar to Eq. (10.2). That is, in any plane normal to the velocity, the pressure varies according to the hydrostatic law. The average pressure is then the pressure at the center of gravity of such an area. However, as $z + p/w$ is constant over the entire sectional area, it is immaterial what point is selected, provided the pressure and the elevation both apply to the same point.

On a horizontal diameter through a pipe, the pressure is the same at every point. As the velocity is much higher in the center than near the walls, it follows that H is also higher there. This is reasonable because the streamlines near the wall lose more energy due to friction than those near the center of the stream. This emphasizes the fact that a flow equation really applies along the same streamline but not between two streamlines, any more than between two streams in two separate channels.

Static pressure.   Even if a fluid is flowing, the pressure measured at right angles to its flow is called the *static pressure*. This is the value given by manometers or piezometer tubes and pressure gauges.

Stagnation pressure.   The center streamline shows that the velocity becomes zero at the stagnation point. If p_o/w denotes the static pressure head at some distance away where the velocity is V_o, while p_s/w denotes the pressure head at the stagnation point, then, applying Eq. (10.8) to these two points, $p_o/w + 0 + V_o^2/2g = p_s/w + 0 + 0$, or the stagnation pressure is:

$$p_s = p_o + w \,\frac{V_o^2}{2g} = p_o + \rho \,\frac{V_o^2}{2} \tag{10.31}$$

The quantity $wV_o^2/2g$ or $pV_o^2/2$ is called the *dynamic pressure*.

Equation (10.31) applies to a fluid where compressibility may be disregarded. Using the preceding subscripts in Eq. (10.18) and rearranging so as to solve this equation for $p_s - p_o$, and expanding by the binomial theorem, the final result is:

$$p_s - p_o = w_o \,\frac{V_o^2}{2g} \left(1 + \frac{w_o}{2p_o k} \frac{V_o^2}{2g} + \cdots \right)$$

For a gas, the acoustic velocity is $c = (gkpv)^{1/2} = (gkRT)^{1/2}$. As

$$\frac{gkp_o}{w_o} = gkp_o v_o = c^2$$

The preceding equation may be written as

$$p_s = p_o + w_o \,\frac{V_o^2}{2g} \left(1 + \frac{V_o^2}{4c^2} + \cdots \right) \tag{10.32}$$

For air at 70°F, $c = 1130$ ft/s. If $V_o = 226$ ft/s, the error in neglecting the compressibility factor (the value in the parentheses) is only 1%. However, for higher values of V_o, the effect becomes much more important. Equation (10.32) is, however, restricted to values of V_o/c less than 1.

Hydraulic gradient and energy gradient

If a piezometer tube is erected at B in Fig. 10.2, the liquid will rise in it to a height BB′ equal to the pressure head existing at that point. If the end of the pipe at E were closed so that no flow would take place, the height of this column would then be BM. The drop from M to B′ when flow occurs is due to two factors: (1) that a portion of the pressure head has been converted into the velocity head which the liquid has at B, and (2) that there has been a loss of head due to fluid friction between A and B.

If a series of piezometers were erected along the pipe, the liquid would rise in them to various levels. The line drawn through the summits of such an imaginary series of liquid columns is called the *hydraulic gradient*. It may be observed that the hydraulic gradient represents what would be the free surface, if one could exist under such conditions of flow.

The hydraulic gradient represents the pressure along the pipe, as at any point the vertical distance from the pipe to the hydraulic gradient is the pressure head at that point, assuming the profile to be drawn to scale. At C this distance is zero, thus indicating that the pressure is atmospheric at both locations. At D the pipe is above the hydraulic gradient, indicating that there the pressure head is –DN, or vacuum of DN.

If the profile of a pipeline is drawn to scale, then not only does the hydraulic gradient enable the pressure head to be determined at any point by measurement on the diagram but it shows by mere inspection the variation of the pressure in the entire length of the pipe. The gradient is a straight line only if the pipe is straight and of uniform diam-

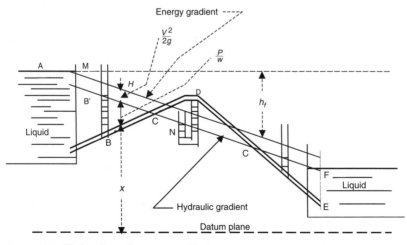

Figure 10.2 Hydraulic and energy gradients.

eter. But, for the gradual curvatures that are often found in long pipelines, the deviation of the gradient from a straight line will be small. Of course, if there are local losses of head aside from those due to normal pipe friction, there may be abrupt drops in the gradient and also any changes in diameter with resultant changes in velocity will make marked changes in the gradient.

If the velocity head is constant, as in Fig. 10.2, the drop in the hydraulic gradient between any two points is the value of the loss of head between those two points, and the slope of the hydraulic gradient is then a measure of the rate of loss. Thus if Fig. 10.2 had different pipe sizes along the hydraulic gradient, then the rate of loss in the larger pipe would be much less than in the smaller pipe. If the velocity changed, the hydraulic gradient might actually rise in the direction of flow.

The vertical distance from the level of the surface at A in Fig. 10.2 down to the hydraulic gradient represents $h_f + V^2/2g$ from A to any point in question. Hence the position of the gradient is independent of the position of the pipe. Thus it is not necessary to compute pressure heads at various points in the pipe to plot the gradient. Instead, values of $h_f + V^2/2g$ from A to various points can be laid off below the horizontal line through A, and this procedure is often more convenient. If the pipe is of uniform diameter, it is necessary to locate only a few points, and often only two are required.

If Fig. 10.2 represents to scale the profile of a pipe of uniform diameter, the hydraulic gradient can be drawn as follows: At the intake to the pipe there will be a drop below the surface at A that should be laid off equal to $V^2/2g$ plus a local entrance loss. At E the pressure is EF, and hence the gradient must end at F. If the pipe discharged freely into the air at E, the line would pass through E. The location of other points, such as B′ and N, may be computed if desired. In the case of a long pipe of uniform diameter, the error is very small if the hydraulic gradient is drawn as a straight line from the surface directly above the intake to a surface directly above the discharge end if the latter is submerged, or to the end of the pipe if there is a free discharge into the atmosphere.

If values of h_f are laid off below the horizontal line through A, the resulting line represents values of H measured above any arbitrary datum plane inasmuch as this line is above the hydraulic gradient a distance equal to $V^2/2g$. This line is the energy gradient. It shows the rate at which the energy decreases, and it must always drop downward in the direction of flow unless there is an energy input from a pump. The energy gradient is also independent of the position of the pipeline.

Other energy gradients could be plotted as in Fig. 10.2, showing the head loss just barely above the minimum pipe diameter for a positive minimum flow in the pipeline.

Method of solution

For the solution of problems of liquid flow, there are two fundamental equations, the equation of continuity and the energy equation in one of the forms from Eq. (10.5) to Eq. (10.9). The following procedure may be employed:

1. Choose a datum plane through any convenient point.

2. Note at what sections the velocity is known or is to be assumed. If at any point the section area is larger than elsewhere, the velocity head is so small that it may be disregarded.

3. Note at what points the pressure is known or is to be assumed. In a body of liquid at rest with a free surface, the pressure is known at every point within the body. The pressure in a jet is the same as that of the medium surrounding the jet.

4. Note whether or not there is any point where all three terms—pressure, elevation, and velocity—are known.

5. Note whether or not there is any point where there is only one unknown quantity.

It is generally possible to write an energy equation so as to fulfill conditions 4 and 5. If there are two unknowns in the equation, then the continuity equation must be used also.

Example 10.3. The procedure is shown by an example based on water flowing from a reservoir (point 1) through a pipe 6 in in diameter. At the end of the pipe is a 3-in nozzle that discharges a stream into the air at point 5. The velocity of the jet will be denoted by V_5, but as the velocity throughout the entire pipe is the same at all points, it will be indicated merely by V. Assume that the friction loss in any length of pipe may be represented by $h_f = kV^2/2g$, where k depends upon the length of the pipe and other factors. Assume also that point 2 along the pipe run is at the same elevation as points 4 and 5. Point 3, however, between points 4 and 5, is 25 ft above points 2, 4, and 5. Point 3 is also 15 ft below the reservoir surface level at point 1. Suppose that conditions are such that the values of k between the reservoir and point 2, between points 2 and 3, between points 3 and 4, and between points 4 and 5 are 2, 4, 4, and 1, respectively, or the total loss in the entire pipe is $11V^2/2g$. Let it be required to find the pressure head at point 3 when flow occurs.

At point 3 there are both an unknown pressure and an unknown velocity; hence the energy equation is not sufficient, as one equation is capable of determining only one unknown. The following procedure should then be used: The datum plane may be taken at any level, but it is usually desirable to take it at the lowest point in the system and thus avoid negative values of z. Let the datum plane be assumed through point 5. At point 1 on the surface of the water the pressure is atmospheric, which is also the case with the stream at point 5. Thus, whatever the barometer pressure may be, it cancels out in the equations and therefore can be neglected. If the reservoir is large in area compared with the cross section of the pipe, the velocity in it is negligible. Thus at point 1 everything is known, and at point 5 there is only one unknown. It is immaterial how far below the surface the intake for the pipe is located.

Applying the energy equation between points 1 and 5, and following the same order of terms for H as in Eq. (10.7):

$$0 + 40 + 0 = 0 + 0 + \frac{V_5^2}{2g} + 11\frac{V^2}{2g}$$

Now V_5 is the velocity of the jet, and V is the velocity in the pipe, but either one may be replaced in terms of the other by the continuity equation. It is not necessary to compute areas, as it is easier and much more accurate to use the ratio of the diameters. The velocities vary inversely to the square of the diameter ratio, and the velocity heads vary inversely to the fourth power of the diameter ratio. Thus $V_5^2 = (6/3)^4 \, V^2 = 16V^2$.

Substituting for V_5 in the preceding equation, $V^2/2g = 40/27 = 1.48$ ft, and so one of the unknowns at point 3 has been determined. It is not desirable to solve for V at this stage, because what is desired is the value of $V^2/2g$; so the numerical value of $2g$ should not be introduced here. Note that $V^2/2g$ is velocity head and is an entity that might be represented by a single symbol such as h_v if desired.

Next apply the flow equation between either points 1 and 3 or points 3 and 5, as only one unknown is now involved. Using the former:

$$0 + 40 + 0 = \frac{p_3}{w} + 25 + 1.48 + 6 \times 1.48$$

From which $p_3/w = 4.64$ ft.

If the rate of discharge is desired, it is now time to introduce the numerical value of $2g$ and solve for V (but not otherwise). It has been found that $V^2/2g = 1.48$; hence $V = (2g \times 1.48)^{1/2} = 8.022 \, (1.48)^{1/2} = 9.75$ ft/s. Hence $Q = 0.196 \times 9.75 = 1.91$ ft^3/s. If one remembers that $(2g)^{1/2} = 8.022$, it will be much better to compute the velocity in the manner shown rather than to multiply the velocity head by $2g$ and then take the square root of the product.

It would have been equally possible to replace V with V_5 and find its value by the energy equation. If we are now interested in it, we can find it also by the continuity equation, as $V_5 = 4 \times 9.75 = 39$ ft^3/s and $V_5^2/2g = 16 \times 1.48 = 23.68$ ft.

Linear acceleration

If a body of liquid in a tank is transported at a uniform velocity, the conditions are those of ordinary hydrostatics. But if it is moved horizontally by a force F_x with a linear acceleration a_x, gravity becomes a significant force. A particle at the surface is acted upon by gravity; so $F_y = G = mg$.

The horizontal force of acceleration is $F_x = ma_x$. By D'Alembert's principle the problem may be transformed into one of static equilibrium if the actual accelerating force is replaced by a fictitious inertia force of the same magnitude but the opposite direction, $-F_x$. The resultant of the gravity force F_y and $-F_x$ is F, and the surface must be normal to the direction of F. Thus tan $0 + a_x/g$. Hence the surface and all planes of equal hydrostatic pressure must be inclined at this angle 0 with the horizontal.

Flow in curved path

The energy equations previously developed applied fundamentally to flow along a streamline or along a stream having a large cross section if certain average values were used. Now conditions will be investigated in a direction normal to a streamline. Consider for example an elementary prism with a linear dimension dr in the plane of the paper rotated on radius r and an area dA normal to the plane of the paper. The mass of this fluid is $p\,dA\,dr$, and the normal component of acceleration is V^2/r. Thus the centripetal force acting upon it is $p\,dA\,dr\,V^2/r$. As the radius increases from r to $r + dr$, the pressure will change from p to $p + dp$. Thus the resultant force in the direction of the center of curvature is $dp\,dA$. Equating these two:

$$dp = p\,\frac{V^2}{r}\,dr \qquad (10.33)$$

where dr = small increment of the radius
r = radius

When the flow is in a straight line for which r is infinity, the value of dp is zero. That is, no difference in pressure can exist transverse to flow in a straight line, which was proved in a different manner.

As dp is positive if dr is positive, the equation shows that pressure increases from the concave to the convex side of the stream, but the exact way in which it increases depends upon the way in which V varies with the radius. In the next two sections will be presented two important practical cases in which V varies in two very different ways.

Forced vortex

A fluid may be made to rotate as a solid body without relative motion theoretically between particles, either by the rotation of a containing vessel or by stirring the contained fluid so as to force it to rotate. Thus an external torque is applied. A common example is the rotation of fluid within the impeller of a centrifugal pump or the rotation of a gas within the impeller of a centrifugal fan or centrifugal compressor.

Cylindrical vortex. If the entire body of fluid rotates as a solid, then in Eq. (10.33) V varies directly with r, that is, $V = rw$, where w is the imposed angular velocity. Inserting this value in Eq. (10.33), we have:

$$dp = pw^2 r \, dr = \frac{w}{g} w^2 r \, dr$$

Between any two radii r_1 and r_2 this integrates as:

$$\frac{p^2}{w} - \frac{p_1}{w} = \frac{w_2}{2g} (r_2{}^2 - r_1{}^2) \qquad (10.34)$$

If p_o is the value of the pressure when $r_1 = 0$, this becomes:

$$\frac{p}{w} = \frac{w^2}{2g} r^2 + \frac{p_o}{w} \qquad (10.35)$$

which is seen to be the equation of a parabola. It is seen that, if the fluid is a liquid, the pressure head p/w at any point is equal to z, the depth of the point below the free surface. Hence the preceding equation may also be written as:

$$z_2 - z_1 = \frac{w^2}{2g} (r_2^2 - r_1^2) \qquad (10.36)$$

and

$$z = \frac{w^2}{2g} r^2 + z_o \qquad (10.37)$$

where z_o is the elevation when $r_1 = 0$. Equations (10.36) and (10.37) are the equations of the free surface—if one exists—or in any case are the equations for any surfaces of equal pressure, which are a series of paraboloids. For an open vessel, the pressure head at any point is equal to its depth below the free surface. If the liquid is confined within a vessel, the pressure along any radius would vary in just the same way as if there were a free surface. Hence the two are equivalent.

The axis of the open vessel is assumed to be vertical. The axis of the closed vessel may be in any direction. However, pressure varies with elevation as well as with the radius, and a more general equation, considering both these factors at the same time, is:

$$\frac{p_2}{w} - \frac{p_1}{w} + z_2 - z_1 = \frac{w^2}{2g} (r_2{}^2 - r_1{}^2) \qquad (10.38)$$

Equation (10.34) is a special case when $z_1 = z_2$, and Eq. (10.36) is a special case when $p_1 = p_2$. If the axis of rotation of the closed vessel were horizontal, the paraboloid that represents the pressure would be some-what distorted because, at a given radius, the pressure at the top would be less than that at the bottom by the amount $z_2 - z_1$. Inserting the value of p/w from Eq. (10.35), but letting $p_o/w = 0$, in the expression for total head, which is the constant in the Bernoulli equation, Eq. (10.9), we have $H = (p/w) + (V^2/2g) = [(rw)V^2/2g] + (V^2/2g)$. As p and V both increase or decrease together, which is just the opposite of the situation in linear flow, it is seen that H cannot be the same for different circular streamlines. In fact, because in this special case $V = rw$, it follows that $H = 2(rw)^2/2g$. That is, H increases as the square of r.

Spiral vortex. So far the discussion has been confined to the rotation of all particles in concentric circles. Suppose there is now superimposed a flow with a velocity having radial components, either outward or inward. If the height of the walls of the open vessel were less than that of a liq-uid surface spread out by some means of centrifugal force, and if liquid were supplied to the center at the proper rate by some means, then it is obvious that liquid would flow outward, over the vessel walls. If, on the other hand, liquid flowed into the tank over the rim from some source at a higher elevation and were drawn out at the center, the flow would be inward. The combination of this approximately radial flow with the cir-cular flow will result in path lines that are some form of spirals.

If the same vessel is arranged with suitable openings near the center and also around the periphery, and if it is provided with vanes, it becomes either a centrifugal pump impeller or a turbine runner. These vanes constrain the flow of the liquid and determine both its relative magnitude and its direction. If the area of the passages normal to the direction of flow is a, then the equation of continuity fixes the relative velocities, since:

$$Q = a_1 v_1 = a_2 v_2 = \text{constant}$$

This relative flow is the flow as it would appear to an observer or a camera revolving with the vessel. The pressure difference due to this superimposed flow alone is found by the energy equation, neglecting friction losses, to be $(p_2/w) - (p_1/w) = (v_1^2 - v_2^2/2g)$. Hence for the case of rotation with flow, the total pressure difference between two points is found by adding together the pressure differences due to the two flows considered separately. That is:

$$\frac{p_2}{w} - \frac{p_1}{w} = \frac{w^2}{2g}(r_2^2 - r_1^2) + \left(\frac{v_1^2 - v_2^2}{2g}\right) \tag{10.39}$$

Of course, friction losses will modify this result to some extent. It is seen that Eq. (10.34) is a special case of Eq. (10.39) when $v_1 = v_2$, either when both are finite or when $v_1 = v_2 = 0$.

For a forced vortex with spiral flow, energy is put into the fluid in the case of a pump and extracted from it in the case of a turbine. In the limiting case of zero flow, when all path lines become concentric circles, energy input from some external source is still necessary for any real fluid in order to maintain the rotation. Thus a forced vortex is characterized by a transfer of mechanical energy from an external source and a consequent variation of H as a function of the radius from the axis of rotation.

Free or irrotational vortex

In this type there is no expenditure of energy whatever from an outside source, and the fluid rotates by virtue of some rotation previously imparted to it or because of some internal action. Some examples are a whirlpool in a river, the rotary flow that often arises in a shallow vessel when liquid flows out through a hole in the bottom (as is often seen when water empties from a bathtub), and the flow in a centrifugal pump case just outside the impeller or in a turbine case as the water approaches the guide vanes. As no energy is imparted to the fluid, then it follows that, neglecting friction, H is constant throughout, that is, $(p/w) + z + (V^2/2g)$ is a constant.

Cylindrical vortex.   In mechanics, force equals mass times acceleration or force equals time rate of change of momentum. If force is multiplied by a distance, the result is torque, and if time rate of change of momentum is multiplied by a distance, the product is called the time rate of change of angular momentum. The angular momentum of a particle of mass m rotating in a circular path of radius r with a velocity V is mVr. As there is no torque applied in the case of a free vortex, it follows that the angular momentum is constant and thus $rV = C$ is a constant, where the value of C determined by knowing the value of V at some radius r. Inserting $V = C/r$ in Eq. (10.33), we obtain:

$$dp = p\,\frac{C^2}{r^2}\,\frac{dr}{r} = \frac{w}{g}\,\frac{C^2}{r^3}\,dr$$

Between any two radii r_1 and r_2, this integrates as

$$\frac{p_2}{w} - \frac{p_1}{w} = \frac{C^2}{2g}\left(\frac{1}{r_1^2} - \frac{1}{r_2^2}\right) = \frac{V_1^2}{2_g}\left[1 - \left(\frac{r_1}{r_2}\right)^2\right] \tag{10.40}$$

If there is a free surface, the pressure head p/w at any point is equal to the depth below the surface. Also, at any radius the pressure varies in a vertical direction according to the hydrostatic law. Hence this equation is merely a special case where $z_1 = z_2$.

As $H = (p/w) + z + (V^2/2g)$ is a constant, it follows that at any radius r:

$$\frac{p}{w} + z = H - \frac{V^2}{2g} = H - \frac{C^2}{2gr^2} = H - \left[\frac{V_1^2}{2g}\left(\frac{r_1}{r}\right)^2\right] \qquad (10.41)$$

Assuming the axis to be vertical, the pressure along the radius can be found from this equation by taking z as a constant. Then, for any constant pressure p, values of z, determining a surface of equal pressure, can be found. If p is zero, the values of z determine the free surface, if one exists.

Equation (10.41) shows that H is the asymptote approached by $(p/w) + z$ as r approaches infinity and V approaches zero. On the other hand, as r approaches zero, V approaches infinity, and $(p/w) + z$ approaches minus infinity. Since this is physically impossible, the free vortex cannot extend to the axis of rotation. In reality, as high velocities are attained as the axis is approached, the friction losses, which vary as the square of the velocity, become of increasing importance and are no longer negligible. Hence the assumption that H is constant no longer holds. The core of the vortex tends to rotate as a solid body as in the central part of a pump impeller.

Spiral vortex. If a radial flow is superimposed upon the concentric flow previously described, the path lines will then be spirals. If the flow goes out through a circular hole in the bottom of a shallow vessel, the surface of liquid takes the form of an empty hole, with an air "core" sucked down the hole. If an outlet symmetrical with the axis is provided, as in a pump impeller, we might have a flow either radially inward or radially outward. If the two plates are a constant distance B apart, the radial flow with a velocity V_r is then across a series of concentric cylindrical surfaces whose area is $0.2\pi rB$. Thus $Q = 2\pi rBV_r$ is a constant, from which it is seen that rVr is a constant. Thus the radial velocity varies in the same way with r that the circumferential velocity did in the preceding discussion. Hence the pressure variation with the radial velocity is just the same as for pure rotation. Therefore the pressure gradient of flow applies exactly to the case of spiral flow, as well as to pure rotation.

If the enclosed impeller case is not constant, then V_r = constant/rB and the pressure variation due to the radial velocity will be somewhat different from that due to rotation alone; B is the diameter of the pipe receiving the flow. The true difference in pressure between any two points may then be determined by adding together the pressure differences caused by these two types of flow considered separately. This procedure is sometimes convenient when friction losses are to be considered and where the losses for the two types of flow are considered to vary in different ways. However, for the case where H is considered as constant for either type of flow, it is possible to combine V_r and V_u, the latter now being used to denote the circumferential velocity, to determine the true velocity V. The total pressure difference may then be determined directly by using this value of V.

In Fig. 10.3 it is seen that, if both V_r and V_u vary inversely to the radius r, the angle $\acute{\alpha}$ is constant. Hence for such a case the path line is the equiangular or logarithmic spiral. For a case where B varies, V_r will not vary in the same way as V_u, and hence the angle $\acute{\alpha}$ is not a constant.

The plotting of such spiral path lines may be of practical value in those cases where it is desired to place some object in the stream for structural reasons but where it is essential that the interference with the stream flow be minimized. If the structure is shaped so as to conform to these streamlines as nearly as possible, it will offer a minimum disturbance to the flow.

Cavitation

According to the Bernoulli theorem, Eq. (10.9), if at any point the velocity head increases, there must be a corresponding decrease in the pressure head. For any liquid there is a minimum absolute pressure, and this is the vapor pressure of the liquid. As has been shown, this value depends both upon the identity of liquid and upon its temperature. If the conditions are such that a calculation results in a lower absolute pressure than the vapor pressure, this simply means that the assumptions upon which the calculations are based no longer apply.

If at any point the local velocity is so high that the pressure in a liquid is reduced to its vapor pressure, the liquid will then vaporize (or boil) at that point and bubbles of vapor will form. As the fluid flows on into a region of higher pressure, the bubbles of vapor will suddenly condense—in other words, they may be said to "collapse." This action produces very high dynamic pressures upon the solid walls adjacent, and as this action is continuous and has a high frequency, the material in that zone will be damaged. Turbine runners, pump impellers, and ship screw propellers are often severely and quickly damaged by such

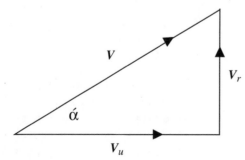

Figure 10.3 If both V_r and V_u vary inversely to r, angle $\acute{\alpha}$ is constant.

action, as holes are rapidly produced in the metal. Not only is cavitation destructive, but it immediately produces a drop in the efficiency of the machine, propeller, or other device.

In order to avoid cavitation, it is necessary that the pressure at every point be above the vapor pressure. To ensure this, it is necessary either to raise the general pressure level, as by placing a centrifugal pump below the intake level so that the liquid flows to it by gravity rather than being drawn up by suction, or so to design the machine that there are no local velocities high enough to produce such a low pressure.

Simulation Laws of Fluid Flow

Definition and uses

It is usually impossible to determine all the essential facts for a given fluid flow by pure theory, and hence much dependence must be placed on experimental investigations. The number of tests to be made can be greatly reduced by a systematic program based on dimensional analysis and specifically on the laws of similitude or similarity, which means certain relations by which test data can be applied to other cases.

Thus, the similarity laws enable us to make experiments with a convenient fluid—such as water, for example—and then apply the results to a fluid which is less convenient to work with, such as air, gas, steam, or oil. Also, in both hydraulics and aeronautics valuable results can be obtained at a minimum cost by tests made with small-scale models of the full-size apparatus. The laws of similitude make it possible to determine the performance of the full-size *prototype,* from tests made with the model. It is not necessary that the same fluid be used for the model and its prototype. Neither is the model necessarily smaller than its prototype. Thus, the flow in a carburetor might be studied in a very large model. And the flow of water at the entrance to a small centrifugal-pump runner might be investigated by the flow of air at the entrance to a large model of the runner.

A few other examples in which models may be used include ships in towing tanks; airplanes in wind tunnels; hydraulic turbines, centrifugal pumps, dam spillways, and river channels; and the study of such phenomena as the action of waves and tides on beaches, soil erosion, and the transportation of sediment. It must be emphasized that the model may not necessarily be different in size from its prototype. In fact, it may be the same device, the variables in the case being the velocity and the physical properties of the fluid.

Geometric similarity

One of the desirable features in model studies is that there be geometric similarity, which means that the model and its prototype be identical in shape and differ only in size. The two may look alike, and again they may bear little resemblance to each other, but the flow patterns should be geometrically similar. If the scale ratio be denoted by L_r, which means the ratio of the linear dimensions of the model to corresponding dimensions in the prototype, then it follows that areas vary as L_r^2 and volumes as L_r^3. Complete geometric similarity is not always easy to attain. Thus, the surface roughness of a small model may not be reduced in proportion unless it is possible to make its surface very much smoother than that of the prototype. In the study of sediment transportation, it may not be possible to scale down the bed materials without having material so fine as to be impractical. Thus, fine powder does not simulate the behavior of sand. Again, in the case of a river, the horizontal scale is usually limited by the available floor space, and this same scale used for the vertical dimensions may produce a stream so shallow that capillarity has an appreciable effect; also, the slope may be such that the flow is laminar. In such cases it is necessary to use a distorted model, which means that the vertical scale is larger than the horizontal scale. If the horizontal scale be denoted by L_r and the vertical scale be L_r' then the cross section area ratio is $L_r L_r'$.

Kinematic similarity

This implies geometric similarity and, in addition, that the ratio of the velocities at all corresponding points is the same. If subscripts m and p denote model and prototype, respectively, the velocity ratio V_r is as follows:

$$V_r = \frac{V_m}{V_p} \qquad (10.42)$$

Its value in terms of L_r will be determined by dynamic considerations, as explained in the following subsection.

As time T is dimensionally L/V, the time scale is

$$T_r = \frac{L_r}{V_r} \qquad (10.43)$$

In a similar manner, the acceleration scale is

$$a_r = \frac{L_r}{T_r^2} \qquad (10.44)$$

Dynamic similarity

If two systems are dynamically similar, corresponding forces must be in the same ratio in the two. The forces are fluid friction, inertia, gravity, pressure, and surface tension. The physical properties involved are viscosity, density, elasticity, and capillarity. In most cases it is impossible to satisfy all requirements for dynamic similarity, but some of these forces either do not apply at all or else are negligible in specific cases. Hence it is possible to confine attention to those which are most important.

If N is the ratio of two types of forces and is therefore a dimensionless number, then, for any two forces considered, dynamic similarity requires that N be the same for both model and prototype; that is, $N_m = N_p$. As stated before, this equality is not confined to model studies alone but is applicable to any two flow systems.

Reynolds number. In the flow of a fluid through a completely filled conduit, gravity does not affect the flow pattern, and neither does the pressure if the fluid be considered as incompressible. It is also obvious that capillarity is of no practical importance; hence, the significant forces are inertia and fluid friction due to viscosity. The same is true of an airplane at speeds below that at which compressibility of the air is appreciable. Also, for a submarine submerged far enough so as not to produce waves on the surface, the only forces involved are those of friction and inertia.

Considering the ratio of inertia forces to friction forces, a parameter is obtained called the *Reynolds number*, or *Reynolds' law*, in honor of Osborne Reynolds, who presented it in a publication of his experimental work in 1882. However, it was Lord Rayleigh 10 years later who developed the theory of dynamic similarity.

From Newton's law, $F = ma$, the inertia forces on a volume represented by L^3 are proportional to $\rho L^3 V/T = \rho L^2 (L/T) V = \rho L^2 V^2$, where T represents time. Therefore, the ratio of these two forces, which is a Reynolds number, is

$$N_R = \frac{L^2 V^2 \rho}{LV\mu} = \frac{LV\rho}{\mu} = \frac{LV}{\nu} \tag{10.45}$$

For any consistent system of units, N_R is a dimensionless number. The linear dimension L may be any length that is significant in the flow pattern. Thus, for a pipe completely filled it might be either the diameter or the radius, and the numerical value of N_R will vary accordingly. General usage in this country prescribes L as the pipe diameter.

If two systems, such as a model and its prototype or two pipelines with different fluids, are to be dynamically equivalent so far as inertia and viscous friction are concerned, they must both have the same value of N_R. For the same fluid in both cases, the equation shows that a high velocity must be used with a model of small linear dimensions. It is also possible to compare the action of fluids of very different viscosities, provided only that L and V are so chosen as to give the same value of N_R.

It is seen that for the velocity ratio of Eq. (10.42) $V_r = (v/L)r$ and from Eq. (10.43) $T_r = L^2/v$, for the same value of N_R.

Froude number.   Considering inertia and gravity forces alone, a ratio is obtained called the *Froude number* or *Froude's law,* in honor of William Froude, who experimented with flat plates towed lengthwise through water in order to estimate the resistance of ships due to wavemaking. The force of gravity on any body with a linear dimension L is

$$wL^3 = \rho g L^3$$

Therefore, the ratio of inertia forces to gravity forces is

$$\frac{\rho L^2 V^2}{\rho g L^3} = \frac{V^2}{gL}$$

Although this is sometimes defined as a Froude number, it is more common to use the square root so as to have V in the first power, as in the Reynolds number. Thus, the Froude number is

$$N_F = \frac{V}{(gL)^{1/2}} \tag{10.46}$$

Systems involving gravity and inertia forces include the wave action set up by a ship, the flow of water in open channels, the forces of a stream on a bridge pier, the flow over a spillway, the flow of a stream from an orifice, and other cases where gravity is a dominant factor.

A comparison of Eqs. (10.45) and (10.46) shows that the two cannot be satisfied at the same time with a fluid of the same viscosity, as one requires that the velocity vary inversely as L, while the other requires it to vary directly as $L^{1/2}$. If both friction and gravity are involved, it is then necessary to decide which of the two factors is more important or more useful. In the case of a ship, the towing of a model will give the total resistance, from which must be subtracted the computed skin friction to determine the wavemaking resistance, and the latter may be even smaller than the former. But, for the same Froude number, the wavemaking resistance of the full-size ship may be determined from this result. A computed skin friction for the ship is then to be added to this value to give the total ship resistance.

In the flow of water in open channels, fluid friction is a factor as well as gravity and inertia, and apparently we face the same difficulty here. However, for flow in an open channel there is usually fully developed turbulence, so that the hydraulic friction loss is exactly proportional to V^2, as will be shown later. The fluid friction is therefore independent of Reynolds number, with rare exceptions, and thus is a function of the Froude number alone.

The only way to satisfy Eqs. (10.45) and (10.46) for both a model and its prototype is to use fluids of very different viscosities in the two cases. Sometimes this can be done, but often it is either impractical or impossible.

For the computation of N_F, the length L must be some linear dimension that is significant in the flow pattern. For a ship it is commonly taken as the length at the waterline. For an open channel it is customary to take it as the depth. From Eq. (10.46), V varies as $-(gL)^{1/2}$, and if g be considered as a constant, as is usually the case, then from Eq. (10.42):

$$V_m : V_p = (L_r)^{1/2} : 1$$

And from Eq. (10.43), the ratio of time for model and prototype is

$$T_m : T_p = (L_r : 1)^{1/2}$$

A knowledge of the time scale is useful in the study of cyclic phenomena such as waves and tides.

As the velocity varies as $(L_r)^{1/2}$ and the cross section area as L_r^2, it follows that

$$Q_m : Q_P = L_r^{5/2} : 1$$

As previously stated, it is usually necessary to use a river model with an enlarged vertical scale. In this case, the velocity varies as L_r' and hence

$$Q_m : Q_p = L_r L_r^{3/2} : 1$$

Mach number. Where compressibility is important, it is necessary to consider the ratio of the fluid velocity (or the velocity of a solid through a stationary fluid) to that of a sound wave in the same medium. This ratio, called the *Mach number* in honor of an Austrian scientist of that name, is

$$N_M = \frac{V}{c} \qquad (10.47)$$

where c is the acoustic velocity (or celerity) in the medium in question. (The ratio of inertia forces to elastic forces represented by the modulus of elasticity is called the *Cauchy number.*) If N_M is less than 1, the flow is *subsonic;* if it is equal to 1, the flow is *sonic;* if it is greater than 1, the flow is *supersonic;* and for extremely high values of N_M, the flow is *hypersonic.*

A certain coefficient might readily be a function of all three parameters, N_R, N_F, and N_M, and then its values could be represented only by a family of curves.

Weber number. In a few cases of flow, surface tension may be important, but normally this is negligible. The ratio of inertia forces to surface tension is called the *Weber number:*

$$N_W = \frac{V}{(\sigma/\rho L)^{1/2}} \qquad (10.48)$$

An illustration of its application is where a very thin sheet of liquid flows over a surface.

Comments on models

In the use of models, it is essential that the fluid velocity not be so low that laminar flow exists when the flow in the prototype is turbulent. Also, conditions in the model should not be such that surface tension is important if such conditions do not exist in the prototype. For example, the depth of water flowing over the crest of a model spillway should not be too low.

While model studies are very important and valuable, it is necessary to exercise some judgment in transferring results to other cases and often a scale effect must be allowed for.

Neither is it always necessary or desirable that these various ratios be adhered to in every case. Thus, in tests of model centrifugal pumps, geometric similarity is essential, but it is desirable to operate at such a rotational speed that the peripheral velocity and all fluid velocities are the same as in the prototype, as only in this way may cavitation be detected. The roughness of a model should be scaled down in the same ratio as the other linear dimensions, which means that a small model should have surfaces that are much smoother than those in its prototype. But this requirement imposes a limit on the scale that can be used if true geometric similarity is to be had. However, in the case of a river model, with a vertical scale larger than the horizontal scale, it may be necessary to make the model surface rough in order to simulate the flow conditions in the actual stream. As any distorted model lacks the proper similitude, no simple rule can be given for this; the roughness should be determined by trial until the flow conditions are judged to be typical of those in the prototype.

Vacuum and barometric pressure

In the case of a siphon, such as a pipe connecting two tanks that rises above the tanks, there is a vacuum at its summit, and if the siphon is to operate, the absolute pressure at the summit should equal or exceed the vapor pressure. An exact solution should consider the friction losses in the lengths of conduit involved, but we shall here confine ourselves to the ideal case of frictionless flow with all loss occurring after exit from the siphon. Thus the absolute pressure at the summit is ideally

$$p_2 = p_a - w\left(z + \frac{V^2}{2g}\right) = p_a - w\,(z + h) \tag{10.49}$$

where p_a is the atmospheric or barometer pressure and where it is assumed that $V^2/2g = h$. With friction considered, the value of p^2 would be increased, so the preceding is on the safe side.

A model of the siphon would have a minimum pressure of:

$$(p_2)_m = (p_a)_m - L_r w (z + h)$$

And if it is to operate at the same barometric pressure as its prototype, so that $(P_a)_m = P_a$, then

$$p_2 = (p_a)_m - (1 - L_r)w(z + h)$$

Thus, the minimum pressure in the prototype is much less than that in the model. Hence, the model might operate, but the full-size siphon would not. In order that they be comparable, the same minimum pressure should exist in each, and, equating $(p_2)_m$ to p^2, we obtain:

$$(p_a)_m = p_a - (1 - L_r)w(z + h) \tag{10.50}$$

Which shows that the entire model siphon should be enclosed in a tank in which there is a partial vacuum to produce a barometric pressure $(P_a)_m$ much less than the normal atmospheric pressure. It is apparent that there is also a limit beyond which it is impossible to go. Combining Eqs. (10.49) and (10.50), we obtain:

$$(p_a)_m = L_r p_a + (1 - L_r)p_2 \tag{10.51}$$

By substituting the minimum allowable pressure p^2 for the prototype, this determines the ideal reduced barometric pressure under which the model siphon must operate.

Flow Measurement

Velocity measurement

This section deals principally with the application of fluid dynamics to those devices whose primary use is the measurement of either velocity or rate of discharge. The measurement of local velocities at different points in the cross section of a stream represents a two- or three-dimensional approach to the problem of determining the rate of discharge. The distribution of velocity over the section is found experimentally and integrated over the area to obtain the rate of discharge. (Velocity measurements also have other uses, such as the measurement of fluctuating velocity components. We place most emphasis on devices and methods which have uses outside of the laboratory as well as within it.) The direct measurement of discharge, on the other hand, is an application of the one-dimensional method of analysis, in which we are concerned not with variations across the section but only in the mean motion of the stream as a whole.

Pitot tube

One means of measuring the local velocity is the *pitot tube* (Fig. 10.4), named after Henri Pitot, who used a bent glass tube in 1730 to measure velocities in the River Seine. It is shown that the pressure at the forward stagnation point of a stationary body in a flowing fluid is $p_2 = p_o + \rho V_o^2/2$, where p_o and V_o are the pressure and velocity, respectively, in the undisturbed flow upstream from the body. If $p_2 - p_o$ can be measured, then the velocity at a point is determined by this relation. The stagnation pressure can be measured by a tube facing upstream, such as in Fig. 10.4. For an open stream or a jet, only this single tube is necessary, as the height h to which the liquid rises in the tube above the surrounding free surface is equal to the velocity head in the stream approaching the tip of the tube.

For a closed conduit under pressure it is necessary to measure the static pressure also. *Static pressure* is simply the difference between the pressure of the fluid flowing outside the pitot tube and the pressure inside the pitot tube; this is also called the differential head h. The

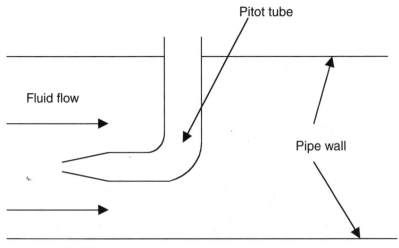

Figure 10.4 Typical pitot tube.

determination of the static pressure is the chief source of error in the use of the pitot tube.

The static pressure may be measured at the wall of the pipe if the wall is smooth and the pipe is straight so that there will be no variation in pressure across the diameter due to centrifugal action. A tube projecting into the stream, as in Fig. 10.4, will give a reading that is too low, owing to the acceleration experienced by the fluid in flowing around the end.

A tube directed downstream, such as that shown in Fig. 10.4 but turned to face in the opposite direction, will give a still lower reading, although not as much as $V_o^2/2g$ below the static pressure. Pitot tubes facing in both directions may be combined into a single instrument, called a *pitometer,* which will give a differential reading greater than h. Such an instrument must be calibrated, as its coefficient cannot be computed by theory.

Obviously, the differential pressure may be measured by any suitable manometer arrangement and does not require an actual column of the fluid concerned. If the fluid were air, for example, this would be impossible.

The formula for the pitot tube follows directly from the stagnation-pressure relation. The difference between the stagnation- and the static-pressure heads is seen to be h, which is the dynamic-pressure head. Hence for a pitot tube pointed directly upstream and in a flow without appreciable turbulence, the equation is

$$V_o = (2gh)^{1/2} \tag{10.52}$$

This equation is not merely an ideal one; it is correct for the conditions specified. It may therefore be used directly for such cases as measuring the velocities in a jet from a good nozzle.

However, in most cases of flow there is appreciable turbulence in the stream, which means that the velocities of individual particles fluctuate in both direction and magnitude. Hence, in general, the velocities of the particles make varying angles with the axial direction of the channel. The effect of this is to cause the instrument to read a value that is higher than the temporal mean axial component of velocity, which is what we generally desire. Hence the preceding equation must be modified to give

$$V_o = C_p \, (2gh)^{1/2} \tag{10.53}$$

where C_p is a pitot-tube coefficient that is slightly less than unity.

The value of C_p decreases with increasing turbulence and ranges from 1.0 down to about 0.97. Experimental evidence from a number of sources indicates that a value of between 0.995 and 0.98 will apply to conditions usually encountered [1–4]. Moody gives a specific expression for a pitot tube used in a pipe, which is $C_p = 1 - 0.15(1 - V/U_{max})$, where V/U_{max} is the ratio of the mean to the maximum velocity in the pipe, called the *pipe factor* [5].

Where conditions are such that it is impractical to measure static pressure at the wall, a combined pitot-static tube may be used. This is actually two tubes, one inside the other. The outer tube is plugged off at the end facing the flowing fluid, but small holes are drilled through it to receive the fluid pressure. These holes open into the annular space between the tubes. The static pressure is measured through two or more of these holes, and it is assumed that the flow follows along the outside of the tube in such a way that the true static pressure is obtained.

For a round-nosed body of revolution with its axis parallel to the flow, the stagnation pressure is found at the tip; the pressure relative to the static pressure then decreases along the surface of such a round-nosed body and even becomes less than the static pressure. But the effect of the stem, at right angles to the stream, is to produce an excess pressure head which diminishes upstream from the stem. If the piezometer orifice in the side of the tube is located where the excess pressure produced by the stem equals the negative pressure caused by the flow around the nose and along the tube, the true static head will be obtained. Tubes constructed with proportions such that one is inside the other are found to be very accurate. A feature of this tube is that, even if not placed exactly in line with the flow, the errors in the static and stagnation values balance each other so that correct velocity heads will be read for deviation angles up to $\pm 15°$.

Still greater insensitivity to angularity may be obtained by guiding the flow past the pitot tube by means of a shroud. The shroud is a tee fitting with curving edges facing the fluid flow; the pitot tube is placed inside it. Such an arrangement, called a *Kiel probe,* is used extensively in aeronautics. The stagnation-pressure measurement is accurate to within 1% of the dynamic pressure for yaw angles up to $\pm 54°$. A disadvantage is that the static pressure must be measured independently, as by a wall piezometer or the static portion of the Prandtl tube.

The Prandtl and Kiel tubes have been seen to be relatively insensitive to direction. Another type is just the reverse and is therefore called a *direction-finding pitot tube.* In the flow past a cylinder whose axis is at right angles to the stream, called a *right-angle pitot tube,* the flow net shows there is a point on either side of the stagnation point where the velocity along the cylinder wall is equal to the undisturbed velocity in the stream. It follows from the Bernoulli theorem that a piezometer hole in the wall at this point will register the true static pressure.

This point depends on a Reynolds number but for ordinary cases has been found experimentally to be at an angle of 39.75° from the center line, giving a total included angle between a and b of 78.5°. If the tube is rotated about its axis until the pressure differential between a and b is 0, then it is known that the direction of the stream is midway between those two points. If the pressure is then read for either a or b alone, the value obtained will be the true static pressure. If the tube is rotated 39.25° from this position, the total pressure is obtained. Hence the direction-finding tube gives not only the direction of the velocity but also correct values of both static and total pressure. The velocity pressure is of course the difference. The pressure measured by a right-angle pitot tube, like that measured by the shroud type, is the pressure head relative to the static head. This type of tube has been used extensively in wind tunnels and in investigations of hydraulic machinery [6].

Piezometer connections

Some of the precautions which must be taken in making piezometer connections for the determination of static pressure have been given. It is emphasized again that the piezometer opening in the side of the channel should be normal to and strictly flush with the surface. Placing anything around it or putting it near an ell fitting will result in error. Thus, Allen and Hooper found that a projection of 0.1 in will cause an error of 16% of the local velocity head [7]. Also, a slight recess will produce a positive error, while a larger recess will produce a negative error, which vanishes as the length of the recess becomes 2 diameters or more. This indicates that, whatever the size of the piezometer orifice, it should be maintained at that size for at least 2 diameters.

The location for a piezometer connection should preferably be in a straight section of sufficient length for normal flow to become established and one that is free of disturbing influences. In the case of a wall piezometer for a pitot tube, the connection should be located upstream from the stem of the tube a distance of at least 10 stem diameters. Furthermore, if the tube is large relative to the pipe, the wall piezometer should be placed 8 tube diameters upstream from the pitot-tube tip to avoid the decrease in pressure caused by the accelerated flow through the reduced section. In order to correct for possible imperfections of the wall and flow conditions, it is desirable to have two or more openings around the periphery of the section. If it is necessary to measure the pressure where there is curved flow, it is essential to take pressure readings for several points spaced uniformly around the periphery and to average them [8].

The diameter of the orifice should be small relative to the channel dimensions; but if it is too small, excessive throttling caused by it may be a source of error, as the observer may believe equilibrium has been attained when this is not the case. Also, any slight leakage from the connecting tubing will then produce a marked effect on the readings. Allen and Hooper did not find any appreciable variation with orifice diameters ranging from $\frac{1}{16}$ to $\frac{27}{32}$ in, however.

Where the pressure is pulsating, the fluctuations may be damped down by throttling for the sake of ease in reading. This will give a true mean value of the pressure, but in most of the fluid metering devices the flow is a function of the square root of the pressure. For a pulsating flow, the square root of the mean pressure is always greater than the mean of the square roots of the pressures, and it is the latter that should be used in the formulas. Hence, the value of the pressure, or the differential pressure, actually used is too large. Therefore, for such a flow the discharge coefficients should be further reduced below the values given. However, since the amount by which they should be reduced cannot well be predicted, it is necessary to eliminate the pulsations if accuracy is to be attained.

Other methods of measuring velocity

The following are a few additional methods of measuring the velocity at a point.

Hot-wire anemometer.   This device is useful not only for indicating the velocity at a point but for measuring the fluctuation in velocity as it responds rapidly to any change. Its operation depends on the fact that the electrical resistance of a wire depends on its temperature; that the temperature, in turn, depends on the heat transfer to the surrounding fluid; and that the coefficient of heat transfer increases with increasing velocity. The hot wire, which is generally of platinum or tungsten, can be connected through a suitable circuit either to carry a constant current, in which case the voltage is a measure of the velocity, or to hold a constant voltage, the resulting current then being a function of the velocity.

Current meter and rotating anemometer.   These two, which are the same in principle, determine the velocity as a function of the speed at which a series of cups or vanes rotate about an axis either parallel to or normal to the flow. The instrument used in water is called a *current meter,* and when designed for use in air it is called an *anemometer.* As the force exerted depends on the density of the fluid as well as on its velocity, the anemometer must be so made as to operate with less friction than the current meter.

If the meter is made with cups which move in a circular path about an axis perpendicular to the flow, it always rotates in the same direction and at the same rate regardless of the direction of the velocity, whether positive or negative, and it even rotates when the velocity is at right angles to its plane of rotation. Thus, this type is not suitable where there are eddies or other irregularities in the flow. If the meter is constructed of vanes rotating about an axis parallel to the flow, resembling a propeller, it will register the component of velocity along its axis, especially if it is surrounded by a shielding cylinder. It will rotate in an opposite direction for negative flow and is thus a more dependable type of meter.

Photographic methods. The camera is one of the most valuable tools in a fluid mechanics research laboratory. In studying the motion of water, for example, a series of small spheres consisting of a mixture of benzene and carbon tetrachloride adjusted to the same specific gravity as the water can be introduced into the flow through suitable nozzles. When illuminated from the direction of the camera, these spheres will stand out in a picture. If successive exposures are taken on the same film, the velocities and the accelerations of the particles can be determined.

In the study of compressible fluids, many techniques have been devised to measure optically the change in density, as given by the interferometer, or the rate at which density changes in space, as determined by the shadowgraph and schlieren methods. From such measurements of density and density gradient it is possible to locate shock waves, which in turn yield velocities. Although of great importance, these photographic methods are too complex to warrant further description here.

Measurement of mean velocity

The preceding devices measure local velocities only, but there are other methods by which the mean velocity can be measured directly. Two of these which are widely used are discussed here.

Salt velocity. This method is based on the increase in electrical conductivity of water when salt is added. Sets of electrodes are installed in a conduit at two sections some distance apart. A single charge of concentrated salt solution is injected above the first station, and the time of passage of the solution between the two stations is obtained from the conductivity measurements, thus yielding the mean velocity [9].

Gibson method. The mean velocity in a penstock leading from a reservoir can be determined by rapidly closing a valve at the lower end and recording a pressure-time diagram at a point just upstream from the valve. The principle that impulse equals change of momentum can be applied to the mass of liquid between the reservoir and the point of measurement. The impulse is given by the area under the recorded pressure-time curve, and from this the initial velocity can be determined [10].

Discharge measurement

Devices for the measurement of discharge can be divided into two categories, those which measure by weight or positive displacement a certain quantity of fluid, and those which employ some aspect of fluid dynamics to measure the mean velocity, which determines the rate of discharge. The ultimate standard must lie in the first of these groups, but it is obvious that such measurements are often impossible—as in the flow of a large river, for example—and more frequently impractical. It is customary, therefore, to employ more convenient instruments which must be calibrated or else a coefficient assumed for them on the basis of experimental work on similar instruments. In order that this may be done intelligently, it is necessary to know the theory underlying each instrument, for only in this way can we make reliable application of the results of experimental work.

Orifices, nozzles, and tubes

Among the devices used for the measurement of discharge are orifices and nozzles. Tubes are rarely so used but are included here because their theory is the same, and experiments upon tubes provide information as to entrance losses from reservoirs into pipelines.

An *orifice* is an opening in the wall of a tank or in a plate which is normal to the axis of a pipe, the plate being either at the end of the pipe or in some intermediate location. An orifice is characterized by the fact that the thickness of the wall or plate is very small relative to the size of the opening. A standard orifice is one with a sharp edge as in Fig. 10.5a or an absolutely square shoulder, so that there is only a line contact with the fluid. Those shown in Figs. 10.5b and 10.5c are not standard because the flow through them is affected by the thickness of the plate, the roughness of the surface, and for 10.5c the radius of curvature. Hence, such orifices should be calibrated if high accuracy is desired [11].

A *nozzle* is a converging tube, such as a cone opening at the smaller end, if it is used for liquids; but for a gas or a vapor, a nozzle may first converge and then diverge again, if the pressure drop through it is sufficiently great, because at the lower pressures the specific volume increases at a faster rate than does the velocity. A nozzle has other important uses, such as providing a high-velocity stream for fire fighting or for power in a steam turbine or a Pelton waterwheel.

A *tube* is a short pipe whose length is not more than 2 or 3 diameters. There is no sharp distinction between a tube and the thick-walled orifices of Figs. 10.5b and 10.5c. It may be straight, or it may diverge.

Characteristics of fluid jets

A *jet* is a stream issuing from an orifice, nozzle, or tube. It is not enclosed by solid boundary walls but is surrounded by a fluid whose

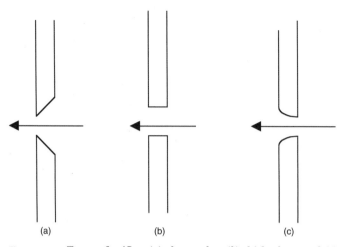

Figure 10.5 Types of orifice: (*a*) sharp edge, (*b*) thick plate, and (*c*) thick plate with curved radius.

velocity is less than its own. The two fluids may be different, or they may be of the same kind. A *free jet* is a stream of liquid surrounded by a gas and is therefore directly under the influence of gravity. A *submerged jet* is a stream of any fluid surrounded by a fluid of the same kind. As it is buoyed up by this surrounding fluid, it is not directly under the action of gravity.

Jet contraction.   Where the streamlines converge in approaching an orifice, as is the case in Fig. 10.5*a*, they continue to converge beyond the orifice until they reach the orifice opening, where they become parallel. For the ideal case, this location is an infinite distance away; but in reality, owing to friction between the jet and the surrounding fluid, this section is usually found very close to the orifice. For a free jet with a high velocity or a submerged jet with any velocity, the friction with the surroundings is such that the jet slows down and diverges beyond the opening. The orifice opening is then a section of minimum area and is called the *vena contracta*. (Of course, if a free jet is discharged vertically downward, the acceleration due to gravity will cause its velocity to increase and the area to decrease continuously so that there may be no apparent section of minimum area. In such special cases, the vena contracta should be taken as the place where marked contraction ceases and before the place where gravity has increased the velocity to any appreciable extent above the true jet velocity.)

By *jet velocity* is meant the value of the average velocity at the vena contracta. The velocity profile across an orifice opening such as that shown in Fig. 10.5*a* is seen to be nearly constant except for a small annular ring around the outside. The average velocity V is thus only slightly less than the center or maximum velocity u_c. As the streamlines are straight at the orifice opening, it follows that at the vena contracta the pressure is constant across the orifice diameter and, except for a minute increase due to surface tension, this pressure must be equal to that in the medium surrounding the jet. (The pressure within a jet is totally different from the force that the jet is capable of exerting upon some object that it may strike.)

In the plane of the orifice (Fig. 10.5a), the cross-sectional area is greater than at a plane immediately upstream; hence, the average velocity must be less. As the streamlines at this point are curved, centrifugal action will cause the pressure to increase from the edge of the orifice to the center. Thus, the pressure across the diameter will vary somewhat at the orifice area plane of the wedge type (Fig. 10.5a); consequently, from the flow equation, the velocity u must also vary.

As the contraction of the jet is due to the converging of the approaching streamlines, it may be prevented altogether by causing the particles of fluid to approach in an axial direction solely, as in the free discharge from the end of a straight pipe or by a construction such as in Figs. 10.5b or 10.5c. Also, if an orifice is placed near a corner of a tank, the streamlines are partially prevented from converging, and the contraction is thus diminished. Roughness of the surface around the orifice reduces the velocity of the fluid approaching from the side and thereby also reduces the contraction.

Jet trajectory. A free liquid jet in air will describe a *trajectory*, or path under the action of gravity, with a vertical velocity component which is continually changing. The trajectory is a streamline; consequently, if air friction is neglected, Bernoulli's theorem may be applied to it, with all the pressure terms 0. Thus, the sum of the elevation and velocity head must be the same for all points of the curve. The energy gradient is a horizontal line at distance $V^2/2g$ above the nozzle, where V is the velocity leaving the nozzle.

The equation for the trajectory may be obtained by applying Newton's equations of uniformly accelerated motion to a particle of the liquid passing from the nozzle to point P, whose coordinates are x, z, in time t. Then $x = V_x t$ and $z = V_z t - \frac{1}{2}gt^2$. Evaluating t from the first equation and substituting it in the second gives

$$Z = \frac{V_z}{V_x} x - \frac{g}{2V_x^2} x^2 \qquad (10.54)$$

This is seen to be the equation of an inverted parabola having its vertex at $x_o = V_x V_z/g$ and $z_o = V_o^2/2g$. Since the velocity at the top of the trajectory is horizontal and equal to V_x, the distance from this point to the energy gradient is evidently $V_x^2/2g$. This may be obtained in another way by considering that $V^2 = V_x^2 + V_z^2$. Dividing each term by $2g$ gives these relations.

If the jet is initially horizontal, as in the flow from a vertical orifice, $V_x = V$ and $V_z = O$. Equation (10.54) is then readily reduced to an expression for the jet velocity in terms of the coordinates from the vena contracta to any point of the trajectory, z now being positive downward,

$$V = x \left(\frac{g}{2z} \right)^{1/2} \qquad (10.55)$$

Example 10.4. If a jet is inclined upward 30° from the horizontal, what must be its velocity to reach over a 10-ft wall at a horizontal distance of 60 ft, neglecting friction?

$$V_x = V \cos 30° = 0.866V$$

$$V_z = V \sin 30° = 0.5V$$

From Newton's laws,

$$x = 0.866\, Vt = 60$$

$$z = 0.5Vt - 16.1t^2 = 10$$

From the first equation, $t = 69.3/V$. Substituting this in the second equation,

$$0.5V\,\frac{69.3}{V} - \frac{1}{2}\,32.2\left(\frac{69.3}{V}\right)^2 = 10$$

from which $V^2 = 3140$ or $V = 56$ ft/s.

The same value may be obtained by substituting for V_x and V_x in Eq. (10.54). The foregoing calculations are for an ideal case. In reality, air friction and turbulence will alter the trajectory.

Jet coefficients

Coefficient of velocity C_v. The velocity that would be attained at the vena contracta if friction did not exist may be termed the ideal velocity V_i. (This is frequently called "theoretical velocity," but this is a misuse of the word *theoretical*. Any correct theory should allow for the fact that friction exists and affects the result. Otherwise, it is not correct theory but merely an incorrect hypotheses.) Owing to friction, the actual average velocity V is less than the ideal velocity, and the ratio V/V_i is called the *coefficient of velocity*. Thus, $V = C_v V_i$.

Coefficient of contraction C_c. The ratio of the area of a jet A at the vena contracta to the area of the orifice or other opening A_o is called the *coefficient of contraction*. Thus, $A = C_c A_o$.

Coefficient of discharge C_d.   The ratio of the actual rate of discharge Q to the ideal rate of discharge Q_i, if there were no friction and no contraction, may be defined as the *coefficient of discharge*. Thus, $Q = C_d Q_i$. By observing that $Q = AV$ and $Q_i = A_o V_i$, it is seen that $C_d = C_c C_v$. Other coefficients which are sometimes more convenient to use will be defined in specific cases in later subsections.

The coefficient of velocity may be determined by a velocity traverse of the jet with a fine pitot tube in order to obtain the mean velocity. This is subject to some slight error, as it is impossible to measure the velocity at the outer edge of the jet. The velocity may also be computed approximately from the coordinates of the trajectory. The ideal velocity is computed by the Bernoulli theorem.

The coefficient of contraction may be obtained by measuring the diameter of the jet with suitable calipers, which must be rigidly supported. Precision may be lacking, especially in the case of small jets.

The coefficient of discharge is the one that can most readily be obtained, and with a high degree of accuracy. It is also the one that is of the most practical value. For a liquid, the actual Q can be determined by some standard methods, such as a volume or weight measurement over a known time. For a gas, one can note the change in pressure and temperature in a container of known volume from which the gas may flow. Obviously, if any two of the coefficients are measured, the third can be computed from them.

Jet with no velocity of approach

The velocity of a fluid approaching an orifice or nozzle or similar device is called the *velocity of approach*. For example, consider a large tank, filled with liquid, with a small orifice on its wall, near the bottom. It is assumed that the area of the tank is so large relative to that of the orifice that the velocity at the surface of the liquid (point 1) is negligible. Let the Bernoulli theorem be written between point 1 at the surface and point 2 at the orifice jet discharge, assuming the pressure is the same at both points. Let h equal the height of the liquid, measured from the surface level to the center of the orifice. Then $0 + h + 0 = 0 + 0 + V_i^2/2g$, from which

$$V_i = (2gh)^{1/2} \tag{10.56}$$

This was one of the earliest rational laws of hydraulics (1643), known as Torricelli's theorem from its discoverer, who was a student of Galileo.

According to the definitions just given, the actual velocity of the jet is

$$V = C_v (2gh)^{1/2} \tag{10.57}$$

and the actual discharge is

$$Q = AV = C_c A_o \ [C_v \ (2gh)^{1/2}] = C_d A_o \ (2gh)^{1/2} \tag{10.58}$$

In case the pressures are not the same at points 1 and 2, then in place of h in the preceding equations we should have $h + p_1/w - p_2/w$.

Submerged orifice

For a submerged orifice, points 1 and 2 are the same as in the preceding case, and point 3 is the surface of the liquid downstream of the submerged orifice. The equation is written between points 1 and 2, realizing that the pressure head on the jet at point 2 is equal to y, the height of point 3 above the orifice center. Let h now equal the height of point 1, the surface level of the liquid in the tank, above point 3, the surface level of the liquid downstream of the orifice. Thus

$$0 + (h + y) + 0 = y + 0 + \frac{V_i^2}{2g}$$

and as the y term cancels, the result is the same as in the preceding case. Thus the actual velocity is

$$V = C_c \ (2gh)^{1/2}$$

For a submerged orifice, the coefficients are practically the same as for a free jet, except that for heads less than 10 ft and for very small orifices the discharge coefficient may be slightly less. It is of interest to observe that if the energy equation is written between points 1 and 3, the result is $h_f = H_1 - H_3 = h$. Therefore, this is one instance where it is impossible to assume no friction loss.

Jet with velocity of approach

In case the velocity of approach is not negligible, as was assumed in the preceding discussion, it is necessary to consider the velocity head at point 1. Thus, for a conical nozzle, writing the Bernoulli equation between a point inside the pipe immediately upstream of the nozzle (point 1) and a point immediately downstream of the nozzle (point 2), and then introducing the velocity coefficient, the actual jet velocity is

$$V = C_v \left[2g \left(\frac{p_1}{w} + \frac{V_1^2}{2g} \right) \right]^{1/2} \tag{10.59}$$

This is seen to be similar in form to Eq. (10.57), and if the velocity at point 1 could be determined in some way, the value of the jet velocity could be computed directly. If Q were measured, then $V_1 = Q/A_1$, or V_1 might be determined by a pitot-tube traverse. But even if V_1 is known only approximately, the error in using such a value cannot be great, as $V_1^2/2g$ is usually small compared with p_1/w. However, V_1 can be eliminated by the continuity equation as $V_1 = AV/A_1 = C_c A_o V/A_1 = C_c (D_o/D_1)^2 V$.

Substituting this value of V_1 in Eq. (10.59) and noting that $C_v C_o = C_d$, the actual jet velocity is

$$V = \frac{C_v}{[1 - C_d^2 (D_o/D_1)^4]^{1/2}} \left(2g \frac{p_1}{w}\right)^{1/2} \tag{10.60}$$

Comparing this with Eq. (10.57), we see that the divisor is the factor that corrects for the velocity of approach where D_o and D_1 are the diameters of the orifice or nozzle tip and the approach pipe, respectively, and h is replaced by p_1/w.

Multiplying the jet velocity by the jet area, which is $A = C_c A_o$, the rate of discharge is

$$Q = \frac{C_d A_o}{[1 - C_d^2 (D_o/D_1)^4]^{1/2}} \left(2g \frac{p_1}{w}\right)^{1/2} \tag{10.61}$$

Although this formula is strictly logical, it is more convenient to employ a coefficient C, so that

$$Q = \frac{CA_o}{[1 - (D_o/D_1)^4]^{1/2}} \left(2g \frac{p_1}{w}\right)^{1/2} \tag{10.62}$$

It is seen that C_d and C are identical in numerical value only when $D_o/D_1 = 0$, but they are very nearly the same in value when D_o/D_1 is small, such as <0.4.

Inasmuch as either C_d or C should be determined by experiment, it is even more convenient to replace both Eqs. (10.61) and (10.62) by

$$Q = KA_o \left(2g \frac{p_1}{w}\right)^{1/2} \tag{10.63}$$

where K is a flow coefficient, determined experimentally for a given ratio of D_o/D_1.

Energy loss and nozzle efficiency

The loss of head in friction in an orifice, nozzle, or tube may be determined by writing the energy equation between some point upstream and the vena contracta of the jet. Letting H now represent the total head upstream while $V^2/2g$ is the velocity head in the jet, the equation is $H - h_f = V^2/2g$. But as $V = C_v (2gH)^{1/2}$ and $V^2/2g = C_v^2 H$, the substitution of this last expression in the preceding energy equation results in two equivalent expressions for the loss of head

$$h_f = (1 - C_v^2) H = \frac{[(1/C_v^2) - 1] V^2}{2g} \tag{10.64}$$

It should be noted that the contraction coefficient is merely the relation between the area of a jet and the orifice and is not involved at all in friction loss.

As the jet from a nozzle is frequently employed for power purposes in a steam turbine or a Pelton wheel, we are interested in its energy efficiency. The efficiency of a nozzle is defined as the ratio of the power in the jet to the power passing a section in the pipe at the base of the nozzle. As the discharge is the same for the two points, the efficiency is merely the ratio of the heads at these two sections. Thus, referring again to the conical nozzle, $e = H_2/H_1$. But $H^2 = V^2/2g$, and $H_1 = p_1/w + V_1^2/2g$, and so from Eq. (10.59) it follows that $H^2 = C_v^2 H_1$. Hence the nozzle efficiency is

$$e = C_v^2 \tag{10.65}$$

Owing to the fact that the velocity in the jet is not uniform across a diameter, the true kinetic energy in the jet is slightly greater than that by $V^2/2g$. Thus, for a good nozzle, the true efficiency may be about 1% more than the value given by Eq. (10.65).

Nozzle performance

The height to which a good fire stream can be thrown by a nozzle is about two-thirds to three-fourths of the effective head at the base of the nozzle. The proportion is higher for large jets than for small ones, for smooth nozzles than for rough ones, and for low than for high base pressures. Tests have shown that air friction and dispersion of the jet reduce this angle to about 30° from the horizontal [12]. The same tests showed that an increase in pressure at the base of the nozzle beyond 175 psi produced no further increase in the throw of a 1.5-in jet. They showed further that the most important feature in improving the performance of fire streams is the elimination of turbulence generated by abrupt changes in cross section or flow direction within the nozzle. The best nozzle form was found to be a curving, converging conical nozzle, rather than the conventional straight conical tip.

It has also been found that the efficiency of Pelton wheels can be improved by several percent by similar means, as a better nozzle design provides a jet with a more uniform velocity and one that has less tendency to disperse before it strikes the wheel.

Example 10.5. A 2-in circular orifice (not standard) in the end of a 3-in-ID pipe discharges a measured flow of 0.60 ft³/s of water when the pressure in the pipe is 10.0 psi. If the jet velocity is determined by pitot tube to be 39.2 ft/s, find the values of the coefficients C_v, C_c, C_d, C, and K.

$$\frac{p_1}{w} = 10 \times \frac{144}{62.4} = 23.1 \text{ ft}$$

$$V_1 = \frac{Q}{A_1} = \frac{0.60}{0.0491} = 12.23 \text{ ft/s}$$

$$\frac{V_1^2}{2g} = 2.32 \text{ ft}$$

Then, from Eq. (10.59):

$$C_v = 0.970$$

$$\text{Area of jet} = \frac{Q}{V} = \frac{0.60}{39.2} = 0.0153 \text{ ft}^2$$

Then,

$$C_c = \frac{A_j}{A_o} = \frac{0.0153}{0.0218} = 0.702$$

From $C_d = C_v C_c$,

$$C_d = 0.970 \times 0.702 = 0.680$$

[This value could also have been obtained from Eq. (10.61).]
From Eq. (10.62):

$$C = 0.640$$

From Eq. (10.63):

$$K = 0.715$$

Values of coefficients

Velocity coefficient. The coefficient of velocity is a measure of the extent to which friction retards the jet velocity of a real fluid. For sharp-edged circular orifices, the velocity coefficient may range from about 0.95 to 0.994. The smaller values are found with small orifices and low heads. For good nozzles with smooth walls, the velocity coefficient is usually from 0.97 to 0.99. For certain tubes and other devices, the velocity coefficients may be much lower than the preceding, as may be observed in Figs. 10.6b and 10.6c. Such tubes have considerable friction loss, not because of the initial contraction of the stream entering, but because of its subsequent reexpansion to fill the tube. Rounding the entrance, as in Fig. 10.6a, greatly reduces the friction.

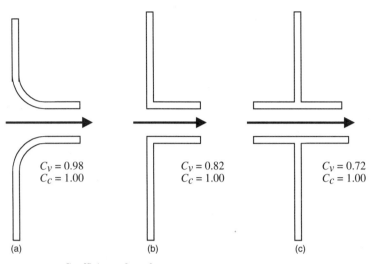

$C_V = 0.98$
$C_C = 1.00$

$C_V = 0.82$
$C_C = 1.00$

$C_V = 0.72$
$C_C = 1.00$

(a)   (b)   (c)

Figure 10.6   Coefficients for tubes.

Contraction coefficient.   This merely determines the size of a stream that may issue from a certain area. Its value may range from approximately 0.50 to 1.00. For the ideal flow through a rectangular sharp-edged orifice, classical hydrodynamics yields a value for C_c of

$$\frac{\pi}{\pi + 2} = 0.611$$

and this is also practically the value for standard sharp-edged circular orifices, for orifice diameters 2.5 in or larger and for heads above about 3 ft. For very small orifices and for heads as low as a few inches, the effects of adhesion, surface tension, and capillarity will increase the coefficient up to 0.72, and even more in extreme cases.

Any device that provides a uniform diameter for a short distance before exit, such as the tubes in Fig. 10.6 or a straight-tip nozzle, will usually eliminate any contraction; thus $C_c = 1.00$. While this increases the size of the jet from a given area, it also tends to produce more friction. The contraction coefficient is very sensitive to small changes in the edge of the orifice or in the upstream face of the plate. Thus, slightly rounding the edge of the orifice in Fig. 10.6*b*, instead of maintaining a perfectly square corner, will increase the contraction coefficient materially.

The tubes in Figs. 10.6*b* and 10.6*c* are flowing full, and, because of the turbulence, the jets issuing from them will have a "broomy" appearance. Because of the contraction of the jet at the entrance to these tubes, the local velocity in the central portion of the stream will be higher than that at the exit from the tubes; hence, the pressure will be lower. If the pressure is lowered to that of the vapor pressure of the fluid, the streamlines will then no longer follow the walls. In such a case, the tube in Fig. 10.6b becomes equivalent to the orifice in Fig. 10.5b, while the tube in Fig. 10.6c tends to converge the exiting fluid to a smaller diameter. If its length is less than its diameter, the reentrant tube is called a *Borda mouthpiece*. Because of the greater curvature of the streamlines for a reentrant tube, the contraction coefficient is lower than for any other type, and the velocity coefficient is also lower. But if the jet springs clear to the atmosphere, the velocity coefficient is as high as for a sharp-edged orifice.

The Borda mouthpiece is of interest because it is one device for which the contraction coefficient can be very simply calculated. For all other orifices and tubes, there is a reduction of the pressure on the wall adjacent to the opening, but the exact pressure values are unknown. However, for the reentrant tube, the fluid is at rest on the wall around the tube; hence, the pressure must be exactly that due to the depth below the surface. The only unbalanced pressure is that on an equal area opposite to the tube, and its value is whA_o. The time rate of change of momentum due to the flow out of the tube is $(W/g)V = wA V^2/g$, where A is the area of the jet. Equating force to time rate of change of momentum, $whA_o = wA V^2/g$. Ideally, $V^2 = 2gh$, and thus ideally $C_c = A/A_o = 0.5$. If it is assumed that $C_v = 0.98$, then the actual values will be $C_c = 0.52$ and $C_d = 0.51$.

It is not easy to compute contraction coefficients in other cases, but they can be determined reasonably well by means of flow nets. However, without taking this much trouble, a little judgment will enable one to make a fair estimate of the contraction.

Discharge coefficient. If this coefficient has to be estimated in a particular case, it is usually better to assume velocity and contraction coefficients separately and calculate the discharge coefficient from them. Thus, it is known that the velocity coefficient for a standard orifice is about 0.98 and its contraction coefficient is 0.62; hence, $C_d = 0.98 \times 0.62 = 0.61$ approximately. For some devices the velocity coefficient is the determining factor, and for some the contraction coefficient is the controlling factor.

Although numerous experiments have been made and their results published, there is as yet no general agreement on precise values for discharge coefficients for orifices and nozzles, so that calibration is recommended if high accuracy is desired. However, for standard sharp-edged circular orifices, the range of values for C_d is 0.59 to 0.68. The values decrease with increasing orifice size and also with increasing head. For conical nozzles, the discharge coefficients are the same or nearly the same as the velocity coefficients. For a nozzle such as a low-turbulence fire nozzle with a 30° inlet, there will be a material jet contraction, in which case the discharge coefficient will be lower than the velocity coefficient; see Refs. [13] and [14].

Venturi tube

The converging tube is an efficient device for converting pressure head to velocity head, while the diverging tube converts velocity head to pressure head. The two may be combined to form a *venturi tube,* named after Venturi, an Italian, who investigated its principle about 1791. It was applied to the measurement of water by Clemens Herschel in 1886. It consists of a tube with a constricted throat which produces an increased velocity accompanied by a reduction in pressure, followed by a gradually diverging portion in which the velocity is transformed back into pressure with slight friction loss. As there is a definite relation between the pressure differential and the rate of flow, the tube may be made to serve as a metering device.

The Herschel type of venturi meter has a converging cone, with a reducing section converging 20° on the inlet side of the throat and an enlarging section diverging 5° on the outlet side of the throat. These angles are measured as deviations from a straight centerline through the length of the pipe. If the converging section is replaced with a simple flange having an entrance such as that shown in Fig. 10.6a, the coefficient is believed to be less sensitive to changes in the smoothness of the surface.

Writing the Bernoulli equation between points 1 and 2, we have, for the ideal case,

$$\frac{p_1}{w} + z_1 + \frac{V_1^2}{2g} = \frac{p_2}{w} + z_2 + \frac{V_2^2}{2g}$$

and as $(p_1/w + z_1) - (p_2/w + z_2) = h$, whether the tube is horizontal or not, we have

$$\frac{V_2^2}{2g} - \frac{V_1^2}{2g} = h$$

By the continuity equation, $V_1 = (A_2/A_1)/V_2$; hence

$$V_2 = \left[\frac{2gh}{1 - (A_2/A_1)^2}\right]^{1/2}$$

As there is some friction loss between points 1 and 2, the true velocity is slightly less than the value given by this expression. Hence, we may introduce a discharge coefficient C, so that the flow is given by

$$Q = A_2 V_2 = \frac{CA_2}{[1 - (D_2/D_1)^4]^{1/2}} (2gh)^{1/2} \tag{10.66}$$

which is equivalent to Eq. (10.62). For a given meter, the dimensions are known and constant. Although C varies with the rate of flow, the variation is very small, and if an average value is assumed, the equation for a particular meter reduces to

$$Q = Mh^{1/2} \tag{10.67}$$

The venturi tube is an accurate device for measurement of flow of all types of fluid and is most valuable for large flow in pipelines. With a suitable recording device, it can integrate the flow rate so as to give the total quantity. Aside from the installation cost, its only disadvantage is that it introduces a permanent frictional resistance in the pipeline. This loss is practically all in the diverging part from points 2 to 3 and is ordinarily from $0.1h$ to $0.2h$.

Values of D_2/D_1 may vary from 0.25 to 0.75, but a common ratio is 0.50. A small ratio gives increased accuracy of the gauge reading but is accompanied by a higher friction loss and may produce an undesirably low pressure at the throat, sufficient in some cases to cause liberation of dissolved air or even vaporization of the liquid at this point.

The angles of convergence and divergence in the original Herschel meter are 20 and 5°, respectively, but in other tubes these angles have been increased to as much as 30 and 14°, respectively. It is believed that a minimum friction loss is obtained for an angle of 5° in the diffuser, but a larger angle will reduce the length and cost of the tube.

For accuracy in use, the venturi meter should be preceded by a straight pipe whose length is at least 5 to 10 pipe diameters. The approach section becomes more important as the diameter ratio increases, and the required length of straight pipe depends on the conditions preceding it. Thus, the vortex formed from two short-radius elbows in planes at right angles, for example, is not eliminated in 30 pipe diameters. Such a condition is alleviated by the installation of straightening vanes preceding the meter [15].

The pressure differential should be obtained from piezometer rings surrounding the pipe, with a number of suitable openings into the two sections. In fact, these openings may sometimes be replaced by very narrow slots extending all the way around the circumference. A pulsating flow gives a differential which is proportional to the square of the instantaneous velocities; thus, a mean velocity computed from the square root of the manometer reading will result in a measured flow greater than the actual flow [16, 17].

Venturi meter coefficients

Unless specific information is available for a given venturi tube, the value of C may be assumed to be about 0.99 for large tubes and about 0.97 or 0.98 for small ones, provided the flow is such as to give reasonably high Reynolds numbers. A roughening of the surface of the converging section from age or scale deposit will reduce the coefficient slightly. Venturi tubes in service for many years have shown a decrease in C of the order of 1 to 2% [18, 19].

For a given venturi tube, C is a function of Reynolds number, which makes it possible to select the proper value of C for any fluid whatever and does not restrict it to the fluid with which the tube has been r calibrated. Values of venturi tube coefficients as functions of both Reynolds number and size are shown in Fig. 10.7. (Figure 10.7 is adapted from a similar graph by W. S. Pardoe [20].) Note that N_R is computed for conditions at the throat. To determine the Reynolds number, it is often convenient to use a "hybrid" set of units and compensate by a correction factor. Thus

$$N_R = \frac{15.28Q}{dv} = \frac{22,740W}{d\mu}$$

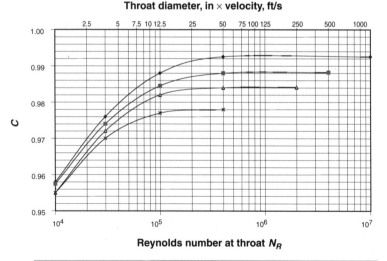

Figure 10.7 Coefficients for venturi tubes with a diameter ratio D_2/D_1 of 0.5. $N_R = D_2 V_2 \rho_2 / \mu_2$. *(Adapted from W.S. Pardoe, "The Coefficients of Herschel-Type Cast-Iron Venturi Meter," Trans. ASME 67:339, 1945.)*

where d = diameter, in
  v = kinematic viscosity, ft^2/s
  μ = fluid viscosity, cP
  Q = rate of flow, ft^3/s
  W = rate of discharge, lb/s

This last form is especially convenient in case of gases. Figure 10.7 is for a diameter ratio of $D_2/D_1 = 0.5$ only, but it is sensibly valid for smaller ratios also. For larger ratios, C decreases slightly. Thus, for $D_2/D_1 = 0.75$ the values are about 1 percent less than shown in Fig. 10.7. Although this diagram is based on venturi tubes with a 20° entrance and –5° exit, the coefficients will be accurate within 1% for similar devices, such as a venturi nozzle with a curved flange entrance.

Occasionally, the precise calibration of a venturi tube has given a value of C greater than 1. Such an abnormal result is sometimes due to improper piezometer openings. Another explanation is that we may apply it to the venturi tube with $p_1/w - p_2/w = h$. The equation then may be written as

$$\frac{V_2^2}{2g} - \frac{V_1^2}{2g} = h - \left[h_f + \frac{(\alpha_2 - 1)\, V_2^2}{2g} - \frac{(\alpha_1 - 1)\, V_1^2}{2g} \right]$$

For the ideal case of frictionless flow, every term in the brackets is zero, resulting in a coefficient of unity. The same result may be produced in a real flow if the sum of all the terms in the brackets is zero. The friction h_f is always positive, but in converging flow the sum of the other two terms can be negative.

As the pressure across any section of a straight pipe is independent of the velocity profile, it follows that $p_1/w - p_2/w$ has the same value for all streamlines. Therefore if V_1 and V_2 indicate values on a streamline in ideal flow,

$$(V_2')^2 - (V_1')^2 = \frac{p_1}{w} - \frac{p_2}{w} = \text{constant}$$

or

$$(V_2')^2 = (V_1')^2 + \text{constant}$$

Thus, graphically, V_2' is the hypotenuse of a right triangle of which V_1' is one side while the other side is proportional to the square root of the pressure drop. Considering a converging velocity profile from a conical nozzle at point 1 into that at point 2 shows that the percentage variation in velocity becomes much less. Assuming a diameter ratio of 0.5, it is thus apparent that α_2 is less than α_1. In cases where h_f is extremely small owing to a very smooth surface of the converging section, but where α_1 may be relatively large, it is possible to have the entire bracket not only zero, but even negative, thereby giving the coefficient C a value greater than unity without violating the law of conservation of energy.

Although such a case is unusual, it has been found at times. A reason for introducing this discussion is that the venturi tube is one case where there may be considerable difference in velocity profiles at the two sections, and it is desirable to observe that what is ordinarily termed "friction" may include other effects as well. It is in cases where true friction is very small that these other factors become appreciable. The foregoing discussion also explains the use of a converging section preceding the working section of every wind tunnel or water tunnel where it is desired to have a uniform velocity distribution.

Flow nozzle

If the discharge cone of a venturi nozzle is omitted, the result is a flow nozzle. This is simply a plate with a sized hole fitted between the flanges of a pipe union. The hole is curved and lipped as in Fig. 10.6a. Again as for a venturi, D_1 is the diameter of the connecting pipe and D_2 is the diameter of the hole. The flow nozzle is simpler than the venturi tube and can be installed between the flanges of a pipeline. It will answer the same purpose, though at the expense of an increased friction loss in the pipe. Although the venturi meter equation [Eq. (10.66)] can be employed for the flow nozzle, it is more convenient and customary to include the correction for velocity of approach with the coefficient of discharge, as in Eq. (10.63), so that

$$Q = KA_2 \, (2gh)^{1/2} \tag{10.68}$$

where K is called the *flow coefficient* and A_2 is the area of the nozzle throat. Comparison with Eq. (10.66) establishes the relation

$$K = \frac{C_v}{[1 - (D_2/D_1)^4]^{1/2}} \tag{10.69}$$

Although there are many designs of flow nozzles, the International Standards Association (ISA) nozzle has become an accepted standard form in many countries. Values of K for various diameter ratios of the ISA nozzle are shown in Fig. 10.8 as a function of Reynolds number. Note that in this case the Reynolds number is computed for the approach pipe rather than for the nozzle throat, which is a convenience as N_R in the pipe is frequently needed for other computations also.

As shown in Fig. 10.8, many of the values of K are greater than unity, which is a result of including the correction for approach velocity with

Pipe diameter, in × velocity, ft/s

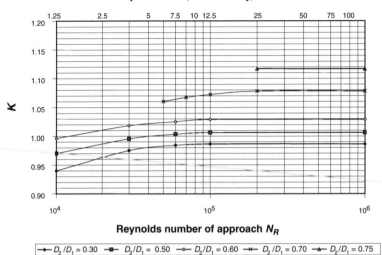

Figure 10.8   Flow coefficients for ISA nozzle. $N_R = D_1 V_1 \rho_1 / \mu_1$.

the conventional coefficient of discharge. There have been many attempts to design a nozzle for which the velocity-of-approach correction would just cancel the discharge coefficient, leaving a value of the flow coefficient equal to unity. Detailed information on these so-called long-radius nozzles may be found in the ASME publications on fluid meters and flow measurement [21–23]. As in the case of the venturi meter, the flow nozzle should be preceded by at least 10 diameters of straight pipe for accurate measurement. Two alternative locations for the pressure taps may be drilled directly into the union flanges holding the flow nozzle plate. The head loss across an ISA flow nozzle is shown in Fig. 10.9 as a percentage of the differential head h.

Orifice meter

An orifice in a pipeline, as in Fig. 10.5a, may be used as a meter in the same manner as the venturi tube or the flow nozzle. It may also be placed on the end of the pipe so as to discharge a free jet. The coefficients are practically identical in the two cases. The difference between the orifice in the present discussion and that in the earlier sections of this chapter is that here the pipe walls are nearer to the edge of the orifice so that there is less contraction of the jet, resulting in a higher value of C_c, and also there is a much larger value of the velocity of approach. For the numerical values of the coefficients to apply, the ratio D_o/D_1 should be less than ⅕, where D_o and D_1 are the diameters of the orifice and the approach pipe, respectively. In the orifice meter the

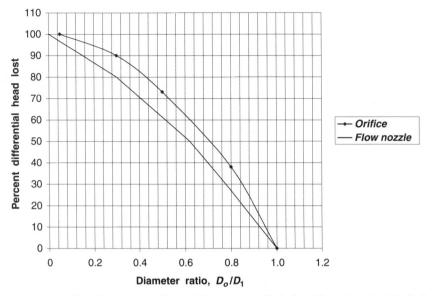

Figure 10.9 Head loss across orifice and flow nozzle. *(Calculated from data in Refs. [24] and [25].)*

ratio is usually much higher than that, and C_c increases with increasing D_o/D_1, while C_v remains nearly constant. Thus, the value $C_d = C_c C_v$ will increase with increasing ratio, as shown in Fig. 10.10. Instead of using Eq. (10.61), however, it is more convenient to use Eq. (10.62). For ease of reference this is given here as

$$Q = \frac{CA_o}{[1 - (D_o/D_1)^4]^{1/2}} \, (2gh)^{1/2} \qquad (10.70)$$

This equation has the additional advantage that C varies little for diameter ratios up to 0.85, as shown in Fig. 10.10. As it is not satisfactory to use higher ratios, the marked variation beyond that point is of no importance. It will be observed that this equation is the same as that for the venturi meter and differs from it only in the numerical value of C. As given here, the equation differs from the form that h replaces p_1/w, but Eqs. (10.62), (10.66), and (10.70) are really identical.

As in the case of the orifice and the conical flow nozzle, it is convenient to reduce the equation to

$$Q = KA_o \, (2gh)^{1/2} \qquad (10.71)$$

Although all three forms of the orifice equation will be found in use, the present trend is toward Eq. (10.71) because of the simplified computation. The relation between C_d, C, and K is shown in Fig. 10.10. (The

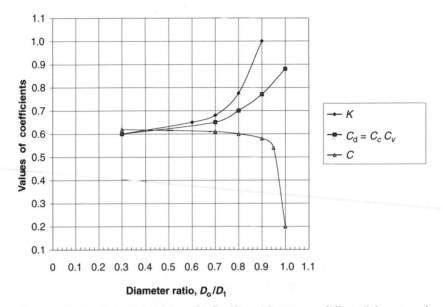

Figure 10.10 Coefficients for sharp-edged orifice with pressure differential measured either at the flanges or at the vena contracta. *(Calculated from data in Refs. [24] and [25].)*

curves shown in Fig. 10.10 were calculated from data given in Refs. [24] and [25].) If the value of anyone of these coefficients is known for a given diameter ratio, corresponding values of the other three may be computed or read from Fig. 10.10.

The orifice meter coefficient is subject to the same influences of finish on the edge of the orifice and the surface of the plate as were discussed for the simple orifice in the subsection "Values of Coefficients." It is also affected by size owing to the relative roughness of the approach pipe. Thus, for a small pipe the relative roughness is greater, even though the character of the surface is exactly the same. This tends to retard the velocity of flow along the pipe walls and thus along the face of the orifice plate. It is this velocity coming in from the side that produces the jet contraction, and any reduction in it will reduce the contraction. Thus, if the coefficient for an orifice in a 15-in pipe is 0.60, it might be 0.61 for the same diameter ratio in a 3-in pipe.

The difference between an orifice meter and a venturi meter or flow nozzle is that for both of the latter there is no contraction, so that A_2 is also the area of the throat and is fixed, while for the orifice, A_2 is the area of the jet and is a variable and is, in general, less than the area of the orifice A_o. For the venturi tube or flow nozzle the discharge coefficient is practically a velocity coefficient, while for the orifice the value of C or K is much more affected by C_c than it is by C_v.

Thus, the coefficient C for an orifice meter is much less than it is for a venturi or a flow nozzle, and it also varies in a different manner with Reynolds number. Consider a pressure gradient between points in a pipe fitted with an orifice plate. Point 1 is upstream of the orifice plate. Point 2 is immediately downstream of the orifice, at the vena contracta. Point 3 is further downstream where the flow is normal, as it was at point 1, though at lower pressure owing to the orifice pressure loss.

This pressure gradient shows that the pressure drops to a minimum at the vena contracta at point 2 and then gradually rises to point 3, which is 4 to 8 pipe diameters downstream, after which it slowly diminishes owing to normal pipe friction. The drop in pressure between points 1 and 2 is due to the acceleration given the stream in passing through the orifice, while the pressure drop between points 1 and 3, which is the loss of head caused by the orifice meter, is due solely to the friction caused by the throttling effect of the orifice. The mechanism of this is that a considerable portion of the kinetic energy of the jet due to the velocity of translation is transformed into kinetic energy of rotation in vortices which are formed from the high fluid shear stresses as the jet expands to fill the pipe. This kinetic energy of rotation is not converted into pressure but is eventually dissipated by viscosity effects and converted into heat.

The pressure differential may be measured between point 1, which is about 1 pipe diameter upstream, and the vena contracta, which is approximately 0.5 pipe diameter downstream. The distance to the vena contracta is not a constant but decreases as D_o/D_1 increases. The differential can also be measured between the two corners on each side of the orifice plate. These *flange taps* have the advantage that the orifice meter is self-contained; the plate may be slipped into a pipeline without the necessity of making piezometer connections in the pipe. In the corner where the orifice joins the pipe, the pressure is a little higher than in the center, owing to the stagnation effect. For a similar reason, the pressure on the downstream corner is a little higher than that at the vena contracta; hence, the differential pressure measured with flange taps is not materially different from that obtained with *vena contracta taps* as previously described. For precise work, however, there exist curves of the flow coefficient obtained with each of these two schemes of measuring the pressure differential [26].

The pressure differential may also be measured between points 1 and 3, but the value so obtained is much smaller. This is undesirable, except where it is necessary to employ some manometer with a limited capacity.

This method has the merit that it is not very important just where the downstream piezometer is located, provided only that it is not too close. The same connections may then be used for different-sized orifices in the same pipe. However, as the roughness of the pipe has an appreciable effect, the entire setup should be calibrated under service conditions. The orifice as a measuring device has the merit that it may be installed in a pipeline with a minimum of trouble and expense. Its principal disadvantage is the greater frictional resistance offered by it as compared with the venturi tube. The friction added by the orifice is plotted in Figs. 10.9 and 10.10 together with that for a flow nozzle. The orifice head loss is given closely by $h_f = [1 - (D_o/D_1)^2]h$. For further information on orifice meters, see Refs. [27] and [28].

Variation of meter coefficients with Reynolds number

The coefficients for venturi meters, flow nozzles, and orifice meters vary with Reynolds number as shown in Figs. 10.7 to 10.10. The curve for an orifice meter shown in Fig. 10.10 covers an unusually wide range of both viscosity and Reynolds number. The fluids used were water and a series of oils up to a very viscous road oil, and for each fluid a number of different velocities were used, so that the curve represents points for many combinations of velocity and viscosity. Although the orifice plate may not be a standard beveled form, the value of C for high Reynolds numbers agrees closely with the value of C in Fig. 10.10 for a diameter ratio of 0.75.

It has been shown that the coefficients for any meter vary with Reynolds number but approach constant values at large Reynolds numbers. Figs. 10.9 and 10.10 show such constant values for each diameter ratio. It is desirable that any meter be used, if possible, at such Reynolds numbers that its coefficient is substantially constant.

An explanation for the form of the curve in Fig. 10.10 and also of those in Fig. 10.9 may be had by considering N_R to be decreased by using a series of fluids of increasing viscosity but keeping all other quantities the same. Starting with a low viscosity and a high N_R, the coefficient is seen to be approximately constant. As N_R decreases, a point is reached at which the viscosity is sufficient to retard effectively the flow of the film of fluid over the upstream face of the orifice plate and thus to reduce the contraction of the jet. Further increases in viscosity continue to reduce the contraction until finally the size of the jet is equal to that of the orifice. Thus, C_c continuously increases from its initial low value at high N_R until it equals 1.0.

With increasing viscosity and hence increasing fluid friction, the velocity coefficient will continuously decrease from a value of 1 for an ideal fluid to a value that approaches 0 as the viscosity approaches infinity. The discharge coefficient, which is roughly the product of these two, at first increases with decreasing N_R, as the contraction coefficient dominates, until it reaches a maximum value. When C_c becomes 1, then the decreasing velocity coefficient causes the discharge coefficient to decrease rapidly with smaller values of N_R.

For the venturi tube, on the other hand, there is no contraction at any stage; hence, its coefficient would be 1.0 for zero viscosity or a value for N_R of infinity and will then decrease continuously with increasing viscosity or decreasing Reynolds number, as in Fig. 10.7.

Example 10.6 A 3-in ISA flow nozzle is installed in 4-in pipe carrying water at 72°F. If a water-air manometer shows a differential of 2 in, find the flow.

solution This is a trial-and-error type of solution. First, assume a reasonable value of K. From Fig. 10.37, for $D_2/D_1 = 0.67$, for the level part of the curve, $K = 1.06$. Then from Eq. (10.68)

$$Q = KA_2\,(2gh)^{1/2}$$

where

$$A_2 = \frac{\pi}{4} \times \frac{3^2}{144} = 0.0491 \text{ ft}^2$$

and

$$h = \frac{\Delta p}{w} = \frac{2}{12} = 0.167 \text{ ft}$$

Thus

$$Q = 1.06 \times 0.0491 \times 8.02 \times (0.167)^{1/2} = 0.171 \text{ ft}^3/\text{s}$$

With this first determination of Q,

$$V_1 = \frac{Q}{A} = \frac{0.171}{0.0872} = 1.96 \text{ ft/s}$$

Then

$$D_1'V_1 = 4 \times 1.96 = 7.84$$

From Fig. 10.37, $K = 1.05$ and

$$Q = \frac{1.05}{1.06} \times 0.171 = 0.1694 \text{ ft}^3/\text{s}$$

No further correction is necessary.

Weir

The weir has long been a standard device for the measurement of water in an open channel. It is an orifice placed at the water surface, so that the head on its upper edge is 0. Consequently, the upper edge is eliminated, leaving only the lower edge, or crest. Consider a weir having a vertical measure H of water flowing over it. The rate of flow is determined by measuring the height H, relative to the crest, at a distance upstream at least 4 times the maximum head that is to be employed. The amount of drawdown at the crest is about $0.15H$.

The upstream face of the weir plate should be smooth, and the plate should be strictly vertical. The crest should have a sharp, square upstream edge, a top width of ⅟₁₆ to ⅛ in, and a bevel on the downstream side, so that the nappe springs clear, making a line contact for all but the very lowest heads. If it does not spring clear, the flow cannot be considered as true weir flow, and the coefficients do not apply. If the crest extends downstream so that it supports the nappe, the weir is broad-crested. The velocity at any point in the nappe can be determined. The approach channel should be long enough so that normal velocity distribution exists, and the surface should be free of waves. It may sometimes be necessary to install baffles to ensure a quiet flow of approach. All deviations from proper weir construction affect the flow in the same way. That is, they increase it over the computed value.

The value of H is measured by a head gauge. The hook gauge is brought up to the surface from below until its point just pierces the water surface, when the scale is read. From this reading must be subtracted the reading when the point is level with the weir crest. The point gauge is lowered to the water surface from above, and the scale is read just as the point touches the surface. The point gauge may be connected into an electric circuit so that a small light shows when contact is made with the water. Both gauges may be used in the open stream, but it is better if they are used in a stilling well at one side of the channel.

Suppressed rectangular weir.   This type of weir is as wide as the channel, and the width of the nappe is the same as the length of the crest. As there are no contractions of the stream at the sides, it is said that end contractions are *suppressed*. It is essential that the sides of the channel upstream be smooth and regular. It is common to extend the sides of the channel downstream beyond the crest so that the nappe is confined laterally. The flowing water tends to entrain air from this enclosed space under the nappe, and unless this space is adequately ventilated, there will be a partial vacuum, and perhaps all the air may eventually be swept out. The water will then cling to the downstream face of the plate, and the discharge will be greater for a given head than when the space is vented. Therefore, venting is necessary if the standard formulas are to be applied.

Weir with end contractions. When the length B of the crest of a rectangular weir is less than the width of the channel, there will be a lateral contraction of the nappe so that its width is less than B. It is believed that end contractions are a source of error, and so this type of weir is not considered so accurate as the preceding. Its chief virtue is that the approach channel need not be of uniform cross section or have smooth sides.

Triangular or V-notch weir. For relatively small flows, the rectangular weir must be very narrow and thus of limited maximum capacity, or else the value of H will be so small that the nappe will not spring clear but will cling to the plate. For such a case the triangular weir has the advantage that it can function for a very small flow and also measure reasonably large flows as well. The vertex angle may be anything from 10 to 90° but is rarely more than the latter.

Basic weir formulas

To derive a flow equation for the weir, consider an elementary area $dA = B\,dh$ in the plane of the crest, and assume the velocity through this area to be equal to $(2gh)^{1/2}$. B is the width of the crest, and H is the height of liquid flowing over the crest or height of the weir. The apparent flow through this area is

$$dQ = B\,dh\,(2gh)^{1/2} = B\,(2g)^{1/2}\,h^{1/2}\,dh$$

and this is to be integrated over the whole area—that is, from $h = 0$ to $h = H$. Performing this integration, we obtain an ideal Q_i which is

$$Q_i = (2g)^{1/2}\,B \int_o^H h^{1/2}\,dh = \frac{2}{3}\,(2g)^{1/2}\,BH^{3/2}$$

But this ideal flow will be decreased slightly by fluid friction and decreased much more by other factors. It is apparent that the area of the stream in the plane of the crest is much less than BH, owing to the drawdown at the top and the crest contraction below. As h is never as small as 0, there is no zero velocity at the top of the nappe; furthermore, the streamlines in general are not normal to the plane of the area. To correct for these discrepancies, it is necessary to introduce an experimental coefficient of discharge C_d. This then gives us the basic weir formula:

$$Q = C_d\,\frac{2}{3}\,(2g)^{1/2}\,BH^{3/2} \tag{10.72}$$

For a given weir, if C_d is assumed to be constant, this equation could be reduced to $Q = KH^{3/2}$; $H^{3/2} = H \times H^{1/2}$.

If V_o, the velocity of approach at the section where H is measured, is appreciable, the limits of integration should then be $H + V_o^2/2g$ and $V_o^2/2g$; so the resulting fundamental equation is

$$Q = C_d \frac{2}{3} (2g)^{1/2} B\left[\left(H + \frac{V_o^2}{2g}\right)^{3/2} - \left(\frac{V_o^2}{2g}\right)^{3/2}\right] \tag{10.73}$$

To be strictly correct, the velocity head should be multiplied by a to correct for the nonuniformity of the velocity. To do this precisely, it would be necessary to measure local velocities over the area, but this is not usually practicable, so a is usually assumed to be anything from 1.0 to 2.2, according to one's judgment as to the velocity distribution. As it is believed that the ratio of surface velocity to mean velocity is an important criterion for this case, it is customary to assume a as 1.0 to 1.3 for H equal to or greater than Z and to increase it up to 2.2 as the ratio H/Z decreases.

Numerous weir formulas have been proposed, but they all differ in the ways in which C_d is evaluated and in the methods of correcting for velocity of approach.

Rectangular weir with end contractions suppressed

Francis weir formula. In 1848 to 1852 at Lowell, Massachusetts, James B. Francis made precise investigations of the flow of water over weirs of large size. His observations showed the flow varied as $H^{1.47}$, but he adopted $H^{3/2}$ for greater convenience. He then selected a constant average value of $C_d = 0.622$ in Eq. (10.72), so that for a suppressed weir with negligible velocity of approach

$$Q = 3.33BH^{3/2} \tag{10.74}$$

(In this and subsequent weir formulas which are not in dimensionless terms, foot-second units are assumed.)

For a suppressed weir with velocity of approach considered, he used the same value for C_d in Eq. (10.73). The Francis formulas have been checked by many others and found to be accurate within 1 to 3% for all heads above 0.3 ft but to be about 7% too low for heads of 0.1 ft and under.

Equation (10.73), with $C_d = 0.622$ inserted, must be solved by trial. An approximate value of Q is first obtained from Eq. (10.74). This value is then divided by the area of the channel where H is measured to obtain an approximate value of V_o. A value of a may be assumed, if desired, and Eq. (10.73) used to find a new and slightly larger Q. A second value of V_o might now be used in a second calculation, but this is seldom necessary.

It is cumbersome to use Eq. (10.73) because of the trial solution required. This may be avoided with sufficient accuracy for most practical purposes by a series of mathematical steps, which will not be reproduced here, but which, if $C_d = 0.622$, will result in an approximate formula

$$Q = 3.33BH^{3/2}\left[1 + 0.259\left(\frac{H}{H+Z}\right)^2\right]$$

Rehbock formula. Small-scale but precise experiments covering a wide range of conditions led T. Rehbock of the Karlsruhe Hydraulic Laboratory in Germany to the following expression for C_d in Eq. (10.72):

$$C_d = 0.605 + \frac{1}{305H} + 0.08\frac{H}{Z} \tag{10.75}$$

This equation was obtained by fitting a curve to the plotted values of C_d for a great many experiments and is purely empirical. Capillarity is accounted for by the second term, while velocity of approach (assumed to be uniform) is responsible for the last term. Rehbock's formula has been found to be accurate within 0.5% for values of Z of 0.33 to 3.3 ft and for values of H of 0.08 to 2.0 ft with the ratio H/Z not greater than 1.0. It is even valid for greater ratios than 1.0 if the bottom of the discharge channel is lower than that of the approach channel, so that backwater does not affect the head.

There are numerous other weir formulas, which will not be given here, as the foregoing are typical and adequate. It is believed that the most accurate weir is the one with no end contractions but with a very deep channel of approach so that the velocity of approach is negligible.

Weir with end contractions

When the width B of a rectangular weir is less than that of the channel, Francis concluded that the effect of each side contraction is to reduce the width of the nappe by $0.1H$. The usual contracted weir will have a contraction at each side, but occasionally a weir is placed against one side wall, so that the contraction on one side is suppressed. If n = the number of contractions, which may be 2, 1, or 0, the Francis formula based on Eq. (10.74) is

$$Q = 3.33\,(B - 0.1nH)\,H^{3/2} \tag{10.76}$$

This formula is an excellent illustration of the limitations of an empirical formula. It is accurate within the limits of the experiments on which it was based. But it is seen that, if H is only large enough relative to B, the formula would indicate zero flow, and, for an even larger value of H, the computed result would be negative. If this equation is to be used, the minimum proportions allowable are $B > 3H$, with the width of each side contraction $2H$ and the height of the crest $2H$, but even larger ratios are desirable. Even with such limitations, it may yield results that are up to 3% too high.

For a weir with two end contractions, Gourley and Crimp, on the basis of tests they conducted, obtained the formula

$$Q = 3.10B\ 1.02H^{1.47} \tag{10.77}$$

This seems to agree well with Eq. (10.76) and has the merit that it has no such limitations on the minimum value of B/H.

Cipolletti weir. In order to avoid correcting for end contractions, the sides of the Cipolletti weir are given as 0.25:1 batter, which is supposed to add enough to the effective width of the stream to offset the lateral contraction. Thus, if B is the width or length of the crest,

$$Q = 3.367BH^{3/2} \tag{10.78}$$

Triangular or V-notch weir

Consider a triangular weir with a vertex θ. H is the height from the bottom of the weir to the water's surface, and θ is the angle of the bottom of the weir. The rate of discharge through an elementary area dA is $dQ = C_d\,(2gh)^{1/2}\,dA$. Now $dA = 2x\,dh$, and $x/(H - h) = \tan\theta/2$. Substituting in the foregoing, the following is obtained for the entire notch:

$$Q = C_d\ 2(2g)^{1/2} \tan \frac{\theta}{2} \int_{o}^{H} (H - h)\ h^{1/2}\ dh$$

Integrating between limits and reducing, the fundamental equation for all triangular weirs is obtained:

$$Q = C_d\ \frac{8}{15}\ (2g)^{1/2} \tan \frac{\theta}{2}\ H^{5/2} \tag{10.79}$$

Note that $H^{5/2} = H^2\ H^{1/2}$

For a given angle θ this may be reduced to

$$Q = KH^{5/2} \tag{10.80}$$

Experimental values of C_d vary from 0.58 to 0.68 for water flowing over V-notch weirs with central angles varying from 10 to 90° [29, 30].

The incomplete contraction at low heads produces a higher value of C_d, but for even lower heads the frictional effects reduce the coefficient continuously. At very low heads, when the nappe clings to the weir plate, the phenomenon can no longer be classed as weir flow.

Neglecting the low-head region, C_d decreases somewhat with increasing head, suggesting that the discharge varies with H to some power slightly less than 2.5. As a fair average for heads of 0.2 to 2.0 ft, it may be found that the curves here shown will give for the 60° weir (in foot-second units)

$$Q = 1.44H^{2.48} \tag{10.81}$$

and for the 90° weir

$$Q = 2.48H^{2.47} \tag{10.82}$$

Submerged weir

If the water surface downstream from the weir is higher than the weir crest, the weir is said to be submerged. Two kinds of submerged flow are recognizable: (1) the plunging nappe, in which the weir nappe is similar to that for free flow, plunging down to the bed of the flume; and (2) the surface nappe, in which the nappe remains on or near the surface, leaving the water beneath relatively undisturbed. The plunging nappe occurs at low values of the submergence ratio $S = H_2/H_1$, while the surface nappe occurs at high ones.

The net flow Q over the submerged weir has been given by Villemonte to be

$$Q = Q_1 (1 - S^n)^{0.385} \tag{10.83}$$

where Q_1 is the free flow which would result from the head H_1 if the weir were not submerged and n is the exponent which appears in the normal weir equation, that is, 3/2 for the rectangular weir and 5/2 for the triangular weir.

This equation may be considered accurate in general only to within 5%, for a practical submergence range of 0 to 0.9. It may, however, be applied to a 90° triangular weir with an accuracy of 0.5% for S less than 0.7. It should be emphasized that the outstanding advantage of the submerged weir is the great saving in head loss over any of the free-fall types.

Sluice gate

The sluice gate is similar to the weir, with the major difference that the water flows under the gate rather than over it. Thus, a rather large head of pressure may be exerted on the passage through the gate due to the depth of the water upstream.

The sluice gate is used to control the passage of water in an open channel. When properly calibrated, it may also serve as a means of flow measurement. As the lower edge of the gate opening is flush with the floor of the channel, contraction of the bottom side of the issuing stream is entirely suppressed. Side contractions will of course depend on the extent to which the opening spans the width of the channel. The complete contraction on the top side, however, owing to the larger velocity components parallel to the face of the gate, will offset the suppressed bottom contraction, resulting in a coefficient of contraction nearly the same as for a slot with contractions at top and bottom.

Flow through a sluice gate differs fundamentally from that through a slot in that the jet is not free but guided by a horizontal floor. Consequently, the final jet pressure is not atmospheric but distributed hydrostatically in the vertical section. Consider a sluice gate with point 1 upstream and point 2 downstream of the gate, where y is the height of water upstream. Writing the Bernoulli theorem with respect to the stream bed as datum from point 1 to point 2 in the free-flow case,

$$\frac{V_1^2}{2g} + y_1 = \frac{V_2^2}{2g} + y_2$$

from which, introducing a velocity coefficient,

$$V_2 = C_v \left[2g\left(y_1 + \frac{V_1^2}{2g} - y_2\right)\right]^{1/2}$$

Following the customary definition of the coefficient of discharge, $C_d = C_c C_v$, there results

$$Q = C_d\, A\, [2g\,(y_1 - y_2) + V_1^2]^{1/2}$$

in which $A = aB$ is the area of the gate opening. As in the case of the flow nozzle and the orifice meter, it is convenient to absorb the velocity of approach into an experimental flow coefficient and to include also in this coefficient the effect of downstream depth. There results the simple discharge equation

$$Q = K_s A(2gy_1)^{1/2} \tag{10.84}$$

Values of K_s fall between 0.55 and 0.60 for free flow but are materially reduced when the flow conditions downstream are such as to produce submerged flow. (Values of discharge coefficients for sluice and other types of gates may be found in Hunter Rouse [31].)

Other methods of measuring discharge

In addition to the foregoing standard devices for measuring the flow of fluids, there exist a number of supplementary devices less amenable to exact theoretical analysis but worthy of brief mention. One of the simplest for measuring flow in a pipeline is the *elbow meter*, which consists of nothing more than piezometer taps at the inner and outer walls of a 90° elbow in the line. The pressure difference, due to the centrifugal effects at the bend, will vary approximately as the velocity head in the pipe. Like other meters, the elbow should have sections of straight pipe upstream and downstream and should be calibrated in place [32].

The rotameter consists of a tapered glass tube in which the metering float is suspended by the upward motion of the fluid around it. Directional notches cut in the float keep it rotating and thus free of wall friction. The rate of flow determines the equilibrium height of the float, and the tube is graduated to read the flow directly. The rotameter is also used for gas flow, but the weight of the float and the graduation must be changed accordingly.

Example 10.7 An air duct of 2- by 2-ft square cross section turns a bend of radius 4 ft to the center line of the duct. If the measured pressure difference between the inside and outside walls of the bend is 1 in of water, estimate the rate of air flow in the duct. Assume standard sea-level conditions in the duct, and assume ideal flow around the bend.

Flow measurement of compressible fluids

Strictly speaking, most of the equations that are presented in the preceding part of this chapter apply only to incompressible fluids; but practically, they may be used for all liquids and even for gases and vapors where the pressure differential is small relative to the total pressure. As in the case of incompressible fluids, equations may be derived for ideal frictionless flow and then a coefficient introduced to obtain a correct result. The ideal conditions that will be imposed for a compressible fluid are that it is frictionless and that there is to be no transfer of heat; that is, the flow is adiabatic. This last is practically true for metering devices, as the time for the fluid to pass through is so short that very little heat transfer can take place. Because of the variation in density with both pressure and temperature, it is necessary to express rate of discharge in terms of weight rather than volume. Also, the continuity equation must now be

$$W = w_1 A_1 V_1 = w_2 A_2 V_2$$

Applying to a fluid meter,

$$V_2 = C_v \left[\frac{2gJ\,(E_1 - E_2)}{1 - (w_2\,A_2/w_1\,A_1)^2} \right]^{1/2}$$

(10.85)

and the rate of discharge is

$$W = Cw_2\,A_2 \left[\frac{2gJ\,(E_1 - E_2)}{1 - (w_2\,A_2/w_1\,A_1)^2} \right]^{1/2}$$

(10.86)

In these equations E_2 is the value of enthalpy after a frictionless adiabatic expansion from p_1 to p_2. It may be obtained from vapor tables or charts and is very convenient for vapors. The discussion of such tables and their use is outside the scope of this text and is strictly in the field of thermodynamics, but for a gas $E_1 - E_2 = c_p(T_1 - T_2)$, where $T_2 = T_1\,[p_2/p_2\,(k - 1)]/k$. For compressible fluids, p must be absolute pressure.

However, Eqs. (10.85) and (10.86) can be replaced by equations involving pressures directly, and in these forms they can be used for either gases or vapors by the selection of the appropriate value of k.

$$V_2 = C_v \left[2g\,\frac{k}{k-1}\,p_1 v_1\,\frac{1 - (p_2/p_1)^{(k-1)/k}}{1 - (A_2/A_1)^2\,(p_2/p_1)^{2/k}} \right]^{1/2}$$

(10.87)

and, multiplying this by $w_2\,A_2$ and replacing W_2 in terms of W_1, the rate of discharge is

$$W = CA_2 \left[2g\,\frac{k}{k-1}\,p_1\,w_1 \left(\frac{p_2}{p_1}\right)^{2/k}\,\frac{1 - (p_2/p_1)^{(k-1)/k}}{1 - (A_2/A_1)^2\,(p_2/p_1)^{2/k}} \right]^{1/2}$$

(10.88)

where W_1 may be replaced by p_1/RT_1 if desired.

Inasmuch as this equation is too complex for convenient use in metering, it is replaced by the simpler equation for an incompressible fluid, but with the insertion of an expansion factor Y.

For an incompressible fluid with specific weight w_1, the differential head is $h = (p_1 - p_2)/w_1$, and from Eq. (10.66) the flow of an incompressible fluid is

$$W' = w_1 Q' = CA_2 \left[\frac{2gw_1\,(p_1 - p_2)}{1 - (D_2/D_1)^4} \right]^{1/2}$$

(10.89)

If the fluid is compressible, the specific weight will decrease from w_1 to w_2 as the pressure drops from p_1 to p_2 and the value of w' determined by Eq. (10.89) will be larger than the true value W as given by Eq. (10.88). Therefore, we use an expansion factor Y such that $W = YW'$. Therefore, the true flow rate for a compressible fluid is

$$W = CYA_2 \left[\frac{2gw_1 (p_1 - p_2)}{1 - (D_2/D_1)^4} \right]^{1/2} \tag{10.90}$$

where C has the same value as for an incompressible fluid at the same Reynolds number.

[Equations (10.88) to (10.90) seem to be inconsistent in that they use the throat area A_2 with the specific weight w_1 in the pipe or inlet. The practical reason is that the diameter of the throat is more accurately known than the diameter of the pipe, and the pressure and temperature in the pipe are more generally known than those in the throat. Note that w_1 may be replaced by p_1/RT_1.]

Values of Y may be found by experiment or by computing the value of W by Eq. (10.88) and finding the relation between it and W' for a variety of conditions. The values so determined may be used for p_2/p_1. For a venturi or nozzle throat where $C_c = 1$,

$$Y = \left\{ \frac{[k/(k-1)] (p_2/p_1)^{2/k} [1 - (p_2/p_1)^{(k-1)/k}]}{1 - (p_2/p_1)} \right\}^{1/2} \left(\frac{1 - (D_2/D_1)^4}{1 - (D_2/D_1)^4 (p_2/p_1)^{2/k}} \right)^{1/2}$$

Charts could be constructed for values of k, but if no such chart is available, it will be necessary to use Eq. (10.88). However, it will be found that Y does not vary a great deal with different values of k. Thus, for $p_2/p_1 = 0.90$ and $D_2/D_1 = 0.65$, values of Y for $k = 1.4$, 1.3, and 1.2 are 0.9309, 0.9259, and 0.9201, respectively. For smaller values of p_2/p_1 and D_2/D_1, the differences will be greater.

Subsonic velocity

In many cases of metering fluids the velocities are lower than the acoustic velocity. This subsection is devoted to this condition.

Pitot tube.   When a pitot tube is used for the measurement of the velocity of a compressible fluid, the usual hydraulic formula may be employed for all velocities less than about one-fifth that of the acoustic velocity in the medium. The error at this arbitrary limit is about 1%. For higher velocities a correction should be made, which applies up to a Mach number of 1.

Venturi tubes and converging/diverging nozzles.   For these devices, p_2 in the equations in the preceding subsection is the pressure at the throat where the velocity is V_2. In the following discussion it is assumed that the initial values of p_1 and T_1 remain constant while P_2 varies.

As p_2/p_1 decreases, the velocity V_2 increases until the pressure p_c at which the velocity V_2 becomes equal to the acoustic velocity c. Any further reduction of pressure in the space into which the fluid discharges will not lower the pressure at the throat or increase the velocity at this section, because a pressure wave, which is a sound wave whose velocity is c, cannot travel upstream against a fluid whose particles may move at a higher velocity. Thus, the throat velocity and the rate of discharge reach a maximum constant value when p_2 becomes as low as p_c, and they remain constant with any further reduction of pressure in the space downstream.

Hence, the curves for Y values terminate at $p_2/p_1 = 0.528$ for the venturi tube, because, as is shown in the next subsection, this is approximately the ratio at which $p_2 = p_c$ when $k = 1.4$.

Square-edged or thin-plate orifices.   Such orifices differ from the venturi tube in that for compressible fluids the diameter of the jet is no longer constant but increases with decreasing values of p_2/p_1 owing to the expansion of the fluid. The preceding equations for V_2 are valid, but this jet velocity is found at the vena contracta where the area is A_2, and this area is unknown. Hence, we must replace A_2 in the equations by A_o. Thus, C_c may initially equal 0.62—for example, for liquids—but for compressible fluids will continuously increase as p_2/p_1 decreases. Hence, it is no longer possible to compute Y by theory, as in the preceding case, but we must resort to experiment. An empirical equation which covers this variation in C is the following [33, 34]:

$$Y = 1 - \left[0.41 + 0.35 \left(\frac{D_o}{D_1} \right)^4 \right] \frac{p_1 - p_2}{kp_1}$$

In the case of an orifice it is probably true that the maximum jet velocity is the acoustic velocity c, but this does not impose a limit on the rate of discharge because the jet area continues to increase with decreasing values of p_2/p_1.

It has been found by experiment that the flow through an orifice increases continuously as the pressure ratio decreases. Thus, H. B. Reynolds reduced p_2 to a value as low as $0.13p_1$ and for air with a negligible velocity of approach found his data to be fitted accurately by the equation

$$W = \frac{0.545 A_o \, (p_1^2 - p_2^2)^{0.48}}{T_1^{1/2}}$$

where A_o = orifice area, in^2
    p = absolute pressure, psi

Also, J. A. Perry experimented with air with a negligible velocity of approach and carried p_2 down to a value as low as $0.05p_1$, at which pressure the flow was 13.7% more than at the critical pressure. His formula for the orifice is

$$W = 0.465 \, \frac{A_o \, p_1}{T_1^{1/2}} \left[1 - \left(\frac{p_2}{p_1} \right)^2 \right]^{1/2}$$

where A_o and p_1 can both involve either square feet or square inches.

Flow at critical pressure

The sonic or acoustic velocity for a perfect gas is

$$c = (gkpv)^{1/2} = (gkRT)^{1/2} \qquad (10.91)$$

Neglecting the velocity of approach,

$$\frac{V_2^2}{2g} = \frac{k}{k-1} \, (p_1 v_1 - p_2 v_2)$$

Substituting c from Eq. (10.91) for V_2, we obtain an expression

$$P_c = \frac{2}{k-1} \left(p_1 \frac{v_1}{v_c} - p_c \right)$$

where p_c is a special value of p_2 called the critical pressure and v_v is the corresponding specific volume.

$$\frac{v_1}{v_c} = \left(\frac{p_c}{p_1} \right)^{1/k}$$

and from this

$$P_c = \frac{2}{k-1} \left(P_c^{1/k} p_1^{(k-1)/k} - p_c \right)$$

This last expression may be rearranged to give

$$P_c = p_1 \left(\frac{2}{k+1} \right)^{k/(k-1)} \qquad (10.92)$$

For air and diatomic gases, $k = 1.4$ at usual temperatures, though it decreases with increasing temperature; for superheated steam k may be taken as about 1.3; and for wet steam k is about 1.13. As long as the velocity is attained in an expansion from p_1, p_c is $0.528p1$ to $0.58p_1$ for the gases and vapors mentioned. For $k = 1.4$, the value is $p_c = 0.528p_1$.

There are two restrictions concerning Eq. (10.92). However, for the practical cases where this expression for critical pressure is employed, the friction involved is very small, and the final result is corrected by a velocity coefficient anyway. The other factor is that V_1 was dropped in this derivation for the sake of simplicity. Using the steps used in deriving Eq. (10.92), an exact expression including velocity V_1 may be found to be

$$p_c^{(k-1)k} = \frac{2}{k+1} p_1^{(k-1)/k} + \frac{k-1}{k+1} \left(\frac{A_2}{A_1}\right)^2 \left[\frac{p_c^{(k+1)/k}}{p_1^{2/k}}\right]$$

which can be solved by trial.

An empirical expression involving velocity of approach in a very simple manner and one which is also surprisingly accurate has been presented by W. H. Church. It is

$$\frac{p_c}{p_1} = \left(\frac{2}{k+1}\right)^{k/(k-1)} + \frac{(D_2 D_1)^5}{5} \tag{10.93}$$

For venturi tubes and nozzles where the velocity in the throat is the acoustic velocity, W may be found by substituting p_c for p_2 in Eq. (10.90) and employing the value of Y for $p_2/p_1 = p_c/p_1$, if Y is known for the particular k value. If Y is not available for the particular k concerned, then Eq. (10.88) may be used by inserting the numerical value of p_c for p_2.

If the velocity of approach is negligible, which makes $A_2/A_1 = 0$ in Eq. (10.87), then, inserting the expression for p_c of Eq. (10.92) for p_2 in Eq. (10.87), the latter reduces to

$$V_2 = c = \left[2g \frac{k}{k+1} p_1 v_1\right]^{1/2} \tag{10.94}$$

for an ideal frictionless flow. In reality, there is a slight discrepancy here, for fluid friction does require that a velocity coefficient be introduced, and this in turn would make V_2 less than the acoustic velocity. The fact is that the velocity is not uniform across the throat diameter, and if the velocity in the main portion of the stream is acoustic, the small portion of the flow near the wall does not reach the acoustic value until a short distance downstream. However, this discrepancy is minor and is covered by the numerical value of a used in computing W, which is the quantity we are really interested in evaluating.

Equation (10.94) gives the ideal velocity at a venturi throat where the pressure is p_c and the specific volume is v_c. The flow is obtained by multiplying by a coefficient C, the throat area A_2, and specific weight w_c, or preferably dividing by specific volume $v_c = v_1 (p_1/p_c) (1/k)$ and replacing p_c in terms of p_1 by Eq. (10.92). The flow is then

$$W = CA_2 \left[2g \frac{p_1}{v_1} \frac{k}{k+1} \left(\frac{2}{k+1} \right)^{2/(k-1)} \right]^{1/2}$$ (10.95)

where C is a coefficient with the same value as for an incompressible fluid at the same Reynolds number and is normally of the order of 0.98.

Equations (10.91) and (10.94) show that for a perfect gas the acoustic velocity is independent of the pressure and depends only on the absolute temperature. Thus, at the throat the value of c is $(gkRT_2)^{1/2}$ or from Eq. (10.94) is $\{2g \ [k/(k+1)] \ RT_1\}^{1/2}$, from which, when the throat velocity is acoustic, $T_2 = T_1 \ 2/(k+1)$. But Eq. (10.95) shows that the rate of discharge is proportional to the initial pressure p_1 inasmuch as $p_1/v_1 = p_1^2/RT_1$.

It is obvious that for any specific gas with a given value of k, Eq. (10.95) may be greatly simplified. Thus, if $k = 1.4$, for example, the equation reduces to

$$W = C \frac{3.88 \, A_2 p_1}{(RT_1)^{1/2}}$$ (10.96)

Furthermore, for air with $R = 53.3$

$$W = C \frac{0.53 A_2 \, p_1}{T_1^{1/2}}$$ (10.97)

In these last two equations, the area units for both A_2 and p_1 must be the same and can be square inches and pounds per square inch, respectively, or they can both involve square feet. If these last three equations are used for an orifice, then the orifice area A_o replaces A_2, and an appropriate value of C must be used.

It must be remembered that Eqs. (10.94) to (10.97) are to be used only when the velocity of approach is negligible and when p_2 at the throat of a venturi tube or converging/diverging nozzle is reduced to a value p_c as obtained from Eq. (10.92). For an orifice p_2 is the pressure in the space into which the jet issues, no matter how low this pressure may become.

Supersonic velocity

Whenever the velocity of a gas or a vapor exceeds that of sound in the same medium, a decided change takes place in the flow pattern, and certain of the previous equations no longer hold.

The preceding subsection shows that the maximum possible actual velocity in the throat of a venturi tube is the acoustic velocity in the medium for the conditions existing at that section. In order that the velocity of a stream may exceed that of sound, it is necessary to turn from the orifice to a nozzle which converges and then diverges, which is essentially a venturi tube. If the terminal pressure is sufficiently low, the velocity in the throat will be that of a sound wave in the medium, whose pressure is p_c and whose other values are E_c, W_c, V_c, and T_c. Thus, the area of the throat determines the maximum capacity of the nozzle for a given initial p_1 and T_1 inasmuch as c is the maximum possible velocity at that section.

In the diverging portion of the nozzle, the gas or vapor can continue to expand to lower pressures, and the velocity can continue to increase. If the terminal area is sufficiently large, the fluid can expand completely to the terminal pressure, no matter how small the latter may become. The velocity of the jet so obtained will then be much greater than the acoustic velocity at the throat. Thus, for a jet from such a nozzle, Eqs. (10.85) and (10.87) give the jet velocity when the terminal pressure of the nozzle has any value whatever. For this case, the venturi throat pressure p_2 should be replaced by the venturi expansion outlet pressure p_3, which is the pressure to which the fluid expands at the end of the nozzle but is not necessarily the same as the pressure in the space into which the jet discharges. However, the rate of discharge is still limited by the throat conditions.

The venturi tube is seen to be a converging/diverging device and thus may enable supersonic velocities to be attained. If the pressure at the throat, point 2, is greater than P_c, a gas or a vapor will flow through the tube in much the same manner as a liquid. That is, the velocity will increase to a maximum value at the throat, which will be less, however, than the acoustic velocity; and then from points 2 to 3 the velocity will diminish, while the pressure increases, as shown by the hydraulic gradient ABD. The pressure at point 3, as represented by the height to D, is only slightly less than p_1 at A. If p_1 remains constant while the pressure at point 3 decreases a little, the hydraulic gradient ABD will change to a similar one but with slightly lower values. As the pressure at point 3 continues to decrease, p_1 remaining constant, the pressure at point 2 continues to diminish and the velocity to increase until the limiting acoustic velocity is reached, when the pressure gradient is ACE. If the pressure is further reduced to H, the pressure gradient is $ACFGH$, the jump from F to G being a pressure shock, similar to the hydraulic jump or standing wave often seen in open channels conveying water. As the pressure at point 3 continues to diminish, the terminal pressure finally becomes H''' so that the pressure history is $ACFH'''$. In contrast with conditions for flow below the acoustic velocity, the velocity in the last case continuously increases from point 1 to its maximum value at point 3, which is much greater than the acoustic velocity, while the pressure continuously decreases from points 1 to 3.

As long as p_1 remains unchanged, the value of p_c, and hence of the throat velocity, remains constant; thus, the rate of discharge through the venturi tube is unaffected by any further decrease in the pressure at point 3. The value of the pressure at point 3 merely determines the velocity that may be attained at that point and the necessary area of the terminal cross section.

If p_1 is increased, the acoustic velocity may be shown to remain unaltered, but, since the density of the gas is increased, the rate of disharge is greater. The orifice and the venturi tube are alike insofar as discharge capacity is concerned. The only difference is that with the venturi tube, or a converging/diverging nozzle, a supersonic velocity may be attained at discharge from the device, while with the orifice the acoustic velocity is the maximum value possible at any point.

It may also be added at this point that, when the velocity of a body through any fluid, whether a liquid or a gas, exceeds that of a sound wave in the same fluid, the flow conditions are entirely different from those for all velocities lower than this value. Thus, instead of streamlines the conditions might be as represented, which is a schlieren photograph of supersonic flow past a sharp-nosed model in a wind tunnel. It could also represent a projectile in flight through still air. A conical compression or shock wave extends backward from the tip, as may be seen by the strong density gradient revealed as a bright shadow in the photograph. A streamline in the undisturbed fluid is unaffected by the solid boundary or by a moving projectile until it intersects a shock-wave front, when it is abruptly changed in direction, proceeding roughly parallel to the nose form. Where the conical nose is joined to the cylindrical body of the model, dark shadows may be seen, representing rarefaction waves. The streamlines are again changed in direction through this region, becoming parallel to the main flow again. The reason why streamlines are unaffected in front of a projectile is that the body travels faster than the disturbance can be transmitted ahead.

Consider a point source of an infinitesimal disturbance moving with velocity V through a fluid. At the instant when this source passes through the point A_o the disturbance commences to radiate in all directions A with the velocity c of a sound wave in this medium. In successive instants the source passes through the points $A_1, A_2,$ and A_3, which last may represent the position at the instant of observation. While the source has covered the distance A_oA_3 with velocity V, the sound wave, traveling at the slower acoustic velocity c, has progressed only as far as radius A_oB_o. Similar termini of disturbances emanating from A_1 and A_2 form a straight-line envelope, which is the shock wave, The angle β is called the *Mach angle,* and it is seen that

$$\sin \beta = \frac{A_o B_o}{A_o A_3} = \frac{c}{V} \tag{10.98}$$

where the dimensionless velocity ratio V/c is the Mach number.

In the case of the finite projectile, the shock wave leaves the tip at an angle with the main flow which exceeds the Mach angle, on account of the conical nose which follows the tip. Appropriate corrections may be applied, however, and the shock-wave angle from such a sharp-nosed object remains an accurate means of measuring supersonic velocities.

Example 10.8   Air enters a converging/diverging nozzle (venturi) at a pressure of 120 psia and a temperature of 90°F. Neglecting the entrance velocity, and assuming a frictionless process, find the Mach number at the cross section where the pressure is 35 psia.

solution

$p_1 = 120$ psia $= 17,280$ psfa
$p_2 = 35$ psia $= 5040$ psfa
$T_1 = 90 + 460 = 550°$ abs
$V_1 = 0$
$N_M = \dfrac{V_2}{c}$

From the subsection "Flow Measurement of Compressible Fluids" [following Eq. (10.86)]

$$T_2 = T_1 \left(\frac{p_2}{p_1} \right)^{(k-1)/k}$$

From air, $k = 1.4$, $c_p = 0.24$ Btu/lb · °F. Thus

$$T_2 = 550 \left(\frac{35}{120} \right)^{(1.4-1)/1.4}$$

$$= 550 \times 0.292^{0.286} = 550 \times 0.703 = 387°$$

Then

$$V_2^2 = 50,000 \times 0.24 \times (550 - 387) = 1,956,000$$

$$V_2 = 1397 \text{ ft/s}$$

To find c, from Eq. (10.91):

$$c = (gkRT_2)^{1/2}$$

$$c = (32.2 \times 1.4 \times 53.3 \times 387)^{1/2} = 965 \text{ ft/s}$$

Thus

$$N_M = \frac{V_2}{c} = \frac{1397}{965} = 1.45$$

Uniform Flow in Open Channels

Open channels

An open channel is one in which the stream is not completely enclosed by solid boundaries and therefore has a free surface subjected only to atmospheric pressure. The flow in such a channel is caused not by some external head but rather by the slope of the channel and of the water surface.

The principal types of open channels are natural streams or rivers; artificial canals; and sewers, tunnels, or pipelines not completely filled. Artificial canals may be built to convey water for purposes of water power development, irrigation or city water supply, drainage or flood control, and numerous others. While there are examples of open channels carrying liquids other than water, there exist few experimental data for such, and the numerical coefficients given here apply only to water at natural temperatures.

The accurate solution of problems of flow in open channels is much more difficult than in the case of pressure pipes. Not only are reliable experimental data more difficult to secure, but there is a wider range of conditions than is met with in the case of pipes. Practically all pipes are round, but the cross sections of open channels may be of any shape from circular to the irregular forms of natural streams. In pipes the degree of roughness ordinarily ranges from that of new, smooth metal or wood stave pipes on the one hand to that of old, corroded iron or steel pipes on the other. But with open channels the surfaces vary from smooth timber to the rough or irregular beds of some rivers. Hence the choice of friction coefficients is attended with greater uncertainty in the case of open channels than in the case of pipes.

Uniform flow was described earlier as it applies to hydraulic phenomena in general. In the case of open channels, uniform flow means that the water cross section and depth remain constant over a certain reach of the channel. This requires that the drop in potential energy owing to the fall in elevation along the channel be exactly consumed by the energy dissipation through boundary friction and turbulence.

Uniform flow will eventually be established in any channel that continues sufficiently far with a constant slope and cross section. This may be stated in another way, as follows: For any channel of given roughness, cross section, and slope, there exists for a given flow one and only one water depth y_0 at which the flow will be uniform. Let A represent the water depth at the slope top. Let B represent the water depth at the point where the slope starts downward off the top. Let points C and D represent the water flow at uniform depth and flow down the slope. Let point E represent the beginning of stabilized flow after reaching a horizontal plane at the slope bottom. Thus, the flow is accelerating in the reach from A to C, becomes established as uniform from C to D, suffers a violent deceleration owing to the change of slope between D and E, and finally approaches a new depth of uniform flow somewhere beyond E.

The Reynolds number does not ordinarily figure prominently in open-channel flow, as the viscous forces are generally far outweighed by roughness considerations. When the occasion calls for it, however, it is usually defined so as to be compatible with that for pipes, or $N_R = 4RV/I_v$, where R is the hydraulic radius.

Hydraulic gradient

It is quite evident that in the case of an open channel the hydraulic gradient coincides with the water surface, because if a piezometer tube is attached to the side of the channel, the water will rise in it until its surface is level with that of the water in the channel.

By hydraulic slope is meant the value of S, which is given by

$$S = \frac{h_f}{L} \tag{10.99}$$

where L is measured along the channel (not the horizontal). Thus S is the slope of the energy gradient. It applies both to closed pipes and open channels. For uniform flow in either pressure pipes or open channels, the velocity is constant along the length of the conduit, and therefore for this special case the energy gradient and the pressure gradient (or hydraulic gradient) are parallel. Consequently, for an open channel with uniform flow, S becomes equal to the slope of the water surface S_w.

In a closed pipe conveying fluid under pressure, there is no relation between the slope of the pipe and the slope of the hydraulic gradient. In uniform flow in an open channel, the water surface must be parallel to the bed of the stream, and consequently S is then also the slope of the bed, S_0; thus we arrive at the important relation, for uniform flow only, that

$$S = S_w = S_0.$$

Equation for uniform flow

A general equation for frictional resistance in a pressure conduit was developed. The same reasoning may now be applied to uniform flow with a free surface. Consider a sloped channel of water flowing over a constant slope of angle θ, which shows the short reach of length L between stations 1 and 2 of a channel in uniform flow with area A of the water section. As the flow is neither accelerating nor decelerating, we may consider the body of water contained in the reach to be in static equilibrium. Summing forces along the channel, the hydrostatic pressure forces F_1 and F_2 balance each other, as there is no change in the depth y between the stations.

The only force in the direction of motion is the gravity component, and this must be resisted by the boundary shear stress τ_0 acting over the area PL, where P is the wetted perimeter of the section. Thus

$$wAL \sin \theta = \tau_0 PL$$

But $\sin \theta = h_f/L = S$. Solving for τ_0, we have

$$\tau_0 = w \frac{A}{P} S = wRS \qquad (10.100)$$

where R is the hydraulic radius. Substituting the value of τ_0, the following is also true:

$$\tau_0 = C_f p \frac{V^2}{2} = wRS$$

This may be solved for V in terms of either the friction coefficient C_f or the conventional friction factor f to give

$$V = \left(\frac{2g}{C_f} RS\right)^{1/2} = \left(\frac{8g}{f} RS\right)^{1/2} \qquad (10.101)$$

Chézy formula

In 1775 Chézy proposed that the velocity in an open channel varied as $(RS)^{1/2}$, which led to the formula

$$V = C\,(RS)^{1/2} \qquad (10.102)$$

which is known by his name. It has been widely used both for open channels and for pipes under pressure. Comparing Eqs. (10.101) and (10.102), it is seen that $C = (8g/f)^{1/2}$. Despite the simplicity of Eq. (10.102), it has the distinct drawback that C is not a pure number but has the dimensions $L^{1/2}ST^{-1}$, requiring that values of C in metric units be converted before being used with English units in the rest of the formula.

As C and f are related, the same considerations that have been presented regarding the determination of a value for f apply also to C. For a small open channel with smooth sides, the problem of determining f or C is the same as in the case of a pipe. But most channels are relatively large compared with pipes, thus giving Reynolds numbers that are higher than those commonly encountered in pipes. Also, open channels are frequently rougher than pipes, especially in the case of natural streams. As the Reynolds number and the relative roughness both increase, the value of f becomes practically independent of N_R and depends only on the relative roughness. In 1897 Bazin published his formula for C, which, in view of the foregoing, is not greatly out of line with more modern theory. His formula, expressed in foot-pound-second units, is

$$C = \frac{157.6}{1 + m/R^{1/2}} \tag{10.103}$$

where m is a measure of the absolute roughness and varies from 0.11 for smooth cement or planed wood to 0.83 for rubble masonry and 2.36 for ordinary earth channels.

Manning formula

One of the best and most widely used formulas for open-channel flow is that of Robert Manning, who published it in 1890. Manning found from many tests that the value of C varied approximately as $R^{1/6}$, and others observed that the proportionality factor was very close to the reciprocal of n, the coefficient of roughness in the classical Kutter formula. The Kutter formula, which was for many years the most widely used of all open-channel formulas, is now of interest principally for its historical value and as an outstanding example of empirical hydraulics. This formula, which may be found in several handbooks, included terms to make C a function of S, based on some river flow data that were later proved to be in error. This led to the Manning formula, which has since spread to all parts of the world. In metric units, the Manning formula is

$$V = \frac{1}{n} R^{2/3} S^{1/2}$$

The dimensions of n are seen to be $TL^{-1/3}$. As it is unreasonable to suppose that the roughness coefficient should contain the dimension T, the Manning equation is more properly adjusted so as to contain $(g)^{1/2}$ in the numerator, thus yielding the dimension of $L^{1/6}$ for n. In order to avoid converting the numerical value of n for use with English units, the formula itself is changed so as to leave the value of n unaffected. Thus, in foot-pound-second units, the Manning formula is

$$V = \frac{1.486}{n} R^{2/3} S^{1/2} \tag{10.104}$$

in which 1.486 is the cube root of 3.281, the number of feet in a meter. Despite the dimensional difficulties of the Manning formula, which have long plagued those attempting to put all of fluid mechanics on a rational dimensionless basis, it continues to be popular because it is simple to use and reasonably accurate. As n is presumed to vary only with the channel roughness, no trial-and-error solutions are involved in its use. Representative values of n for various surfaces are given in Table 10.1.

It was learned that ε is a measure of the absolute roughness of the inside of a pipe. The question naturally arises as to whether ε and n may be functionally related to one another. Such a correlation may be accomplished by use of the Prandtl-Karman equation for rough pipes. Written in terms of R, recalling that for a circular pipe $D = 4R$,

$$\frac{1}{f^{1/2}} = 2 \log_{10}\left(14.8\,\frac{R}{\varepsilon}\right) \tag{10.105}$$

The Manning formula may be put in this form by comparing Eqs. (10.102) and (10.104) from which

$$C = \left(\frac{8g}{f}\right)^{1/2} = \left(\frac{1.486}{n}\right)R^{1/6}$$

TABLE 10.1   Values of n in Manning's Formulas

Nature of surface	n	
	Minimum	Maximum
Neat cement	0.010	0.013
Wood stave pipe	0.010	0.013
Plank flumes, planed	0.010	1.014
Vitrified sewer pipe	1.010	0.017
Metal flumes, smooth	0.011	0.015
Concrete, precast	0.011	0.013
Cement mortar	0.011	0.015
Plank flumes, unplaned	0.011	0.015
Common clay drainage tile	0.011	0.017
Concrete, monolithic	0.012	0.016
Brick with cement mortar	0.012	0.017
Cast iron	0.013	0.017
Cement rubble	0.017	0.030
Riveted steel	0.017	0.020
Canals and ditches, smooth earth	0.017	0.025
Metal flumes, corrugated	0.022	0.030
Canals		
Dredged in earth, smooth	0.025	0.033
In rock cuts, smooth	0.025	0.035
Rough beds and weeds on sides	0.025	0.040
Rock cuts, jagged and irregular	0.035	0.045
Natural streams		
Smoothest	0.025	0.033
Roughest	0.045	0.060
Very weedy	0.075	0.150

or

$$\frac{1}{(f)^{1/2}} = \frac{1.486R^{1/6}}{n(8g)^{1/2}}$$
(10.106)

Equating the right sides of Eqs. (10.105) and (10.106) provides the desired correlation.

Values of ε vs. n vary proportionately in the range of $\varepsilon = 0.01$ to 0.40 for the range of $n = 0.015$ to 0.03, respectively. Values of ε vs. n must also be regarded in light of the components making up the equation being plotted. The values of ε, for example, were originally determined for artificially roughened pipes of circular section. We have presumed to extend Eq. (10.105) to open channels of noncircular section with roughness often many times greater than in corresponding pipes. Thus it is not surprising to find a value of n of 0.35, for example, representing a roughness projection in a wide, shallow stream almost equal to the depth. Powell found that Eq. (10.105) did not fit experimental data taken by himself and others on channels of definite roughness. He proposed an equation that, reduced to the form of Eq. (10.105), is shown in Fig. 10.55 and is plotted as a dashed line for a value of $R = 0.2$ ft, which was near the middle of the experimental range observed. Powell's equation in this form is valid only for rectangular channels.

Despite the discrepancies in the correlation functions from different sources, the salient feature of the curves is at once apparent. A threefold increase in n (as between the practical values of 0.01 to 0.03) represents something between a hundredfold and a thousandfold increase in E. That is why the value of E can be in error by 100% or so without causing an error of more than a few percent in the velocity or rate of discharge.

Velocity distribution in open channels

It was noted that the velocity in a channel approaching a weir might be so badly distributed as to require a value of 1.3 to 2.2 for the kinetic energy correction factor. In unobstructed uniform channels, however, the velocity distribution not only is more uniform but is readily amenable to theoretical analysis. Vanonil has demonstrated that the Prandtl universal logarithmic velocity distribution law for pipes also applies to a two-dimensional open channel, i.e., one that is infinitely wide. This equation may be written

$$\frac{u - u_{\text{mas}}}{(gyS)^{1/2}} = \frac{2.3}{k} \log_{10} \frac{y'}{y}$$

in which y is the depth of water in the channel, u is the velocity at a distance y' from the channel bed, and K is the von Karman constant, having a value of about 0.40. This expression can be integrated over the depth to yield the more useful relation

$$u = V + \frac{1}{K} (gyS)^{1/2} \left(1 + 2.3 \log_{10} \frac{y'}{y}\right) \qquad (10.107)$$

which expresses the distribution law in terms of the mean velocity V. If this equation were plotted together with velocity measurements that are made on the centerline of a rectangular flume 2.77 ft wide and 0.59 ft deep, the filament whose velocity u is equal to V would lie at a distance of $0.632y$ beneath the surface.

Velocity measurements made in a trapezoidal canal, reported by O'Brien, yield the distribution contours, with the accompanying values of the correction factors for kinetic energy and momentum. The filament of maximum velocity is seen to lie beneath the surface, and the correction factors for kinetic energy and momentum are greater than in the corresponding case of pipe flow. Despite the added importance of these factors, however, the treatment in this section will follow the earlier procedure of assuming the values of α and β to be unity, unless stated otherwise. Any thoroughgoing analysis would, of course, have to take account of their true values.

The velocity distribution in a natural stream becomes important when it is desired to gauge the flow by using a current meter. The average velocity in any vertical section is usually found at about $0.62y$ below the surface. This result, gained from a great many river measurements, checks closely with the value of $0.632y$. A slightly better average may be found by taking the mean of the velocities at $0.2y$ and $0.8y$.

It is interesting to note that for a logarithmic velocity distribution the average of the velocities at $0.2y$ and $0.8y$ is exactly equal to that at $0.6y$. The surface velocity at the center is ordinarily about 10% above the average velocity, but it may be greatly affected by wind and is not considered an accurate indicator.

Most efficient cross section

Any of the open-channel formulas given here show that, for a given slope and roughness, the velocity increases with the hydraulic radius. Therefore, for a cross section of a given area of water, the rate of discharge will be a maximum when R is a maximum, which is to say when the wetted perimeter is a minimum. Such a section is called the *most efficient cross section* for the given area. Or for a given rate of discharge the cross-sectional area will be a minimum when the design is such as to make R a maximum (and thus P a minimum). This section would be the most efficient cross section for the given rate of discharge.

Of all geometric figures, the circle has the least perimeter for a given area. Hence a semicircular open channel will discharge more water than one of any other shape, assuming that the area, slope, and surface roughness are the same. Semicircular open channels are often built of pressed steel and other forms of metal, but for other types of construction such a shape is impractical. For wooden flumes the rectangular shape is usually employed. Canals excavated in earth must have a trapezoidal cross section, with side slopes less than the angle of repose of the bank material. Thus there are other factors besides hydraulic efficiency that determine the best cross section. For given side slopes and area, however, it is possible to determine the shape of the water cross section so that the wetted perimeter, and thus the cost of the channel grading and lining, will be minimized.

Consider a trapezoidal channel having side walls sloped outward from the channel bottom's flat floor. This slope angle ϕ is measured from a vertical line that begins at the channel's floor, for both side walls. Then:

$$A = By + y^2 \tan \phi \quad \text{and} \quad P = B + 2y \sec \phi$$

where B = floor width
y = water depth
$y \sec \phi$ = side wall dimension

Substituting for B in terms of A gives A

$$P = \frac{A}{y} - y \tan \phi + 2y \sec \phi$$

Differentiating with respect to y and equating to zero,

$$\frac{A}{y^2} = \frac{By + y^2 \tan \phi}{y^2} = 2 \sec \phi - \tan \phi$$

From which

$$B = 2y(\sec \phi - \tan \phi) \tag{10.108}$$

It will be noticed that, for the rectangular channel ($\phi = 0$), the most efficient section will have a width of twice the depth.

The hydraulic radius of any cross section of greatest efficiency may now be evaluated from

$$R = \frac{A}{P} = \frac{y^2 (2 \sec \phi - \tan \phi)}{2y(2 \sec \phi - \tan \phi)} = \frac{y}{2} \tag{10.109}$$

To find the side slope giving the greatest possible efficiency for a trapezoidal section, we set P as a function of A (which is constant) and the variable ϕ, then differentiate and equate to zero. From the preceding relations for most efficient section, $P = 2[A(2 \sec \phi - \tan \phi)]^{1/2}$. The differentiation is simpler if we first square P; then

$$\frac{dP^2}{d\phi} = 4A (2 \tan \phi \sec \phi - \sec^2 \phi) = 0$$

from which we can see that $\sec \phi = 2 \tan \phi$, $\sin \phi = \frac{1}{2}$, and $\phi = 30°$. The second derivative is positive, indicating that this is a true minimum. This, together with the relation of Eq. (10.108), demonstrates that the most efficient of all trapezoidal sections is the half-hexagon. A simicircle having its center at the middle of the water surface can always be inscribed within a cross section of greatest efficiency. This may be proved for the trapezoidal section by making lines *OM, ON,* and *OQ* from the t point 0 at the center of the water surface perpendicular to the sides of the channel. Letting the lengths x, y, and z be the lengths of these perpendicular lines, the area and wetted perimeter are given by

$$A = xz + \tfrac{1}{2}yB \qquad \text{and} \qquad P = 2z + B$$

As $R = A/P = y/2$ for the most efficient section, these equations may be solved to give $x = y$.

Circular sections not flowing full

Frequently the civil engineer encounters flow in circular sections, as in sewers, which are not under pressure and must therefore be treated as open channels. The maximum rate of discharge in such a section occurs at slightly less than full depth. Let the following variables represent this analysis:

D = diameter of circular section
r = radius of circular section
y = depth of water in circular section
$2\theta = 360°$ – angle of water surface width in section (measured at
    radius center of section)

Thus:

$$A = \frac{D^2}{4} (\theta - \sin \theta \cos \theta) = \frac{D^2}{4} (\theta - \tfrac{1}{2} \sin 2\theta)$$

$$P = D\theta$$

where θ is expressed in radians. This gives

$$R = \frac{A}{P} = \frac{D}{4} \frac{1 - (\sin\theta\cos\theta)}{\theta} = \frac{D}{4}\left(1 - \frac{\sin\theta}{2\theta}\right)$$

To find the maximum rate of discharge, as by the Manning formula, we must have $AR^{2/3}$ at a maximum. This expression may be established from the preceding relations, then differentiated with respect to θ, set equal to zero, and solved, giving $\theta = 151.15°$. The corresponding value of the depth y is $0.938\,D$. The maximum velocity will be found to occur at $0.81\,D$. (This derivation is based upon a roughness coefficient that remains constant as the depth changes. Actually the value of n has been shown to increase by as much as 28% from full to about one-quarter full depth, where it appears to be a maximum. This effect causes the actual maximum discharge and velocity to occur at water depths of about 0.97 and 0.83 full depth, respectively. Despite the foregoing analysis, circular sections are usually designed to carry the design capacity flowing full, as the conditions producing maximum flow would frequently include sufficient backwater to place the conduit under slight pressure.

The rectangle, trapezoid, and circle are the simplest geometric sections from the standpoint of hydraulics, but other forms of cross section are often used, either because they have certain advantages in construction or because they are desirable from other standpoints. Thus oval- or egg-shaped sections are common for sewers and similar channels where there may be large fluctuations in the rate of discharge. It is desirable that, when a small quantity is flowing, the velocity be kept high enough to prevent the deposit of sediment, and that, when the conduit is full, the velocity should not be so high as to cause excessive wear of the lining.

Specific energy and alternate depths of flow

Totally separate from the concept of energy gradient or energy difference between two sections of a stream is the matter of the energy at a single section with reference to the channel bed. This is called the *specific energy* at that section, and its value is given by the following equation:

$$E = y + V^2 \tag{10.110}$$

If q denotes the flow per unit width of a wide rectangular channel, then $V = q/y$ and

$$E = y + \frac{q^2}{2gy^2} \tag{10.111}$$

or

$$q = y[2g\,(E - y)]^{1/2} \tag{10.112}$$

Plotting Eq. (10.112) in dimensionless form as y/E vs. q/q_{max} will show that the maximum discharge for a given specific energy occurs when the depth is between 0.6 and 0.7E. This may be established more exactly by differentiating Eq. (10.112) with respect to y and equating to zero. Thus

$$\frac{dq}{dy} = (2g)^2 \left[(E - y)^{1/2} - \frac{1}{2}\frac{y}{(E-y)^{1/2}} \right] = 0$$

from which

$$Y_c = \frac{2}{3E} \tag{10.113}$$

where Y_c is the critical depth for the given specific energy. The maximum rate of discharge for a given specific energy may now be determined by substituting from Eq. (10.113) into Eq. (10.112);

$$q_{max} = g^{1/2} \left(\frac{2}{3}\,E \right)^{3/2} = (gy_c^3)^{1/2} \tag{10.114}$$

Plotting Eq. (10.112) will also show that any rate of discharge less than the maximum can occur at two different depths for a given value of specific energy. If the flow is on the upper limb of the curve, it is said to be *upper-stage* or *tranquil;* if on the lower limb, it is called *lower-stage* or *shooting.*

The velocity and rate of discharge occurring at the critical depth are termed V_c and $q_c = q_{max}$, the critical velocity and flow, respectively. On account of the greater area, the velocity of upper-stage flow is slower than the critical and is called subcritical velocity; likewise, supercritical velocity occurs at lower-stage conditions. Combining Eqs. (10.110) and (10.113) yields a simple expression for critical velocity,

$$\frac{V_c^2}{2g} = \frac{y_c}{2} \quad \text{or} \quad V_c = (gy_c)^{1/2} \tag{10.115}$$

Thus a condition of upper- or lower-stage flow may readily be tested by determining whether the velocity head is less than or greater than half the depth, respectively. This critical velocity bears no relation to the other critical velocity, which separates laminar from turbulent flow, but unfortunately custom has given us a duplication of names.

All the foregoing treatment of alternate depths has been based on a specific energy that has been assumed constant while different depths and rates of discharge were considered. The case may now be put differently, and we may ask about the possible depths corresponding to different specific energies for a given rate of discharge. Equation (10.111) may be plotted on a specific energy diagram for three successively increasing constant rates of discharge per unit width, q, q', and q''. In this diagram the critical depth appears as the depth of minimum specific energy for a given flow. As this has the same meaning as the depth producing maximum discharge for a given specific energy, y_c may be evaluated from Eq. (10.114) for any value of q,

$$y_c = \left(\frac{q^2}{g}\right)^{1/3} \tag{10.116}$$

This result may be obtained independently by differentiating Eq. (10.111) with respect to y and equating to zero. It may be observed that the depth, which may be plotted vertically to determine the curve, is also represented by the horizontal distance from the vertical axis to the 45° line. It is also seen that the upper limb of such a curve corresponds to subcritical flow, while the lower limb refers to the alternate condition of supercritical flow.

We may summarize much of the foregoing into some axioms of open-channel flow related to conditions at a given section in a rectangular channel:

1. A flow condition, i.e., a certain rate of discharge flowing at a certain depth, is completely specified by any two of the variables y, q, V, and E, except the combination q and E, which yields in general two alternate stages of flow.

2. For any value of E there exists a critical depth, given by Eq. (10.113), for which the flow is a maximum.

3. For any value of q there exists a critical depth, given by Eq. 8 (10.106), for which the specific energy is a minimum.

4. When flow occurs at critical depth, both Eqs. (10.113) and (10.115) are satisfied and the velocity head is one-half the depth.

5. For any flow condition other than critical, there exists an alternate stage at which the same rate of discharge is carried by the same specific energy. The alternate depth may be found from either the discharge curve or the specific-energy diagram by extending a vertical line to the alternate limb of the curve. Analytically, the alternate depth is found by solving Eq. (10.111), which may be reduced from a cubic to a quadratic equation by dividing by $y - y_1$, where y_1 is the alternate depth which is known.

Channel slope and alternate depths of flow

It will now be emphasized again that uniform flow occurs at a depth which depends only on the rate of discharge, the shape and roughness of the cross section, and the slope of the stream bed. If, for a given roughness and shape, the channel slope is such that the given flow is tranquil, the slope is said to be *mild*. If the flow is supercritical, or shooting, the slope is termed *steep*. Thus, the hydraulic steepness of a channel slope is determined by more than its elevation gradient. A steep slope for a channel with a smooth lining could be a mild slope for the same flow in a channel with a rough lining. Even for a given channel, the slope may be mild for a low rate of discharge and steep for a higher one.

The slope of a channel which will just sustain the given rate of discharge in uniform flow at critical depth is termed the *critical* slope B_e. Evidently, for a given cross section and rate of discharge, the uniform flow is upper-stage if $S < S_e$ and lower-stage if $S > S_c$. When the flow is near critical, a small change in specific energy results in a large change in depth. Because of the undulating stream surface with flow at or near critical depth, it is undesirable to design channels with slopes near the critical.

Critical depth in nonrectangular channels

For simplicity of explanation, the treatment of critical depth, is confined to wide rectangular channels. We shall now consider an irregular section (no measurable channel walls or smooth channel surfaces) of area A carrying a flow Q. Thus, Eq. (10.111) becomes

$$E = y + \frac{Q^2}{2gA^2} \tag{10.117}$$

Differentiating with respect to y,

$$\frac{dE}{dy} = 1 - \frac{Q^2}{2g}\left(\frac{2}{A^3}\frac{dA}{dy}\right)$$

This may now be set equal to zero and solved for the value of the critical depth for the given flow. As A may or may not be a reasonable function of y, it is helpful to observe that $dA = B\,dy$ and thus $dA/dy = B$, the width of the water surface. Substituting this in the preceding expression results in

$$\frac{Q^2}{g} = \left(\frac{A^3}{B}\right)_{y=yc} \tag{10.118}$$

as the equation which must be satisfied for critical flow. For a given cross section the right-hand side is a function of y only. A trial-and-error solution is generally required to find the value of y which satisfies Eq. (10.118). We may next solve for V, the critical velocity in the irregular channel I, by observing that $Q = AV$. Substituting this in Eq. (10.118) yields

$$\frac{V_c^2}{g} = \frac{A_c}{B_c} \quad \text{or} \quad V_c = \left(\frac{gA_c}{B_c}\right)^{1/2} \tag{10.119}$$

If the channel is rectangular, $A_c = B_c y_c$, and the preceding is seen to reduce to Eq. (10.115).

It has already been pointed out that the cross section most commonly encountered in open-channel hydraulics is not rectangular but trapezoidal. As repeated trial-and-error solutions of Eq. (10.119) become very tedious, practicing hydraulic engineers avail themselves of numerous tables and curves which have been prepared for finding the critical depth in trapezoidal channels of any bottom width and side slopes.

Nonuniform Flow in Open Channels

Occurrence of nonuniform flow

As a rule, uniform flow is found only in artificial channels of constant shape and slope, although even under these conditions the flow for some distance may be nonuniform. But with a natural stream the slope of the bed and the shape and size of the cross section usually vary to such an extent that true uniform flow is rare. Hence the application of the equations given to natural streams can be expected to yield results that are only approximations of the truth. In order to apply these equations at all, the stream must be divided into lengths within which the conditions are approximately the same.

In the case of artificial channels that are free from the irregularities found in natural streams, it is possible to apply analytical methods to the various problems of nonuniform flow. In many instances, however, the formulas developed are merely approximations, and we must often resort to trial solutions and even purely empirical methods. For the treatment of many types of flow, see Bakhmeteff [35].

In the case of pressure conduits, we have dealt with uniform and nonuniform flow without drawing much distinction between them. This can be done because in a closed pipe the area of the water section, and hence the mean velocity, is fixed at every point. But in an open channel these conditions are not fixed, and the stream adjusts itself to the size of cross section that the slope of the hydraulic gradient requires.

Energy equation for gradually varied flow

The principal forces involved in flow in an open channel are inertia, gravity, hydrostatic force due to change in depth, and friction. The first three represent the useful kinetic and potential energies of the liquid, while the fourth dissipates useful energy into the useless kinetic energy of turbulence and eventually into heat owing to the action of viscosity. To compose an energy relation of nonuniform flow, assume the following variables:

E = energy of flow, $y + \alpha(V^2/2g)$
$H = E + z$
S = overall energy gradient slope from point 1 to point 2
S_w = hydraulic gradient slope from point 1 to point 2
V = velocity of fluid flow downslope
y = liquid depth at any point, measured from bed to surface
z = dimension from bed at point 1 to a datum reference point
α = angle of bed slope to horizontal datum line

The total energy of the elementary volume of liquid is proportional to

$$H = z + y + \alpha \frac{V^2}{2g} \tag{10.120}$$

where $z + y$ is the "potential energy" above the arbitrary datum, and $\alpha(V^2/2g)$ is the kinetic energy, with V being the mean velocity in the section. Once again, the depth y does not represent a real energy possessed by every particle in the volume element. It exists in the equation for the same reason that the term p/w exists in the standard Bernoulli equation: because we are concerned with a continuous-flow process. Quite evidently, the only true potential energy possessed by any particle is its total height above the datum.

The value of α will generally be found to be higher in open channels than in pipes. It may range from 1.05 to 1.40, and, in the case of a channel with an obstruction, the value of α just upstream may be as high as 2.00 or even more. As the value of α is not known unless the velocity distribution is determined, it is often omitted, but an effort should be made to employ it if accuracy is necessary. Differentiating Eq. (10.120) with respect to x, the distance along the channel, the rate of energy dissipation is found to be (with $\alpha = 1$)

$$\frac{dH}{dx} = \frac{dz}{dx} + \frac{dy}{dx} + \frac{1}{2g}\frac{d(V^2)}{dx} \tag{10.121}$$

The slope of the energy gradient is defined as $S = -dH/dx$, while the slope of the channel bed $S_0 = -dz/dx$, and the slope of the hydraulic gradient or water surface is given by $S_w = -dz/dx - dy/dx$.

In an open stream on a falling grade the effect of gravity is to tend to produce a flow with a continually increasing velocity along the path, as in the case of a freely falling body. The gravity force is opposed by the frictional resistance. The frictional force increases with velocity, while gravity is constant; so eventually the two will be in balance, and uniform flow will occur. When the two forces are not in balance, the flow is nonuniform. There are two types of nonuniform flow. In one the changing conditions extend over a long distance, and this may be called *gradually varied flow*. In the other the change may take place very abruptly and the transition be confined to a short distance. This may be designated as a *local nonuniform phenomenon*. Gradually varied flow can occur at either the upper or lower stage, but the transition from one stage to the other is ordinarily abrupt. Other cases of local nonuniform flow occur at the entrance and exit of a channel, at changes in cross section, at bends, and at obstructions such as dams, weirs, or bridge piers.

The energy equation between two sections (1) and (2) a distance L apart is

$$z_1 + y_1 + \alpha_1 \frac{V_1^2}{2g} = z_2 + y_2 + \alpha_2 \frac{V_2^2}{2g} + h_f \qquad (10.122)$$

As $z_1 - z_2 = S_0 L$ and $h_f = SL$, the energy equation may also be written in the form (with $\alpha_1 = \alpha_2 = 1$)

$$y_1 + \frac{V_1^2}{2g} = y_2 + \frac{V_2^2}{2g} + (S - S_0) L \qquad (10.123)$$

The Manning equation for uniform flow, Eq. (10.125), can be applied to nonuniform flow with an accuracy that is dependent on the length of the reach taken. Thus a long stream will be divided into several reaches, such that the change in depth is roughly the same within each reach. Then, within a reach, the Manning formula gives

$$S = \left(\frac{n V_m}{1.486 R_m^{2/3}} \right)^2 \qquad (10.124)$$

where V_m and R_m are the means of the respective values at the two ends of the reach. With S_0 and n known and the depth and velocity at one end of the reach given, the length L to the end corresponding to the other depth can be computed from Eq. (10.123).

Critical velocity and celerity of gravity waves

A channel laid on a certain slope will carry a certain rate of discharge at critical velocity in uniform flow. The concepts of critical velocity and depth and alternate stages of flow, however, are of greater significance in cases of nonuniform flow because of the relation between the critical velocity and the celerity or velocity of propagation of the gravity wave.

Consider a solitary wave, progressing to the left in an open channel, with celerity (wave velocity) c. We may replace this situation with the equivalent steady-flow case in which the wave stands still while the flow enters at velocity $V_1 = -c$. Writing the energy equation, Eq. (10.122), between points 1 and 2 (with $z_1 = z_2$, $\alpha_1 = \alpha_2 = 1$, and neglecting friction), and keeping the same variable definitions, we have

$$y_1 + \frac{V_1^2}{2g} = y_2 + \frac{V_2^2}{2g}$$

By continuity, $y_1 V_1 = y_2 V_2$. Substituting for V_2 and rearranging terms results in

$$\frac{V_1^2}{2g} = \frac{y_2 - y_1}{1 - (y_1/y_2)^2}$$

If we now let $y_1 = y$, the undisturbed depth, and $y_2 = y + \Delta y$, where Δy is the wave height, and drop terms of order higher than the first, we have for the wave celerity approximately

$$C = (gy)^{1/2} \left(1 + \frac{3}{2} \frac{\Delta y}{y}\right)^{1/2} \approx (gy)^{1/2} \left(1 + \frac{3}{4} \frac{\Delta y}{y}\right) \qquad (10.125)$$

Equation (10.125) is valid only for waves in shallow water, i.e., for waves of great length and moderate amplitude relative to their depth. For so-called deep-water waves, as might be encountered in the ocean, for example, but still presuming small amplitudes, a more accurate equation is

$$C = \left(\frac{g\lambda}{2\pi} \tanh \frac{2\pi y}{\lambda}\right)^{1/2} \qquad (10.126)$$

where λ is the wavelength from crest to crest of a wave train. For the background of this equation, and further information on waves in open channels, see Rouse [36].

As the wave height y in Eq. (10.125) becomes small compared to the depth, or the wavelength λ in Eq. (10.126) becomes great compared to the depth, either of these equations reduces to

$$C = (gy)^{1/2} \qquad (10.127)$$

which is the same as the critical velocity given by Eq. (10.115) in terms of the critical depth y_c. From Eq. (10.127) it is seen that the celerity of a wave increases as the depth of the water increases. But this is the celerity relative to the water. If the water is flowing, the absolute speed of travel will be the resultant of the two velocities. When the stream is flowing at its critical depth y_c, the stream velocity and the wave celerity will be equal. This means that when the surface is disturbed due to any cause, the wave so produced *cannot travel upstream*. Quite evidently this also applies to any portion of the stream in supercritical flow, at depths less than y_c. If the disturbance is a permanent one, such as produced by an obstruction or a change in the channel, the wave remains stationary and is therefore called a *standing wave*. When the depth is greater than y_c, and consequently the velocity of flow is less than the critical, the wave celerity c is then greater than critical, owing to the increase in y. Consequently, any surface disturbance will be able to travel upstream against the flow. Hence the entire stream picture is dependent on whether the stream velocity is smaller or greater than the critical velocity. The situation is analogous to that of the acoustic velocity. The standing wave that exists because of a permanent disturbance, when the flow velocity is above the critical, will be at such a direction that $\sin \beta = c/V = (gy)^{1/2}/V$, where β is the angle between the direction of flow and the wave front [37].

Surface curvature in gradually varied flow

As there are some 12 different circumstances giving rise to as many different fundamental types of varied flow, it is helpful to have a logical scheme of type classification. In general, any problem of varied flow, no matter how complex it may appear, with the stream passing over dams, under sluice gates, down steep chutes, on the level, or even on an upgrade, can be broken down into reaches such that the flow within any reach either is uniform or falls within one of the given nonuniform classifications. The stream is then analyzed one reach at a time, proceeding from one to the next until the desired result is obtained.

For simplicity, the following treatment is based on channels of rectangular section. The section will be considered sufficiently wide and shallow that we may confine our attention to a section 1 ft wide, through which the velocity is essentially uniform. It is important to bear in mind that all the following development is based on a constant value of the discharge per unit width q and on one value of the roughness coefficient n. Commencing with Eq. (10.121), we may observe that, as $V = q/y$,

$$\frac{1}{2g} \frac{d(V^2)}{dx} = \frac{1}{2g} \frac{d}{dx}\left(\frac{q^2}{y^2}\right) = -\frac{q^2}{g} \frac{1}{y^3} \frac{dy}{dx}$$

Substituting this, plus the S and S_0 terms derived earlier, in Eq. (10.121) yields

$$-S = -S_0 + \frac{dy}{dx}\left(1 - \frac{q^2}{gy^3}\right); \qquad \frac{dy}{dx} = \frac{S_0 - S}{1 - (q^2/gy^3)} = \frac{S_0 - S}{1 - (V^2/gy)} \qquad (10.128)$$

Evidently, if the value of dy/dx as determined by Eq. (10.128) is positive, the water depth will be increasing along the channel; if negative, it will be decreasing. Looking first at the numerator, S may be considered as the slope, such as would be obtained from Eq. (10.124), which would carry the given discharge at depth y with uniform flow. Let y_0 denote the depth for uniform flow on the bed slope S_0. Then, by Eq. (10.125), written for the unit width flow,

$$q = AV = y\,\frac{1.486}{n}\,y^{2/3}S^{1/2} = y_0\,\frac{1.486}{n}\,y_0^{2/3}\,S_0^{1/2}$$

This demonstrates that, for constant q and n, $S/S_0 = (y_0/y)^{10/3}$ and, consequently, for $y > y_0$, $S < S_0$ and the numerator is positive. Conversely, for $y < y_0$, $S > S_0$, and the numerator is negative.

To investigate the denominator of Eq. (10.128) we observe that, for critical flow, $y_c = V_c^2/g$ by Eq. (10.126). When the depth is greater than critical, the denominator is positive. Conversely, when $y < y_c$, the denominator is negative. The term V^2/gy is seen to have the dimensions of a Froude number squared. It is sometimes called the Froude number or the kinetic flow factor. Evidently, a Froude number greater than 1 corresponds to supercritical flow and a negative denominator, while $N_F < 1$ means tranquil flow and a positive denominator. The foregoing analyses have been combined into a series of backwater scenarios, as they are called, although the term *backwater* applies strictly only to those cases where an obstruction in the stream causes retarded flow. Reviewing the variables, y_0 = water depth from surface to bed, y_c = water depth from surface to weir bottom or overflow crest, and y = water depth from surface to sloped channel bottom. The surface profiles are classified according to slope and depth as follows: If S_0 is positive, the bed slope is termed *mild* (M) when $y_0 > y_c$, *critical* (C) when $y_0 = y_c$, and *steep* (S) when $y_0 < y_c$; if $S_0 = 0$, the channel is *horizontal* (H); and if S_0 is negative, the bed slope is called *adverse* (A). If the stream surface lies above both the normal (uniform flow) and critical depth lines, it is of type 1; if between these lines, it is of type 2; and if below both lines, it is of type 3. The 12 scenarios of surface curvature are labeled accordingly.

It may be noted further that, as even a hydraulically steep slope varies but a few degrees from the horizontal, it makes little difference whether the depth y is measured vertically or perpendicular to the bed.

It will be observed that some of the curves are concave upward, while others are concave downward. While the mathematical proof for this is not given, the physical explanation is not hard to find. In the case of the type 1 curves, the surface must approach a horizontal asymptote as the velocity is progressively slowed down owing to the increasing depth. Likewise, all curves that approach the normal or uniform depth line j

must approach it asymptotically, because uniform flow will prevail at sections sufficiently remote from disturbances, as we learned in the section on occurrence of nonuniform flow. The curves that cross the critical depth line must do so vertically, as the denominator of Eq. (10.128) becomes zero in this case, with a vertical water surface (actually indicating a limiting condition in which the assumptions underlying the theory become invalid). The critical slope curves, for which $y_0 = y_c$, constitute exceptions to both the foregoing statements, as it is not possible for a curve to be both tangent and perpendicular to the critical uniform depth line.

Many of the examples show a rapid change from a depth below the critical to a depth above the critical. This is a local phenomenon, known as the *hydraulic jump*, which will be discussed in detail following the examples of gradually varied flow.

The qualitative analysis of backwater scenarios has been restricted to rectangular sections of great width. These equations are, however, applicable to any channel of uniform cross section, if y_0 is the depth for uniform flow and y_c is the depth that satisfies Eq. (10.118). The surface profiles can even be used qualitatively in the analysis of natural stream surfaces as well, provided that local variations in slope, shape, roughness of cross section, etc., are taken into account. The step-by-step integration method for the solution of nonuniform flow problems is not restricted to uniform channels and is therefore suited to backwater computations for any stream whatever.

Examples of backwater scenarios

The M_1 scenario.   This is a mild slope case. The most common case of backwater is where the depth is already above the critical and is increased still further by a dam. Referring to the specific energy equation, Eq. (10.116), as the depth increases, the velocity diminishes without any abrupt transitions, so that a smooth water surface plane is obtained. In the case of flow in an artificial channel with a constant bed slope, the backwater curve would be asymptotic at infinity to the surface for uniform flow, as noted before. But the problems that are usually of more important interest are those concerned with the effect of a dam on a natural stream and the extent to which it raises the water surface at various points upstream.

For an artificial channel where the conditions are uniform, save for the variation in water depth, this problem may be solved by use of Eqs. (10.123) and (10.124). Usually the solution commences at the dam, where conditions are assumed to be known, and the lengths of successive reaches upstream, corresponding to assumed increments of depth, are computed. A tabular type of solution is the most helpful, with column headings corresponding to the various elements of Eqs. (10.123) and (10.124), the last column being ΣL, which sums up the length from the dam to the point in question. It is important, if accuracy is desired, to keep the depth increment small within any reach; a depth change of 10% or less is fairly satisfactory. The smaller the depth increment used in this step integration procedure, the greater the overall distance to some point of specified depth or elevation. When the depth increments are infinitesimally small, the effect is that of integrating the equation of nonuniform flow. This has been done numerically, and tables are available that greatly lessen the work of making repeated solutions of backwater curves [38].

Of course, this type of solution is not restricted to the Manning formula. Equation (10.124) may be replaced with a similar relation based on a constant value of the Chézy C or on a value that varies with R in accordance with Eq. (10.124).

For a natural stream, the solution is not so direct, because the form and dimensions of a cross section cannot be assumed and then the distance to its location computed. As there exist various slopes and cross sections at different distances upstream, the value of L in Eq. (10.123) must be assumed and then the depth of stream at this section can be computed by trial.

In view of the fact that for an irregular stream so many approximations must be made as to the actual conditions, the refinements of Eq. (10.123) are not always justified and it is frequently as satisfactory to apply a simple equation for uniform flow $V = C \, (RS)^{1/2}$. In order to do this, the stream must be divided into various reaches within which the flow may be assumed to be fairly uniform. Then, for each reach, average values V_m and R_m are used, and the value of S is determined by trial.

The M_2 scenario.   This is also a mild slope case, one in which the flow falls off of a smooth plane edge. This scenario, representing accelerated flow at the upper stage on a slope that is flatter than critical, exists, like the M_1 slope, because of a control condition downstream. In this case, however, the control is not an obstruction but the removal of the hydrostatic resistance of the water downstream, as in the case of the free overfall. As in the mild M_1 slope, the surface will approach the depth for uniform flow at an infinite distance upstream. Practically, because of slight wave action and other irregularities, the distinction between the M_2 or drop-down curve and the curve for uniform flow disappears within a finite distance. Since friction produces a constant diminution in energy in the direction of flow, it is obvious that at the point of outfall the total energy H must be less than at any point upstream. As the critical depth is the value for which the specific energy is a minimum, then this should apparently be the depth of the stream at this point. However, the value for the critical depth is derived on the assumptions that the water is flowing in straight lines, so that there is no centrifugal effect to alter the pressure over the section, and also that the pressure on the stream bed is equal to the hydrostatic depth. These are the conditions in many cases, but in the free fall, if the underside of the nappe is open to atmospheric pressure, then the pressure on the channel bottom at the outfall point will be atmospheric pressure and independent of the depth of the water. Owing to the convergence of the stream filaments between the upper and lower surfaces of the flow, however, the pressure within the nappe at the crest is greater than atmospheric.

When these changed conditions are considered, it is found that the true energy can be even less than the minimum value found for the preceding case, and thus the depth at the end is not y_c but is smaller than y_c. By treating the free overfall as a sharp-crested weir of zero height, it has been shown that

$$y_c = 0.715\, y_c \qquad (10.129)$$

in the case of a horizontal stream bed [39]. Certain experiments have yielded results that agree very closely with this value and furthermore show that the normal critical depth y_c is not attained until the distance upstream is between 4 and 12 times y_c [39, 40]. The free overfall then becomes an easy means of measuring the discharge with good accuracy from a laboratory flume.

If, instead of a free overfall, the conduit discharges into water whose surface is higher than the level, then of course the surface at discharge must rise to that value (neglecting losses), but it can never be lower than y_c.

The M₃ scenario.   This occurs because of an upstream control, as by the sluice gate. The bed slope is not sufficient to sustain lower-stage flow, and, at a certain point determined by energy and momentum relations, the water surface will pass through a hydraulic jump to upper-stage flow unless this is made unnecessary by the existence of a free overfall before the M_3 crest reaches critical depth.

The S scenarios.   These are steep slope cases. They may be analyzed in much the same fashion as the M scenarios, having due regard for downstream control in the case of upper-stage flow and upstream control for lower-stage flow. Thus a dam or an obstruction on a steep slope produces an S_1 scenario, which approaches the horizontal asymptotically but cannot so approach the uniform depth line, which lies below the critical depth. Therefore this curve must be preceded by a hydraulic jump. The S_2 scenario shows accelerated lower-stage flow, smoothly approaching uniform depth. Such a curve will occur whenever a steep channel receives flow at critical depth, as from an obstruction (as shown) or reservoir. The sluice gate on a steep channel will produce the S_3 scenario, which also approaches smoothly the uniform depth line.

The C scenarios.   These scenarios, with the anomalous condition at $y_e = y_0$, have already been discussed. Needless to say, the critical slope profiles are not of frequent occurrence.

The H and A scenarios.   These scenarios have in common the fact that there is no condition of uniform flow possible. The smooth-plane fall drop-down curves are similar to the M_2 scenario, but even more noticeable. The value of y_e given in Eq. (10.129) applies strictly only to the smooth-plane scenario, but is approximately true for the M_2 scenario also. The sluice gate on the horizontal and adverse slopes produces H_3 and A_3 scenarios that are like the sluice gate scenario but cannot exist for as long as the M_3 scenario before a hydraulic jump must occur. Of course it is not possible to have a channel of any appreciable length carry water on a horizontal or an adverse grade.

The hydraulic jump

By far the most important of the local nonuniform flow phenomena is that which occurs when supercritical flow has its velocity reduced to subcritical. We have seen in these example scenarios that there is no ordinary means of changing from lower- to upper-stage flow with a smooth transition, because the theory calls for a vertical slope of the water surface. The result, then, is a marked discontinuity in the surface, characterized by a steep upward slope of the profile, broken throughout with violent turbulence, and known universally as the *hydraulic jump*.

The specific reason for the occurrence of a hydraulic jump can perhaps best be explained by reference to the M_3 scenario. The mild slope is insufficient to sustain the lower-stage flow that has been imposed by the sluice gate. The specific energy is decreasing as the depth is increasing (proceeding to the left along the lower limb of the specific energy diagram). Were this condition to progress until the flow reached critical depth, with minimum specific energy for the given flow, an increase in specific energy would be required as the depth increased from the critical to the uniform upper-stage flow depth downstream. But this is a physical impossibility, as we know that for any $y < y_0$ the energy gradient is falling faster than the bed slope. The hydraulic jump, therefore, forms before the critical depth is reached, conserving a portion of the excess specific energy and permitting flow at uniform depth to proceed directly from the jump.

The hydraulic jump can also occur from an upstream condition of uniform lower-stage flow to a nonuniform S_1 curve downstream when there is an obstruction on a steep slope, or again from a nonuniform upstream condition to a nonuniform downstream condition, as illustrated by the H_3-H_2 or the A_3-A_2 combinations. In addition to the foregoing cases, wherein the channel bed continues at a uniform slope, a jump will form when the slope changes from steep to mild, as on the apron at the base of the spillway from a dam. This is an excellent example of the jump serving a useful purpose, for it dissipates much of the destructive energy of the high-velocity water, thereby reducing downstream erosion. The same type of jump is seen to occur at the base of the free overfall shown in the scenarios previously discussed.

The equation for the height of the hydraulic jump will be derived for the case of a horizontal channel bottom (the H_3-H_2 combination). It is assumed that the friction forces acting are negligible because of the short length of channel involved and because the shock losses are large in comparison. The key to the solution is the law of conservation of momentum, which states that the time rate of change of momentum of a flowing stream must be equal to the components of all forces acting. As the channel is horizontal, the gravity forces will have no component in the direction of flow. Hence we have

$$\frac{w}{g}\, Q\, (V_2 - V_1) = w \int_{A1} h_1\, dA - w \int_{A2} h_2\, dA$$

where h_1 and h_2 are the variable distances from the surface to flow of area at points 1 and 2, defined as points in the downstream flow from a slope. Point 1 is at the lower end of the sloped stream. Point 2 is downstream, where the flow approaches increasing water depth. At point 2, the opposing force of the increasing depth equals the force of the flow, stopping the water flow. It is seen that each integral is equal to hA, the static moment of the area about an axis lying in the surface, where h is the vertical distance from the surface to the center of gravity of the area. Hence, after rearranging, we have

$$\frac{w}{g}\, QV_2 + wh_2 A_2 = \frac{w}{g}\, QV_1 + wh_1 A_1 \qquad (10.130)$$

That is, the momentum plus the pressure on the cross-sectional area is constant, or, dividing by w and observing that $V = Q/A$,

$$f_m = \frac{Q^2}{Ag} + Ah = \text{constant} \qquad (10.131)$$

In the case of a rectangular channel, this reduces to a unit width basis,

$$f_m = \frac{q^2}{y_1 g} + \frac{y_1^2}{2} = \frac{q^2}{y_2 g} + \frac{y_2^2}{2} \qquad (10.132)$$

As the loss of energy in the jump does not affect the "force" quantity f_m, the latter is the same after the jump as before and therefore any vertical line on the f_m curve serves to locate two conjugate depths y_1 and y_2.

Thus the line for the initial water level y_1 intersects the f_m curve at a, giving the value of f_m that must be the same after the jump. The vertical line ab then fixes the value of y_2. This depth is then transposed to the specific energy diagram to determine the value cd of $V_2^2/2g$. The value of $V_1^2/2g$ is the vertical distance ef, and the head loss caused by the jump is the drop in energy gradient from 1 to 2.

When the rate of flow and the depth before or after the jump are given, it is seen that to solve Eq. (10.132) for the other depth results in a cubic equation. This may readily be reduced to a quadratic, however, by observing that $y_2^2 - y_1^2 = (y_2 + y_1)(y_2 - y_1)$ and substituting the known depth in the resulting expression,

$$\frac{q^2}{g} = y_1 y_2 \frac{y_1 + y_2}{2} \qquad (10.133)$$

In the case of a hydraulic jump on a sloping channel, it is simply necessary to add the sine component of the weight of the water to the general momentum equation. Although there exist more refined methods for doing this [41], a good approximation may be obtained by assuming the jump section to be a trapezoid with bases y_1 and y_2 and altitude of about $6Y_2$.

The problem of determining where a hydraulic jump will occur is a combined application. In the case of supercritical flow on a mild slope, for instance, the tail water depth y_2 is determined by the uniform flow depth y_0 for that slope. The rate of flow and the application of Eq. (10.133) then fix y_1, and the length of the M_3 curve required to reach this depth from the upstream control may be computed from Eq. (10.123). Similarly, in the case of subcritical flow on a steep slope, the initial depth is equal to y_0, the tail water depth is given by Eq. (10.133), and the length of the S_1 curve to the jump from the downstream control is computed from Eq. (10.123). For application of the hydraulic jump to design problems, and for analysis of the jump in circular and other nonrectangular sections, the reader is referred to more extensive treatises on the subject [42].

Example 10.9   A wide rectangular channel, of slope $S = 0.0003$ and roughness $n = 0.020$, carries a steady flow of 50 ft^3/s per foot of width. If a sluice gate is so adjusted as to produce a depth of 1.5 ft in this channel, determine whether a hydraulic jump will form downstream, and, if so, find the distance if from the gate to the jump.

First find the normal depth y_0 for uniform flow. From Eq. (10.125), with $R = y$ for a wide channel,

$$q = Vy = \frac{1.486}{n} y^{2/3} S^{1/2} y$$

or

$$50 = \frac{1.486}{0.02} y^{5/3} (0.0003)^{1/2}$$

from which $y_0 = (39)^{3/5} = 9$ ft. This will be the depth downstream from the jump, if there is one. To find the depth preceding the jump y_2, we write Eq. (10.133),

$$\frac{q^2}{g} = y_2 y_0 \frac{y_2 + y_0}{2}$$

$$\frac{50^2}{32.2} = y_2 \times 9 \frac{y_2 + 9}{2}$$

from which $y_2 = 1.625$ ft. Since the depth of 1.5 ft at the sluice gate is less than this, the conditions are proper for an M_3 curve and a jump will form.

The total change in depth over the M_3 curve is $1.625 - 1.5$ or 0.125 ft. Since this is less than 10% of the depth, the whole distance may be considered in one reach.

$y_1 = 1.50$ ft	$V_1 = \dfrac{50}{1.5} = 33.3$ ft/s	$R_1 = 1.5$ ft
$y_2 = 1.625$ ft	$V_2 = \dfrac{50}{1.625} = 30.8$ ft/s	$R_2 = 1.625$ ft

Therefore $V_m = 32.05$ ft/s; $R_m = 1.562$ ft. Then, from Eq. (10.124),

$$S = \left(\frac{nV_m}{1.486 R_m^{2/3}} \right)^2$$

$$S = \left(\frac{0.02 \times 32.05}{1.486 \times (1.562)^{2/3}} \right)^2 = 0.321^2 = 0.1031$$

From Eq. (10.123),

$$1.5 + \frac{33.3^2}{64.4} = 1.625 + \frac{30.8^2}{64.4} + (0.1031 - 0.0003) L$$

from which $L = 8.52/0.1028 = 83$ ft.

If the gate were lowered further, y_1 would be reduced, while y_2 would remain the same; so the jump would move farther downstream. If the gate were raised, the jump would move upstream, but if it were raised above 1.625 ft, no distinct jump could form. Instead the sluice gate would become submerged.

The hydraulic drop and broad-crested weir

When water from a reservoir enters a canal in which the depth is greater than critical, there will be a drop in the surface owing to the velocity head and also to any friction loss at entrance. Water flowing over a dam in a channel is such a case. But if the uniform depth is less than the critical value, the water level at entrance will drop to the critical depth and no farther, no matter how low the water level may be in the stream below. The maximum flow in the canal is then limited by this factor. When the channel dam immediately slopes downstream, the surface curve will have a point of inflection at about the entrance section; and although a steep slope curve downstream may be changed with different channel conditions, the portion of this curve upstream from this point of inflection remains unaltered.

A very short flume becomes a *broad-crested weir* and can be used to measure the flow in a channel. If the surface of a channel dam is horizontal and extends downstream for 5 or more times the depth of the water upstream of the dam, this dam is by definition a broad-crested weir, and if the flow along it is in parallel lines so that centrifugal effect does not alter the normal hydrostatic pressure variation, then the flow is at critical depth. This enables the rate of discharge per unit width to be given by Eq. (10.114). Practically, however, the curvilinear flow at entrance and outfall prevents the "true" critical depth from forming at any one section for all flow rates. In addition, the flow at or near critical depth results in a wavy surface, which makes it difficult to measure the depth with any accuracy. While the depth may oscillate considerably in the region of the critical value, however, the rate of discharge for a constant specific energy varies but little when conditions are near critical. Thus it is possible to obtain a fairly accurate measurement of the flow rate by measurement of the upstream head, as in the case of the ordinary weir.

The head from the velocity of approach must be added to the measured head to give the total specific energy related to the horizontal weir crest. The value of E so obtained may be used in Eq. (10.114) to determine the rate of discharge per unit width. Owing to friction, however, the true rate of discharge is less than that computed in this way. For the broad-crested weir described previously, with the upstream edge rounded, the actual rate of discharge varies from 90 to 92.5% of the value given by Eq. (10.114) as the head varied from 0.50 to 1.15 ft and greater [43]. Among the advantages claimed for the broad-crested weir are that it can operate submerged without appreciable change in coefficient, thereby conserving head, and that there is no sharp crest to wear round and alter the coefficient.

Flow around channel bends and obstructions

The flow around the bend of a channel provides an application of the fundamentals of flow in a curved path. As there is no torque applied to the fluid, the flow should follow the laws of the free vortex—and indeed it would, were it not for the effect of friction on the walls and bottom of the channel.

In the case of a real liquid in an open channel, it is necessary to differentiate between the behavior at subcritical and supercritical velocities. Subcritical flow in a rectangular channel has been investigated experimentally and has been found to conform fairly well to ideal conditions, especially within the first part of the bend [44]. As the flow continues around the bend, the velocity distribution becomes complicated by the phenomenon of spiral flow, which for open channels is analogous to the secondary counterrotating currents found at bends in closed pipes.

The existence of spiral flow is observed near the short radius of the bend. The water surface is superelevated at the outside wall for the cylindrical free vortex. The element EF is subjected to a centrifugal force mV^2/r, which is balanced by an increased hydrostatic force on the left side due to the superelevation of the water surface at C above that at D. The element GH has exactly the same hydrostatic force inward, but the centrifugal force outward is much less because the velocity is decreased by friction near the bottom. This results in a cross flow inward along the bottom of the channel, which is balanced by an outward flow near the water surface, hence the spiral. This spiral flow is largely responsible for the commonly observed erosion of the outside bank of a river bend, with consequent deposition and building of a sand bar near the inside bank.

It is generally not possible to perceive a convex surface profile, given by the free vortex theory. For most practical purposes the water surface may be supposed to be a straight line from A to B, raised at the outside wall and depressed at the inside, with the slope given by the ordinary superelevation formula used for highway curves, $\tan \theta = V^2/gr$, where r is the radius of the curve to the center of the channel.

With supercritical flow the complicating factor is the effect of distrubance waves generated by the very start of the curve. These waves, one from the outside wall and one from the inside, traverse the channel, making an angle β with the original direction of flow. The water surface along the outside wall around a bend in a rectangle will rise from the beginning of the curve, reaching a maximum at point where the wave from the inside wall reaches the outside wall. The wave is then reflected back to the inside wall, and the outer surface falls again, and so on around the bend. The maximum rise is approximately twice the value computed by the free vortex or superelevation formulas.

Several schemes to lessen the surface rise from wave effect have been investigated [45]. The bed of the channel may be banked, for instance, so that all elements of the flow are acted upon simultaneously, which is not possible when the turning force comes from the walls only. As in a banked railway curve, this requires a transition section preceding the main curve. Another method is to introduce a counterdisturbance calculated to offset the disturbance wave caused by the curve. Such a counterdisturbance can be provided by a section of curved channel of twice the radius of the main section, by a spiral transition curve, or by diagonal sills on the channel bed, all preceding the main curve. When an obstruction or transition causes a rise in the floor of the channel, the water surface will be lowered or raised according to whether the flow is upper- or lower-stage, respectively. In an upper-stage case, where water flows over a smooth sloped weir (sloped equally both upstream and downstream), it may be seen that a rise in the channel bottom represents a decreased specific energy E_2, a corresponding increased velocity head, and a decreased depth, given by the specific energy diagram for flow over the weir. Quite evidently, if the weir rises any higher than required to produce critical depth with minimum specific energy, these relations are no longer possible. An M_1 backwater curve will result, building up specific energy until there is enough to produce critical depth over the obstruction, as in the case of a broad-crested weir.

If the channel is constricted laterally, the specific energy remains constant but the discharge q per unit width increases. An important application of the foregoing theories is the Parshall flume [46], which involves a constriction in the sides and a depression in the floor as well. This construction causes critical depth to occur near the beginning of the drop in floor level. When the flume operates submerged, a hydraulic jump forms downstream of the sloping section; but this does not affect conditions upstream, and the flume will measure flow with an error of 3% or less for submergence ratios of up to 0.77.

Example 10.10 The theory of the hydraulic drop serves to determine the capacity of a flume leading from a reservoir. Consider a rectangular flume 15 ft wide, built of unplaned planks ($n = 0.014$), leading from a reservoir in which the water surface is maintained constant at a height of 6 ft above the bed of the flume at entrance. The flume slopes 1 ft in its length of 1000 ft, and the depth at the downstream end is held constant at 4 ft by some control further downstream. The flume is then said to be submerged. Assuming an entrance loss of $0.2V_1^2/2g$, find the capacity.

For a first, approximate answer, we will consider the entire flume as one reach. The equations to be satisfied follow.

Energy at entrance:

$$y_1 + \frac{1.2V_1^2}{2g} = 6.0 \tag{10.134a}$$

Energy equation (10.123) for the entire reach:

$$y_1 + \frac{V_1^2}{2g} = 4.0 + \frac{V_2^2}{2g} + (S - 0.001)L \tag{10.134b}$$

In which S is given by Eq. (10.124),

$$S = \left(\frac{nV_m}{1.486R_m^{2/3}} \right)^2 \tag{10.134c}$$

The procedure is to make successive trials of the upstream depth y_1. This determines corresponding values of V_1, q_1, V_2, V_m, R_m, and S. The trials are repeated until the value of L from Eq. (10.134b) is close to 1000 ft. The solution is conveniently shown in Table 10.2.

The capacity is then $Q = qB$, or $36.2 \times 15 = 543.0$ ft³/s. A more accurate result can be obtained by dividing the flume into reaches in which the depth change is about 10% of the depth. Thus, commencing at the downstream end, with $q = 36.2$ ft²/s, the distance may be calculated to the point where $y = 4.4$ ft, thence to the depth of 4.8 ft, and so on. This procedure will normally lead to a value of ΣL between the assumed end depths, which is greater than the length of the flume. The value of y_1 is then decreased, with a consequent increase in the discharge. The approximate procedure shown is seen to be conservative, i.e., the true capacity is greater than that calculated.

Unsteady Flow

Introduction

Most texts are confined to steady flow, as the majority of cases of engineering interest are of this nature. However, there are a few cases of unsteady flow that are very important. It has been explained that tur-

TABLE 10.2 Solving for Variables in Determining the Capacity of a Flume Leading from a Reservoir

Trial y_1, ft	V_1, Eq. (10.134a), ft/s	$q = y_1V_1$, ft³/s	$V_2 = q/4$, ft/s	V_m, ft/s	R_1, ft	R_m, ft	S, Eq. (10.134c)	L, Eq. (10.134b)	
5.0	7.32	36.3	9.15	8.24	3.0	2.7	0.00160	890	Increase y_1
5.2	6.55	34.0	8.50	7.52	3.07	2.74	0.00131	2400	Decrease y_1
5.03	7.20	36.2	9.05	8.13	3.01	2.71	0.00155	1030	Close enough

bulent flow is unsteady in the strictest sense of that word but that, if the mean temporal values are constant over a period of time, it is called *mean steady flow*. Attention will here be directed to cases where the mean temporal values continuously vary.

There are two main types of unsteady flow. The first is where the water level is steadily rising or falling, as in a reservoir or a ship lock, so that the rate of flow varies continuously but change takes place slowly. The second is where the velocity in a pipeline is changed rapidly by the fast closing or opening of a valve. In the first case of slow changes, the flow is subject to the same forces as have been previously considered. The fast changes involved in the second type require the consideration of elastic forces.

Unsteady flow also includes such topics as oscillations in connected reservoirs and in U tubes and such phenomena as tidal motion and flood waves. Likewise, the field of machinery regulation by servomechanisms is intimately connected with unsteady motion. However, all these topics are considered to be beyond the scope of the present text.

Discharge with varying head

Under this condition the rate of discharge will continuously vary, and the problem may be either to find the total change in volume in a given time interval or to find the length of time necessary for a certain change in level. Let V_L be the total volume under consideration, while into it there is an inflow at a rate Q_1 and an outflow at a rate Q_2. It follows that in any time dt the total change in volume is

$$dV_L = Q_1\,dt - Q_2\,dt$$

If S is the area of the surface of the volume while dz is the change in level of the surface, then $dV_L = S\,dz$. Equating these two expressions for dV_L,

$$S\,dz = Q_1\,dt - Q_2\,dt \tag{10.135}$$

Either Q_1 or Q_2 or both may be variable and functions of z, or one of the two may be either constant or zero. For example, if liquid is discharged through an orifice or a pipe of area A under a differential head z, $Q_2 = C_d (2gz)^{1/2}$, where C_d is a numerical discharge coefficient and z is a variable. If the liquid flows out over a weir or a spillway of length B, $Q_2 = CBz^{3/2}$, where C is the appropriate coefficient. (For steady flow, z would be the constant H.) In either case z is the variable height of the liquid surface above the appropriate datum. In like manner, Q_1 may be some function of z.

Equation (10.135) is general, and if it is possible to express S, Q_1, and Q_2 as functions of z, the time t may be found by

$$t = \int_{z_2}^{z_1} S\, dz/(Q_1 - Q_2) \qquad (10.136)$$

In many practical cases involving natural reservoirs, the surface area is not a simple mathematical function of z, but values of it may be known for various values of z. In such a case, Eq. (10.136) may be solved graphically by plotting values of $S/(Q_1 - Q_2)$ against simultaneous values of z. The area under such a curve to some scale is the numerical value of the integral. It may be observed that instantaneous values for Q have been expressed in the same manner as for steady flow. This is not strictly correct, as for unsteady flow the energy equation should also include an acceleration head. The introduction of such a term renders the solution much more difficult. In cases where the value of z does not vary rapidly, no appreciable error will be involved by disregarding this acceleration term. Therefore the equations will be written as for steady flow.

Velocity of pressure wave

Unsteady phenomena with rapid changes taking place frequently involve transmission of pressure in waves or surges. The velocity of a pressure wave is

$$c = \left(\frac{gE_v}{w}\right)^{1/2} = \left(\frac{E_v}{\rho}\right)^{1/2} \qquad (10.137)$$

where E_v is the volume modulus of the medium. For water a typical value of E_v is $144 \times 300{,}000$ lb/ft², and thus the velocity of a pressure wave is $c = 4720$ ft/s. But for water in an elastic pipe this value is modified by the stretching of the pipe walls with E_v being replaced by K such that

$$K = \frac{E_v}{1 + (D/t)\,(E_v/E)}$$

where D and t are the diameter and wall thickness of the pipe, respectively, and E is the modulus of elasticity of the pipe material. As the ratios D/t and E_v/E are dimensionless, any consistent units may be used in each.

The velocity of a pressure wave in water in an elastic pipe is then

$$c = \left(\frac{g}{w}K\right)^{1/2} = \frac{4720}{[1 + (D/t)\,(300{,}000 \times 144/E)]^{1/2}} \qquad (10.138)$$

Values of the modulus of elasticity for steel, cast iron, and wood are about 30,000,000, 15,000,000, and 1,500,000 psi, respectively. Inasmuch as only ratios are involved in Eq. (10.138), the 144 may be omitted and these values used directly for E since the 300,000 is also in psi. For normal pipe dimensions the velocity of a pressure wave usually ranges between 2000 and 4000 ft/s, but it must always be less than 4720 ft/s.

Water hammer

In the preceding unsteady flow cases, changes in velocity were presumed to take place so slowly that instantaneous conditions could be treated as steady flow. But if the velocity of a liquid in a pipeline is abruptly decreased by a valve movement, the phenomenon encountered is called *water hammer*. This is a very important problem in the case of hydroelectric plants, where the flow of water must be rapidly varied in proportion to the load changes on the turbine.

Instantaneous closure. Although it is physically impossible to close a valve instantaneously, such a concept is useful as an introduction to the study of real cases. Assume that a valve immediately downstream of point N in a pipe is closed instantaneously. The lamina of water next to the valve will be compressed by the rest of the column of water flowing against it. At the same time the walls of the pipe surrounding this lamina will be stretched by the excess pressure produced. The next lamina layer will then be brought to rest, and so on. It is seen that the volume of water in the pipe does not behave as a rigid, incompressible body but that the phenomena are affected by the elasticity of both the water and the pipe. Thus the cessation of flow and the resulting pressure increase progress along the pipe as a wave action with the velocity c.

After a short interval of time the volume of water BN will have been brought to rest, while the water in a pipe length will still be flowing with its initial velocity and initial pressure. When the pressure wave reaches the inlet at M, the entire mass in the length L will be at rest but will be under an excess pressure throughout. There will be a transient hydraulic gradient parallel to the original hydraulic gradient, but at a height p'/w above it.

It is impossible for a pressure to exist at M that is greater than that due to the depth producing the flow at M, and so it instantly drops down to the value it would have for zero flow. But the entire pipe is now under an excess pressure, so the water in it is compressed and the pipe walls are stretched. Then some water starts to flow back into the reservoir, and a wave of pressure unloading travels along the pipe from M to N. At the instant this unloading wave reaches N, the entire mass of water will be under the static pressure equal to the pressure initiating flow at M. But the water is still flowing back into the reservoir, and this reverse velocity must now be instantaneously stopped. This will produce a drop in pressure at N that ideally will be as far below the static pressure as the pressure an instant before was above it. Then a wave of rarefaction travels back up the pipe from N to M. Ideally there would be a series of pressure waves traveling back and forth over the length of the pipe and alternating equally between high and low pressures about the value for zero flow.

The time for a round trip of the pressure wave from N to M and back again is

$$T_r = 2(L/c) \tag{10.139}$$

where L is the length of pipe, and so for an instantaneous valve closure the excess pressure remains constant for this length of time, before it is affected by the return of the unloading pressure wave; and in like manner the pressure defect during the period of rarefaction remains constant for the same length of time. At a distance x from the inlet, such as at B, the time for a round trip of a pressure wave is only $2x/c$, and hence at any point, such as at B, the time duration of the excess or deficient pressure will be less. At the inlet M, where $x = 0$, the excess pressure is only an instantaneous value.

The mass of water in the pipe is $pAL = wAL/g$, and the force required to change the velocity by an amount V is $Ap'A$. Applying Newton's law, $p'A = pAL\,V\,It$, where t is time. The time required to reduce the entire mass of water by the amount V is $t = L/c$, and therefore

$$\Delta p' = \rho c\,\Delta V = \frac{w}{g}\,c\,\Delta V \tag{10.140}$$

If there is complete valve closure, so that the velocity is reduced to zero, then the maximum water hammer excess pressure is

$$p' = \rho cV = \frac{w}{g}\,cV \tag{10.141}$$

It will be observed that the pressure increase is independent of the length of the pipe and depends solely upon the acoustic velocity and the change in the velocity of the water.

Consider now conditions at the valve as affected by both pipe friction and damping. When the pressure wave from N has reached a midpoint B in the pipe length L, the water in BN will be at rest and for zero flow the hydraulic gradient should be a horizontal line. There is thus a tendency for the gradient to flatten out for the portion BN. Hence, instead of the transient gradient having the slope imposed by friction, it will approach a horizontal line starting from the transient value at B. Thus the pressure head at N will be raised to a slightly higher value than NS shortly after the valve closure.

This slight increase in pressure head at the valve over the theoretical value Vc/g has been borne out by tests. Consider a pressure wave profile on an xy plot, with the energy value p/w the y axis and time the x axis. The pressure wave makes a square wave pattern, with the time dimension ab on the top half of the first wave and ef on the bottom half. Both ab and ef are defined by $2L/c$. The line ab slopes upward owing to this adverse pressure gradient, and for the same reason ef may slope slightly downward, as all conditions are now reversed. But damping acts to oppose this, and if the pipe friction is very small, damping may flatten out ab and ef so that they are practically horizontal or may even converge toward the x axis. Also, owing to damping, the drop de will be less than bd, and the waves will be of decreasing amplitude about the line of zero flow until the final equilibrium pressure is reached.

All the preceding analysis assumes that the wave of rarefaction will not cause the minimum pressure at any point to drop down to or below the vapor pressure. If it should do so, the water would separate and produce a discontinuity.

Rapid closure. It is physically impossible for a valve to be closed instantaneously, so we will now consider the real case where the valve is closed in a finite time that is more than zero but less than $2L/c$. An actual case resembles a sine waveform more than a square waveform, with $\Delta p/w$ the energy amplitude on the y axis. The shape of the curve during the time t' depends entirely upon the operation 1 of the valve and its varying effect upon the velocity in the pipe. But the maximum pressure rise is still the same as for instantaneous closure.

The only differences are that it endures for a shorter period of time and the vertical lines of the square wave are changed to the sloping lines of a sine wave. If the time of valve closure were exactly $2L/c$, the maximum pressure rise at the valve would still be the same but the curves of the sine wave would all end in sharp points for both maximum and minimum values, as the time of duration would be reduced to zero. These references to square and sinusoidal pressure shock waves are based on an experimental pipe with the following data: $L = 3060$ ft, internal diameter $= 2.06$ in, $c = 4371$ ft/s, $V = 1.11$ ft/s, $V_c/g = 151$ ft, $2L/c = 1.40$ s, static head $= 306.7$ ft, head before walve closure $= 301.6$ ft, $h_f = 5.1$ ft. For the square wave, the time of closure is 1 s, and it will be noted that the actual pressure rise is more than 151 ft. For the sine wave, the time of closure is 3 s.

No matter how rapid the valve closure may be, so long as it is not the imaginary instantaneous case, there will be some distance from the intake, such as xo, within which the valve closure time is more than $2xo/c$. Thus in any real case the maximum pressure rise cannot extend all the way to the reservoir intake. In the actual case the maximum pressure rise will be constant at the instantaneous value p' (or p') for a distance from the valve up to this point a distance xo from the intake. From this point the excess pressure will diminish to zero at the intake.

Slow closure.   The preceding discussion has assumed a closure so rapid (or a pipe so long) that there is insufficient time for a pressure wave to make the round trip before the valve is closed. Slow closure will be defined as closure in which the time of valve movement is greater than $2L/c$. In this case the maximum pressure rise will be less than in the preceding cases because the wave of pressure unloading will reach the valve before the movement is completed and will thus stop any further increase in pressure.

Thus the pressure of any shock wave generated in a pipeline would continue to rise if it were not for the fact that at a time $2L/c$ a return unloading pressure wave reaches the valve and stops the pressure rise at a value of about 53 ft, as contrasted with about three times that value if this were not the case. Subsequent pressure changes as elastic waves travel back and forth are very complex and require a detailed step-by-step analysis that is beyond the scope of this text. In brief, the method consists of assuming that the valve movement takes place in a series of steps each of which produces a pressure p' proportional to each V. Other texts contain details of computing successive pressures for slow valve closure and further explanation of much of this condensed treatment [47–49].

Tests have shown that for slow valve closure, i.e., in a time greater than $2L/c$, the excess pressure produced decreases uniformly from the value at the valve to zero at the intake.

Surge chamber

In a hydroelectric plant it is necessary to decrease the flow of water to a turbine very rapidly in the event of a sudden drop in the load. In such a case a relief valve may be opened so as to bypass water around the machine, and then this relief valve may be slowly closed. Alternatively, an air chamber may be used to absorb the shock. Some excess water may flow into this device and compress the air to a higher pressure. One disadvantage is that water under pressure absorbs air, so it is necessary to renew the air supply periodically by the operation of a small air compressor. Also, an air chamber has distinct size limitations.

Both these objections are overcome by a surge chamber, which permits a relatively large quantity of water to flow into it and raise the hydraulic gradient to provide a decelerating force. Water hammer is characterized by pressure waves that travel at high velocity, but the physical movement of the water itself is relatively minor. With a surge chamber there are appreciable movements of the water, but these surges are less rapid than pressure waves, and the pressure fluctuations are much less severe. The surge chamber fulfills another desirable function in that, in the event of a sudden demand for increased flow, it can provide some excess water while the entire mass of water in a long pipeline is being accelerated. The topography usually makes it impractical to provide a surge chamber next to the powerhouse, but the two should be as close together as physical conditions permit.

Valve opening

When a valve at the terminal end of a pipe is quickly opened, elastic waves travel back and forth until damping produces an equilibrium condition similar to that in valve closure. The analysis is similar to that for closure, except that the signs are reversed. Thus the initial pressure wave is one of rarefaction. Also, the pressure drop is limited to the difference between the initial valves on the two sides of the valve and can have no such magnitude as is possible in the case of water hammer. A further difference is that with valve closure the velocity of flow changes very rapidly and is accompanied by rapid and large pressure changes; with valve opening the pressures remain constant at both ends of the pipe, and the velocity of flow changes gradually.

In view of the foregoing, a very close approximation for valve opening may be achieved by assuming an incompressible liquid in a rigid pipe. Such an assumption for instantaneous closure would indicate an infinite pressure rise, but no such result is possible for an instantaneous opening. Therefore in this simplified case we can employ the energy equation, but with a term added for the acceleration head.

Assume a pipe of length L, and let the head NX (previously explained as a distance upstream from the pipeline valve) be denoted by h. Consider the friction loss in the pipe to be represented as $h_f = kV^2/2g$, where k is assumed to be constant, although actually it may vary with the velocity unless the pipe is very rough. With these assumptions the energy equation written between X and N becomes

$$h - (1 + k) \frac{V^2}{2g} = 1/g \int_0^L \frac{\partial V}{\partial t}\, ds \qquad (10.142)$$

where the term on the right is the acceleration head. As the velocity varies only with time, this may be written as

$$\frac{L}{g} \frac{dV}{dt} = h - (1 + k) \frac{V^2}{2g}$$

The equilibrium value may be denoted by V_0 and $(1 + k)V_0^2 = 2gh$. Inserting this in the preceding equation after multiplying by $2g$,

$$dt = \frac{2L}{1 + k} \frac{dV}{V_0^2 - V^2}$$

From which

$$t = \frac{l}{(1 + k) V_0} \log_e \frac{V_0 + V}{V_0 - V} \tag{10.143}$$

This equation indicates that equilibrium will be attained only after an infinite time, but it will be found that the equilibrium value will be approached very closely within a short time. It must be remembered that this is an idealized case. In reality there will be elastic waves and damping, so that true equilibrium will be reached in a finite time.

References

1. White, W. M., and Rheingans, W. J., "Photoflow Method of Water Measurement," *Trans. ASME* 57:273, 1935.
2. Cole, E. S., "Pitot Tube Practice," *Trans. ASME* 57:281, 1935.
3. Discussion in *Trans. ASME* 58(2), 1936.
4. Hubbard, C. W., "Investigations of Errors of Pitot Tubes," *Trans. ASME* 61:477, 1939.
5. Moody, L. F., "Some Pipe Flow Characteristics of Engineering Interest and an Approximate Method of Discharge Measurement," *La Houille Blanche* 5:313, May–June 1950.
6. Binder, R. C., and Knapp, R. T., "Experimental Determinations of the Flow Characteristics in the Volutes of Centrifugal Pumps," *Trans. ASME* 58, 1936.
7. Allen, C. M., and Hopper, L. J., "Piezometer Investigation," *Trans. ASME* 54(9), 1932.
8. Prandtl-Tietjens, *Applied Hydro and Aeromechanics,*" McGraw-Hill, New York, 1934, p. 227.
9. Allen, C. M., and E. A. Taylor, "The Salt Velocity Method of Water Measurement," *Trans. ASME* 45:283, 1923.
10. Gibson, N. R., "The Gibson Method and Apparatus for Measuring the Flow of Water in Closed Conduits," *Trans. ASME* 45:343, 1923.
11. ASME, "Flow Measurement by Means of Standardized Nozzles and Orifice Plates, of Information on Instruments and Apparatus," *ASME Power Test Codes*, American Society of Mechanical Engineers, New York, 1940, p. 45.
12. Rouse, Hunter, Howe, J. W., and Metzler, D. E., "Experimental Investigations of Fire Monitors and Nozzles, *Trans. ASCE* 117:1147–1188, 1952.
13. Marks, L. S., *Mechanical Engineers' Handbook*, 5th ed., McGraw-Hill, New York, 1954, pp. 239–243.
14. Greve, F. W., *Flow of Liquids through Vertical Circular Orifices and Triangular Weirs*, Purdue Engineering Experimental Station Research Series 95, 1945, p. 21.
15. Pardoe, W. S., "The Effect of Installation on the Coefficients of Venturi Meters," *Trans. ASME* 65:337, 1945.
16. Beitler, S. R., "The Effect of Pulsation on Orifice Meters," *Trans. ASME* 61:309, 1939.
17. Beitler, S. R., Lindahl, E. J., and McNichols, H. B., "Developments in the Measuring of Pulsating Flows with Inferential-Head Meters," *Trans. ASME* 65:337, 1943.
18. Allen, C. M., and Hooper, L. J., "Venturi Weir Measurements," *Mech. Eng.*, June 1935, p. 369.
19. Pardoe, W. S., *Mech. Eng.*, January 1936, p. 60.

20. Pardoe, W. S., "The Coefficients of Herschel-Type Cast-Iron Venturi Meter," *Trans. ASME* 67:339, 1945.
21. ASME, "Flow Measurement," *ASME Power Test Codes,* American Society of Mechanical Engineers, New York, 1940, pt. 5, chap. 4, p. 28.
22. Bean, H. S., Beitler, S. R., and Sprenkle, R. E., "Discharge Coefficients of Long-Range Nozzles When Used with Pipe-Wall Pressure Taps, *Trans. ASME* 63:439, 1941.
23. Folsom, R. G., "Nozzle Coefficients for Free and Submerged Discharge," *Trans. ASME* 61:233, 1939.
24. Beitler, S. R., and Bucher, P., "The Flow of Fluids through Orifices in 6 in Pipes," *Trans. ASME* 52(30), 1930.
25. Bean, H. S., Buckingham, E., and Murphy, P. S., "Discharge Coefficients of Square-Edged Orifices for Measuring the Flow of Air," U.S. National Bureau of Standards Research Paper 49, ASME Fluid Meters Report, 1931.
26. ASME, "Flow Measurement," *ASME Power Test Codes,* American Society of Mechanical Engineers, New York, 1940, pt. 5, chap. 4, p. 46.
27. Ambrosius, E. E., and Spink, L. K., "Coefficients of Discharge of Sharp-Edged Concentric Orifices in Commercial 2 in, 3 in, and 4 in Pipes for Low Reynolds Numbers Using Flange Taps, *Trans. ASME* 69:805, 1947.
28. Hodgson, J. L., "The Laws of Similarity for Orifice and Nozzle Flows," *Trans. ASME* 51(FSP—51-42):302–332, 1929.
29. Lenz, Arno T., "Viscosity and Surface Tension Effects on V-Notch Weir Coefficients, *Trans. ASME* 108:759–802, 1943.
30. Barr, James, "Experiments upon the Flow of Water over Triangular Notches," *Engineering,* April 8 & 15, 1910.
31. Rouse, Hunter, *Engineering Hydraulics,* John Wiley & Sons, New York, 1950, pp. 536–543.
32. Lansford, W. M., *The Use of an Elbow in a Pipe Line for Determining the Rate of Flow in the Pipe,* University of Illinois Engineering Experimental Station Bulletin 289, December 1936.
33. ASME, "Flow Measurement," *ASME Power Test Codes,* American Society of Mechanical Engineers, New York, 1940, p. 47.
34. Buckingham, Edgar, U.S. National Bureau of Standards Research Paper 459.
35. Bakhmeteff, B. A., *Hydraulics of Open Channels,* McGraw-Hill, New York, 1932.
36. Rouse, H., *Fluid Mechanics for Hydraulic Engineers,* cited in R. L. Daugherty and A. C. Ingersoll, *Fluid Mechanics,* McGraw-Hill, New York, 1954.
37. Ippen, A. T., "Mechanics of Supercritical Flow," *Trans. ASCE* 116:274, 1951.
38. Von Seggern, M. E., "Integrating the Equation of Non-uniform Flow," *Trans. ASCE* 115:71–106, 1950.
39. Rouse, H., "Discharge Characteristics of the Free Overfall," *Civil Eng.,* cited in R. L. Daugherty and A. C. Ingersoll, *Fluid Mechanics,* McGraw-Hill, New York, 1954.
40. O'Brien, M. P., "Analyzing Hydraulic Models for Effects of Distortion," *Eng. News-Record.,* cited in R. L. Daugherty and A. C. Ingersoll, *Fluid Mechanics,* McGraw-Hill, New York, 1954.
41. Ellms, R. W., "Computation of the Trailwater Depth of the Hydraulic Jump in Sloping Flumes," cited in R. L. Daugherty and A. C. Ingersoll, *Fluid Mechanics,* McGraw-Hill, New York, 1954.
42. Stevens, J. C., "The Hydraulic Jump in Standard Conduits," cited in R. L. Daugherty and A. C. Ingersoll, *Fluid Mechanics,* McGraw-Hill, New York, 1954.
43. Woodburn, J. G., *Texts of Broad-Crested Weirs,* cited in R. L. Daugherty and A. C. Ingersoll, *Fluid Mechanics,* McGraw-Hill, New York, 1954.
44. Mockmore, C. A., "Flow around Bends in Stable Channels," cited in R. L. Daugherty and A. C. Ingersoll, *Fluid Mechanics,* McGraw-Hill, New York, 1954.
45. Knapp, R. T., "Design of Channel Curves for Supercritical Flow," cited in R. L. Daugherty and A. C. Ingersoll, *Fluid Mechanics,* McGraw-Hill, New York, 1954.
46. Parshall, R. L., "The Parshall Measuring Flume," cited in R. L. Daugherty and A. C. Ingersoll, *Fluid Mechanics,* McGraw-Hill, New York, 1954.
47. Durand, W. F., "Hydraulics of Pipe Lines," cited in R. L. Daugherty and A. C. Ingersoll, *Fluid Mechanics,* McGraw-Hill, New York, 1954.

48. Quick, R. S., "Comparison and Limitation of Water Hammer Theories," cited in R. L. Daugherty and A. C. Ingersoll, *Fluid Mechanics,* McGraw-Hill, New York, 1954.
49. Angus, R. W., "Graphical Analysis of Water Hammer in Branched and Compound Pipes," cited in R. L. Daugherty and A. C. Ingersoll, *Fluid Mechanics,* McGraw-Hill, New York, 1954.

Industrial Chemistry

Electrolysis

Using electrical conductance to help determine whether certain solutes should be classified as electrolytes or nonelectrolytes and referring to electricity in the decomposition of certain compounds indicate that electrical energy often may be used to advantage in chemistry. In fact, both in the laboratory and in the chemical industries, electricity is one of the most important tools at the chemist's command. Following a study of the nature of the chemical changes that may be brought about through the agency of electrical energy, numerous important commercial applications are considered in some detail here.

Electrolytic cells

The process of *electrolysis* may be defined as the transformation of chemical substances as the result of passage of direct electric current. In order to bring about such transformations, the use of specialized apparatus is necessary (or at least convenient). A simple form of electrolytic cell is illustrated by Fig. 11.1. The cell as a whole consists of a vessel containing a solution of an electrolyte in which are immersed two electrodes (or poles). The electrodes are connected to storage batteries or to a generator, which serve as sources of the electrical energy needed to bring about the desired electrolytic transformation. The electrode through which the flow of electrons enters the cell is known as the *cathode* and is designated as the *negative* (–) electrode. By convention, the other electrode, which is called the *anode,* may be thought of as that through which electrons are withdrawn from the solution, and this electrode is designated as *positive* (+). Throughout the following sections, the terms *cathode* and *anode* are used to designate the negative and positive electrodes, respectively.

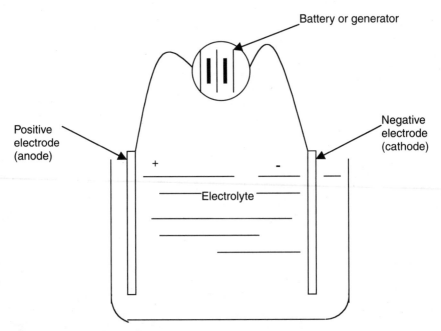

Figure 11.1 Electrolytic cell.

It must be recognized that Fig. 11.1 is only a highly simplified representation of an electrolytic cell. Actual commercial practice usually employs a connected series of such cells constructed in a manner to meet the needs of each specific operation. Frequently, the vessel containing the electrolyte is made of metal and serves as one of the electrodes. Other modifications are shown in connection with commercial applications of electrolysis.

Electrolysis of zinc chloride solution

The occurrence of a simple electrolysis may be demonstrated conveniently by passing a direct current through an aqueous solution of zinc chloride. A rectangular glass vessel (Fig. 11.2) is filled partly with zinc chloride solution. A polished strip of brass serves as the cathode, and a carbon rod constitutes the anode. In this particular case, the use of a carbon rather than a metal rod is advisable because most metals react chemically with one of the products of the electrolysis, thus complicating the situation with secondary reactions not pertinent to the primary process of electrolysis. Further, it is convenient to surround the anode with a glass housing in which the gaseous electrolysis product may be collected and from which it may be continuously withdrawn through a suitable exit tube.

Immediately upon connecting the cell to a source of direct current, a deposit of gray metallic zinc appears on the surface of the cathode and bubbles of chlorine gas appear at the surface of the anode. A simple chemical test for chlorine may be made by leading this gas into an aqueous sodium iodide solution, whereupon the solution assumes a yellow color caused by displacement of iodine by chlorine. Accordingly, it is concluded that the products of the electrolysis of a zinc chloride solution are elemental zinc and elemental chlorine, and the next problem is that of explaining the mechanism by which these products may be produced.

The solution in the cell contains zinc ions (Zn^{2+}) and chloride ions (Cl^-). The chloride ions are attracted to the anode (+ electrode), where they give up electrons and form molecules of elemental chlorine.

$$\text{At anode:} \quad 2Cl^- - 2e^- \rightarrow Cl_2$$

The battery acts as a pump and forces these electrons through the outside circuit and into the cathode, at the surface of which these electrons are acquired by the zinc ions in the solution surrounding the cathode. That is,

$$\text{At cathode:} \quad Zn^{2+} + 2e^- \rightarrow Zn$$

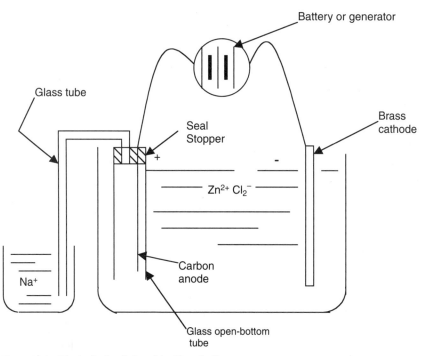

Figure 11.2 Electrolysis of zinc chloride solution.

Obviously, these reactions involve loss and gain of electrons and therefore are to be recognized as oxidation-reduction reactions. The sum of the changes that occur in the regions of the anode and cathode serves to represent the overall chemical change that occurs in the cell as a whole:

Anode: $\qquad 2Cl^- - 2e^- \rightarrow Cl_2$

Cathode: $\qquad Zn^{2+} + 2e^- \rightarrow Zn$

Cell: $\qquad Zn^{2+} + 2Cl^- \rightarrow Zn + Cl_2$

To avoid confusion, equations representing electrolysis should always be labeled in this way.

Some broad aspects of electrolysis

The electrolysis of aqueous zinc chloride solution serves to illustrate a number of features common to all situations of this general type. These are summarized here along with some necessary explanatory remarks and illustrations.

1. *The occurrence of electrolysis requires the utilization of an outside source of electrical energy.* The chemical changes that occur during electrolysis involve an accompanying transformation of electrical energy into chemical energy. Electrolysis may be looked upon as the work required to transfer electrons from the anode through the wires that constitute the outside electrical circuit and finally into the cathode. The magnitude of the outside energy supply required is dependent on the particular material being electrolyzed and is different for different electrolytes.

2. *The chemical changes that occur during electrolysis are forced and not spontaneous.* Depending on their particular structures, the atoms of each kind of element exhibit a tendency to either gain or lose electrons. Thus, chlorine atoms tend to gain electrons and become chloride ions, and zinc atoms tend to lose electrons and become zinc ions. However, as has already been shown, the electrolysis of zinc chloride solution produces exactly the reverse of these changes. Once a compound has been formed by virtue of natural tendencies of the character referred to here, the ease with which the compounds may be decomposed by electrolysis depends on the magnitude of the tendencies responsible for the initial compound formation.

3. *The quantities of matter transformed at the two electrodes are chemically equivalent.* In other words, the quantity of chlorine gas evolved during the electrolysis of a zinc chloride solution must be exactly that quantity which would combine with the zinc deposited on the cathode to form zinc chloride. In terms of the electrons involved, the

number of electrons lost at the anode must be identical to the number gained at the cathode. This requirement also follows from the fact that these reactions are oxidation-reduction changes. Although it is true that one may represent the anode and cathode reactions by what may appear to be independent chemical equations, it must always be borne in mind that a reaction cannot occur at one electrode without the simultaneous occurrence of another reaction at the other electrode.

In this connection, it should also be recognized that the total quantity of matter transformed during electrolysis is proportional to the quantity of electrical energy supplied from the battery or generator. Thus, the total weight of zinc deposited on the cathode when one electrolyzes a zinc chloride solution is related quantitatively to the total quantity of electricity supplied from outside the cell.

4. *The passage of the current through the solution involves the transport of ions.* If one should construct an electrolytic cell consisting of every essential part except the electrolytic solution between the electrodes, the electrical circuit would be incomplete and current could not flow. If one now places a solution of an electrolyte between the two electrodes, the circuit is thereby completed and the current flows through the entire circuit. Since the solution is a conductor of the second class, which may act as a conductor of electricity if transformations of matter occur, the process by which the current passes through the solution must be fundamentally different from that involved in the flow of electrons through the connecting metal wires and through the electrodes. To clarify this problem, it is helpful to develop a mental picture of what occurs in the solution during electrolysis of zinc chloride.

Consider first the changes that occur in the region of the cathode. Zinc ions gain electrons and are deposited in the form of metallic zinc on the surface of the cathode. Since positive ions are thus removed, the solution immediately surrounding the cathode might contain, momentarily, an excess of negative ions. However, the electrical neutrality of the solution is maintained by the migration of zinc ions from the solution farther from the cathode. At the same time, chloride ions are removed in the region of the anode and leave an excess of positive zinc ions; this causes the chloride ions in the remainder of the solution to migrate toward the anode. Thus, as electrolysis proceeds, positive ions are transported toward the cathode while at the same time negative ions are transported toward the anode, and the entire solution becomes progressively more dilute with respect to the ions involved in the reactions that occur at the two electrodes. It is this shifting or transporting of ions that is commonly referred to as the "passage of the current through the solution."

Relative speed of ions. The motion of cations and anions in opposite directions through the solution between the electrodes is analogous to the drawing together of two boats by means of a rope between them.

Assuming that the applied pull is the same at both ends of the rope, it is evident that both boats move at different speeds if they differ in weight. Similarly, ions that differ greatly in size do not move at the same speed. For example, during the electrolysis of hydrochloric acid solution, the very small hydrogen ions move about five times as fast as the larger chloride ions. Accordingly, five-sixths of the ability of hydrochloric acid solution to act as a conductor of electricity is attributed to the hydrogen ions and one-sixth to the chloride ions.

Faraday's laws of electrolysis

It has already been pointed out that the quantities of chemical changes occurring at the anode and cathode must be chemically equivalent. However, nothing really definite has been indicated with regard to the relationship between the quantities of chemical change and the quantity of electricity that flows through the cell. This relationship was established as a result of investigations carried out more than 100 years ago by the English scientist Michael Faraday. By carefully controlled experiments, Faraday was able to prove that the extent to which chemical changes occur during electrolysis is independent of concentration, temperature, and rate of flow of the current and is dependent *only* on the quantity of electricity that flows. The results of these studies may be summarized in generalizations that have come to be known as *Faraday's laws*.

1. *The quantity of chemical change that occurs at the electrodes during electrolysis is directly proportional to the quantity of electricity that flows through the cell.* The unit of measurement of quantities of electricity is the *coulomb* (C), which is that quantity required to deposit (on a cathode) exactly 0.001118 g of silver. Since the atomic weight of silver is 107.088, the number of coulombs required to deposit 1 gram-atomic weight of silver is 107.88 + 0.001118 = 96,500 C. This number of coulombs is called 1 *farad* (F) and represents the quantity of electricity involved in the chemical transformation of 1 gram-atomic weight of *any* element having a valence of 1. The relation between the faraday and the quantities of divalent, trivalent, and so on elements deposited is embodied in Faraday's second law.

2. *The quantities of the various elements liberated at the electrodes during electrolysis are in the ratio of their equivalent weights.* Thus, if a given quantity of electricity, 96,500 C for example, is passed successively through solutions $Ag^+NO_3^-$, $Cu^{2+}(NO_3)_2^-$, and $Bi^{3+}(NO_3)_3^-$, with each solution in a separate container, 107.88 g of silver (i.e., 1 gram-equivalent weight) is deposited on the cathode from the first solution. From the second solution, one-half of 1 gram-atomic weight of copper $(63.57/2 = 31.785$ g), or 1 gram-equivalent weight of copper, is deposited

on the cathode. Similarly, one-third of 1 gram-atomic weight of bismuth (i.e., 1 gram-equivalent weight = 209/3 = 69.667 g) is deposited on the cathode. If one should electrolyze separate solutions containing chloride ions and sulfide ions by using 1 F of electricity, the quantities of these nonmetals liberated would be 35.457 g of chlorine and 16 g of sulfur, or 1 equivalent weight in each case.

Electrochemical equivalent. The electrochemical equivalent of any substance is that quantity which is liberated by the passage of 1 C of electricity. It has already been shown that 1 C is required for the deposition of 0.001118 g of silver; hence, 0.001, 118, or 1.118×10^{-3} g, is the electrochemical equivalent. Similarly, 1 F of electricity liberates 1 gram-equivalent weight of copper (31.785 g); the electrochemical equivalent of copper is therefore $31.785/96,500 = 3.294 \times 10^{-4}$ g, and the electrochemical equivalent of bismuth or any other element may be calculated in an entirely analogous manner.

Determination of Avogadro's number. Faraday's second law is the basis for one of the most accurate methods of determining the best value for the Avogadro number. As stated earlier, the passage of 96,500 C of electricity results in the deposition of 1 gram-atomic weight of silver (i.e., 6.02×10^{23} atoms of silver). Since the silver was in solution in the form of univalent silver ions (Ag^+), 1 electron was required for the discharge of each silver ion, and the total number of electrons required was therefore 6.02×10^{23}. From this it follows that 96,500 C is that quantity of negative electricity which is carried by Avogadro's number of electrons. Knowing that the charge borne by 1 electron is 1.602×10^{-19} C, the number of electrons necessary to give 96,500 C of negative electricity must be $96,500/1.602 \times 10^{-19} = 6.02 \times 10^{23}$, which is the accepted value for Avogadro's number.

Discharge of ions from aqueous solutions

During the electrolysis of a solution of zinc chloride, both of the ions of the *solute* (i.e., Zn^{2+} and Cl^-) have their ionic charges neutralized at the cathode and anode, respectively. These ions may therefore be said to have been *discharged* and liberated in the form of elemental zinc and chlorine. To determine whether the ions corresponding to the solute are always discharged at the electrodes during electrolysis, it is necessary to inquire into the behavior of other electrolytes.

Electrolysis of water. When a direct current is passed through a dilute solution of sulfuric acid, the products liberated at the electrodes are the same as would result from the electrolysis of water alone. The advantage in using dilute acid solution rather than pure water lies in the fact that one may thus avoid the high resistance (to passage of the current) offered by the weak electrolyte water. The products of electrolysis are hydrogen, liberated at the cathode, and oxygen, liberated at the anode. If this electrolysis is carried out in a cell of the type by the Hoffman apparatus, it may be observed readily that hydrogen and oxygen are produced in a 2:1 volume ratio. These products result from the discharge of hydrogen and hydroxyl ions, and the reactions involved are represented by the following equations:

Anode: $4OH^- - 4e^- \rightarrow O_2 + 2H_2O$

Cathode: $4H^+ + 4e^- \rightarrow 2H_2$

Cell: $4H_2O \rightarrow 2H_2 + O_2 + 2H_2O$

or

$$2H_2O \rightarrow 2H_2 + O_2$$

The dilute acid solution contained hydrogen ions supplied by the solute and also by the slight ionization of water. Particular attention should be directed to the fact that the solution contained two anions (i.e., a very low concentration of hydroxyl ions) and a relatively high concentration of sulfate ions. Despite the fact that the sulfate ions were present at much greater concentration than the hydroxyl ions, only the latter were discharged at the anode. From this fact alone, it may be concluded that not all anions are discharged on electrolysis of aqueous solutions and that in such cases the anion of the solvent is involved in the reaction at the anode.

Electrolysis of sodium bromide solution. When an aqueous solution of sodium bromide is electrolyzed, hydrogen gas is liberated at the cathode and bromine gas at the anode. These products are to be anticipated because of the marked chemical similarity of bromine and chlorine and because sodium, even if it were liberated by the discharge of sodium ions, would react immediately with water to liberate hydrogen. In effect, then, the electrolysis of an aqueous solution of sodium bromide involves the discharge of the anion of the solute and the cation of the solvent.

Anode: $2Br^- - 2e^- \rightarrow Br_2$

Cathode: $2H^+ + 2e^- \rightarrow H_2$

Cell: $2H^+ + 2Br^- \rightarrow H_2 + Br_2$

In view of these results, one may recognize that not all cations are discharged during the electrolysis of aqueous solutions and that the cation of the solvent may be discharged at the cathode in place of the cation of the solute.

Generalizations about the discharge of anions and cations from aqueous solutions. From the foregoing cases taken together, along with the results of many similar experiments, it becomes possible to generalize broadly with respect to the behavior of anions and cations during the passage of the direct current through aqueous solutions of electrolytes. These general rules may be formulated as follows:

- All common cations may be discharged except Mg^{2+}, Al^{3+}, the ions of the alkali metals (Li^+, Na^+, K^+, Rb^+, Cs^+), and those of the alkaline earth metals (Ca^{2+}, Sr^{2+}, and Ba^{2+}).

- The only common anions that may be discharged are Cl^-, Br^-, I^-, OH^-, and S^{2-}.

Thus, as is typical in science, extensive experimentation leads to generalizations that may be used thereafter to predict the behavior of substances not previously investigated. Numerous examples of the utility of these general rules are cited in the remainder of this section.

The presence of a nondischargeable anion or cation has an important bearing on the nature of the solution remaining in the cell following the occurrence of electrolysis. In the electrolysis of sodium bromide solution, for example, the discharge of hydrogen ions (from water) occurs in the region surrounding the cathode; hence, the solution becomes basic since it contains the ions corresponding to the strong base Na^+OH^-. Conversely, the electrolysis of an aqueous solution of copper nitrate solution would result in the formation of an acidic solution in the region of the anode since OH^- would be discharged, leaving the nondischargeable NO_3^- ions and the H^+ ions (from water). These two typical examples are in conformity with two additional generalizations, which may be stated as follows:

- The electrolysis of an aqueous solution containing a nondischargeable cation produces a basic solution in the region of the cathode.

- The electrolysis of an aqueous solution containing a nondischargeable anion produces an acidic solution in the region of the anode.

Electrolysis of aqueous solutions of salts

On the basis of the preceding generalizations, it is possible to predict the products of electrolysis of aqueous solutions of simple salts. It is not possible, however, to predict the specific experimental conditions under which these electrolyses may be carried out most advantageously. Usually, the optimum experimental conditions can be ascertained only by means of experiments relating to any particular case.

Suppose one desires to know which products would result from the electrolysis of an aqueous solution of nickel acetate. According to the general rules set forth, nickel ions are discharged at the cathode and hydroxyl ions, not acetate ions, are discharged at the anode.

Anode:
$$4OH^- - 4e^- \rightarrow O_2 + 2H_2O$$

Cathode:
$$2Ni^{2+} + 4e^- \rightarrow 2Ni$$

Cells:
$$2Ni^{2+} + 4OH^- \rightarrow 2Ni + O_2 + 2H_2O$$

Furthermore, it is also apparent that the region of the anode would become progressively more acidic, owing to an accumulation of acetic acid, as the electrolysis proceeds. In a similar manner, one would predict that the electrolysis of strontium nitrate solution would lead to the formation of oxygen gas and hydrogen gas at the anode and cathode, respectively.

Anode:
$$4OH^- - 4e^- \rightarrow O_2 + 2H_2O$$

Cathode:
$$4H^+ + 4e^- \rightarrow 2H_2$$

Cell:
$$4H_2O \rightarrow 2H_2 + O_2 + 2H_2O$$

or

$$2H_2O \rightarrow 2H_2 + O_2$$

The products are the same as those resulting from the electrolysis of water since neither the cation or anion of the solute is dischargeable. During the course of this electrolysis, the region of the anode becomes acidic $(H^+NO_3^-)$ and the region of the cathode becomes basic $[Sr^{2+}(OH)_2^-]$.

Similar predictions may be made readily with regard to the electrolysis of aqueous solutions of other salts.

Electrolysis of aqueous solutions of acids

Since all soluble acids furnish hydrogen ions, the electrolysis of aqueous solutions of acids yields hydrogen gas at the cathode. Consequently, the only question that remains is whether the anion of the particular acid is dischargeable. If not, oxygen gas is liberated. There are two possible types of behavior. One is illustrated by hydriodic acid:

Anode: $2I^- - 2e^- \rightarrow I_2$

Cathode: $2H^+ + 2e^- \rightarrow H_2$

Cell: $2H^+I^- \rightarrow H_2 + I_2$

The other is illustrated by phosphoric acid:

Anode: $4OH^- - 4e^- \rightarrow O_2 + 2H_2O$

Cathode: $4H^+ + 4e^- \rightarrow 2H_2$

Cell: $4H_2O \rightarrow 2H_2 + O_2 + 2H_2O$

or

$$2H_2O \rightarrow 2H_2 + O_2$$

Electrolysis of aqueous solutions of bases

The soluble bases are very limited in number and involve nondischargeable cations. Accordingly, when an aqueous solution of a base is electrolyzed, hydrogen gas is evolved at the cathode and, of course, the discharge of hydroxyl ions at the anode results in the liberation of gaseous oxygen. In such cases, the products are the same as those obtained by the electrolysis of water. Since a commercial application of the electrolysis of the strong base sodium hydroxide is described in the next section, no further discussion of this topic seems necessary at this juncture.

Electrolysis of fused salts

Although many important commercial processes involve the electrolysis of aqueous solutions, this fact should by no means be interpreted as implying that one is restricted to the use of aqueous solutions. Electrolyses may be effected by employing solvents other than water (i.e., *nonaqueous* solvents) or in the absence of any solvent. In any event, if one wished to produce a metal such as sodium, the electrolysis would have to be carried out in the absence of water. This may often be accomplished by the use of fused salts. Since most solid salt crystals consist of ions, application of heat sufficient to melt the salt results in the formation of a molten liquid that still contains ions and is therefore capable of behaving as an electrolyte. Because of their comparatively low melting temperatures, chlorides (and less frequently, hydroxides) are commonly employed. The high temperatures required to melt inorganic salts are disadvantageous, and, wherever possible, a second salt is added for the purpose of lowering the melting temperature of the salt that it is desired to electrolyze. Thus the melting temperature of sodium chloride is lowered by the addition of a relatively small quantity of barium chloride. This lowering of the melting temperature may be considered analogous to the lowering of the freezing temperature of a pure solvent by the addition of a nonvolatile solute, and the added salt is usually referred to as a *flux*.

Electrolyses involving fused salts are more difficult to carry out than those in which aqueous solutions are used. The high temperatures usually required and the generally reactive character of the products necessitate elaborate equipment specially designed to fulfill the specific requirements in any given case. A number of typical cases in which fused salts are employed are considered in the next section.

Industrial Electrochemical Processes

The development of processes for the production of chemicals by electrolysis on an industrial scale has been slowed in some measure by the unavailability of cheap electricity. Consequently, the extensive use of electrochemical processes has been limited largely to regions such as the Niagara Falls area. However, the relatively recent and widespread development of federal power projects such as Boulder Dam has already resulted in more extensive use of electrochemical processes, and there is every indication that this phase of the chemical industries will be continuingly expanded in various regions in the United States.

The following subsections discuss a few chemical processes that involve the electrolysis of aqueous solutions or fused salts. It is not intended to provide here an exhaustive treatment of the subject but rather to select a few typical cases that serve to acquaint the student with the nature, the scope, and the importance of these industries.

Production of active metals

It has already been indicated that the alkali metals, alkaline-earth metals, magnesium, and aluminum may be prepared by electrolysis only in the absence of water, and the same is true of the metal beryllium. Although zinc is more commonly produced by other methods, it may also be produced by electrolysis.

Alkali metals.   All the alkali metals, Li, Na, K, Rb, and Cs, are very light metals that exhibit a silvery metallic luster and have low melting temperatures. These are relatively costly metals, as the cheapest one, sodium, sells at about $0.17 per pound, and the next in order of cost, potassium, sells at about $10.00 per pound. All the alkali metals are produced commercially by the electrolysis of their fused chlorides in electrolytic cells, which must be designed specially for each metal. Since sodium is the only alkali metal produced industrially on a large scale, its production is outlined here in some detail.

Practically all of the sodium produced in the United States involves the use of the Downs electrolytic cell, which consists of a massive graphite anode surrounded by two or more iron cathodes. The electrolyte is an aqueous solution of sodium choloride. The NaCl salt is continuously added. As electric current is passed between the electrodes, chlorine gas is collected in a hood over the graphite anode and piped off to further processing for marketing. The electrolysis of sodium chloride proceeds as follows:

Anode: $2Cl^- - 2e^- \rightarrow Cl_2$

Cathode: $2Na^+ + 2e^- \rightarrow 2Na$

Cell: $2Na^+Cl^- \rightarrow 2Na + Cl_2$

The sodium liberated at the cathode rises to the surface in the annular compartment surrounding the cathode and flows into a container where it is collected under oil and subsequently withdrawn. The chlorine liberated at the anode is sold as elemental chlorine or for use in other chemical processes.

Sodium is chiefly used in the preparation of compounds such as sodium peroxide (Na_2O_2) and sodium cyanide (NaCN). Sodium is also used in many preparative procedures for the formation of compounds that do not contain sodium (etraethyl lead, dyes, organic medicinals, etc.). Electric lamps containing sodium and neon are known as *sodium vapor lamps*. Such lamps produce a soft, penetrating, yellow light and is used in the illumination of highways. Although the other alkali metals are used to only a very limited extent, each has certain rather highly specialized uses (e.g., the use of cesium in photoelectric cells and in certain types of radio tubes).

Alkaline-earth metals.   Of these metals, Ca, Sr, and Ba, only calcium is produced commercially in appreciable quantities. These three metals are more difficult to produce than the alkali metals since the chlorides of the alkaline-earth metals melt at relatively high temperatures. Furthermore, when these metals are liberated at the cathode, they tend to become colloidally dispersed throughout the molten electrolyte. Accordingly, it is necessary to design the electrolytic cells in such manner to permit the immediate collection of the elemental metal. The type of cell used in the production of calcium serves as a suitable illustration. Molten calcium chloride is placed in a cylindrical vessel around the walls of which are placed several graphite anodes and through the bottom of which extends a massive iron cathode. Directly above the cathode is placed a bar of calcium in contact with the surface of the electrolyte and cooled by water that flows through pipes surrounding the bar. As calcium is liberated at the cathode, it rises through the electrolyte, comes in contact with the bar of calcium, and adheres to it.

Anode: $\qquad 2Cl^- - 2e^- \rightarrow Cl_2$

Cathode: $\qquad Ca^{2+} + 2e^- \rightarrow Ca$

Cell: $\qquad Ca^{2+} + 2Cl^- \rightarrow Ca + Cl_2$

The bar of calcium is slowly withdrawn at a rate equal to that at which the calcium metal is collected. Of course, chlorine is produced at the anode.

The development of extensive uses for the alkaline-earth metals has been retarded by their high cost of production. Nevertheless, these metals are very useful in the chemical laboratory and are used commercially to a limited extent as reducing agents in the production of other metals.

Magnesium. The method most widely used in the production of magnesium involves the electrolysis of fused magnesium chloride in an iron vessel (the walls of which serve as the cathode) in which anodes made of carbon are suspended. The magnesium liberated at the cathode is lighter than the electrolyte and therefore rises to the surface, where it is removed mechanically.

Although magnesium metal is used in the manufacture of signal flares, in the removal of gases from radio tubes, and in the production and purification of other metals, the single most important use for this metal lies in the fabrication of alloys. In recent years, many industrial products previously made from heavy metals such as copper and iron have been replaced by alloys of the light metals such as magnesium, beryllium, and aluminum. The increased speed and efficiency of modern transportation facilities (airplanes, automobiles, streamlined trains, etc.) are possible largely through the availability of alloys of the light metals. With the present extensive expansion of transportation by air, it is not unreasonable to suggest that magnesium and aluminum may eventually come to rival the traditional industrial importance of iron.

Zinc. Although zinc is a fairly active metal, its chemical activity is markedly less than that of the other metals considered in this section. Only about one-fourth of the zinc produced in this country is prepared by electrolysis. The remainder is produced by nonelectrolytic methods (described subsequently). In the electrolytic process, zinc ores are treated (*leached*) with dilute sulfuric acid, which extracts the zinc as zinc sulfate. The resulting aqueous solution is electrolyzed between a carbon anode and a cathode consisting of a thin sheet of pure zinc. Oxygen is liberated at the anode. As the electrolysis proceeds, the solution becomes progressively dilute with respect to zinc ion and progressively concentrated with respect to H_2SO_4. Consequently, the aqueous electrolyte may finally be used to treat a new batch of zinc ore.

Zinc is used chiefly as a protective coating for other metals and in the manufacture of alloys such as bronzes, brass, and bearing metals. Lesser quantities of zinc are used in the manufacture of dry-cell batteries and as a reducing agent both in small-scale laboratory use and in the industrial production of less active metals such as silver.

Aluminum. It is significant that chemical industries are characterized by ever-increasing quality of products accompanied by ever-decreasing prices. This trend is very well illustrated by the history of aluminum production. This metal was first prepared in Denmark in 1825 by the reduction of aluminum chloride by potassium at elevated temperatures.

$$AlCl_3 + 3K \rightarrow Al + 3KCl$$

The aluminum so produced sold for $158 per pound owing to the high cost of potassium and to the expense involved in obtaining dry aluminum chloride. Substitution of the cheaper metal sodium for the potassium lowered the price of aluminum to around $25 per pound, and subsequent improvements in methods for the production of sodium made possible the sale of aluminum at $5 per pound. Even so, these improved methods resulted in a total production of less than 100 lb of aluminum during the years in which these methods were used.

As an undergraduate student at Oberlin College in Ohio, Charles Martin Hall became interested in the problem of producing aluminum efficiently and at low cost. Using makeshift equipment and working under many handicaps, he successfully devised the electrolytic process that has come to bear his name. Hall first produced aluminum in 1886, and its price soon thereafter dropped to about $0.20 per pound. In the Hall process, the cell proper consists of an iron box lined with carbon, which serves as the cathode; a series of carbon rods extending into the vessel serves as the anode. The vessel is charged with the mineral *cryolite,* sodium aluminum fluoride (Na_3AlF_6), which is melted by the heat generated by the electric current from the generator. To the melted cryolite is then added aluminum oxide (Al_3O_3) obtained from the ore known as *bauxite* ($Al_2O_3 \cdot H_2O$ and $Al_2O_3 \cdot 2H_2O$). The oxide dissolves in the sodium aluminum fluoride and serves as the electrolyte. Thus, the molten cryolite serves as the solvent and aluminum oxide as the solute. During electrolysis, aluminum liberated at the cathode settles to the bottom of the cell and is withdrawn at suitable intervals. Of the oxygen liberated at the anode, some escapes as such while a part reacts with the carbon anodes to form oxides of carbon. As the dissolved aluminum oxide is electrolyzed, provision is made for addition of more Al_2O_3 at regular intervals so that the cells may be operated continuously.

Using the Hall process exclusively, the aluminum industry in the United States alone produces more than 150,000 tons of aluminum each year, and it is impossible to estimate the magnitude of probable future production. The commercial product obtained directly by electrolysis has a purity greater than 99%. It is of interest to note that a few months following the discovery of the Hall process an identical method was discovered independently by the French chemist Paul-Louis-Toussaint Héroult.

The uses of aluminum are similar to those of the light metals, some of which have previously been discussed under magnesium. Aluminum is also used in fabricating wire cables for electric-power transmission lines; in the manufacture of kitchen utensils, furniture, and paint; and in producing a wide variety of other useful articles. In connection with the general topic of alloys, attention is called to several important alloys containing aluminum.

Electrolytic refining of copper

A large percentage of the metallic copper produced is in the form of copper wire for the conduction of electricity. For this purpose, copper must be rendered very pure since the electrical conductivity of copper increases with increased purity. Copper produced by the usual chemical methods is not sufficiently pure for this purpose and is therefore subjected to an electrolytic refilling process. Crude copper in the form of large slabs serves as the anode, and a thin sheet of pure copper serves as the cathode in an electrolytic cell in which an aqueous solution of copper sulfate containing an excess of dilute sulfuric acid is used as the electrolyte. The chief impurities in the crude copper anode consist of Zn, Ni, Fe, Ag, Au, Pt, and Pd. During electrolysis the anode dissolves to form Cu^{2+}, Zn^{2+}, Ni^{2+}, and Fe^{2+}, and the applied voltage from the generator is so regulated that only the copper is deposited on the cathode, leaving the Zn^{2+}, Ni^{2+}, and Fe^{2+} (which are present in relatively small amounts) in solution. At the same time, the less active metals, Ag, Au, Pt, and Pd, do not dissolve to form ions and hence simply settle to the bottom of the cell, where they collect in the form of a sludge. The copper deposited on the cathode is usually at least 99.9% pure.

Each industrial chemical process has as its objective the economical production of a particular primary product. It is frequently true that, in attaining this objective, one or more by-products may become available. If these by-products can be disposed of at a profit, this serves to decrease the overall cost of operation and to permit the sale of the primary product at a lower, more favorable, competitive price. Thus, the cost of electrolytically refined copper is dependent on the recovery and sale of the by-products—silver, gold, platinum, and palladium. These *precious metals* are recovered in large quantities from accumulated anode sludges. Fully one-fourth of the total production of silver, about one-eighth of the gold, and lesser quantities of platinum and palladium are obtained as by-products of the electrolytic refining of copper.

Electroplating industries

Everyone is aware of the fact that if objects made of iron, steel, or other metals or alloys are allowed to remain in contact with air and atmospheric moisture, their surfaces become tarnished, rusted, or corroded. The practice of placing a protective metallic coating on such objects by electrolysis constitutes one of the oldest of the electrochemical industries. The metal comprising the protective coating usually consists of less active metals such as silver, gold, platinum, chromium, nickel, or copper. Electroplating is sometimes done merely for the purpose of improving the appearance of the finished product. In the process of electroplating silver on the surface of table cutlery, for example, the object to be plated is made the cathode of the electrolytic cell. A sheet of pure silver constitutes the anode, which dissolves at the same rate that silver is deposited on the cathode. The electrolyte consists of potassium argentocyanide $[KAg(CN)_2]$.

In this as in most other electroplating operations, it is desired to obtain a firm adherent deposit of metal, which can thereafter be burnished or polished. To secure such a deposit, careful regulation of acidity, temperature, current density, concentration of electrolyte, and *rate* of deposition must be maintained throughout the entire electrolysis. Failure to control one or more of these variables usually results in flaky or crumbly nonadherent deposits.

In a similar manner, copperplating is accomplished using a copper anode and a solution of potassium cuprocyanide $[KCu(CN)_2]$ as the electrolyte. Nickel is usually plated from nickel ammonium sulfate solutions $[Ni(NH_4)_2(SO_4)_2]$, and the electrolyte used in chromium plating consists of a mixture of chromic acid (H_2CrO_4) and chromic sulfate $[Cr_2(SO_4)_3]$. The case of chromium serves as a good example of the problems frequently encountered in the electroplating industries. In order to place a bright, attractive, yet highly protective coating on objects such as automobile bumpers and wheels, it is desirable to coat the original iron or steel object with nickel and chromium. Accordingly, nickel is plated directly on the iron, and chromium (which does not adhere readily to iron) is plated on the nickel.

Electrolytic production of nonmetallic elements

With the single exception of fluorine, the nonmetallic elements are usually more conveniently prepared by nonelectrolytic methods, some of which are described in detail in a later section. However, certain nonmetals are produced in sufficient quantity by electrolytic methods to warrant their consideration here.

Hydrogen and oxygen. From the preceding sections, it should be apparent that oxygen and hydrogen may often appear as by-products of industrial electrochemical processes in which nondischargeable anions and cations are involved. Whether these elements are collected and used depends, in any case, on the costs involved and the competition offered by other methods of production. Since oxygen can be produced so readily and at such low cost from liquid air, little oxygen is produced commercially by other methods. A considerable quantity of hydrogen is produced as the primary product, together with by-product oxygen, by the electrolysis of aqueous sodium hydroxide solution. The electrolytic cell employs an iron cathode and a nickel anode. The products, hydrogen and oxygen, are the same as those resulting from the electrolysis of water, dilute sulfuric acid, and so on. Large quantities of hydrogen are also produced as a by-product of the electrolysis of aqueous sodium chloride solution.

Halogens. An important industrial method for the production of chlorine is perhaps more significant in terms of the primary product, sodium hydroxide. Of course, chlorine, bromine, and iodine may be formed by the electrolysis of either aqueous or fused chlorides, bromides, or iodides. However, with the exception of chlorine, these elements are more commonly produced by other methods.

Fluorine can be produced only by electrolysis, and then only with difficulty. The complications involved in its production arise from the extreme chemical activity of this element. Of all the known elements, fluorine has the greatest affinity for electrons; hence, it is the most powerful oxidizing agent (electron acceptor). Accordingly, fluorine cannot be prepared by the chemical oxidation of fluorine compounds, since this would require an oxidizing agent more powerful than fluorine. This element displaces oxygen from water, attacks all but the very unreactive metals such as platinum, and forms compounds with all elements except the inert gases. Because of its generally corrosive character, the design of apparatus for the preparation, storage, and use of fluorine presents many difficult problems. The element is prepared by the electrolysis of fused potassium hydrogen fluoride (KHF_2) in a copper cell that serves as a cathode and which is fitted with graphite anodes. Although cells made of copper are fairly resistant to attack by fluorine, they last for only a short time. More durable cells may be made of silver, platinum, or gold.

The importance of fluorine lies largely in the use of several compounds of that element. Calcium fluoride, or *fluorspar* (CaF_2), is used as a flux and, like sodium fluoride (NaF), as an insecticide. Ammonium fluoride (NH_4F) is used as a disinfectant; hydrofluoric acid is used in the etching of glass; and organic compounds are used as commercial refrigerants.

Electrolytic production of compounds

Only a small number of compounds are produced directly by electrolysis. To illustrate this type of process, the electrolytic production of sodium hydroxide is described in detail. Then it is shown how this process may be modified to permit the formation of two other valuable commercial chemicals.

Sodium hydroxide. The electrolysis of an aqueous solution of sodium chloride yields chlorine at the anode and hydrogen at the cathode. Discharge of hydrogen ions (from water) leaves sodium and hydroxyl ions in the region surrounding the cathode. Thus, sodium hydroxide is produced in the solution about the cathode. The use of this method for the production of both chlorine and sodium hydroxide depends on devising some means of keeping the chlorine and sodium hydroxide from coming into contact, since they react to form other compounds. Of the numerous cells designed to accomplish this purpose, one used rather widely is the Nelson cell.

The walls of the vessel that serves as the container for the salt solution (brine) and as the cathode consist of perforated sheets of iron or of wire gauze covered on the *inside* with layers of insulating material. The anodes, composed of graphite, are immersed in the solution in the center of the cell. The thickness of the asbestos paper and the height of the liquid in the cell are such that the aqueous solution seeps through the paper and reaches the cathode just fast enough to sweep the hydroxyl ions to the back of the cathode as rapidly as they are left behind by the discharge of hydrogen ions. The liquid at the back of the cathode takes no further part in the electrolysis—and drops off the cathode. At the same time, the chlorine gas liberated at the anode rises into the space above the solution, from which the gas is withdrawn by slight suction. Thus, the chlorine is collected inside the asbestos walls where it cannot come into contact with the sodium hydroxide. In the operation of the Nelson cell, provision is made for continuous operation by adding fresh brine automatically.

In this cell, not all of the dissolved sodium chloride is electrolyzed. Consequently, the solution that seeps through the perforated cathode contains sodium hydroxide together with some unchanged sodium chloride. The solution is concentrated by evaporation, whereupon most of the less soluble sodium chloride crystallizes and the very soluble sodium hydroxide remains in solution. This concentrated solution of sodium hydroxide (caustic soda) may be sold as such, or the remainder of the water may be driven off by heating to form solid sodium hydroxide. If a purer product is desired, the solid is dissolved in alcohol, which does not dissolve the remaining traces of sodium chloride. Pure sodium hydroxide is then secured by filtration, followed by evaporation of the alcohol.

Another somewhat more complicated cell for the production of chlorine and sodium hydroxide by the electrolysis of sodium chloride solution is the Castner-Kellner cell, which employs a liquid mercury cathode.

Sodium hypochlorite.   If, in the electrolysis of sodium chloride solution, provision is made for the intimate mixing of the chlorine and sodium hydroxide at low temperatures, the following reaction occurs:

$$Cl_2 + 2Na^+OH^- \rightarrow Na^+ClO^- + Na^+Cl^- + H_2O$$

The resulting aqueous solution containing sodium chloride and sodium hypochlorite is known as *Javelle water* and is used as an antiseptic (Dakin's solution). Treatment of Javelle water with a strong acid results in liberation (in solution) of the weak and unstable hypochlorous acid

$$2Na^+ClO^- + H_2^+SO_4^{2-} \rightarrow 2HClO + Na_2^+SO_4^{2-}$$

which decomposes readily to liberate oxygen and is used extensively as a bleaching agent.

Sodium carbonate.   When sodium hydroxide is produced by the electrolysis of sodium chloride solution saturated with carbon dioxide, sodium carbonate is produced as follows:

$$CO_2 + H_2O \rightleftarrows H_2CO_3$$

$$Na^+OH^- + H_2CO_3 \rightarrow H_2O + Na^+HCO_3^-$$

If the quantity of available carbon dioxide is limited, the normal carbonate results.

$$2Na^+OH^- + H_2CO_3 \rightarrow 2H_2O + Na_2^+ CO_3^{2-}$$

Although this electrolytic method has many advantageous features in comparison with other methods, it suffers the competitive disadvantage of high electrical power costs.

Battery Cells

Everyone is familiar with the use of storage batteries in automobiles, the use of the *dry cell* in flashlights, and so forth. In connection with the study of electrolysis, attention was called to the possibility of employing storage batteries as the source of electrical energy necessary for the occurrence of electrochemical transformations. In view of the fact that the energy supplied by a battery cell is of chemical origin, it is interesting to inquire how the energy originating in chemical reactions can be used to perform useful work. Such an inquiry serves also to establish the relationship between the chemical changes that occur during electrolysis and those that take place in battery cells.

Relationship between electrolytic cells and battery cells

Perhaps the best approach to understanding the operation of battery cells may be had by reconsidering a simple case of electrolysis. The nonspontaneous changes that occur during the electrolysis of an aqueous solution of zinc chloride may be represented by the following equations:

Anode:	$2Cl^- - 2e^- \rightarrow Cl_2$
Cathode:	$Zn^{2+} + 2e^- \rightarrow Zn$
Cell:	$Zn^{2+} + 2Cl^- \rightarrow Zn + Cl_2$

These changes are caused by the imposition of an outside force, which may take the form of electrical energy supplied by a battery cell. Thus, energy is used and work is done in electrolyzing the zinc chloride solution. Since energy can be neither created nor destroyed, it follows that the electrical energy used during electrolysis must have been transformed into chemical energy possessed by the products of electrolysis. This fact leads directly to the question of whether this chemical energy stored up in the products of electrolysis (zinc and chlorine) may be retransformed into electrical energy. In other words, it is proposed to electrolyze zinc chloride solution in a suitable cell and thereafter to use this cell as a source of energy (i.e., as a battery).

The feasibility of this suggestion may be ascertained by electrolyzing a zinc chloride solution until an appreciable quantity of the products of electrolysis has been accumulated and thereafter connecting the electrodes to a voltmeter, replacing the battery shown in Fig. 11.2 with the voltmeter. The cell in question is capable of functioning as a battery that delivers about 2.1 volts.

As one continues to draw current from this battery, it can be shown that zinc dissolves off the electrode and goes into solution as Zn^{2+}. Hence, at that electrode, which shall be designated as the negative (−) terminal of the battery, electrons are made available by the occurrence of the oxidation reaction,

$$(-) \text{ Terminal:} \qquad Zn - 2e^- \rightarrow Zn^{2+}$$

which represents the normal tendency toward spontaneous loss of electrons by zinc in its effort to approach the electronic structure of the nearest inert gas, argon. Correspondingly, the reaction at the positive (+) terminal of the battery involves the reduction reaction,

$$(+) \text{ Terminal:} \qquad Cl_2 + 2e^- \rightarrow 2Cl^-$$

which corresponds to the spontaneous reaction tendency normally exhibited by chlorine, as atoms of this element seek additional electrons that enable each chlorine atom to acquire an extra nuclear structure like that of an inert gas. Thus, electrons flow from the negative terminal through the outside circuit (including the voltmeter) and over to the positive terminal. This is just the opposite of the direction of flow of electrons during electrolysis, and the oxidation-reduction reactions representing the changes that occur in the battery cell,

$$(-) \text{ Terminal:} \qquad Zn - 2e^- \rightarrow Zn^{2+}$$

$$(+) \text{ Terminal:} \qquad Cl_2 + 2e^- \rightarrow 2Cl^-$$
_____

$$\text{Battery:} \qquad Zn + Cl_2 \rightarrow Zn^{2+} Cl_2^-$$

are also seen to be the reverse of those that occur during the electrolysis of zinc chloride solution.

Although a battery cell depending on these reactions of zinc and chlorine could be used as a source of electrical energy, it is rather obvious that a battery involving the use of gaseous chlorine would hardly be practical in everyday use. Hence, before going any further into the problem of the general characteristics of battery cells, a battery more nearly suitable for actual use is considered.

Daniell cell

One may construct a Daniell cell quite simply by immersing a strip of zinc metal in an aqueous solution of zinc sulfate contained in a porous unglazed porcelain cup and similarly immersing a strip of copper in an aqueous solution of copper sulfate in a second container. When these two units are then placed in a container partly filled with a solution of an electrolyte such as sodium chloride and the two terminals are connected to a voltmeter, as shown by replacing the generator in Fig. 11.2 with a voltmeter, it is observed that a voltage of 1.102 volts is produced.

It is immediately apparent that the Daniell cell differs from the zinc-chlorine battery in that the electrode materials (i.e., zinc and copper) of the former are both metals that normally exhibit a tendency to lose electrons. If the Daniell cell is to function as a battery, both metals cannot lose electrons—one must lose and the other must gain electrons. In this particular case, the issue can be decided, qualitatively at least, in terms of the order of activity of the metals. From Table 11.1, it should be recalled that zinc is much more active chemically than copper; hence zinc might be expected to lose electrons more readily than copper if the metals are in contact with solutions of their ions at the same concentration. It may be inferred correctly that the reactions that occur when the Daniell cell serves as a source of electrical energy are as follows:

$$(-)\ \text{Terminal:} \qquad Zn - 2e^- \rightarrow Zn^{2+}$$

$$(+)\ \text{Terminal:} \qquad Cu^{2+} + 2e^- \rightarrow Cu$$

$$\text{Battery:} \qquad Zn + Cu^{2+} \rightarrow Zn^{2+} + Cu$$

TABLE 11.1   Order of Activity of Metals

Potassium
Barium
Calcium
Sodium
Magnesium
Aluminum
Manganese
Zinc
Chromium
Iron
Cadmium
Nickel
Tin
Lead
Hydrogen
Bismuth
Copper
Silver
Mercury
Platinum
Gold

From this it may be seen that the reactions that occur at the two terminals of a battery employing metals as both terminals are dependent on the relative tendencies of the two metals toward loss of electrons. Consequently, any further study of battery cells should be based on some suitable *quantitative* expression of these tendencies.

Electrode potentials

The atoms of all the elements have *some* tendency toward loss of electrons. In the case of the nonmetallic elements, this tendency is extremely small in relation to their tendency to gain electrons. Consider, for example, the tendencies exhibited when a strip of metallic zinc is placed in contact with a solution containing zinc ions. Atoms of zinc tend to detach themselves from the surface of the metal and go into solution as zinc ions by losing electrons. At the same time, zinc ions in the solution tend to acquire electrons at the surface of the metal and become zinc atoms. Thus, in the case of zinc, there are two opposing tendencies, the larger of which is that toward loss of electrons. An equilibrium between these two tendencies becomes established, and since the tendency toward loss of electrons predominates, the strip of zinc possesses an excess of electrons and may be said to have acquired a negative electrical potential (i.e., there will be a difference of potential between the metal and the solution). For zinc or for any other substance in contact with a solution of its ions, the magnitude of this difference in potential depends on both temperature and concentration. This must be true since the return of ions from the solution to the metal depends on the velocity with which the ions move through the solution and on the number of collisions that occur between the ions and the surface of the metal. The higher the temperature, the greater the speed of the ions, and the greater the concentration, the greater the number of collisions per unit of time.

If, on the other hand, a strip of copper is placed in contact with a solution of copper ions, the same tendencies are exhibited. Although copper atoms do tend to lose electrons and become copper ions, this is opposed by a greater tendency for copper ions to gain electrons and become copper atoms. Hence, in this case, the copper electrode acquires a potential with respect to the solution that is less negative than in the case of zinc. Accordingly, the electrode potential of copper is given a positive sign, whereas that of zinc is given a negative sign, from which it follows that the sign of an electrode potential indicates which of these two tendencies predominates.

Measurement of electrode potentials. It is not practical to attempt to measure the absolute potential difference between each element and a solution of its ions. Rather, a suitable reference electrode is chosen and other electrode potentials are measured in relation to this standard. Ordinarily, the hydrogen electrode is used as the reference. This electrode consists of a platinum foil immersed at 25°C in a solution that is 1.0 N with respect to hydrogen ions and constructed to permit hydrogen gas (under a pressure of 1 atm) to be bubbled into the solution. The potential of this electrode is arbitrarily assumed to have the value of zero. To measure the potential of the zinc electrode, for example, the hydrogen electrode constitutes one-half of the complete cell. The other half of the cell consists of the zinc strip in contact with a 1-molal solution of zinc ions. Thus, one actually measures not the absolute potential of the zinc electrode but rather the net effect due to the potential difference between zinc and its ions and that between elemental hydrogen and hydrogen ions.

However, since one is interested primarily in comparing the potential of one electrode with that of another, the constant error involved in assuming the potential of the hydrogen electrode to be zero does not detract from the utility of the electrode potential values obtained in this manner. By making a series of such measurements using a variety of electrodes in comparison with the hydrogen electrode, an extensive list of electrode potential data may be compiled. Such a list is given in Table 11.2.

In terms of the data of Table 11.2, the voltage produced by the Daniell cell is seen to be the *sum* of the negative potential of the zinc electrode and the positive potential of the copper electrode (i.e., 0.758 + 0.344 = 1.102 V). The voltage produced by a battery involving a cadmium and a lead electrode would be the *difference* between the two potentials (i.e., 0.397 − 0.12 = 0.277 V), since both of these electrodes are negative with respect to solutions of their ions.

Construction of battery cells

Any consideration of the requirements to be fulfilled in the construction of battery cells should recognize first that the substances used as electrodes may be, but need not be, produced by electrolysis. In the earlier discussion of the zinc-chlorine battery, both of the substances involved at the two terminals were considered to be the products of a previously conducted electrolysis. However, the zinc-chlorine battery could just as well have been constructed by the use of zinc and chlorine produced by entirely nonelectrolytic methods. You should recall that, in connection with the description of the Daniell cell, no specifications were made with regard to the origin of any of the chemicals involved. This freedom to select suitable materials regardless of their origin or past history follows from the fact that the changes that occur during

TABLE 11.2   Standard Electrode Potentials at 25°C

Element or ion	Reaction	Potential, V
Potassium	$K \leftrightarrow K^+ + 1e^-$	−2.922
Barium	$Ba \leftrightarrow Ba^{2+} + 2e^-$	−2.90
Calcium	$Ca \leftrightarrow Ca^{2+} + 2e^-$	−2.763
Sodium	$Na \leftrightarrow Na^+ + 1e^-$	−2.712
Magnesium	$Mg \leftrightarrow Mg^{2+} + 2e^-$	−2.40
Aluminum	$Al \leftrightarrow Al^{3+} \, 3e^-$	−1.70
Manganese	$Mn \leftrightarrow Mn^{2+} + 2e^-$	−1.10
Zinc	$Zn \leftrightarrow Zn^{2+} + 2e^-$	−0.758
Chromium	$Cr \leftrightarrow Cr^{2+} + 2e^-$	−0.557
Sulfide ion	$S^{2+} \leftrightarrow S + 2e^-$	−0.51
Iron	$Fe \leftrightarrow Fe^{2+} + 2e^-$	−0.44
Cadmium	$Cd \leftrightarrow Cd^{2+} + 2e^-$	−0.397
Nickel	$Ni \leftrightarrow Ni^{2+} + 2e^-$	−0.22
Tin	$Sn \leftrightarrow Sn^{2+} + 2e^-$	−0.13
Lead	$Pb \leftrightarrow Pb^{2+} + 2e^-$	−0.12
Sulfide ion*	$S^{2+} \leftrightarrow S + 2e^-$	−0.07
Hydrogen	$\frac{1}{2}H_2 \leftrightarrow H^+ + 1e^-$	0.00
Stannous ion	$Sn^{2+} \leftrightarrow Sn^{4+} + 2e^-$	+0.13
Bismuth	$Bi \leftrightarrow Bi^{3+} \, 3e^-$	+0.20
Copper	$Cu \leftrightarrow Cu^{2+} + 2e^-$	+0.344
Hydroxyl ion	$2OH^- \leftrightarrow \frac{1}{2}O_2 + H_2O + 2e^-$	+0.40
Iodide ion	$2I^- \leftrightarrow I_2 + 2e^-$	+0.535
Ferrous ion	$Fe^{2+} \leftrightarrow Fe^{3+} + 1e^-$	+0.770
Silver	$Ag \leftrightarrow Ag^+ + 1e^-$	+0.779
Mercury	$Hg \leftrightarrow Hg^{2+} + 2e^-$	+0.86
Platinum	$Pt \leftrightarrow Pt^{4+} + 4e^-$	+0.863
Bromide ion	$2Br^- \leftrightarrow Br_2 + 2e^-$	+1.065
Chloride ion	$2Cl^- \leftrightarrow Cl_2 + 2e^-$	+1.359
Chromic ion	$2Cr^{3+} + 7H_2O \leftrightarrow Cr_2O_7^{2+} + 14H^+ + 6e$	+1.36
Gold	$Au \leftrightarrow Au^{3+} + 3e^-$	+1.36
Manganous ion	$Mn^{2+} + 4H_2O \leftrightarrow MnO_4^- + 8H^+ + 5e^-$	+1.52
Fluoride ion	$2F^- \leftrightarrow F_2 + 2e^-$	+2.80

* Saturated solution of hydrogen sulfide in water; $[S] = 1 \times 10^{-15}M$.

the production of current by a battery are entirely spontaneous (i.e., zinc has a certain tendency to lose electrons, which is the same irrespective of the origin of the zinc).

Since suitable arrangement of two substances having different tendencies toward loss of electrons results in a battery, the number of possible batteries would seem to be almost without limit. Certain broad requirements must be recognized and enumerated before considering the relative practicality of the many batteries that might conceivably be constructed:

1. *A battery requires two half cells, each of which must involve two oxidation states of an element.*   Thus, in the Daniell cell, one of the half cells consists of copper metal (oxidation number = 0) in contact with

copper ions (oxidation number = +2) and the other of zinc metal (oxidation number = 0) and zinc ions (oxidation number = +2). Although often convenient, one component having zero oxidation number is not required. For example, a battery could be constructed to involve ferrous (Fe^{2+}) and ferric (Fe^{3+}) ions, thus meeting this requirement.

2. *The two solutions surrounding the terminals must be either in contact or connected by a solution of an electrolyte.* Examples of suitable arrangements may be found in preceding discussions and in those that follow. The function of the electrolyte between the two terminals is exactly the same as that discussed in connection with the study of electrolysis, namely, the maintenance of electrical neutrality in the region of the electrodes. This is accomplished by the transport of ions through the solution.

3. *The negative terminal of the battery is the one involving the substance having the greater tendency to lose electrons.* Electrons must therefore be acquired at the other terminal, which is designated as the positive terminal.

4. *The voltage delivered by a battery is determined by the innate tendencies of the active materials to lose electrons and by the concentrations of these materials.* In Table 11.2, the listed numerical values representing the various electrode potentials were obtained experimentally under comparable conditions of concentration and temperature. The relative position of any electrode included in that list would be changed if the experimental measurements were made using a sufficiently different concentration of ions surrounding the electrode. The influence of a change in concentration on these equilibria is wholly in accord with facts presented earlier in connection with the study of reversible reactions and chemical equilibria.

Practical batteries. In the manufacture of batteries for the performance of useful work, convenience, cost, and durability are the major considerations. Since so-called storage batteries generally involve the use of aqueous solutions of electrolytes, very active metals such as sodium, potassium, and calcium are excluded, because these metals displace hydrogen from water. The very active nonmetal fluorine is also excluded, because it displaces oxygen from water. The gaseous elements and compounds are impractical as well, because provision for the use of gases in batteries for everyday application would be very difficult. With these exclusions, it follows that practical batteries usually employ electrodes consisting of the solid elements (both metals and nonmetals) and their solid compounds in contact with electrolytes in aqueous solutions whose concentration may be varied over wide ranges.

If one wishes to construct a battery to deliver a given voltage, the initial selection of the electrode materials may be made by reference to a tabulation of electrode potentials, such as is given in Table 11.2. For example, if a 2.0-V cell is desired, the use of manganese and mercury terminals would be indicated since this combination (see Table 11.2) would produce 1.10 + 0.86 = 1.96 V. By suitably adjusting the concentration of the electrolytes surrounding the two terminals, exactly the required voltage could be obtained. If one wished to consider electrodes other than the substances listed in Table 11.2, reference should be made to more extensive tables of data.

By considering the changes that occur while a battery is being used as a source of energy, it may be seen that the negative terminal is dissolved. Also, the concentration of the ions corresponding to the positive terminal is progressively decreased as these ions are converted to the corresponding neutral atoms that are liberated at the surface of the positive terminal. Hence, the useful life of a battery would appear to be limited by the available supply of the materials that react (spontaneously) at the two terminals. This is not the case if the battery is capable of being charged. While a battery is delivering current, it is said to be undergoing a process of *discharge.* Reversal of the reactions that occur during discharge constitutes the process of *charging* the battery, which is accomplished by electrolysis. In effect, the production of the zinc-chlorine battery consisted of the charging of that battery. If the battery was then used as a source of current until all or nearly all of the zinc and chlorine was used up, the cell could be charged again by repeating the process of electrolysis. To charge a battery capable of providing a given voltage, it is necessary only to connect the terminals to the corresponding terminals of another battery (or more likely a generator) capable of producing a higher voltage. Examples of batteries that cannot be recharged conveniently are the Daniell cell and the common dry cell. The familiar lead storage battery owes much of its utility to the ease with which it may be recharged.

Gravity cell

The gravity or crowfoot battery is merely a special case of the Daniell cell discussed previously and involves the identical reactions. The gravity cell is constructed by surrounding thin sheets of copper (which serve as the positive terminal) with a saturated solution of copper sulfate together with an excess of solid crystals of $CuSO_4 \cdot 5H_2O$. The solution surrounding the negative terminal consists of a *dilute* solution of zinc sulfate having a specific gravity so much less than that of the *concentrated* copper sulfate solution that the two solutions do not mix appreciably—hence the name *gravity cell.* A heavy zinc electrode (the negative terminal) is fashioned in the form of a crowfoot in order to present an extensive surface area and is immersed in the dilute zinc sulfate solution. When the two terminals are connected, electrons flow from the zinc to the copper electrode. The initial voltage supplied amounts to 1.102 V. However, as the battery continues to deliver current, the zinc electrode dissolves and the surrounding solution becomes progressively more concentrated; hence, the zinc electrode becomes less negative with respect to the solution, and the voltage drops off in proportion. It is neither convenient nor practical to attempt to recharge a gravity cell.

Lead storage battery

This battery is the one commonly used in automobiles and in numerous other applications that require relatively small voltages. The essential parts of the lead storage battery are described here. As constructed commercially, a hard rubber or plastic case contains dilute sulfuric acid of initial specific gravity of 1.250 to 1.275, which serves as the electrolyte. The negative terminal consists of a sheet (or sheets) of lead formed to permit a coating of spongy elemental lead to be held on the surface exposed to the electrolyte. The positive terminal is constructed similarly and consists of grids of lead packed with lead dioxide (PbO_2). During discharge, the cell reactions are as follows:

(–) Terminal: $\qquad\qquad Pb - 2e^- \rightarrow Pb^{2+}$

(+) Terminal: $\qquad\qquad \underline{Pb^{4+} + 2e^- \rightarrow Pb^{2+}}$

Battery: $\qquad\qquad\quad Pb + Pb^{4+} \rightarrow 2Pb^{2+}$

However, since these reactions occur in the presence of sulfuric acid, the lead ions resulting from the primary reactions indicated here are converted to insoluble lead sulfate. Accordingly, the complete description of the process of discharge is represented by the following reaction:

$$Pb + PbO_2 + 2H_2SO_4 \underset{\text{charge}}{\overset{\text{discharge}}{\rightleftarrows}} 2PbSO_4 + 2H_2O$$

As the battery delivers current, the forward reaction results in using up some of the available H_2SO_4 and in the formation of water, which dilutes the remaining sulfuric acid. Consequently, the process of discharge is accompanied by a decrease in the specific gravity of the electrolyte, and this is the basis of the test usually applied to determine the condition of a storage battery. Ordinarily, a storage battery should be charged when the specific gravity of the electrolyte decreases to a value in the neighborhood of 1.100.

During discharge, both electrodes become coated with insoluble lead sulfate, and if this coating is allowed to become excessive, the lead sulfate interferes mechanically with further reaction. This deposit of lead sulfate is removed during the charging process, which involves the reversal of the reactions that occur during discharge. The lead storage battery may be charged by connecting it to a 20-V dc source. The ensuing electrolysis not only results in the dissolving of the lead sulfate and the restoration of the initial specific gravity of the electrolyte, but also regenerates spongy lead on the one electrode and lead dioxide at the other. At the same time, electrolysis results (to a limited extent) in the discharge of hydrogen and hydroxyl ions with liberation of hydrogen and oxygen. Consequently, water thus lost by electrolysis as well as that lost by evaporation must be replaced periodically by the addition of pure water to maintain the proper concentration of the electrolyte. When completely charged, the lead storage cell delivers a voltage of about 2.1 V. An ordinary 12-V lead storage battery consists of a combination of six such units connected in series.

Edison storage cell

Another example of a practical battery, but one less commonly used, is the Edison cell. The active electrode materials consist of iron and nickel dioxide in contact with an electrolyte consisting of aqueous potassium hydroxide solution. Owing to its marked instability, the NiO_2 changes spontaneously to Ni_2O_3 with liberation of oxygen. The reactions responsible for the flow of current are as follows:

(–) Terminal: $$Fe - 2e^- \rightarrow Fe^{2+}$$

(+) Terminal: $$2Ni^{3+} + 2e^- \rightarrow 2Ni^{2+}$$

Battery: $$Fe + 2Ni^{3+} \rightarrow Fe^{2+} + 2Ni^{2+}$$

In the presence of potassium hydroxide, however, the insoluble hydroxides of iron and nickel are produced. The processes of discharge and charge may be represented as follows:

$$Fe + Ni_2O_3 + 3H_2O \underset{\text{charge}}{\overset{\text{discharge}}{\rightleftharpoons}} Fe(OH)_2 + 2Ni(OH)_2$$

Dry cell

Perhaps the most familiar application of the dry cell is its use in flashlights. This battery consists of a zinc case that serves as the negative terminal. The inner walls are covered with a thin coating of porous paper or cloth, which keeps the zinc from coming into direct contact with the contents of the cell. A carbon rod placed in the center of the cell constitutes the positive terminal, and the remaining space is filled with a pasty mass consisting of a mixture of ammonium chloride and manganese dioxide together with a small amount of zinc chloride. The cell is closed at the top by a coating of pitch or wax, which prevents evaporation of water from the moist electrolyte and which also prevents escape of any gases produced within the cell. The *primary* reactions that occur during discharge are as follows:

(–) Terminal: $$Zn - 2e^- \rightarrow Zn^{2+}$$

(+) Terminal: $$2NH_4^+ + 2e^- \rightarrow 2NH_3 + H_2$$

Battery: $$Zn + 2NH_4^+ \rightarrow Zn^{2+} + 2NH_3 + H_2$$

Since the gases, NH_3 and H_2, cannot escape, they must be used in reactions within the cell. Ammonia molecules combine with zinc ions (produced at the negative terminal) to form the complex ion $[Zn(NH_4)_3]^{2+}$. The hydrogen is consumed by reaction with manganese dioxide, in a *slow* reaction:

$$2MnO_2 + H_2 \rightarrow Mn_2O_3 + H_2O$$

If a dry cell is used continuously for an appreciable period of time, the foregoing reaction is so slow that the hydrogen is not used as rapidly as it is formed. Consequently, gaseous hydrogen collects on the surface of the carbon rod and thus tends to insulate the positive terminal to such an extent that ammonium ions are less readily able to approach the surface of the carbon rod. This results in a decrease in the voltage delivered by the cell. However, if a period of rest is permitted, the accumulated hydrogen is removed slowly by reaction with manganese dioxide, and the interference by gaseous hydrogen is thereby eliminated.

The best commercial dry cells have a rating of about 1.5 V, and such batteries cannot be charged conveniently. The dry cell is rather different from most common batteries in that it is made up largely of solid materials.

Metals and Alloys

Although no definite line of demarcation can be drawn between the metallic and the nonmetallic elements, approximately three-fourths of the known chemical elements are usually considered to be predominantly metallic in character. The objective of the present discussion is to show certain chemical and physical properties that make these metals a class of elements possessing common characteristics. Wholly aside from the fact that metals are far more numerous than nonmetals, the many and varied industrial uses of metals make it important to understand the properties of the metallic state. Such knowledge has been largely responsible for the development of industrial applications that have contributed so much to the progress of modern industry.

Occurrence of metals

The availability of an element (whether metallic or nonmetallic) depends on both its abundance and its mode of existence in nature. Of two metals on or near the earth's surface, one may be far more abundant than the other and yet far less accessible because it does not occur in high concentration at one or more places. In other words, a very abundant element may be relatively inaccessible because it is fairly uniformly distributed throughout the earth's surface. On the other hand, a metal that exists on the earth only to a limited extent may be readily available because existing supplies are found in rich deposits consisting principally of that metal or one or more of its compounds. These ideas are strikingly illustrated by comparing the histories of magnesium and lead. Although magnesium is over 1000 times more abundant than lead, its extensive commercial use is a relatively modern development, whereas lead was known and used by the Babylonians and the Romans.

Early use of lead, as well as metals such as tin, copper, silver, gold, or mercury, was also possible by virtue of the fact that these metals exist in nature, either in an uncombined condition or in chemically combined forms from which these metals may be secured by relatively simple procedures.

Regardless of their chemical character, the various forms in which both nonmetals and metals occur in nature are usually called *minerals*. A mineral that is of such a character that it serves as an economical source for the commercial production of a metal or a nonmetal is referred to as an *ore*.

Native metals.   Only a relatively small number of the metals are capable of existence in nature in the uncombined condition—for example, copper, silver, gold, antimony, bismuth, and platinum. These so-called noble metals are those which exhibit such low degrees of chemical activity that they can exist in nature without entering into chemical combinations resulting from reactions with water, atmospheric oxygen, or carbon dioxide.

Types of ores.   In addition to the native metals, which are relatively unimportant, a wide variety of minerals is found in nature, and these minerals provide the chief ores from which the various metals are extracted commercially. Although these ores encompass many different types of chemical compounds, the great majority of the more important metal ores are notable for their chemical simplicity, as shown by the following list. (Note that the names assigned to minerals have little, if any, significance. These names are not systematic and provide no indication of the chemical character of the substance involved. However, since these trivial names are in common use, it is best to become familiar with both the mineral names and the corresponding chemical names. For example, the names *cuprous sulfide* and *chalcocite* should both be associated with the formula Cu_2S.)

Sulfides.   The simple or (less frequently) complex sulfides of a number of heavy metals are employed extensively as sources of these metals. For example, the minerals *chalcocite* (CU_2S), *chalcopyrite* ($CuFeS_2$), *cinnabar* (HgS), *galena* (PbS), and *sphalerite* (ZnS) are typical of the useful sulfide ores of heavy metals.

Oxides.   The minerals *hematite* (Fe_2O_3), *bauxite* ($Al_2O_3 \cdot H_2O$ and $Al_2O_3 \cdot 2H_2O$), *cassiterite* (SnO_2), *pyroluiite* (MnO_2), and *franklinite* ($ZnO \cdot Fe_2O_3$) serve as examples of common oxide ores. Because of their importance as sources of iron and aluminum, oxide ores are perhaps of greater importance than any other single type.

Carbonates.   Among many carbonate minerals that provide the corresponding metals, the basic carbonate *malachite* [$Cu_2(OH)_2CO_3$] and the normal carbonates *siderite* ($FeCO_3$), *cerussite* ($PbCO_3$), *strontianite* ($SrCO_3$), and *dolomite* ($CaCO_3 \cdot MgCO_3$) are representative ores.

Other types.   Without going into detail regarding each type of compound, the list of minerals in Table 11.3 shows the wide variety in the types of compounds that are known and used. This tabulation is intended to be illustrative rather than exhaustive.

Extraction of metals from their ores

The science of *metallurgy* is concerned with the extraction of metals from their ores and with their subsequent purification. Since metals exist in nature in such a wide variety of types of chemical combination, it is apparent that many different kinds of chemical treatment may be required. Other factors that complicate the subject of metallurgy include the varying degrees of chemical activity exhibited by the metals and the fact that both rich and low-grade ores must be used. Although it is somewhat difficult to generalize with respect to a problem involving so many variables, it is convenient to consider the subject of metallurgy as involving a number of fairly well defined steps.

TABLE 11.3　Some Common Minerals Other Than Sulfides, Oxides, and Carbonates

Type of compound	Mineral name	Formula	Chemical
Halide	Rock salt	NaCl	Sodium chloride
	Carnallite	$KCl \cdot MgCl_2 \cdot 6H_2O$	Potassium magnesium chloride hexahydrate
	Sylvite	KCl	Potassium chloride
	Horn silver	AgCl	Silver chloride
Sulfate	Epsomite	$MgSO_4 \cdot 7H_2O^*$	Magnesium sulfate heptahydrate
	Barite	$BaSO_4$	Barium sulfate
	Gypsum	$CaSO_4 \cdot 2H_2O$	Calcium sulfate dihydrate
	Anglesite	$PbSO_4$	Lead sulfate
Silicate	Willemite	Zn_2SiO_4	Zinc orthosilicate
	Asbestos	$CaMg_3Si_4O_{12}$	Calcium magnesium silicate
	Mica	$KAl_3H_2Si_3O_{12}$	Potassium aluminum hydrogen silicate
	Beryl	$Be_3Al_2Si_6O_{18}$	Beryllium aluminum silicate
Miscellaneous	Wolfram	$FeWO_4$	Ferrous tungstate
	Chalcolite	$Cu(UO_2)_2 \cdot 8H_2O$	Copper uranyl phosphate octahydrate
	Chromite	$Fe(CrO_2)_2$	Ferrous chromite
	Crocoite	$PbCrO_4$	Lead chromate

* In the pure condition, this compound is known as *Epsom salt*.

Concentration processes. On the assumption that the ore has been mined and is available for processing, the first problem to be faced is that of eliminating undesirable components of the natural ore. Most ores contain a certain amount of rock or other earthy materials that do not contain any of the desired metal, and this condition is particularly prevalent among low-grade ores which contain only a small percentage of the desired metal. The method used in concentrating the ore depends on the character of the naturally occurring material. If the density of the desired component is greater than that of the worthless impurities, the latter often may be removed merely by stirring the finely powdered crude ore with water, thus washing away the lighter fractions and leaving behind a concentrate containing the metal sought. A few minerals are magnetic, and this property is sometimes used to effect a gross separation from nonmagnetic impurities. Most commonly, however, ores are concentrated by the process of *flotation*. This process, which is most frequently employed in the concentration of sulfide ores, involves agitation (by a stream of air) of the finely ground ore in water to which pine oil, "cresylic" (tolylic) acids, or other organic materials have been added. The sulfides are carried to the surface by adherence to the resulting froth, while sand particles and other heavy materials settle to the bottom. Removal of the material thus brought to the surface followed by removal of adhering water provides a concentrate much richer in the desired metal than the original crude ore. For example, crude low-grade copper ores containing as little as 0.6% Cu are subjected to concentration by flotation, and the resulting concentrate (after one flotation) contains as much as 20% Cu. As the richer ore deposits become exhausted, it becomes increasingly necessary to resort to the concentration of low-grade ores by flotation or by other suitable procedures.

Roasting.   The process of roasting consists of heating the finely powdered ore (either a concentrate or the crude high-grade ore) in the presence of air. This treatment may serve one of three purposes, depending on the nature of the ore:

1. Sulfides are converted to the oxide of the metal and sulfur dioxide by reaction with atmospheric oxygen:

$$2MS + 3O_2 \rightarrow 2MO + 2SO_2$$

2. Carbonates are converted to oxides with liberation of carbon dioxide:

$$MCO_3 \rightarrow MO + CO_2$$

3. Water, including that held in chemical combination, is removed when the ore is heated.

Smelting.   After an ore has been subjected to an appropriate treatment as outlined here—or, as is sometimes possible, without any pretreatment—the ore is subjected to a process known as *smelting*. It is in this procedure that the metal is actually obtained in the elemental condition. Although the details differ considerably among the many metallurgical processes in everyday use, the overall characteristics usually conform to one of the following general methods.

Electrolytic reduction.   Although perhaps not properly classified as a smelting process, the production of metals by electrolysis may be included at this juncture. Examples of the production of active metals by the electrolysis of either fused salts or aqueous solutions are included in the discussion of industrial electrochemical processes earlier in this chapter.

Reduction by carbon.   Naturally occurring oxide ores or those produced by the roasting process are frequently reduced by carbon at elevated temperatures. Depending largely on the temperature employed, an oxide such as germanium dioxide may be reduced with direct formation of the metal and carbon dioxide:

$$GeO_2 + 2C + heat \rightarrow Ge + 2CO$$

$$GeO_2 + 2CO + heat \rightarrow Ge + 2CO_2$$

In general, reduction by carbon is not suited to the production of metals of a high degree of purity, since the reduced metal usually contains carbon as an impurity.

Reduction by active metals. As previously indicated, the active metal aluminum was first produced by the reduction of anhydrous aluminum chloride by the more active metal potassium:

$$AlCl_3 + 3K \rightarrow Al + 3KCI$$

However, the most useful application of reduction by active metals is illustrated by the *Goldschmidt process,* which involves the high-temperature reduction of oxides or sulfides by active metals such as aluminum or magnesium. This method is used in the case of ores that are difficult to reduce (e.g., the oxides of manganese and chromium). In the reduction of manganese dioxide, an intimate mixture of powdered MnO_2 and powdered Al is placed in a clay crucible which is then embedded in sand. Into a depression at the top of the charge is placed an ignition mixture, consisting of powdered magnesium and either sodium peroxide, barium peroxide, or potassium chlorate, into which extends a piece of magnesium ribbon. When the magnesium ribbon is ignited, the heat generated is sufficient to initiate the combustion of the ignition mixture, and this in turn initiates the strongly exothermal reaction in the charge proper:

$$3MnO_2 + 4Al \rightarrow 3Mn + 2Al_2O_3 + heat$$

Reactions of this type are so strongly exothermal that the metal is melted and settles to the bottom of the crucible, while the aluminum oxide floats on top of the melt. Goldschmidt reduction is particularly useful when one wishes to prepare metals free from carbon, but it is a relatively expensive method.

The common practice of *thermite welding* is another application of the Goldschmidt process. Iron rails are often welded together by this procedure. The ends to be welded together are surrounded by a packing of sand and clay. Molten iron is produced by the reaction

$$Fe_2O_3 + 2Al \rightarrow 2Fe + Al_2O_3 + heat$$

This iron flows through a small opening in the bottom of the crucible and into the space between the ends of the two rails. The temperatures produced by these reactions are often as high as 3500°C.

Refining operations

When metals are produced by any method, the extent to which they must be purified depends on the applications intended. For many purposes, the crude metals containing appreciable quantities of impurities may be employed; for other applications, extensive refining may be necessary. Since this problem involves so many variables, it seems sufficient merely to indicate the general character of the refining processes most commonly utilized. Purification by electrolysis has already been discussed, and this is unquestionably one of the most important and useful methods. Some metals, notably iron and lead, may be purified by oxidation of the impurities by gaseous oxygen (from air), followed by the removal of the oxidized impurities. Still other metals, such as mercury and zinc, are sufficiently volatile that they may be purified by distillation.

Physical properties of metals

Although the metallic elements exhibit many differences in physical characteristics, the class as a whole is characterized by certain fairly well defined properties. All metals are similar in that they are (1) conductors of electricity, (2) conductors of heat, (3) opaque to light, and (4) reflectors of light. This latter property is usually described by stating that metals exhibit a *metallic luster,* which is silvery in appearance in all cases except copper and gold. Aside from these properties, metals exhibit a wide range of *densities,* from the extremely dense (such as lead and gold) to the relatively light alkali metals (beryllium, aluminum, etc.). All the metals except mercury (a liquid) are solids at ordinary temperatures, and they exhibit considerable variation in hardness. Thus, although most metals are rather hard substances, some metals (e.g., sodium and lead) are so soft that they may be cut by an ordinary knife. Most of the metals possess the properties of *malleability* (i.e., they are capable of being rolled or hammered into thin sheets) and *ductility* (i.e., they are capable of being drawn into wires).

It must be recognized that, in general, the physical properties of any metal are dependent on (1) the crystalline structure of the metal, (2) the presence of impurities, and (3) the mode of production and the mechanical treatment to which the metal may have been subjected.

Chemical properties of metals

At this point, you should recall and review numerous important items of information concerning the chemical properties of the metals, including the relative activity of metals; the union of metals with oxygen; reactions between metals and acids, water, and bases; base-forming properties; behavior as reducing agents; and so forth.

The atoms of the metals in general tend to lose electrons and form positive ions. For this reason, the metals are said to be *electropositive;* the most electropositive metals are those whose atoms possess a small number of electrons in the outermost shell (1 or 2 e^-) and hence those which can lose these electrons most readily. With respect to base-forming properties, the strongest bases are formed by the most electropositive metals. Although the metals do exhibit a predominant tendency to form ionic compounds, they may also form covalent compounds. This tendency is most prevalent among metals whose atoms contain three or more electrons in the outermost shell and among the metalloids.

Oxides of the metals. Although each metal usually combines with oxygen to form an oxide having a composition in conformity with the type formula for the particular periodic group involved, it is generally true that a metal forms oxides of other types as well. For those metallic elements which are capable of existence in various states of oxidation, corresponding oxides may be formed (e.g., CU_2O and CuO; PbO and PbO_2; MnO, Mn_2O_3, MnO_2, Mn_3O_4, and Mn_2O_7; etc.). Oxides such as Fe_3O_4 and Mn_3O_4 may be looked upon as *double oxides* in which the valence of the metals becomes more apparent when the formulas are written $FeO \cdot Fe_2O_3$ and $MnO \cdot Mn_2O_3$, respectively. Corresponding to hydrogen peroxide (H_2O_2), which has long been used as an antiseptic, some of the metals also form *peroxides;* perhaps the most common are sodium peroxide (Na_2O_2) and barium peroxide (BaO_2). These oxides are characterized by the presence of an oxygen-to-oxygen bond, as shown by the following graphic formulas:

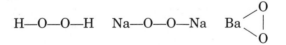

and by the fact that they react with sulfuric acid to form hydrogen peroxide:

$$BaO_2 + H_2SO_4 \rightarrow BaSO_4 + H_2O_2$$

Some oxides having formulas of the type MO_2 are sometimes erroneously called peroxides, for example, PbO_2 and MnO_2. Since these oxides do not possess the properties of true peroxides, it is better simply to designate these as *dioxides*.

Hydrides of the metals. Combination of metals with hydrogen occurs much less readily than union with oxygen and often leads to products of indefinite and variable composition. Some of the hydrides of the metals are covalent, others are ionic or saltlike in character, and some of the heavier metals appear to absorb, adsorb, or form alloys with hydrogen. In general, the metals exhibit much greater regularity in their behavior toward oxygen than toward hydrogen.

Alloys

Products obtained by melting two or more metals together and allowing the resulting mixture to cool and solidify are called *alloys*. By far the great majority of the metallic objects encountered in everyday use consist of alloys rather than pure metals. In fact, iron and aluminum are the only metals used extensively in relatively pure form. Some of the more common alloys such as bronze have been known since antiquity and have contributed markedly to the development of modern civilization. Since alloys are of greater practical importance than pure metals, you should acquire at least some familiarity with the nature, properties, and uses of alloys.

Nature of alloys. In view of the fact that alloys are *mixtures,* it follows that they may be made up of any desired number of metals and may have any desired composition. For simplicity, this discussion is limited largely to the consideration of alloys consisting of *two* metals (i.e., binary alloys). As a result of extensive study, it has been found that when two metals are melted together and the melt is allowed to cool and solidify, the resulting solid consists of one (or any combination) of the following: (1) mixtures of pure crystals of the two component metals, (2) solid solutions, or (3) intermetallic compounds. In any event, the character of the solid alloy is dependent on the specific properties of the metals concerned and on the proportions in which they are mixed.

The formation of *mixed crystals* results from limited solubility or insolubility of one solid metal in the other. Lead and tin and lead and antimony are examples of pairs of metals that form alloys consisting of intimate mixtures of tiny pure crystals of each metal. The formation of *solid solutions* results when the liquid metals are miscible in all proportions and are capable of solidification to compositions that are essentially the same as those of the melts. Many of the most common and useful alloys consist of homogeneous solid solutions of one metal in the other (e.g., alloys of copper and zinc, gold and silver, nickel and chromium, and copper and nickel). *Intermetallic compound* formation occurs in many cases and usually leads to alloys that are too hard and brittle to be of much practical value. However, these very properties may be utilized to advantage in the manufacture of bearing metals and the like, where resistance to abrasion is desired. Typical formulas for intermetallic compounds—Cu_3Sn, $AuMg$, $NaCd_2$, Mn_4Sn, Ca_3Mg_4, Ag_3Bi— show that these compounds often have compositions different from those to be expected on the basis of the normal valences of the metals involved.

The eutectic. The manner in which the properties of one metal are influenced by the presence of another and the sequence of events when a solution of one metal in another is allowed to solidify can best be understood by the consideration of a specific example. Pure lead melts at 327°C, and pure antimony melts at 631°C. If a little antimony is dissolved in a large quantity of lead and the resulting solution is cooled slowly, the crystals that appear first consist of pure lead. However, the freezing temperature of the lead that crystallizes from the melt is below 327°C. Similarly, the crystals of pure antimony that separate from a solution of a small quantity of lead in a large quantity of antimony are found to appear at a temperature below 631°C. Thus, each metal lowers the freezing temperature of the other, and it would seem that for any pair of metals there is some particular composition that exhibits a *minimum* freezing temperature, since the extent of the lowering of the freezing temperature (analogous to the influence of solutes on the freezing temperature of water) depends on the quantity of the second metal added. The relationship between composition and freezing temperature may be illustrated by considering a typical case in which two metals, antimony and lead, are to be melted together. The melting temperature (freezing temperature) of pure antimony is 640°C. If lead is added, the melting temperature of antimony is lowered as the percentage of lead is progressively increased. Similarly, the melting temperature of pure lead is 320°C. The point of intersection of two curves plotting temperature versus metal mixture ratios represents the lowest melting temperature of lead and antimony that can be realized. A composition of approximately 10% antimony and 90% lead has a melting temperature of 247°C. This is the lowest melting temperature of any mixture of these two metals. This is known as the *eutectic* of the lead-antimony mixture. The composition of the eutectic is referred to as the *eutectic composition,* and the melting temperature of the eutectic (in this case, 247°C) is called the *eutectic temperature.*

If a mixture consisting of 50% lead and 50% antimony is melted at 650°C (i.e., above the melting temperature of antimony), and the resulting melt is allowed to cool slowly, the first solid appears when the temperature reaches 500°C. These first crystals consist of pure antimony; since antimony is thereby removed from solution, the solution becomes increasingly richer in lead, and the temperature of solidification is progressively lowered until the eutectic temperature (247°C) is reached. At this temperature, crystals of both lead and antimony separate from the solution, and the composition of the mixture of the two kinds of crystals is the same as that of the liquid (viz., the eutectic composition). Once the eutectic temperature is reached, crystallization continues to complete solidification of the mixture without any further change in composition.

As is true in most cases, the eutectic of lead and antimony consists of a homogeneous mixture of the crystals of the two metals. This uniform fine texture of the eutectic and its low melting temperature render the eutectic composition particularly useful for many purposes. Although other compositions are commonly used, it is frequently true that the best alloys are those having either exactly or very nearly the composition of the corresponding eutectics.

Properties and uses of alloys

At first glance, it might be assumed that the properties of an alloy formed from two metals would be intermediate between the properties of the component metals. Such is not the case. In general, the properties of an alloy are quite different from those of either of the component metals. The preceding section has already shown that the melting temperature of an alloy consisting principally of antimony, but containing some lead, is lower than that of pure antimony. However, the melting temperature of an alloy may be higher (or lower) than that of either component. It is generally true that alloys are harder than the pure component metals and are poorer conductors of electricity. When very small amounts of arsenic are alloyed with copper, the electrical conductivity of the copper may be lowered as much as 50%.

It is commonly found that alloys are much more resistant to corrosion than pure metals. It is largely for this reason that alloys are so widely used in the construction of industrial chemical equipment that must be used in contact with corrosive chemicals.

The colors of alloys cannot usually be predicted from a knowledge of the colors of the constituent metals. Thus, 5-cent coins used to be made of copper and nickel and yet were devoid of the color characteristic of copper. Certain alloys of silver and gold are green in color, while a certain alloy of beryllium and copper exhibits the yellow color characteristic of gold.

In view of the foregoing facts, it is evident that alloys possessing an almost unlimited variety of properties may be formed. Vast industries concerned almost entirely with the fabrication of alloys have been developed throughout the world. By experimentation, alloys tailored to fit almost any need are produced, and accomplishments recorded thus far represent only a beginning. With hitherto little-known metals becoming more available as a result of improved methods of production, rapid advancement in our knowledge of and uses for alloys is to be anticipated.

The compositions of a few common alloys are listed in Table 11.4, together with an indication of their outstanding properties and common uses. Since alloys of iron are discussed later, they are not included here.

Heavy Metals

Although the various phases of metallurgical operations may be considered in general terms, a true appreciation of the nature of the many problems involved can be had only by studying of specific cases. Consequently, this section discusses the metallurgy of the so-called heavy metals and related topics. The metallurgy of the light metals has

TABLE 11.4 Common Alloys

Name	Percentage composition*	Properties	Uses
Brass	Cu, 73–66; Zn, 27–34	Malleable and ductile	Sheets, tubes, wires, etc.
Shot metal	Pb, 99; As, 1	Harder than Pb yet readily fusible	Bullets and shot (cast and molded)
Solder	Pb, 67; Sn, 33	Low melting point	Plumbing
Magnalium	Al, 95–70; Mg, 5–30	Light	Scientific instruments
Monel metal	Ni, 68; Cu, 28; Fe, 1.9	Resistant to corrosion; easily machined and readily polished	Pump cylinders, valves, piston rods, etc.
Type metal	Pb, 60–56; Sn, 10–40; Sb, 4.5–3	Low melting point; expands upon solidification	Cast type
Babbitt metal	Sn, 89; Sb, 7.3; Cu, 3.7	Hard; readily polished	Bearings
German silver	Cu, 55; Zn, 24; Ni, 20	Hard; readily polished	Substitute for silver
Duralumin	Al, 90; Cu, 4; Mg, 05; Mn, 5.5	Light; high tensile strength	Airplane and automobile parts, etc.

* It should not be assumed that the composition of commercial alloys will necessarily correspond to the percentages indicated or that other metals may not be present in small amounts. For example, the properties of brass differ with variation in the copper-zinc ratio. Similarly, the properties of commercial brass are often profoundly changed by the presence of small quantities of other metals. Since alloys are usually fabricated from metals that have not been subjected to elaborate purification processes, the resulting alloys are frequently contaminated with metals present in the primary ores.

already been discussed, and the metallurgy of iron is discussed in the next section.

Copper

People have been using this metal for more than 5000 years, and it presently ranks as one of the most useful heavy metals. Of approximately 2 million tons of copper produced each year, the United States provides 0.5 million tons.

Occurrence. Small amounts of "native" copper are found in the Lake Superior region and elsewhere in the United States. The principal source of this metal, however, consists of compound ores found in Montana, Utah, Arizona, and New Mexico. Of a rather wide variety of copper-bearing minerals, the most important are *chalcopyrite* ($CuFeS_2$), *chalcocite* (CU_2S), *malachite* [$Cu_2(OH)_2CO_3$], and *bornite* (Cu_5FeS_4). Most other countries produce important quantities of copper; Chile is one of the chief sources.

Metallurgy. The metallurgy of copper is notable for its simplicity. This fact and the widespread occurrence of native copper are undoubtedly responsible for the early use of this metal. Low-grade copper ores are concentrated by flotation, and the resulting concentrates (or the better-grade ores) are roasted in order to accomplish a partial conversion of sulfides and a complete conversion of carbonates to oxides. During this treatment, arsenic, present as an impurity, is volatilized as arsenious oxide (As_2O_3). The roasted ore, which contains some iron sulfide as an impurity, is then heated with a suitable flux (sand or limestone) in a blast furnace. During this treatment, the sulfides of copper and iron are melted, and a slag containing the gross impurities separates and is removed. The molten mixture of sulfides of iron and copper (known as *copper matte*) is transferred to a special type of Bessemer converter, more sand is added, and a blast of air is passed through the resulting mixture. This treatment partly converts sulfides to oxides, and the iron oxide so formed reacts with the sand to form a slag consisting principally of ferrous silicate. The cuprous oxide reacts with unchanged cuprous sulfide to form molten elemental copper.

$$Cu_2S + 2Cu_2O \rightarrow 6Cu + SO_2$$

The slag of ferrous silicate floats on top of the molten copper and is poured off, while the molten copper is poured into molds and allowed to cool. This product is known as *blister copper* and ordinarily contains about 2% impurities, among which are several noble metals. Since greater purity is usually required, the blister copper is refined by electrolysis to provide substantially pure metallic copper. The pure metal has a reddish color, is soft, malleable, and ductile, and is a good conductor of heat and electricity. It melts at 1083°C and at room temperature has an absolute density of 8.95.

Uses. The single largest use for pure copper is in fabricating electrical wire and cable. Minor and yet important uses include the manufacture of electrical instruments, roofing, coverings for the bottoms of ships, and so forth. Second only to the production of wires and cables, the manufacture of alloys accounts for the utilization of large quantities of copper (e.g., the various types of brasses, bronzes, and coins). For many years, domestic 5-cent coins were made of an alloy of 75% copper and 25% nickel. However, in order to conserve supplies of both of these important metals, the government in 1942 authorized the issuance of "nickel" coins containing 35% silver, 56% copper, and 9% manganese—that is, much less copper and no nickel at all.

In recent years, alloys of copper and beryllium have been found to have many useful properties (e.g., hardness, resistance to corrosion, and resistance to fatigue), and it seems probable that these alloys will become increasingly important. In addition to its uses as a metal, copper is also used to form many compounds, some of which are of considerable importance commercially. Table 11.5 gives examples of copper compounds and their uses.

Mercury

The chemical inactivity of mercury, its high density, and the fact that it is a liquid at ordinary temperatures make this metal uniquely useful in laboratory work. Mercury has been in use since the Middle Ages.

Occurrence. Although small quantities of elemental mercury are found in nature, the mineral *cinnabar* (HgS) constitutes the chief source of this important metal. The richest known deposits of cinnabar are located in Spain and Italy, and these countries have for many years produced the bulk of the world's supply of mercury. Less extensive

TABLE 11.5   Important Compounds of Copper

Name	Formula	Uses
Basic copper acetate	$Cu_2(OH)_2(C_2H_2O_2)_2$	Paint pigment
Paris green	$Cu_4(AsO_3)_2(C_2H_2O_2)_2$	Insecticide
Cupric oxide	CuO	Oxidizing agent
Copper sulfate pentahydrate	$CuSO_4 \cdot 5H_2O$	Insecticide, medicinal battery cells

deposits of cinnabar are found in California, Texas, Oregon, Mexico, Peru, Japan, and China.

Metallurgy. Owing to the instability of mercuric sulfide and mercuric oxide, the metallurgy of mercury is relatively simple. Low-grade ores that have been concentrated by flotation or high-grade ores may be treated in one of two ways. The simplest procedure involves roasting the sulfide to produce mercury vapor and sulfur dioxide:

$$HgS + O_2 \rightarrow Hg + SO_2$$

The mercury vapor is condensed to liquid mercury, and the sulfur dioxide escapes as a gas. Mercuric oxide is not formed since it is unstable at the temperatures employed in the roasting process.

An alternative method involves heating the sulfide in the presence of calcium oxide:

$$2HgS + 2CaO \rightarrow 2CaS + 2Hg + O_2\uparrow$$

In this process, the mercury vapor is condensed and the oxygen is permitted to escape.

The first stage in the purification of mercury involves filtration through chamois skin to remove the gross insoluble impurities. If further purification is required, the liquid metallic mercury (containing other dissolved metals as impurities) is sprayed into an aqueous solution of mercurous nitrate containing a low concentration of nitric acid. The impurities react with mercurous nitrate to form water-soluble salts and liberate elemental mercury. The mercury is subsequently separated and dried. If mercury of a high degree of purity is required, the metal may be purified further by distillation under reduced pressure in a laboratory apparatus having a sealed boiling flask and a lower condensing flask to receive the vapor. The pure liquid metal has a silvery color, hence the common name *quicksilver.* It boils at 356.66°C, freezes at –38.832°C, and has a density of 13.596 at 0°C.

Uses. Mercury is used extensively in thermometers, barometers, and other scientific instruments. Because of its low vapor pressure, its high density, and the fact that it is a poor solvent for many common gases, mercury is useful as a confining liquid for work involving the collection and analysis of gases. The metal is also used in mercury vapor lamps and arc lights, as a substitute for water in the boilers of heat engines, and so forth. In addition, mercury is used in the form of its alloys (*amalgams*) and in various forms of chemical combination, some of which are listed in Table 11.6.

TABLE 11.6   Important Compounds of Mercury

Common name	Chemical name	Formula	Uses
Calomel	Mercurous chloride	Hg_2Cl_2	Medicinal (liver stimulant)
Bichloride of mercury (corrosive sublimate)	Mercuric chloride	$HgCl_2$	Antiseptic
Ammoniated mercury	Mercuric amidochloride	$Hg(NH_2)Cl$	Medicinal ointment (for skin diseases)
Fulminate of mercury	Mercuric cyanate	$HG(OCN)_2$	Explosive (percussion caps)
Vermillion	Mercuric sulfide*	HgS	Paint pigment

* A red form of mercuric sulfide which is formed by grinding mercury and sulfur together in the presence of potassium sulfide.

Amalgams. Liquid mercury is a fairly good solvent for all metals except iron and platinum. The alloys that are formed by dissolving metals in mercury are called *amalgams,* many of which involve intermetallic compounds of the type MHg_x, where x is quite variable. Aside from their many applications in scientific work, amalgams are used in dentistry and in connection with the metallurgy of silver and gold.

Silver

Like copper, the metal silver has been known since ancient times and prized as a precious metal. Silver has long been used in the fabrication of jewelry, ornaments, and coins.

Occurrence. In addition to native silver, this metal occurs in nature in the form of alloys with gold, mercury, and other metals. Practically all the useful ores of silver are sulfides, with the single exception of *cerargyrite,* or "horn silver" (AgCl). Among the common sulfide ores are *argentite* (Ag_2S), *parargyrite* (Ag_3SbS_3), and *proustite* (Ag_3AsS_3). It is commonly true that sulfides of silver occur in nature along with the sulfides of other heavy metals, and attention has already been called to the fact that important quantities of silver are produced as a by-product of the electrolytic refining of copper.

Metallurgy. As is true for any metal, the procedure employed in the metallurgy of silver must be adapted to the particular form in which the metal exists in its ores. Native silver may be separated by the process of *amalgamation.* The crude ore containing metallic silver is treated with mercury to form an amalgam, which is then separated from the undesired earthy components of the ore. The amalgam is heated in a retort, and the silver remains behind as a residue, while the mercury is distilled from the retort, condensed, and used again. This same process may be used in the extraction of silver from horn silver, but in this case amalgamation is preceded by the following reaction:

$$2AgCl + 2Hg \rightarrow 2Ag + Hg_2Cl_2$$

The silver thus liberated amalgamates with the excess mercury.

Sulfide ores are usually of low silver content and must be concentrated by flotation before further processing is possible. Following concentration, the ores may be treated by any one of several *leaching* processes, two of which are described briefly here:

1. The sulfides are converted to sulfates by roasting the ore, and the resulting silver sulfate is separated by leaching with water. Metallic silver then is precipitated by addition of copper:

$$Ag_2SO_4 + Cu \rightarrow 2Ag\downarrow + CuSO_4$$

   The silver is separated by filtration, and the copper sulfate that is recovered from the filtrate constitutes an important by-product.

2. The sulfide ores are commonly roasted with sodium chloride to form AgCl, which is then dissolved by leaching the ore with an aqueous solution of sodium or potassium cyanide. The chloride is dissolved because of the formation of soluble sodium silver cyanide [NaAg(CN)$_2$]. The cyanide solution is treated with finely divided zinc, which displaces the silver from the complex cyanide.

Sulfide ores of silver and lead are also treated by the Parkes process, which is described later, in discussion of the metallurgy of lead.

Whatever process is used in the production of crude metallic silver, the product most likely contains gold, copper, and other metals as impurities. Refining is accomplished by electrolysis from aqueous solutions containing silver nitrate and nitric acid as the electrolyte, with crude silver as the anode and pure silver as the cathode (i.e., a method wholly analogous to that used for copper). Silver is also refined by the process of *cupellation,* which involves the formation of an alloy with lead, followed by oxidation (in a suitable furnace) of many of the impurities by means of atmospheric oxygen. Following separation of the oxidized impurities, the resulting silver is pure enough for some uses; for other purposes, the copper and gold that remain as impurities must be removed by electrolysis, or the silver may be leached with sulfuric acid and the silver displaced from the resulting silver sulfate solution by the addition of copper.

Of all the metals, silver is the best conductor of heat and electricity. Although both malleable and ductile, silver possesses a mechanical strength which, together with its resistance to corrosion, makes it a useful metal for many purposes.

Uses.   Much silver is used in the form of alloys. Alloys of copper and silver are much harder than pure silver and equally or more resistant to corrosion. Sterling silver and silver coins are alloys of this type. Small amounts of silver are used in the electroplating of a wide variety of objects, in the silvering of mirrors, and so forth. This metal is of real value in the form of compounds such as silver nitrate ($AgNO_3$), commonly known as *lunar caustic,* which is used in medicine (to cauterize wounds) and in the manufacture of indelible inks. Silver bromide ($AgBr$) is an extremely important compound because of its use in photography.

Photography.   No discussion of the uses of silver and its compounds could be complete without reference to photographic processes. That photography is responsible for the utilization of a considerable fraction of the annual production of silver is shown by the fact that the Eastman Kodak Company alone uses more than 300,000 lb of silver annually.

The use of silver bromide (or, less commonly, silver chloride or iodide) depends on the fact that these salts, when exposed to light, undergo decomposition to an extent that is dependent on the intensity of the light. The nature of the chemical changes involved in the simplest type of photographic process may best be indicated in terms of the following steps:

Production of the film or plate.   A thin layer of gelatin containing a colloidal dispersion of silver bromide is placed on a film (made of cellulose nitrate or cellulose acetate) or a glass plate. After drying, the film (or plate) is ready for use.

Exposure.   The film or plate is placed in the camera, and the image of the object to be photographed is focused on the film by means of a suitable lens. Since the object to be photographed has light and dark areas, the light reflected from these areas differs in intensity. Accordingly, the low-intensity reflection from a dark area results in less activation of the silver bromide on the surface of the film than is produced by the more intense light reflected from light areas. Hence, over the entire surface of the film, the silver bromide is activated to varying degrees, and a *latent* or potential image is produced.

Developing.   The latent image on the film is rendered visible by immersing the exposed film in a solution of a reducing agent (the developer), which is usually an organic compound. Reduction of silver bromide to elemental silver occurs most rapidly in those areas on the film which were exposed to the most intense light. Thus, those parts of the film which are rendered darkest owing to the formation of elemental silver are those corresponding to the lightest parts of the object photographed. If the reducing action of the developer is allowed to proceed indefinitely, all of the silver bromide is reduced; therefore, it is necessary to stop the reaction as soon as the image becomes plainly visible.

Fixing. The film is removed from the reducing solution, unchanged silver bromide is removed by immersing the film in a solution of sodium thiosulfate or *hypo* ($Na_2S_2O_3$) solution, and the film is washed with water and dried. Because the directions as well as the light and dark areas of the object photographed are reversed (for the reason previously indicated), the resulting image is called the *negative.*

Printing. The formation of the finished print or *positive* is accomplished by causing light to pass through the negative and strike the surface of paper which has been coated in much the same manner as is involved in the production of the original film or plate. Because the light must pass through the negative before striking the paper, the light and dark areas on the finished print are the reverse of those on the negative and therefore correspond to those of the object photographed. The process of developing and fixing the print is essentially the same as that involved in the treatment of the negative.

Lead

Known and used by people for thousands of years, lead is today one of the cheapest and most useful of the heavy metals. The world's annual production of lead amounts to more than 1.5 million tons, of which nearly 50% is produced in the United States from ore deposits found in Missouri, Oklahoma, Utah, and Idaho.

Occurrence. The single most important ore of lead is the sulfide, *galena* (PbS). Minerals of considerably less commercial value include *anglesite* ($PbSO_4$), *cerussite* ($PbCO_2$), and *wulfenite* ($PbMoO_4$).

Metallurgy. After concentration of sulfide ores of lead (usually by flotation), the concentrate is subjected to an incomplete process of roasting. This treatment converts lead sulfide partly to lead monoxide and lead sulfate

$$4PbS + 7O_2 \rightarrow 2PbO + 2PbSO_4 + 2SO_2$$

while some of the lead sulfide remains unchanged. At the same time, sulfides of copper, iron, zinc, arsenic, bismuth, and so forth, which are present as impurities, are converted to the corresponding oxides. This partly roasted ore is then mixed with limestone (which forms a slag with the silicon dioxide present in the ore), iron ore, and coke. This mixture is heated in a furnace while a blast of air is forced through the mixture, and the following changes take place:

$$Fe_2O_3 + C \rightarrow 2FeO + CO$$

$$PbO + C \rightarrow Pb + CO$$

$$PbO + CO \rightarrow Pb + CO_2$$

$$2PbO + PbS \rightarrow 3Pb + SO_2$$

$$PbS + FeO + C \rightarrow Pb + FeS + CO$$

$$PbSO_4 + FeO + 5C \rightarrow Pb + FeS + 5CO$$

These and undoubtedly several other reactions occur during the smelting of the very complex charge placed in the furnace. As indicated by the preceding reactions, elemental lead may be liberated as a product of five (or more) separate reactions. The lead so produced is drained from the furnace and is then ready for purification. The chief impurities present in the crude product (known as *lead bullion*) are Cu, Ag, Au, As, Sb, and Bi.

Gross impurities are removed by melting the crude lead and forcing a stream of air through the molten metal. Most of the impurities are oxidized, and the resulting oxides float to the surface of the molten lead and are skimmed off. The remaining partly purified lead may be purified further by the *Parkes process,* which involves melting the impure lead and adding zinc. The zinc dissolves in lead to only a slight extent, while the impurities such as copper, silver, and gold are much more soluble in zinc than in lead. After the addition of zinc, the molten mixture is stirred and allowed to cool, whereupon the zinc containing the dissolved noble metals rises to the surface and solidifies. After removal of the resulting crust, the lead is treated successively with fresh portions of zinc until the lead is substantially pure. The zinc crusts containing silver, gold, and so forth are heated in a retort; the zinc that distills from the retort is recovered and used again, while the noble metals are reclaimed from the solid residues remaining in the retort. Less commonly, the crude lead bullion is refined by the *Betts process,* which is an electrolytic refining process analogous to that used in refining copper. Fluorosilicic acid (H_2SiF_6) and lead fluorosilicate ($PbSiF_6$) serve as the electrolyte. Large bars of lead bullion serve as anodes, and the cathodes consist of thin sheets of pure lead. During electrolysis, lead is deposited at the cathode and the noble metals collect as a sludge or "mud" in the region of the anode, while the active metals dissolve and remain in the electrolyte. Of the two methods of refining, the electrolytic method produces the product of greater purity.

Pure lead is a soft, bluish-gray metal which is malleable, ductile, and of very low tensile strength. Lead is a very dense metal (density = 11.34) which melts at 327.50°C and boils at about 1600°C. The hardness of lead is usually increased markedly by even small quantities of metallic impurities.

Uses. The chief uses of lead are found in (1) the construction of storage batteries, (2) the manufacture of a wide variety of alloys, (3) the construction of "lead chamber" sulfuric acid manufacturing plants, and (4) the production of many useful compounds. Some of the more common and useful compounds of lead are listed in Table 11.7.

Tin

The mineral *cassiterite* (SnO_2) is the only important ore of tin. About 165,000 tons of tin are produced annually, most of which comes from cassiterite deposits found in Malaysia, Indonesia, Bolivia, Congo, Nigeria, and Thailand. Although the United States consumes nearly one-half of the tin produced, no significant amounts are produced from the very scarce and low-grade domestic ores. A tin smelter located on the Texas Gulf Coast produces tin from cassiterite concentrates imported from Bolivia, Indonesia, and Congo.

Metallurgy. Concentrated cassiterite ores are roasted largely for the purpose of removing arsenic and sulfur (as As_2O_3 and SO_2) and converting metallic impurities into their oxides. These oxides are then removed by leaching with dilute sulfuric acid, in which the SnO_2 is

TABLE 11.7   Important Compounds of Lead

Common name	Chemical name	Formula	Uses
Lithage	Lead monoxide	PbO	Manufacture of glass, pottery, cement
	Lead dioxide	PbO_2	Manufacture of storage batteries
Minium (red lead)	Plumbous plumbate*	Pb_3O_4	Paint pigment
Sugar of lead	Lead acetate trihydrate	$Pb(C_2H_3O_2)_2 \cdot 3H_2O$	Medicinal
Sublimed white lead	Basic lead sulfate	$Pb_2O(SO_4)$	Paint pigment (white)
White lead	Basic lead carbonate	$Pb_2(OH)_2CO_2$	Paint pigment (white)
Chrome yellow	Lead chromate	$PbCrO_4$	Paint pigment
Chrome red	Basic lead chromate	Pb_2OCrO_4	Paint pigment
	Lead arsenate	$Pb_3(AsO_4)_2$	Insecticide
	Lead tetraethyl	$Pb(C_2H_5)_4$	Antiknock additive in motor fluels

* So-called because this compound is actually the lead salt of orthoplumbic acid (H_4PbO_4)— i.e., $Pb_2(PbO_4)$.

insoluble. The prepared concentrate is thereafter reduced by carbon in a reverberatory furnace:

$$SnO_2 + 2C \rightarrow Sn + 2CO$$

The molten tin is removed, cast into blocks (*block tin*), and subsequently purified by a number of refining processes, the most effective of which is an electrolytic process similar to that used in refining lead.

Uses.   The most important uses for tin are concerned with the manufacture of tinplate (sheet iron coated with tin), a variety of alloys, tinfoil, and pipe which, being very resistant to corrosion, is used to carry water and other liquids.

Zinc

The history of zinc is interesting in that it was long used in the form of alloys before it came to be recognized as an element in 1746. For hundreds of years prior to that date, however, alloys of copper and zinc (brasses) were produced by the smelting of ores containing compounds of the two metals.

Occurrence.   Zinc occurs in a rather wide variety of combinations in nature. The chief ores of this metal are *sphalerite* or zinc blende (ZnS) and *franklinite,* which is a rather complex mixture consisting largely of the oxides of zinc and iron together with variable quantities of oxides of manganese. Other zinc-bearing minerals include *smithsonite* or zinc spar ($ZnCO_3$), *zincite* (ZnO), *willemite* (Zn_2SiO_4), and *calamine* ($Zn_2H_2SiO_5$ or $Zn_2SiO_4 \cdot H_2O$). In the United States, the most important zinc-producing areas are the franklinite deposits in New Jersey and the rich deposits of sphalerite in a region comprising parts of Oklahoma, Kansas, and Missouri.

Metallurgy.   The most modern method for the production of zinc is the electrolytic process, which has already been described.

An older metallurgical process, which is still used extensively, involves a rather extended treatment of high-grade ore or concentrates obtained by a flotation process. In either case, the finely divided ore is roasted to convert sulfides and carbonates to oxides, which are then reduced by means of carbon at temperatures of 1200 to 1300°C. Since zinc boils at 907°C, the liberated metal distills from the earthenware retort and may be condensed in suitable receivers. If the temperature of the condenser is kept below the melting temperature of zinc (419.3°C), the metal is obtained in the form of zinc dust which, in addition to metallic impurities, contains approximately 5% zinc oxide. If, however, the zinc vapors are condensed at a temperature above

419.3°C, the metal is obtained in liquid form and is subsequently poured into molds and cooled. The crude metal obtained in this manner is known commercially as *zinc spelter.*

The principal impurities in zinc spelter are cadmium, iron, lead, arsenic, and copper. In addition to purification by electrolysis, zinc spelter may be refined by redistillation or by suitable chemical methods chosen in relation to the desired degree of purity.

Metallic zinc is characterized by a brilliant white luster which tarnishes readily to produce the familiar dull gray appearance. Between 100 and 150°C, the metal is both malleable and ductile. Pure zinc melts at 419.3°C and boils at 907°C. The absolute density of cast zinc is 6.94, while that of rolled zinc is 7.14.

Uses.   The major uses of metallic zinc are in the manufacture of alloys and in the use of zinc as a protective coating on other metallic products, notably iron and steel. Lesser quantities are employed in the manufacture of dry-cell batteries, sinks, gutters, cornices, weather strips, and so forth. The use of zinc in connection with the metallurgy of lead has already been described. In chemical laboratory work, zinc is one of the most widely used reducing agents.

Zinc oxide (ZnO), which is produced by burning zinc vapor in atmospheric oxygen, is by far the most important compound of zinc. Under the name of *zinc white,* the oxide is used as a paint pigment. It is also used as a base in the manufacture of enamels and glass, and as a "ruler" in the fabrication of automobile tires and other kinds of rubber goods. Zinc sulfide (ZnS) is also an important white paint pigment which is used either as such or in the form of *lithopone,* which is a mixture of zinc sulfide and barium sulfate. This widely used pigment is prepared by the metathetical reaction between zinc sulfate and barium sulfide, a reaction in which both of the products are insoluble:

$$Zn^{2+}SO_4^{2-} + Ba^{2+}S^{2-} \rightarrow \underline{ZnS} + \underline{BaSO_4}$$

Another useful compound is zinc chloride ($ZnCl_2$), which is used as a wood preservative, as a soldering fluid to remove oxides from metallic surfaces, and in the production of parchment paper.

Other heavy metals

The preceding examples and the treatment of the metallurgy of iron are typical of the processes involved in the production of metals of relatively high density. The remaining heavy metals that are used industrially are not considered in detail here. However, Table 11.8 gives a limited selection of information on the mode of occurrence, metallurgy, and uses of these metals.

TABLE 11.8   Some Important Heavy Metals

Metal	Mode of occurrence in nature	Nature of metallurgical process	Uses
Cadmium	Associated with zinc minerals*	By-product of refining zinc spelter	Manufacturing alloys; protective coating for other metals
Bismuth	Native and as bismite $(Bi_2O_3 \cdot H_2O)$	Melting of native metal followed by separation of earthy impurities	Producing low-melting alloys
Chromium	Chromite $[Fe(CrO_2)_2]$	Reduction of oxide by carbon[†]	Electroplating; manufacturing corrosion-resistant alloy steels
Manganese	Pyrolusite (MnO_2)	Reduction of oxide by carbon or aluminum[‡]	Manufacturing alloys with iron and other metals
Nickel	Pyrrhotite, a mixture of sulfides of Ni, Fe, and Cu	Reduction of oxide by carbon	Manufacturing alloys; electroplating; catalyst
Platinum	Native, and as sperrylite $(PtAs_2)$	Dissolution of native Pt in aqua regia, precipitation as $(NH_4)2PtCl_6$, and decomposition by heat	Catalyst; adsorbent for gases; manufacturing jewelry, electrical instruments, electrodes, crucibles, etc.

* The mineral greenockite (CdS) occurs in nature mixed with sphalerite.
[†] Unless the iron is first removed, reduction of chromite by carbon produces the iron-chromium alloy ferrochrome, which is used in the manufacture of chromium steel.
[‡] Goldschmidt process.

Iron and Steel

Of all the many and varied aspects of the chemistry of metals, the industrial importance of iron and the various products derived from it are traditionally singled out for special emphasis. It is certainly true that the many rich deposits of high-grade iron ores, the relative ease and low cost with which the metal may be extracted, and its many useful properties have made iron the cornerstone of industrial development. Although modern trends in the metallurgical industries indicate rather clearly that certain of the light metals may eventually come to rival the dominant position of iron, many years will elapse before it is relegated to a position of secondary importance.

Ores of iron

Iron occurs in a wide variety of chemical combinations in nature. Of these, oxides, sulfides, and carbonates are the compounds used as commercial sources of iron. *Hematite* (Fe_2O_3) is by far the most important. Other iron ores used to a lesser extent include *limonite* $[(Fe_2O_3)_2 \cdot 3H_2O]$, *magnetite* or magnetic oxide of iron (Fe_3O_4), *siderite* $(FeCO_3)$, and *pyrite* (FeS_2).

Most of the countries having any marked degree of industrial development have readily available deposits of iron ores. In the United States, the Lake Superior region (Minnesota, Michigan, and Wisconsin) furnishes about 65% of the domestic production while the remainder is obtained from ores located in Alabama, New York, Colorado, California, Virginia, Tennessee, and Ohio. In the 1940s and 1950s, in the Lake Superior region, the famous Mesabi iron range employed the open-pit method of mining and was one of the world's most important iron-producing areas. From this and other nearby ranges, hematite ore was produced at a rate of 90 million tons per year. So rich were these deposits that only the high-grade ores were utilized. However, these pockets of almost pure hematite were exhausted by 1960, and it then became necessary to resort to the use of lower quality ore and refined ore imported from South America. This fact in itself may have had an important bearing on the future of the iron and steel industry as its competitive position was rendered increasingly more difficult as a result of increased costs brought about by the necessity of processing the poorer ores.

Metallurgy of iron

From the standpoint of the chemical changes involved, the production of iron from an oxide ore (or from a carbonate or sulfide ore after roasting) may be represented in terms of a few simple reactions. The raw materials required are the ore, limestone, and coal or coke. The carbon in coal or coke is first changed to carbon dioxide, which in turn is passed over layers of hot coke to convert the dioxide to carbon monoxide.

$$CO_2 + C \rightarrow 2CO$$

$$3Fe_2O_3 + CO \xrightarrow{450°C} 2Fe_3O_4 + CO_2$$

$$Fe_3O_4 + CO \xrightarrow{600°C} 3FeO + CO_2$$

$$FeO + CO \xrightarrow{800°C} Fe + CO_2$$

The iron is liberated in a spongy condition and absorbs from 4 to 4.5% carbon before it melts at about 1150°C. During the progressive increase in temperature, the limestone decomposes to form quicklime:

$$CaCO_3 \xrightarrow{800–900°C} CaO + CO_2$$

which combines with the silicon dioxide (sand) always present in the ore to form *liquid* calcium silicate:

$$CaO + SiO_2 \xrightarrow{1000–1200°C} CaSiO_3$$

This molten calcium silicate and the other more readily fusible silicates that are always present in the crude iron ores form a slag which floats on the surface of the molten iron. The slag is removed, allowed to solidify, and used in the manufacture of cement or as a road-building material after being mixed with asphalt or road tar.

Operation of the blast furnace

The chemical reactions discussed in the preceding subsection occur in a device known as a *blast furnace*. The blast furnace is a tall, slightly conical tower about 90 ft in height and 20 to 25 ft in diameter at the widest region. The furnace is built of heavy steel plate and lined with firebrick. Hollow bronze brick, through which water is circulated, is inserted in the walls to keep the firebrick from melting. A double-cone charging device is located at the top, while a pit for the temporary collection of iron and slag is provided at the bottom. At the sides and near the bottom are provided water-jacketed openings (tuyeres) through which a blast of air heated to 425 to 600°C is sent into the furnace. The charge, which is introduced at the top of the furnace, consists of definite quantities of iron ore, coke, and limestone. The coke burns to form carbon dioxide and carbon monoxide near the openings of the tuyeres, and these gases pass upward to heat the charge.

When the rising gases have heated the charge to about 450°C, the carbon monoxide begins to change the iron ore into Fe_3O_4. At about 600°C, the Fe_3O_4 is converted to FeO, and as the charge sinks and gets hotter (750 to 800°C), the ferrous oxide is reduced to spongy elemental iron, which absorbs carbon from the glowing coke until a temperature of about 900°C is reached, after which no more carbon is absorbed. Within the temperature range of 800 to 900°C, the limestone decomposes to form CO_2 and CaO, and the latter reacts with sand (present in the crude ore) to form the readily fusible calcium silicate ($CaSiO_3$). This, together with other melted silicates (which are present in all crude iron ores), constitutes the slag, the formation of which is complete at 1100 to 1200°C. At these temperatures, both the slag and the iron are molten, and the slag forms a liquid layer which floats on the surface of the molten iron. In a 24-h period of operation, slag is removed at the *cinder notch* about 15 times, while iron is removed at the *iron notch* every 6 h. The molten iron is allowed to solidify in sand molds or clay-lined molds or is used immediately in the production of steel without ever being allowed to solidify. The larger portion of the iron produced in the United States never cools from the time it is first heated in the blast furnace until it is ready to be sent out from the steel mills in the form of rails, structural steel, tinplate, and so forth. The gas escaping from the blast furnace through the large downcomer pipe contains 22 to 25% carbon monoxide and has a heating value of about 90 to 95 Btu/ft^3. This gas is led through a dust catcher (the *cyclone*) and is then used as a fuel. About one-third of this gas is burned in a series of four hot-blast stoves for preheating the dry air that goes into the blast furnace at the tuyeres; the remaining two-thirds is used in gas engines to generate power for the mills surrounding the blast furnace. The operation of a single blast furnace for a 24-h period requires 4000 to 5000 tons of air, 500 to 600 tons of limestone, and 1500 to 1600 tons of iron ore, and produces about 1000 tons of iron.

Varieties of iron—their properties and uses

Molten iron from the blast furnace may be drained into sand molds or clay-lined molds and allowed to cool. The resulting ingots weigh about 150 lb each and are known as *pig iron* or *cast iron*. This relatively impure product contains some slag in addition to the following impurities: 2 to 4.5% carbon, 1 to 2% silicon, 0.1 to 0.3% sulfur, 0.1 to 2.0% phosphorus, and 0.5 to 1% manganese. This product, because of the presence of the indicated impurities, has a melting temperature lower than that of pure iron, is hard and brittle, and exhibits low tensile strength and ductility.

Cast iron. When scrap iron is added to molten pig iron and the resulting liquid is allowed to cool in suitable molds, cast iron is formed. The rust (Fe_2O_3) on the surface of the scrap iron partly oxidizes (at the temperature of the molten mixture) some of the impurities in the crude pig iron:

$$3S + 2Fe_2O_3 \rightarrow 4Fe + 3SO_2$$

It should also be recalled that iron absorbs carbon during its formation in the blast furnace. These two elements combine to form iron carbide (Fe_3C), which is also known as *cementite*. In the production of cast iron, the character of the product is dependent on the extent to which the iron carbide is decomposed. If the mold is chilled so that the iron cools rapidly, most of the iron carbide remains as such and *white cast iron* is produced. This product is hard and brittle, and cannot be machined. If, on the other hand, the molten iron is cooled slowly, *gray cast iron* is obtained; in this process, part of the Fe_3C decomposes to form iron and carbon. This variety is much softer and tougher than white cast iron and can be machined and drilled successfully. If molten pig iron to which scrap iron has been added is maintained at 400 to 600°C for several days until practically all of the cementite has decomposed, the resulting product has a high content of uncombined carbon and is known as *malleable cast iron*. This material expands on cooling and is very useful in the manufacture of stoves, machinery bedplates, radiators, and other articles not subjected to shock during use. Thus, it is evident that the essential difference between white, gray, and malleable cast iron depends on the extent to which carbon is present in the uncombined form, and this factor also governs the properties of the three varieties of cast iron.

Wrought iron. In 1784, an English ironmaster, Henry Cort, devised a method for the conversion of pig iron into a much more useful product known as *wrought iron*. This conversion is accomplished by removal of most of the impurities from pig iron by oxidation in a *puddling* or reverberatory furnace. The furnace is lined with magnesium oxide and is charged with about 0.25 ton of pig iron together with some nearly pure hematite ore, which serves as a source of oxygen. On heating, carbon is oxidized to carbon dioxide, which escapes as a gas, while silicon, phosphorus, and manganese are oxidized to the corresponding oxides. Some of the manganese reacts with sulfur to form manganous sulfide. The acidic oxides of silicon and phosphorus combine with the basic lining of the furnace to form magnesium silicates and phosphates, which together with the oxides and sulfides of manganese constitute a molten slag. As the impurities are removed, the melting temperature of the iron is raised until it finally becomes partly solidified; whereupon it is rolled into large balls or *blooms* and taken out of the furnace to be rolled into sheets. This rolling process squeezes out most of the adhering slag, also lengthens the crystals of the iron, and incorporates any remaining slag in longitudinal "streamers." The resulting wrought iron is the purest form of commercial iron but still usually contains about 1% slag (the presence of which is desirable since it improves the tensile strength) and about 0.1% carbon.

This form of iron has great tensile strength, can be forged and welded, is ductile and malleable, and has a relatively high melting temperature. Wrought iron has a somewhat fibrous structure, owing to small pockets of slag, and is therefore rather difficult to machine. Some wrought iron is used in the manufacture of high-quality tool steel, but its chief uses are in the manufacture of wire, chains, rails, anchors, bolts, nails, rivets, grate bars, pipes, ornamental gates, and the like.

Production of steel

Steel is the name given to iron which contains 0.04 to 1.7% carbon, small percentages of manganese, and only very small amounts of impurities such as sulfur and phosphorus, and which is capable of being hardened by quenching (i.e., rapid cooling by immersion in water, oil, or other suitable liquid). The carbon content of steel must be kept below

2% in order to prevent excessive brittleness. Steel is produced either from pig iron or from wrought iron by several processes, each of which is considered briefly.

Bessemer process. The essential features of this process were introduced by the American William Kelley, in 1852. However, an improved form of this process was discovered and patented by the English engineer Sir Henry Bessemer, in 1855; hence the name *Bessemer process.* The process employs a pear-shaped vessel 12 to 20 ft in height, 10 to 16 ft in diameter, and mounted on trunnions in order that it may be turned on its short axis for loading and unloading the charge. This vessel is known as a *converter* and is lined to a thickness of about 2 ft with an acidic refractory material consisting largely of silica (SiO_2) and silicates bound together with fireclay. One of the trunnions is hollow and leads to a series of tuyeres in the bottom of the converter through which a blast of air is admitted.

The converter is charged with approximately 20 tons of molten pig iron, preferably having a low phosphorus content, and with the converter in a vertical position, a blast of air is blown through the charge. Silicon, manganese, and carbon burn away (in the order named) so violently that the temperature of the charge is raised considerably. Phosphorus is burned to form the oxide (P_2O_5), but this does not enter the slag and hence is not removed. In about 10 or 15 min the "blow" is completed, and the charge remaining in the converter consists of nearly pure iron. Before the melt is poured from the converter, the required amounts of manganese, carbon, and silicon are added in the form of *spiegeleisen,* which is an iron-manganese alloy of high carbon content. The manganese removes sulfur and combined oxygen; silicon removes trapped air bubbles; and at the same time the carbon content is brought up to the desired percentage. The resulting molten steel is transferred to molds and cast in the form of ingots which may weigh as much as 4 tons.

The Bessemer process is relatively less expensive but does not produce a product of high quality. It is not possible to exercise control over the composition of the product because the conversion occurs so quickly. Furthermore, this process does not effect the removal of phosphorus. The phosphorus pentoxide that is formed during the blow is reduced to phosphorus upon addition of carbon and hence remains as an impurity in the final product. Provision for the removal of phosphorus may be made by the use of the so-called basic Bessemer process, which employs a converter lined with magnesia (MgO), but this practice entails other disadvantages. In the United States, the *acid* Bessemer process is used exclusively and accounts for about 15% of the steel produced in this country. Steel so produced is used largely as structural steel, as reinforcement for concrete, and in the tinplate industries.

Despite its disadvantages and limitations, the Bessemer process opened a new era of industrial progress. Its use made cheap steel available and was largely responsible for the rapid expansion of railroad building during the latter part of the nineteenth century.

Open-hearth process. Only a few years after the invention of the Bessemer process, open-hearth gas-fired furnaces were adapted to the production of steel. The open-hearth furnace is built of brick and steel and contains a shallow saucerlike basin (hearth) capable of carrying a charge of 70 to 100 tons of pig iron. This hearth may be lined either with an acidic (SiO_2) or basic (MgO) lining, but since American practice is limited to the basic process, the other possibility is not discussed further here. In the basic process, 70 to 100 tons of pig iron together with rusty scrap iron or steel, limestone, and oxides of iron constitute the charge. Air and gas are introduced through separate flues and are passed through a heated checkerbrick structure before coming into contact with the charge. Combustion of the gases heats the charge, and the hot gases pass out through and heat the checkerbricks on the opposite side of the furnace. Then the current of the gases is reversed, so that the incoming gas flow passes through the newly heated bricks. Thus, frequent reversal of direction of flow of gases maintains a high temperature and conserves fuel.

In the molten charge, silicon, phosphorus, sulfur, manganese, and carbon are converted to the corresponding oxides by reaction with iron oxide (Fe_2O_3). Oxides of carbon and sulfur pass off as gases, while the oxides of phosphorus and silicon combine with calcium oxide (formed by the decomposition of the limestone) to form phosphates and silicates:

$$3CaO + P_2O_5 \rightarrow Ca_3(PO_4)_2$$

$$CaO + SiO_2 \rightarrow CaSiO_3$$

At the same time, manganous oxide reacts with silicon dioxide to form manganous silicate:

$$MnO + SiO_2 \rightarrow MnSiO_3$$

These phosphates and silicates form a slag. Thus, *all* the impurities are removed from the molten iron in a period of 7 to 8 h. Near the end of the period of treatment, samples of the molten iron are removed, solidified, and sent to the laboratory for accurate and rapid quantitative analysis. On the basis of information supplied by the laboratory, it becomes possible to calculate accurately the quantity of spiegeleisen which must be added in order to produce a steel of the desired composition. Thereafter, the molten steel is transferred to molds, where the steel is cast into ingots.

In comparison with the Bessemer process, the open-hearth process has certain important advantages. It provides accurate control of temperature owing to the use of an outside source of heat. The composition of the final product can be predetermined by analysis and thereby controlled. Complete removal of phosphorus is accomplished, and this permits use of low-grade iron ores, which are usually of relatively high phosphorus content. None of these advantages is possessed by the Bessemer process.

More than three-fourths of the steel produced in this country is produced by the open-hearth process. The uses are innumerable and include practically every commercial application except those in which special alloys must be employed.

Crucible process.   This process utilizes either wrought iron or open-hearth steel. The iron (or steel) is melted in graphite-clay crucibles in gas or electric furnaces, and pure carbon is introduced to bring the carbon content up to 0.8 to 1.5%. The steel thus produced is a uniform product of high quality and relatively high cost. It is used in making knives, razor blades, and tools requiring definite carbon content.

Cementation process.   This process, now little used, consists of heating wrought iron or low-carbon steel in powdered charcoal or leather dust for 6 to 11 days in a closed boxlike furnace at 650 to 700°C. At these temperatures, carbon diffuses slowly into the surface of the steel, thus producing a thin coat of high-carbon steel over a core of low-carbon steel. This is essentially a case-hardening procedure, and steel produced in this manner is used largely in the manufacture of tools.

Electric-furnace process.   One of the highest grades of steel now produced is made in electric furnaces in which steel from the Bessemer or open-hearth process is subjected to further refining. The steel produced in this relatively new process is characterized by its high density and the absence of occluded gases. The chief advantage of this process is its provision for more precise control of temperature.

Influence of impurities on the properties of iron

By inference from previous discussions of other commercial processes, one may gain the impression that the objective of all industrial chemical operations is the production of products of high purity. This is not so in the case of iron. The properties of pure iron are not such as to render this metal particularly useful, and the problem therefore becomes one of producing iron containing just enough of the right impurities to impart a desirable set of mechanical properties. The influence of these impurities can be discussed only in general terms, since the properties imparted by any one impurity may be modified by the presence of one or more other impurities.

Carbon added to iron may exist in the iron in the combined form (i.e., Fe_3C) or as elemental graphitic or flake carbon. Combined carbon increases hardness and mechanical strength in cast iron and steel; graphitic carbon decreases both strength and hardness.

Silicon acts as a softener in cast iron, increases fluidity, lessens shrinkage, and decreases strength. Cast iron containing 15 to 20% silicon is very resistant to the action of acids. Since steel and wrought iron contain no more than traces of silicon, it has no appreciable influence so far as these products are concerned.

Sulfur tends to change graphitic carbon into combined carbon (Fe_3C), thus indirectly increasing the hardness, brittleness, and shrinkage of cast iron (and to a lesser extent, of steel). Presence of more than 0.1% sulfur renders iron very weak and dangerously brittle when *hot*.

Phosphorus increases the fluidity and softness of cast iron while decreasing the shrinkage and strength. In steel, phosphorus decreases ductility and slightly increases the hardness and tensile strength of low-carbon steels. In general, however, a high phosphorus content causes steel to fracture easily when subjected to strain or deformation when *cold*.

Manganese is known as a *cleanser* of iron, inasmuch as it combines with any unchanged iron oxides or sulfides, thus removing oxygen and sulfur. When amounts in excess of this cleansing requirement are added, the manganese begins to act as a hardener, 2% making the iron quite hard. Steel having a manganese content of 2 to 6% and a carbon content of less than 0.5% is so brittle that it can be powdered under a hand hammer. However, when more than 6% manganese is present, this brittleness disappears until 12% manganese is reached. At this composition, the original toughness is restored.

Alloy steels

With the development of the automobile industry, there arose the need for special steels that would withstand shocks and strains, resist corrosion, retain hardness at high temperatures, and so forth. These needs were met by the fabrication of alloy steels having the desired properties, and as these alloys became available in quantity, they rapidly came into use in all phases of industry. Metals commonly alloyed with steel include nickel, tungsten, chromium, vanadium, niobium, tantalum, molybdenum, cobalt, copper, and manganese. These metals may be added singly or in combination. The number and variety of these alloy steels have become so vast that space does not permit any discussion of individual cases. For purposes of illustration, a few of the better known alloys are listed in Table 11.9, together with their composition and a brief description of their properties and uses.

Corrosion of metals and alloys

The corrosion or rusting of iron, steel, and other metals is a matter of great economic importance. Millions of dollars are spent each year to replace metals damaged by corrosion. The nature of the chemical changes that occur during corrosion is not fully understood, and several theories have been advanced in an effort to provide an explanation that might lead to the development of new and better methods of control. Of the older theories, one held that corrosion of iron, for example, consists simply of the direct union of iron and oxygen; another held that corrosion results from the reaction between iron and an acid (usually carbonic acid). The presently popular *electrolytic* theory of corrosion is probably the best yet advanced, although not entirely satisfactory. This theory is too involved for complete discussion here; for present purposes it is sufficient to note that the changes that occur

TABLE 11.9   **Alloy Steels**

Name	Alloyed metals, %	Properties	Uses
Stainless steel	Cr, 18; Ni, 8	Corrosion-resistant	Manufacture of ornamental parts of cars, etc.
Nickel steel	Ni, 2–5	Stainless; very hard, yet elastic	Gears, shafting, wire rope, guns
Invar steel	Ni, 36	Nonexpanding	Clocks, pendulums, tapelines, etc.
Manganese steel	Mn, 10–18	Very hard; resists abrasive wear	Safes, railroad frogs, teeth on elevator dredges
Tungsten steel	W, 10–20;	Remains hard at	Cutting tools,
Chrome-vanadium steel	Cr, 4–8	high temperatures	high-speed drills
	Cr, 1–9; V, 0.15–0.2	Fatigue-resistant; high tensile strength	Auto parts; springs, axles, etc.
Chrome-nickel steel	Ni, 1–5; Cr, 1–2	Great hardness; high tensile strength	Armor plate for battleships and tanks
Molybdenum steel	Mo, 0.3–3	Hard; heat-resistant	Axles
Chrome steel	Cr, 2–4	Very hard; shock-resistant	Files, ball bearings, safes, etc.

during rusting are likened to those which occur in battery cells. If the supply of oxygen is unlimited, the cell reactions are as follows:

(–) Terminal: $2Fe - 4e^- \rightarrow 2Fe^{2+}$

(+) Terminal: $O_2 + 4e^- + 2H_2O \rightarrow 4OH^-$

Battery: $2Fe + O_2 + 2H_2O \rightarrow 2Fe(OH)_2$

The ferrous hydroxide is subsequently oxidized (wholly or in part) to ferric hydroxide [$Fe(OH)_3$].

However, if oxygen is not available, the corrosive action becomes analogous to the interaction of iron and an acid, and the battery cell reactions are as follows:

(–) Terminal: $Fe - 2e^- \rightarrow Fe^{2+}$

(+) Terminal: $2H^+ + 2e^- \rightarrow H_2$

Battery: $Fe + 2H^+ \rightarrow Fe^{2+} + H_2\uparrow$

As a result of extensive experimentation and as an outgrowth of the various theories that have been advanced, methods have been devised whereby corrosion may be caused to occur less rapidly or in some cases prevented entirely. Any one or any combination of the following methods may be employed:

- Elimination of contact with oxygen by the application of protective coatings of grease, metals, plastics such as Bakelite, or rust-inhibiting paints. Paint pigments having rust-inhibiting properties include Pb_2OCrO_4, Cr_2O_3, $ZnCrO_4$, Pb_3O_4, and $PbSO_4$.

- Protection from contact with acids. For example, before use in steam boilers, water should be treated with a base to neutralize any acid present.

- Removal of electrolytes. Water containing dissolved electrolytes corrodes metals more rapidly than pure water.

- Removal of impurities. The presence of certain impurities has been found to accelerate the rate of corrosion of metals and alloys.

- Removal of mechanical strains. Rate of corrosion is generally increased by deformation or mechanical strains in metals. It is a familiar fact that a bent iron nail rusts most readily at the bend.

Metallic protective coatings

The practice of placing protective coatings of other metals on the surface of iron or steel objects by electrolysis has already been discussed.

The same end may be accomplished by methods that do not involve the use of electric current.

Galvanizing. The process of galvanizing consists of placing a protective coating of *zinc* upon the surface of iron. Before application of the zinc, the iron must be cleaned of rust or scale by treatment with dilute sulfuric acid, a process known as *pickling,* which incidentally produces important quantities of *copperas,* hydrated ferrous sulfate ($FeSO_4 \cdot 7H_2O$), as a by-product.

$$FeO + H_2SO_4 + 6H_2O \rightarrow FeSO_4 \cdot 7H_2O$$

The iron is then immersed in molten zinc, withdrawn, and allowed to cool, whereupon the zinc crystallizes in the characteristic spangled design. To provide protection from rusting, the quantity of zinc retained on the surface of the iron should be not less than 2 oz/ft^2.

Zinc is frequently applied to the surface of iron or steel by other methods. In the *Schoop process,* molten zinc is sprayed onto the surface by a blast of air in a manner similar to that used in spraying paint. In addition to metals, objects made of wood, leather, paper, and the like may be coated with zinc by this process. In the process of *sherardizing,* iron is covered with a thin layer of zinc dust at temperatures of 700 to 800°C. Under these conditions, zinc penetrates the surface of the iron to a considerable extent.

Iron coated with zinc is effectively protected from corrosion. Zinc itself does not corrode appreciably, since the surface oxidation of zinc produces a thin and very adherent film of zinc oxide, which in turn protects the underlying zinc metal. Should the zinc coating become imperfect by cracking or abrasion, iron becomes exposed to a moist atmosphere containing carbon dioxide. This is equivalent to having the two metals, zinc and iron, immersed in a dilute carbonic acid solution. This combination constitutes a galvanic battery cell, and since zinc is the more active of the two metals, zinc dissolves (corrodes) preferentially. Consequently, the iron is protected from corrosion so long as the zinc coating is present.

Tinplate. Immersion of pickled iron or steel in molten tin results in the material known as *tinplate,* from which cans, caps, pails, pans, and so forth are made. The objective of tin-plating is to obtain a very thin coating of tin that is *free from cracks, holes, or other imperfections.* The reason for this latter requirement becomes clear if one considers the nature of the galvanic battery cell involving the metals iron and tin. Since iron is more active than tin, an imperfection in a coating of tin results in preferential corrosion of iron; hence, an *imperfect* tin plating, rather than providing protection from corrosion, actually *promotes* corrosion and is worse than no coating at all.

Other metal coatings. In addition to zinc and tin, and aside from those metals applied by electrolysis, other metals applied as protective coatings include lead and cadmium. Both of these metals are applied by immersion as in the case of zinc or tin. Lead was formerly used as a protective coating on wire, cables, steel shingles, and the like, but other metals are now used due to its toxicity. Cadmium is (less frequently) used as a coating on automobile parts, laboratory apparatus, and the like. Although objects plated with cadmium present an attractive appearance, the extensive use of this metal is retarded by its relatively high cost, limited production, and toxicity. Plastic materials have largely replaced the use of metals in such applications.

Some Nonmetallic Elements

You should recall from the periodic arrangement of the elements that only a relatively small number of the stable chemical elements may be considered as predominantly nonmetallic in character. This fact, however, should not lead one to assume that these elements are of only secondary importance.

This section summarizes the chemical and physical properties which are shared in some degree by all of the nonmetallic elements, and the following sections briefly delineate the chemistry of several additional nonmetallic elements—carbon, silicon, nitrogen, phosphorus, and sulfur. A later section discusses the properties of the elements of the halogen family.

Physical properties of nonmetals

Of the predominantly nonmetallic elements, seven are solids under ordinary atmospheric conditions (B, C, Si, P, S, Se, and I), only one is a liquid (Br), and the remainder are gases (H, N, O, F, Cl, He, Ne, A, Kr, and Xe). The physical properties of these elements present far more striking contrasts than do those of the metals. Thus, among the nonmetals one encounters the extremely volatile helium, which *boils* at −267°C (i.e., just 5°C above absolute zero), and the nonvolatile element carbon, which *melts* at about 3500°C. Similarly, the densities and other physical properties of these elements differ tremendously, as is made more evident by an inspection of a table of physical properties of the elements.

Allotropy. Particularly among the nonmetals that are solids under ordinary atmospheric conditions, it is frequently found that the same element is capable of existing in different physical forms. Thus, the element carbon may exist as the crystalline diamond or the amorphous wood charcoal. Although these elements exhibit striking differences in physical properties, their chemical properties are usually not significantly different. If, for example, one should burn samples of diamond and wood charcoal in an excess of oxygen, the reaction product in both cases would be carbon dioxide, and the two samples of carbon dioxide would be found to be identical in all respects. The only important difference involved is the magnitude of the accompanying energy change. These different physical forms of the same element are known as *allotropic forms* or *allotropic modifications*. Although allotropy is also encountered among the metals, the best-known examples of this phenomenon are to be found among the nonmetals, and several examples are discussed in the following sections.

Chemical properties of nonmetals

With the exception of the inert gases, most of the nonmetallic elements have four or more electrons in their outermost shells. The atoms of these elements tend to gain electrons to form negative ions; as a consequence, the nonmetallic elements in their various forms of chemical combination usually exist as simple or complex anions. Electrons gained by these atoms are held by the attraction of the positive nucleus, and the firmness with which these electrons are held differs from one nonmetal to another. Since an element that has gained one or more electrons may be said to be in an *electronegative* condition, it follows that to the extent that nonmetals differ in their tendencies toward acquisition of electrons, these elements exhibit different degrees of electronegativity. Thus, just as the metals differ in their tendencies toward loss of electrons, so the nonmetals differ in the opposite direction.

In this connection, it should be recalled that the nonmetals in general (as well as many of the less electropositive metals and the metalloids) show pronounced tendencies toward compound formation as a result of sharing electrons. Another property shared generally by all the nonmetals (except the inert gases) is that of variable valence. You need only consider the various states of oxidation exhibited by elements such as nitrogen and sulfur to appreciate the extent to which variable valence is encountered among typical nonmetals. Finally, recall that the nonmetals include those elements whose binary compounds with oxygen (e.g., SO_3 and P_2O_5) or hydrogen (e.g., HCl, HBr, and H_2S) react with water to form acidic solutions. Thus, the nonmetals are properly designated as the acid-forming elements.

Carbon

Three of the principal sources of carbon are coal, petroleum, and natural gas. Thus, the greater part of the world's available fuel resources has consisted of compounds of carbon. Similarly, all forms of plant and animal life involve carbon compounds, many of which are exceedingly complex. The element carbon is unique in the extent to which it combines with a few other elements to form a wide variety of compounds. Several hundred thousand compounds are known, and many more are being produced each day in chemical laboratories throughout the world.

Allotropic forms of carbon.   In the solid state, the element carbon exists in three different allotropic modifications—*amorphous carbon* and the two crystalline forms known as *diamond* and *graphite*. Amorphous carbon includes numerous common products such as wood charcoal, bone black, coke, lamp black, and carbon black. Each of these varieties of crystalline and amorphous carbon possesses properties that render it useful for a variety of purposes.

Diamond.   The chief sources of the diamond are the Kimberley region in South Africa, Brazil, and the East Indies. The diamond crystal is cubic, its density is 3.5, and although it is the hardest substance known to occur in nature, it is brittle and easily shattered. The diamond has an unusually high melting temperature (3500°C), does not act as a conductor of electricity, and is so inert chemically that the combustion of carbon in oxygen to form carbon dioxide does not occur until a temperature of about 800°C is reached. Through the study of the crystalline structure of the diamond by means of X rays it has been found that the carbon atoms in the crystal are arranged in tetrahedral form, each atom being equidistant from the four other carbon atoms surrounding it.

The chief uses of diamond depend largely upon the properties of hardness and ability to refract light. Diamonds that are transparent and largely free from impurities that impart discoloration are used primarily as gemstones. Diamonds that are variously colored (mostly black) are used in polishing, glass cutting, and the construction of the cutting edges of rock drills, rock-cutting saws, and the like. Synthetic diamonds are also manufactured for industrial uses.

Graphite.   This crystalline form of carbon is found in Ceylon, Siberia, Madagascar, and in various regions in central Europe and the United States, Canada, and Mexico. The melting temperature of graphite is the same as that of diamond, and these two forms of elemental carbon are also similar in their low degree of chemical activity. In contrast to diamond, however, graphite has a relatively low density (225) and is a good conductor of electricity. The crystalline structure of graphite is markedly different from that of diamond and involves carbon atoms arranged in planes of hexagonal ring structure. When diamond is heated to 3500°C, it vaporizes rapidly; on cooling, the vapors condense (solidify) in the form of graphite.

Although much graphite is readily available in nature, large quantities of graphite that is superior to the naturally occurring material are manufactured artificially by a process invented by the American chemist Edward G. Acheson. In the *Acheson process,* powdered anthracite coal is heated between graphite electrodes in an electric furnace for about 20 h. A core of granulated carbon is placed in the center of the charge between the two electrodes since the powdered coal alone is a very poor conductor. To exclude air, the charge in the furnace is covered with a layer of sand and carbon. Utilizing a current of about 200 A at 40,000 to 50,000 V, the temperatures produced within the charge are sufficient to volatilize the impurities and to convert the carbon from the coal into graphite of about 99.5% purity.

Graphite is used in the manufacture of electrodes for high-temperature furnaces, crucibles, pencil "lead", pigments, and the like. Colloidal suspensions of graphite in water, oil, or grease are commonly employed as lubricants. The planar leaflike structure of graphite is such that the layers in the crystal lattice may easily slide over one another, and the greaselike properties of graphite make this material particularly useful in the elimination of friction.

As indicated previously, the so-called amorphous varieties of carbon embrace numerous common materials. It is perhaps improper to designate all these materials as truly amorphous. Some of these materials contain less well-defined crystalline structures comparable to that of graphite but with a much less well-ordered and regular orientation of carbon atoms. Nevertheless, these materials have long been known as forms of amorphous carbon.

Wood charcoal. In addition to a rather wide variety of other products, wood charcoal is produced by heating wood, nutshells, and so forth, in the presence of a limited supply of oxygen. This operation is commonly known as *coking* or *destructive distillation.* After all the volatile products have been removed, the residue amounts to about 20% of the original weight of wood and consists of about 97% carbon. The remaining 3% is made up of nonvolatile minerals present in the wood. The chief use of charcoal is found in the iron and steel industry but, in addition, significant quantities are used as a fuel in the form of briquettes or in powdered form. Some wood charcoal is employed as a material for filtering rainwater and as a low-temperature absorbent for certain vapors. Coconut charcoal is used to absorb poisonous gases in gas masks.

Bone black. By heating "green" bones, one obtains a nonvolatile residue known as *bone black* or *animal charcoal.* This product contains only about 10% finely divided carbon uniformly distributed throughout a porous mass of (principally) calcium phosphate. Because the carbon in this product has the ability to adsorb many colored compounds, it serves as an excellent decolorizer for sugar syrups, as well as many other colored and turbid solutions.

Coke. Just as the destructive distillation of wood yields wood charcoal and volatile products, so the heating of bituminous coals results in a wide variety of useful volatile materials and amorphous carbon in the form known as *coke.* The chief uses of coke are in connection with the iron and steel industries, but important quantities of this form of carbon are also used as a fuel in other industries and in the home.

Lamp black. A relatively pure and very finely divided form of amorphous carbon known as *lamp black* is produced by burning light oils (which are rich in carbon) in a limited supply of air. The soot is collected in settling chambers on coarse cloth screens through which the smoke must pass. Lamp black is used in the manufacture of inks, paints, crayons, and the like.

Carbon black. This material is also nearly pure carbon and is produced by the incomplete combustion of natural gas. This process is exceedingly wasteful. Although burning 1000 ft^3 of natural gas should theoretically yield 30 lb of carbon black, a yield of only about 1.5 lb is ordinarily obtained. This finely divided variety of carbon is used in very large quantities as a filler in the manufacture of automobile tires and other rubber goods. Carbon black is also used in the manufacture of printing inks and certain types of lacquers.

Silicon

Compared to carbon, the related element silicon is relatively unimportant, so far as uses for the uncombined element are concerned. Most of the silicon produced commercially is used in the metallurgical and glass industries. In metallurgy, it is ued in the manufacture of a useful iron-silicon alloy known as *ferrosilicon.* Silicon is also used as an additive in organic products such as plastic and rubber compounds, and elemental silicon is a fundamental material in semiconductor and microprocessor manufacturing.

Preparation and properties of silicon. Elemental silicon of about 98% purity may be produced by the reduction of silicon dioxide by aluminum.

$$3SiO_2 + 4Al \rightarrow 3Si + 2Al_2O_3$$

The crude silicon is then dissolved in molten aluminum and cooled, and the aluminum is dissolved by diluted hydrochloric acid.

There remains elemental silicon in a crystalline form similar to that of diamond.

Silicon is a very hard but brittle element having a melting temperature of 1422°C and a density of 2.40. This element is fairly reactive toward the halogens and solutions of strong bases such as potassium hydroxide. Silicon reacts less readily with oxygen to form silicon dioxide and with other elements similarly to form a class of binary compounds known as *silicides*.

Nitrogen

In relation to the vital processes of all living forms of matter, only a few elements are more widely utilized than nitrogen. Directly or indirectly, all forms of plant and animal life are dependent on the availability of nitrogen in the form of its compounds. Although large deposits of nitrogen compounds are found in nature—for example, the extensive deposits of sodium nitrate (commonly known as *Chile saltpeter*) found in Chile, Argentina, Peru, Bolivia, California, and elsewhere—these supplies are inadequate to permit the maintenance of soil fertility and the continuation of presently existing forms of life. It is therefore necessary to utilize nitrogen in the atmosphere in some way.

Fixation of atmospheric nitrogen.   Nitrogen gas constitutes about four-fifths of the atmosphere; hence, tremendous supplies of this element are available. Elemental nitrogen may be converted into compound forms by several methods. Although animals cannot utilize atmospheric nitrogen directly, certain plants can do so. The bacteria that live in the nodules on the roots of legumes such as peas, beans, and clover are able to convert atmospheric nitrogen into forms that can be utilized by the plants with which these bacteria are associated. Some atmospheric nitrogen also becomes "fixed" through the agency of electrical discharges (lightning) during thunderstorms. In these discharges, atmospheric nitrogen and oxygen combine to form oxides of nitrogen, which dissolve in the falling drops of rain and are thereby imparted to the soil. It is a surprising fact that the quantity of nitrogen made available in this manner amounts to more than 5 lb per acre per year. In addition to the foregoing methods, several artificial processes for the fixation of atmospheric nitrogen have been devised and are used extensively. These are discussed in some detail later.

Preparation of nitrogen.   The preparation of pure nitrogen on a laboratory scale is best accomplished by the formation and subsequent decomposition of ammonium nitrite. These reactions are effected in aqueous solution by warming the reactants *cautiously*.

$$NH_4^+Cl^- + Na^+NO_2^- \rightleftarrows Na^+Cl^- + NH_4^+NO_2^-$$

$$NH_4^+NO_2^- + heat \rightarrow N_2 + 2H_2O$$

Pure nitrogen is also formed when gaseous ammonia is passed over heated copper oxide and the resulting gaseous mixture is dried.

$$3CuO + 2NH_3 \rightarrow 3Cu + 3H_2O + N_2$$

Nitrogen may also be produced directly from air by the fractional distillation of liquid air or by passing air over heated copper. In the latter case, oxygen is removed by combination with the copper to form copper oxide, but the resulting nitrogen is not entirely pure since it is contaminated by the inert gases present in the original sample of air.

Properties and uses of nitrogen.   Pure gaseous nitrogen is tasteless, odorless, colorless, and does not support combustion. This gas is slightly lighter than air, 1 L weighing 1.25048 g under standard conditions of temperature and pressure. The very slight solubility of nitrogen in water is shown by the fact that at 25°C, 100 g of water dissolves only 0.0019 g of nitrogen. Solid nitrogen melts at −209.8°C, and the pure liquid boils at −195.8°C.

Chemically, the element nitrogen is somewhat unreactive. Under appropriate conditions, however, it combines with hydrogen to form several hydrides, the most important of which is the gas ammonia (NH_3). Nitrogen also combines with oxygen to form several different oxides (e.g., N_2O, NO, and N_2O_3) and with other nonmetals and metals to form a class of binary compounds known as *nitrides* (e.g., S_4N_4, AIN, and Mg_3N_2).

The large-scale uses of nitrogen are those involving the commercial production of nitrogen compounds such as nitric acid, ammonia, and cyanamide. Although appreciable quantities of nitrogen are used in a process known as *nitriding* which is employed in the hardening of steel, gaseous ammonia has come into use recently as the source of nitrogen required for this process.

Phosphorus

It is an interesting fact that the two elements most necessary in the maintenance of soil fertility are the related elements nitrogen and phosphorus, which occupy adjacent positions in Group V of the periodic table. Phosphorus occurs in nature only in the combined form, chiefly as the mineral *phosphorite* [$Ca_3PO_4)_2$]. Impure calcium phosphate, known as *phosphate rock,* is mined extensively in Tennessee, Florida, Montana, and Idaho. Large deposits of this mineral are also found in Morocco and Tunisia in North Africa.

Allotropic forms of phosphorus. Solid phosphorus exists in two distinct allotropic modifications and is also commonly encountered in a form consisting of a mixture of the two. *White* (or *yellow*) *phosphorus* is a translucent, waxlike solid which melts at 44°C, boils at about 290°C, and has a density of 1.83. When vaporized, the resulting gas consists of tetraatomic molecules (P_4) up to a temperature of about 1500°C, whereupon these molecules partly dissociate into (and exist in equilibrium with) diatomic molecules (P_2). White phosphorus is insoluble in water but is soluble in solvents such as ethyl ether and carbon disulfide. Great care should always be exercised in handling this form of phosphorus since it is highly flammable and very poisonous. Skin burns caused by phosphorus are exceedingly painful and very slow to heal.

When white phosphorus is heated above 250°C in the absence of air and in the presence of a trace of iodine, which serves as a catalyst, the white allotrope is incompletely converted to *violet phosphorus*. Since this conversion is seldom complete, the resulting amorphous material consists of the white and violet allotropes and, because of its brownish-red appearance, is known as *red phosphorus*.

Relatively little is known about the nature of violet phosphorus. It is best prepared by heating white phosphorus under high pressure or by crystallization from molten lead. Violet phosphorus has a metallic appearance, is nonflammable, and has a density of 2.69.

Because it is a mixture, the physical properties of red phosphorus are variable. Thus, the density ranges from 2.10 to 2.34, depending on the completeness of the transformation from the white to the violet allotrope. The vapor formed when red phosphorus is heated is identical with that formed by white phosphorus; in either case, condensation of these vapors produces white phosphorus. The red modification is much less active chemically than the white variety and is insoluble in those solvents which dissolve white phosphorus.

Preparation and uses of phosphorus. The preparation of elemental (white) phosphorus is carried out in an electrically heated furnace into which a mixture of phosphate rock, sand, and coke is fed continuously by means of a screw conveyer.

$$2Ca_3(PO_4)_2 + 10C + 6SiO_2 \rightarrow P_4 + 10CO\uparrow + 6CaSiO_3\downarrow$$

The phosphorus vapor distills out of the furnace and is liquefied in a condenser, after which the liquid is filtered and cast into sticks in molds immersed in water. The molten calcium silicate is withdrawn from the bottom of the furnace and solidified as a slag.

The greater part of the elemental phosphorus produced in this country is used in the manufacture of a type of alloy known as *phosphor bronze*. In chemical warfare, smoke screens have been produced by the ignition (by reaction with atmospheric oxygen) of white phosphorus liberated from shells, grenades, or bombs. Dense clouds of white smoke were formed owing to oxidation of phosphorus to form finely divided particles of P_2O_3 and P_2O_5. Burning white phosphorus is also useful as an antipersonnel weapon. Another important application for phosphorus lies in its use in the production of phosphorus sesquisulfide (P_4S_3), which is used in the manufacture of matches and in the production of other compounds of phosphorus. The chief use of the red modification of this element is also in the manufacture of matches.

Matches. For many years, matches were made by dipping sticks of wood into a paste made of glue, lead dioxide, and white phosphorus. Long exposure to white phosphorus sometimes causes horrible bone diseases. For this reason and because of the constant fire hazard, the use of white phosphorus in the manufacture of matches is prohibited by law. The modern strike-anywhere match has a head consisting of a mixture of potassium chlorate, paraffin, glue, and finely ground glass. The tip of the match head also contains some phosphorus sesquisulfide (P_4S_3). Ignition of the P_4S_3 by friction initiates combustion, which then extends to the $KClO_3$, the paraffin, and finally to the wood. In the case of the safety match, a side of the matchbox is coated with a mixture of red phosphorus, ground glass, and glue. The head of the match is a mixture of potassium chlorate (the source of oxygen), antimony trisulfide (Sb_2S_3, the combustible material), and glue. Thus, the head of the match contains no phosphorus or sesquisulfide, and such a match can be ignited readily only by friction against the side of the box bearing the red phosphorus which ignites the materials in the match head.

Sulfur

The occurrence of sulfur in the form of simple and complex sulfides of the heavy metals has already been discussed. Of greater consequence, however, are the vast underground deposits of nearly pure elemental sulfur found in the Gulf Coast areas of Louisiana and Texas. Less extensive and less pure deposits of elemental sulfur are also found in a number of other countries, principally Sicily, Spain, Chile, Mexico, and Japan.

Mining of sulfur. In Sicily, sulfur occurs as crystalline deposits disseminated in porous limestone and gypsum. After underground or open-pit surface mining, the crude sulfur-bearing ores are crushed and then treated in the *Calcarone furnace*. Such a furnace consists of a large vertical cylindrical masonry structure having a sloping bottom. The charge is ignited at the top, and the rate of burning is regulated so as to use the minimum amount of sulfur to produce the heat necessary to melt the remainder. Nevertheless, 30 to 35% of the available sulfur is burned to sulfur dioxide. The molten sulfur flows down the sloping bottom to an opening, where it is cast into molds. This rather impure product is the common commercial material known as *brimstone*. This crude sulfur must be purified by distillation to remove impurities such as arsenic and selenium.

The crude sulfur-bearing ores of Japan are charged into furnaces, and the sulfur is distilled out; the resulting vapors are liquefied and cast into "mats" of solid sulfur.

Frasch process.   In the United States, elemental sulfur is made available by the ingenious *Frasch process*. A well is drilled into the sulfur-bearing rocks, which occur at an average depth of about 1000 ft but which may be at depths as great as 1600 to 2000 ft. A 10-in surface casing is cemented into the cap rock to block off the loose overlying formations. The well is then equipped with three concentric pipes, the first of which is a 6-in hot-water (steam) pipe, the second is a 3-in sulfur line, and the third is a 1-in compressed-air line. Steam heated to 160 to 170°C and forced under pressure through the 6-in pipe into the well serves to melt the sulfur. Compressed air under a pressure of 500 psi is sent into the well through the 1-in pipe. Air bubbles become trapped in the molten sulfur and render the liquid sulfur-water mixture lighter and therefore easier to force to the surface (through the 3-in pipe) by the pressure supplied by the compressed air. At the surface, the molten sulfur flows into metering tanks and then is pumped through steam-jacketed pipes to a central solidification vat. Such vats hold 400,000 to 750,000 tons of sulfur of an average purity of 99.5% and a maximum purity of 99.95%. A single well may deliver as much as 400 to 600 tons per day.

The development of the Frasch process, first used successfully in 1891, has made this country substantially independent of foreign sources of sulfur. An inspection shows the growth of this country as a producer of sulfur (in relation to the world production) and the rapid decline in the quantity of sulfur imported.

Allotropic forms of sulfur.   Solid sulfur exists in two crystalline modifications. *Rhombic* sulfur consists of S_8 molecules, is stable at temperatures below 95.5°C, has a specific gravity of 1.96, and is soluble in carbon disulfide. At 95.5°C, rhombic sulfur changes slowly, with absorption of heat, into the *monoclinic* form. Molten rhombic sulfur consists of S_8 molecules and exists as a pale-yellow, thin, and limpid liquid known as λ *sulfur*. When the temperature is raised, λ sulfur is slowly converted to dark and viscous μ *sulfur,* which consists of S_6 and S_4 molecules and which is considered to be the amorphous variety of sulfur. In carbon disulfide, λ sulfur is soluble and μ sulfur is insoluble; thus, a separation of the two forms is permitted. When liquid sulfur that has been heated extensively above its melting temperature is chilled suddenly, there is obtained a mass of *plastic* sulfur, which owes its plasticity to its high content of μ sulfur. If sulfur is sublimated, the vapors condense in the form of very fine and usually very pure crystals known as *flowers of sulfur.*

Uses of sulfur.   In the elemental form, sulfur is used (1) in the vulcanization of rubber; (2) in the manufacture of black powder; (3) as a fungicide (in powder form), particularly in growing grapes; and (4) as lime-sulfur spray for fruit trees. Although the preceding uses require considerable amounts of this element, the great bulk of the world's production of sulfur goes into the manufacture of sulfur dioxide, which may be used as such or converted to other industrially important sulfur compounds, some of which are discussed later.

Some Binary Compounds of Nonmetallic Elements

The preceding sections discuss the preparation, properties, and uses of several typical nonmetals. You should recognize that these elements are capable of forming many binary, ternary, and complex compounds and that the task of becoming familiar with all these individual substances would be a formidable one. Because only a few of these compounds warrant consideration from the standpoint of their practical utility, the following sections are concerned with the study of the more important *binary* compounds of the elements treated. Other binary and ternary compounds of nonmetals are encountered.

Oxides of carbon

The element carbon forms three compounds with oxygen, the most common being the dioxide (CO_2). The others are the very poisonous carbon monoxide (CO) and the relatively less well known carbon suboxide (C_3O_2).

Carbon monoxide.   The preparation of pure gaseous carbon monoxide may be accomplished by the endothermal reaction between carbon dioxide and hot carbon at elevated temperatures:

$$CO_2 + C + 41,500 \text{ cal} \rightarrow 2CO$$

Much carbon monoxide is produced in this manner when coal or coke is used as a fuel under conditions such that the supply of oxygen is limited and carbon is present in excess. The carbon dioxide first formed reacts with the excess carbon to form the monoxide. This gas is also formed in limited quantities by the incomplete combustion of liquid fuels in internal-combustion engines.

For laboratory-scale use, the preparation of carbon monoxide is best accomplished by the thermal decomposition of formic acid or sodium formate:

$$HCOOH + heat \rightarrow CO + H_2O$$

$$HCOONa + heat \rightarrow CO + NaOH$$

From the first of the preceding equations, it is evident that carbon monoxide is the anhydride of formic acid.

Properties of carbon monoxide. Carbon monoxide is a colorless, odorless, and tasteless gas which is slightly lighter than air and only very slightly soluble in water. Carbon monoxide has a pronounced tendency toward combination with oxygen (to form carbon dioxide) and burns in oxygen with a bright, blue flame. Because of its affinity for oxygen, carbon monoxide is an excellent reducing agent for the reduction of the metals. Recall that carbon monoxide is used in connection with the metallurgy of iron. Similarly, oxides of metals such as copper and lead are reduced when heated in the presence of carbon monoxide:

$$CuO + CO \rightarrow Cu + CO_2$$

$$PbO + CO \rightarrow Pb + CO_2$$

Carbon monoxide is an extremely toxic gas. Inhalation of air containing 1 volume of carbon monoxide in 300 volumes of air causes death in a few minutes. The hemoglobin (red coloring matter) of the blood forms an unstable union with oxygen and carries this oxygen to all parts of the body. Carbon monoxide, however, forms a very *stable* compound with hemoglobin and thus renders it incapable of performing its function as an oxygen carrier. Accordingly, death from carbon monoxide poisoning is the result of an inadequate supply of oxygen in the various parts of the body. Many lives are lost each year because of failure to provide for the elimination of carbon monoxide resulting from the incomplete burning of natural gas in home heating units, from allowing automobile engines to operate in closed garages, from explosions in mines, and so forth.

Preparation of carbon dioxide. The commercial production of carbon dioxide by the reaction between steam and either methane or hot carbon is common practice. Carbon dioxide produced industrially as a by-product of the combustion of fuels such as coal and coke is reclaimed by reaction with aqueous sodium carbonate solution:

$$CO_2 + Na_2^+CO_3^{2-} + H_2O \rightarrow 2Na^+HCO_3^-$$

The resulting sodium hydrogen carbonate is heated to liberate carbon dioxide gas:

$$2NaHCO_3 \rightarrow Na_2CO_3 + CO_2 + H_2O$$

and the sodium carbonate is used over again. Carbon dioxide is also formed as a by-product of the manufacture of lime by the thermal decomposition of carbonaceous rocks

$$CaCO_3 + heat \rightarrow CO_2 + CaO$$

and as a by-product of certain fermentation processes used in the production of alcoholic beverages.

In the laboratory, carbon dioxide is prepared most conveniently by treating a carbonate with a strong acid

$$Na_2^+CO_3^{2-} + H_2SO_4 \rightarrow H_2CO_3 + Na_2^+SO_4^{2-}$$

and collecting (over water) the carbon dioxide resulting from the decomposition of the unstable carbonic acid:

$$H_2CO_3 \rightarrow H_2O + CO_2$$

Properties of carbon dioxide.   Carbon dioxide is a colorless, odorless gas, which does not support combustion and which is about 1.5 times as heavy as air. Under a pressure of 1 atm, the pure gas solidifies at $-78.47°C$. At room temperature and at atmospheric pressure, carbon dioxide dissolves in water to the extent of about 1 volume of gas to 1 volume of water. With increasing pressure, the solubility increases in accordance with Henry's law until a pressure of about 6 atm is reached, after which the increase in solubility is less pronounced.

That carbon dioxide is an unusually stable compound is shown by the fact that it is not appreciably decomposed by heat except at very high temperatures. Above $2000°C$, partial decomposition occurs in accordance with the equation:

$$2CO_2 \rightleftarrows 2CO + O_2$$

Carbon dioxide is reduced to the monoxide by reaction with hot carbon. By reaction with water, carbon dioxide (carbonic anhydride) reacts to a limited extent to form carbonic acid

$$H_2O + CO_2 \leftrightarrow H_2CO_3$$

in an equilibrium the forward reaction of which is favored by an increase in pressure and by the slight ionization of the resulting very weak electrolyte:

$$H_2CO_3 \leftrightarrow H^+ + HCO_3^- \qquad K_i = 3 \times 10^{-7}$$

$$HCO_3^- \rightleftarrows H^+ + CO_3^{2-} \qquad K_i = 6 \times 10^{-11}$$

Uses of carbon dioxide. Large quantities of carbon dioxide are consumed in the manufacture of carbonates of sodium by the Solvay process as well as in the production of other chemicals. A more familiar large-scale use for this simple compound is in the production of carbonated beverages. Soda water is produced by dissolving carbon dioxide in water containing suitable flavoring agents, through the application of pressures of 4 to 8 atm (60 to 120 psi). When the carbonated liquid is withdrawn into an open glass, the pressure is reduced to 1 atm, the solubility decreases, and the excess gas escapes, producing the effect of effervescence.

Fire extinguishers. Because carbon dioxide does not support combustion, this gas finds extensive application in the manufacture of several different kinds of fire extinguishers. In one type, a small, stoppered canister of sulfuric acid is mounted inside a much larger cylinder. Inversion of the apparatus permits the stopper to fall off of the acid canister. This permits the acid to mix with the sodium carbonate (or sodium hydrogen carbonate) solution contained in the main vessel and to generate carbon dioxide gas:

$$Na_2^+CO_3^{2-} + H_2^+SO_4^{2-} \rightleftarrows Na_2^+SO_4^{2-} + H_2O + CO_2\uparrow$$

Some of the gas dissolves in the aqueous solution, and the gas pressure developed within the apparatus forces the mixture out through the nozzle, which is directed onto the flame. If some heavy molasses or similar gluelike material is added to the sodium carbonate solution and some aluminum sulfate is added to the sulfuric acid, the solution discharges from the nozzle in the form of a foam in which the carbon dioxide gas is trapped. This heavy blanket of foam containing carbon dioxide settles over and around the burning object. The aluminum sulfate hydrolyzes to form aluminum hydroxide, which, at the temperature of the flame, dehydrates to produce a crust of aluminum oxide over the flame. Both effects serve to extinguish the flame. The common Foamite fire extinguisher is an example of this type.

Dry ice. If a commercial cylinder of liquid carbon dioxide is inverted and some of the liquid is allowed to run out into a cloth bag or a special perforated metal container, only a part of the liquid evaporates. That which evaporates takes heat from the remainder and causes it to solidify in the form of a white solid resembling snow. This material, compressed into blocks, is sold commercially as *dry ice*. Since this solid vaporizes directly to a gas at −78.47°C, evaporation produces no liquid residue. For this reason alone, the advantages of this dry refrigerant are obvious.

When dry ice is mixed with a volatile liquid such as ether or acetone, a freezing mixture providing temperatures in the range of −90 to −100°C (depending upon the pressure) results. Such freezing mixtures are commonly employed in scientific laboratory work where the maintenance of low temperatures is desired.

Carbides

Although the general term *carbide* applies to the binary compounds of the element carbon, this term is used in systematic nomenclature only when carbon is the more electronegative of the two elements involved. Thus, CO_2 is called *carbon dioxide* and not *oxygen carbide* since oxygen is more electronegative than carbon. Although carbon forms binary compounds with most of the nonmetals, metalloids, and metals, only a few of the more common members of this class are considered here.

Carbon disulfide. Carbon disulfide (CS_2) is produced commercially by the direct union of sulfur vapor and hot carbon at considerably elevated temperatures in an enclosed electric furnace. The reaction must be carried out in the absence of air. Solid sulfur and carbon (usually in the form of coke) are fed continuously into the furnace, which is heated electrically to a temperature sufficient to vaporize the sulfur and to maintain the carbon at a temperature such that combination occurs when the sulfur vapor comes into contact with the solid carbon. Since carbon disulfide is a low-boiling-point liquid (46°C), it passes out of the furnace as a vapor, which is conducted into a condenser. Although the resulting pale-yellow liquid may be purified by distillation, the crude product is used directly for many purposes.

The yellow color of commercial carbon disulfide is due to the presence of small amounts of dissolved sulfur. The rather disagreeable odor of this material is also due to the presence of impurities. Carbon disulfide is a highly flammable liquid which is very poisonous and which is only very slightly soluble in water.

The large-scale uses of carbon disulfide center mainly about its properties as a solvent. Many fats, oils, waxes, and resins are abundantly soluble in this liquid. Despite the disadvantages attendant upon its volatility, flammability, and toxicity, carbon disulfide is used extensively as a solvent and in processes for the manufacture of rubber products, lacquers, varnishes, cellophane, and so forth. Because of its toxicity, this compound is used to some extent as an insecticide and as a poison for rodents.

Silicon carbide. Silicon carbide (SiC), more familiarly known by the trade name Carborundum, is produced by heating a mixture of coke, sawdust, sand, and salt to about 3000°C in an electric furnace:

$$SiO_2 + 3C \rightarrow SiC + 2CO$$

The salt is added to the charge placed in the furnace in order to render the charge a better conductor and to permit the partial removal of any iron present. This latter is accomplished owing to the formation of iron chloride, which is volatile at the temperature of the furnace. The sawdust renders the entire charge more porous and thus facilitates the escape of volatile products. Because the charge as a whole is not a particularly good conductor, a core of pure carbon is placed in the center of the charge. When the heated charge is cooled, a mass of crystals of silicon carbide forms in the region of the carbon core.

The product thus obtained consists of sharp iridescent crystals which are extremely inactive chemically. The most notable property of this material, however, is its hardness. Carborundum is almost as hard as diamond and is generally as an *abrasive* (i.e., in the manufacture of grinding stones and wheels, polishing papers and cloths, etc.). A lesser use lies in the incorporation of coarse Carborundum crystals into concrete or terrazzo floors to render them slipproof.

Calcium carbide. The calcium compound having the formula CaC_3 is that which is implied when the common term *carbide* is used. That is, just as the term *salt* commonly denotes sodium chloride, so the term *carbide* is applied to calcium carbide. This compound is formed commercially by the reaction between quicklime and coke in an electric furnace at 2800 to 2900°C:

$$CaO + 3C \rightarrow CaC_2 + CO$$

The equipment is designed so as to permit continuous and automatic operation, and the resultant molten calcium carbide is withdrawn from the furnace and allowed to cool, whereupon it solidifies to a dense gray solid.

Calcium carbide is an important chemical in the manufacture of acetylene

$$CaC_2 + 2H_2O \rightarrow C_2H_2 + Ca(OH)_2$$

and in the manufacture of calcium cyanamide

$$CaC_2 + N_2 \rightarrow CaNCN + C$$

which, in turn, is used in the production of ammonia, sodium cyanide, and so forth.

Oxides of silicon

The nonmetal silicon forms two oxides, SiO and SiO_2. The monoxide is relatively unimportant, while silicon dioxide is a compound of major value in many chemical processes and products.

Silicon monoxide. When silicon dioxide is reduced by carbon at high temperatures, silicon monoxide is formed:

$$SiO_2 + C \rightarrow SiO + CO$$

Silicon monoxide is a light-brown solid which finds limited use as a pigment.

Silicon dioxide. Because silicon dioxide is so readily available in nature, there is only infrequent need to produce this compound in either the laboratory or the plant. Many familiar materials consist of silicon dioxide (or *silica*) in varying degrees of purity. Thus, flint, agate, amethyst, quartz, onyx, opal, granite, petrified wood, sand, and sandstone are all materials that consist entirely or largely of silica. You may recognize that all these materials have a variety of common uses and should recall, from preceding discussions, cases in which silicon dioxide is used in certain commercial chemical processes.

One of the most significant properties of silica is its resistance to chemical attack. This compound is insoluble in water and is unreactive toward all acids except hydrofluoric, with which the following reaction occurs:

$$SiO_2 + 4HF \rightarrow SiF_4 + 2H_2O$$

With strong bases such as sodium hydroxide, silica reacts to form *silicates:*

$$SiO_2 + 2Na^+OH^- \rightarrow Na_2^+SiO_3^{2-} + H_2O$$

This shows that SiO_2 is the anhydride of silicic acid (H_2SiO_3). When silica is fused with dry sodium carbonate, sodium sulfate, or any salt formed from an acid having a volatile anhydride, the latter is released and silicates are produced:

$$Na_2^+CO_3^{2-} + SiO_2 \rightarrow Na_2^+SiO_3^{2-} + CO_2$$

$$Na_2^+SO_4^{2-} + SiO_2 \rightarrow Na_2^+SiO_3^{2-} + SO_3$$

Finally, note the marked contrast between the properties (both chemical and physical) of the related compounds CO_2 and SiO_2.

Ammonia

Of the simple hydrogen compounds of the elements, only water is more widely useful than ammonia. In the gaseous and liquid states and in solutions, ammonia serves a variety of needs in the chemical laboratory, the home, and the industries.

Production of ammonia.   For small-scale laboratory use, ammonia may be prepared by any of the following methods.

Treatment of an ammonium salt with a strong base.   This is perhaps the best available laboratory method and is illustrated by the following reactions:

$$(NH_4)_2{}^+SO_4{}^{2-} + 2Na^+OH^- \rightarrow Na_2{}^+SO_4{}^{2-} + 2NH_3 + 2H_2O$$

$$2NH_4{}^+Cl^- + Ca^{2+}(OH)_2{}^- \rightarrow Ca^{2+}Cl_2{}^- + 2NH_3 + 2H_2O$$

Dissociation of an ammonium salt.   Gaseous ammonia is liberated by heating the normal ammonium salt of a (relatively) nonvolatile acid; for example:

$$(NH_4)_2SO_4 + heat \rightarrow NH_4HSO_4 + NH_3$$

Ammonium phosphate would serve equally well. If, however, one used the ammonium salt of a volatile acid such as hydrochloric:

$$NH_4Cl + heat \rightarrow NH_3 + HCl$$

Ammonia would be liberated together with hydrogen chloride gas, and the two gases would recombine to form the original salt, ammonium chloride.

Hydrolysis of the nitride of a metal.   The nitrides of metals such as magnesium react with water to produce ammonia and the hydroxide of the metal; thus:

$$Mg_3N_2 + 6HOH \rightarrow 3Mg(OH)_2 + 2NH_3$$

Volatilization of ammonia from aqueous solutions.   When an aqueous solution of ammonia (ammonium hydroxide) is warmed, some of the dissolved ammonia volatilizes in the gaseous form. This result is to be anticipated in light of the manner in which the solubility of gases in liquids is generally influenced by an increase in temperature. In the use of this method, the quantity of heat supplied should be such that a minimum quantity of water is vaporized.

None of the preceding methods would be suitable for industrial use. For many years the bulk of the ammonia produced commercially was obtained as a by-product of the destructive distillation of coal. More recently, however, processes that are far more satisfactory have been devised.

Haber process. This method, which is sometimes called the *synthetic ammonia process,* is based on the following reactions:

$$N_2 + 3H_2 \rightleftarrows 2NH_3 + 24,400 \text{ cal}$$

The manner in which this equilibrium is influenced by changes in temperature and pressure should be reviewed. In practice, ammonia is produced by the Haber process at temperatures ranging from 400 to 600°C and at pressures between 200 and 1000 atm. Catalysts that are suitable for use in this process include a mixture of the oxides of iron, potassium, and aluminum; iron oxide alone; mixtures of iron and molybdenum; the metals platinum, osmium, uranium; and a number of others as well.

Cyanamide process. Although this process employs relatively inexpensive raw materials, it cannot compete with the Haber process under normal conditions. Formation of ammonia by the cyanamide process is based on the reaction between calcium cyanamide and steam:

$$CaNCN + 3H_2O \rightarrow CaCO_3 + 2NH_3$$

One of the chief disadvantages involved in this process is the large quantity of energy that must be employed in the production of calcium carbide and in its conversion to calcium cyanamide.

Uses of ammonia. The various uses of ammonia include the use of the compound both as such and in the form of other compounds made from ammonia. In the liquid state, much ammonia is used as the refrigerant liquid in commercial refrigeration plants and in the manufacture of ice. Some liquid ammonia is used both in the laboratory and commercially as a solvent, and its solvent properties are in many respects similar to those of water. Great quantities of ammonia are used in the manufacture of nitric acid, sodium hydrogen carbonate, normal sodium carbonate, aqueous ammonia (or ammonium hydroxide), ammonium salts for use as fertilizers, and many other useful chemicals.

Oxides of nitrogen and phosphorus

The generally abnormal character of an "introductory element" is well illustrated by the case of nitrogen. With respect to oxide formation, for example, this element exhibits more diversity than any other element in Group V of the periodic arrangement.

The related elements nitrogen and phosphorus combine with oxygen to form those oxides which are to be expected on the basis of the position of these elements in the periodic table. In addition, nitrogen forms several other oxides (Table 11.10). Although each of these compounds has its own specific uses or participates in certain reactions that render these oxides of scientific interest, the emphasis here is on the relationship between these oxides (acid anhydrides) and the corresponding acids.

Nitrogen dioxide may be considered as being, at the same time, the anhydride of both nitrous and nitric acids. This view is based on the fact that nitrogen dioxide reacts with *cold* water as follows:

$$2NO_2 + H_2O \rightarrow HNO_2 + HNO_3$$

However, if hot water is used, the reaction is as follows:

$$3NO_2 + H_2O \rightarrow 2HNO_3 + NO$$

The two oxides of phosphorus are, in fact, dimeric, and their formulas accordingly may be written more properly as P_4O_6 and P_4O_{10}.

Oxides of sulfur

Of the two common oxides of sulfur, the dioxide is by far the more important. Sulfur trioxide (SO_3), the anhydride of sulfuric acid, is a strong oxidizing agent which finds limited uses other than in the manufacture of H_2SO_4. Solutions of sulfur trioxide in concentrated sulfuric acid are known as *fuming sulfuric acid* or *oleum,* and such solutions are useful in the laboratory and in a few commercial processes.

TABLE 11.10   Oxides of Nitrogen and Phosphorus

Oxide		Related acid	
Formula	Name	Formula	Name
N_2O	Nitrous oxide*		
NO	Nitric acid		
N_2O_3	Nitrogen trioxide	HNO_2	Nitrous acid
NO_2	Nitrogen dioxide		
N_2O_5	Nitrogen pentoxide	HNO_3	Nitric acid
P_2O_3	Phosphorus trioxide	H_3PO_3	Phosphorus acid
P_2O_5	Phosphorus pentoxide	H_3PO_4	Phosphoric acid

* Laughing gas

Sulfur dioxide.   The best laboratory method for the preparation of sulfur dioxide (SO_2) is based on the metathetical reaction between a sulfite and a strong acid; for example:

$$Na^+HSO_3^- + H_2^+SO_4^{2-} \rightarrow Na^+HSO_4^- + H_2O + SO_2$$

or

$$Na_2^+SO_3^{2-} + H_2^+SO_4^{2-} \rightarrow Na_2^+SO_4^{2-} + H_2O + SO_2$$

Less convenient but otherwise equally satisfactory is the preparation involving the reduction of *hot* concentrated sulfuric acid by means of copper, carbon, or sulfur. The oxidation-reduction reaction involved may be illustrated by the following equations:

$$Cu - 2e^- \rightarrow Cu^{2+}$$

$$S^{6+} + 2e^- \rightarrow S^{4+}$$

$$\overline{Cu + S^{6+} \rightarrow Cu^{2+} + S^{4+}}$$

$$Cu + 2H_2SO_4 \rightarrow CuSO_4 + SO_2 + 2H_2O$$

If sulfur is employed as the reducing agent, the equations are as follows:

$$S - 4e^- \rightarrow S^{4+}$$

$$2S^{6+} + 4e^- \rightarrow 2S^{4+}$$

$$\overline{S + 2S^{6+} \rightarrow 3S^{4+}}$$

$$S + 2H_2SO_4 \rightarrow 3SO_2 + 2H_2O$$

Commercially, sulfur dioxide is produced by roasting of sulfides; for example:

$$2PbS + 3O_2 \rightarrow 2PbO + 2SO_2$$

$$4FeS_2 + 11O_2 \rightarrow 2Fe_2O_3 + 8SO_2$$

or by the direct union of elemental sulfur and oxygen:

$$S + O_2 \rightarrow SO_2$$

Pure sulfur dioxide is a colorless gas which has an irritating odor and which is soluble in water. The gas may be liquefied readily and is usually stored and transported in the liquid state in steel cylinders.

This substance is a strong reducing agent, and therefore a poor oxidizing agent.

The chief uses of sulfur dioxide concern the following processes:

- The hydration of SO_2 to form sulfurous acid, which, in turn, is used to produce sulfites, notably $Ca(HSO_3)_2$, which is used in large quantities in the pulp and paper industries

- The oxidation of SO_2 to SO_3, which is then converted to sulfuric acid

- The bleaching of silk, straw, wool, and other fabrics (This use depends on the fact that sulfur dioxide forms unstable colorless compounds with the colored pigments in these fibers; because of the relative instability of these colorless compounds, the bleaching action of sulfur dioxide is seldom permanent.)

- The removal of excess chlorine or hypochlorous acid from fabrics which have been bleached with chlorine:

$$HClO + SO_2 + H_2O \rightarrow HCl + H_2SO_4$$

- The operation of refrigeration units in which liquid sulfur dioxide is used as the refrigerant liquid

Halogens and Compounds of the Halogens

All the elements of the halogen family except astatine occur in nature in the form of their compounds. Because of their marked chemical activity, they are never found in nature in the uncombined state.

In compound form, *fluorine* is a fairly abundant element which constitutes approximately 0.1% of the solid crust of the earth. The chief fluorine-bearing minerals are *fluorspar* (CaF_2), *cryolite* (Na_3AlF_6), and *fluorapatite* $\{CaF_2[Ca_3(PO_4)_2]_3\}$.

Ordinary rock salt (sodium chloride) is the chief commercial source of *chlorine*. In the United States, the salt wells in Michigan and the salt mines of Texas and Louisiana are rich sources of this material. Chlorine is also obtained from the minerals *carnallite* ($MgCl_2 \cdot KCl \cdot 3H_2O$) and *kainile* ($MgSO_4 \cdot KCl \cdot 6H_2O$). Salt deposits containing combined chlorine are widely scattered over all parts of the world and are among the most abundant, cheap, and useful raw materials of the chemical industries. Oceanic waters have a high salt content, and the waters of salt lakes such as the Great Salt Lake in Utah contain approximately 20% sodium chloride.

Although less abundant than chlorine, the occurrence of *bromine* parallels that of chlorine. Thus, bromine is found in salt deposits in the form of the bromides of sodium, potassium, magnesium, and so forth, and in seawater. Although in this country much bromine is produced from salt wells in Michigan, seawater is the most important source of this element.

Iodine also occurs in seawater and in numerous seaweeds including kelp, from which iodine has been extracted on a commercial scale. The greater part of the iodine of commerce is obtained from sodium iodate $(NaIO_3)$ and sodium periodate $(NaIO_4)$, both of which are associated with the enormous deposits of sodium nitrate in Chile.

Both chlorine and iodine are found in, and are essential to, the human body. The gastric juices normally contain 0.25 to 0.4% hydrochloric acid, while the chloride-ion concentration of the blood is approximately 0.25%. Iodine occurs in the thyroid gland in the form of a complex organic compound which is essential to the gland's proper functioning. Iodine deficiency in the human body is largely responsible for the development of goiter and cretinism. For this reason, iodine, frequently in the form of sodium iodide added to table salt (i.e., iodized salt), is added to the diet as a preventive. In the United States, waters in the coastal regions contain adequate supplies of iodine, while in the inland states, iodine deficiency is much more prevalent.

Methods of preparation

A review of the means whereby the elements of the halogen family may be prepared either in the laboratory or on an industrial scale provides further insight into the interrelationships of these elements.

Fluorine.   The distinguished chemist Henri Moissan first prepared fluorine by the electrolysis of a solution of potassium fluoride in liquid hydrogen fluoride. Because of the extreme chemical activity of this element, the electrolytic cell employed had to be made of platinum. At the present time, fluorine is produced in the laboratory and commercially by the electrolysis of fused potassium hydrogen fluoride (KHF_2) in the manner described in the section on electrolysis.

Chlorine.   The preparation of chlorine in the laboratory is best accomplished by the oxidation of hydrogen chloride using any of a wide variety of oxidizing agents (e.g., HNO_3, $KMnO_4$, $K_2Cr_2O_7$, MnO_2, and $KClO_3$). The HCl may be used in the form of concentrated hydrochloric acid or may be prepared by the reaction between sodium chloride and concentrated sulfuric acid. (This procedure is made possible by the fact that concentrated sulfuric acid is not a sufficiently strong oxidizing agent to oxidize HCl. Hence, a stronger oxidizing agent must be used to produce Cl_2 from the HCl liberated in the reaction between NaCl and H_2SO_4.) The overall reactions are illustrated by the following equations:

$$KClO_3 + 6HCl \rightarrow KCl + 3H_2O + 3Cl_2$$

$$2KMnO_4 + 16HCl \rightarrow 2MnCl_2 + 2KCl + 8H_2O + 5Cl_2$$

$$MnO_2 + 4HCl \rightarrow MnCl_2 + 2H_2O + Cl_2$$

Or, in case the HCl is produced as needed:

$$NaCl + H_2SO_4 \rightarrow HCl + NaHSO_4$$

$$NaCl + NaHSO_4 \rightarrow HCl + Na_2SO_4$$

The oxidation of hydrogen chloride by atmospheric oxygen,

$$4HCl + O_2 \rightarrow 2Cl_2 + 2H_2O$$

at elevated temperatures and in the presence of suitable catalysts, was once used as a commercial procedure for the production of chlorine. However, this process cannot now be operated economically in competition with methods involving electrolysis. As has been described previously, chlorine is now produced commercially almost exclusively by the electrolysis of aqueous solutions of chlorides or of fused chlorides.

Bromine. The laboratory preparation of bromine may be accomplished in a manner analogous to that described for chlorine. Thus, treatment of a bromide with a mixture of sulfuric acid and manganese dioxide results in the liberation of elemental bromine:

$$2NaBr + 2H_2SO_4 + MnO_2 \rightarrow MnSO_4 + Na_2SO_4 + 2H_2O + Br_2$$

Although this equation accounts for the liberation of bromine, this product also arises by virtue of the occurrence of another reaction. Although not a sufficiently strong oxidizing agent to liberate chlorine from hydrogen chloride, concentrated sulfuric acid does act on hydrogen bromide, with a resultant *partial* oxidation to bromine:

$$2HBr + H_2SO_4 \rightarrow Br_2 + SO_2 + 2H_2O$$

Hence, when one treats sodium bromide with concentrated sulfuric acid and manganese dioxide, the hydrogen bromide liberated by the initial reaction,

$$2NaBr + H_2SO_4 \rightarrow 2HBr + Na_2SO_4$$

may be oxidized to bromine either by the concentrated H_2SO_4 or by the MnO_2.

The commercial production of bromine from salt-well brines or from seawater depends on the fact that chlorine is capable of displacing bromine from its salts. Bromine is extracted from seawater by a process involving the following steps:

Seawater is rendered slightly acidic by the addition of sulfuric acid, after which the water solution is sprayed into towers containing gaseous chlorine. If M is used to represent a metal that forms a soluble bromide, the reaction may be represented as follows:

$$2MBr + Cl_2 \rightarrow 2MCl + Br_2$$

The liberated bromine is carried from the tower by means of a stream of air and into an absorber containing sodium carbonate solution. By reaction with sodium carbonate, the bromine is converted to sodium bromide and sodium bromate:

$$3Na_2CO_3 + 3Br_2 \rightarrow 5NaBr + NaBrO_3 + 3CO_2$$

When bromine is needed, this solution is acidified with sulfuric acid, whereupon hydrobromic and bromic acids are formed:

$$5NaBr + NaBrO_3 + 3H_2SO_4 \rightarrow 3Na_2SO_4 + 5HBr + HBrO_3$$

Owing to the oxidizing action of bromic acid, bromine is liberated in accordance with the following equation:

$$5HBr + HBrO_3 \rightarrow 3Br_2 + 3H_2O$$

The resulting bromine may be removed in a current of steam and subsequently separated from the water by condensation, or it may be used directly in the manufacture of ethylene dibromide:

$$C_2H_4 + Br_2 \rightarrow C_2H_4Br_2$$

Although seawater contains only 67 parts per million of bromine in solution, this seemingly rather involved process is both efficient and economical. One cubic mile of seawater contains approximately 300,000 tons of bromine, and the process described permits the recovery of about 80% of the available bromine.

Iodine. Because of the close chemical similarity between chloride, bromine, and iodine, it is to be anticipated that iodine might be prepared in a manner analogous to that used in the preparation of bromine. Such is the case. In the laboratory, iodine is most conveniently produced by treating an iodide with sulfuric acid and manganese dioxide. Just as in the case of bromine, the initial reaction is that between the iodide and sulfuric acid:

$$NaI + H_2SO_4 \rightarrow HI + NaHSO_4$$

$$NaI + NaHSO_4 \rightarrow HI + Na_2SO_4$$

The hydrogen iodide is oxidized by either or both of the oxidizing agents present, concentrated H_2SO_4 or MnO_2:

$$4HI + MnO_2 \rightarrow MnI_2 + I_2 + 2H_2O$$

or

$$2HI + H_2SO_4 \rightarrow I_2 + SO_2 + 2H_2O$$

In addition to relatively small amounts of iodine that are extracted from the ashes of kelp, the great bulk of iodine produced commercially is obtained from sodium iodate. This compound occurs along with sodium nitrate (Chile saltpeter), and the iodine is extracted by treating the iodate with sodium hydrogen sulfite:

$$2NaIO_3 + 5NaHSO_3 \rightarrow I_2 + 2Na_2SO_4 + 3NaHSO_4 + H_2O$$

The resulting solution is evaporated to dryness and heated further, whereupon the iodine sublimates and thus may be separated from the other products of the reaction.

Uses of the halogens

Owing to the extreme chemical activity of *fluorine,* the chief uses of this element arise in connection with the production of its compounds. Thus, sodium fluoride is commonly used as a flux, as an insecticide, and as a wood preservative. Ammonium fluoride is sometimes used as a disinfectant. The compound dichlorodifluoromethane (CCl_2F_2) is used as an insecticide propellent and is an important commercial refrigerant (Freon), particularly well suited for use in domestic refrigeration units.

Chlorine is by far the most useful of the elements of the halogen family. This fact becomes evident when one learns that the annual production of chlorine in the United States alone amounts to approximately 250,000 tons. Much of this chlorine is used in the bleaching of fabrics, wood pulp, and so forth. Considerable quantities of this element are used in treating water supplies for the purpose of destroying harmful bacteria. Many of the poisonous gases used in the world wars were chlorine compounds, such as mustard gas [$(ClC_2H_4)_2S$], chloropicrin

(CCl_3NO_2), phosgene ($COCl_2$), and so forth. Other important and large-scale uses of chlorine are found in the manufacture of compounds such as chloroform ($CHCl_3$), carbon tetrachloride (CCl_4), bleaching powder ($CaOCl_2$), and a host of other compounds that are useful in the home, industry, and the laboratory.

That *bromine* is relatively a less important element from the standpoint of its applications may be seen from the fact that the annual production of bromine in this country is only about one-twentieth that of chlorine. At the present time, the chief use of bromine is in the manufacture of ethylene dibromide ($C_2H_4Br_2$), which is used in the production of high-test gasolines. Bromine is also used in the production of other bromine compounds. Thus, potassium bromide is used medicinally as a sedative and silver bromide is used in photographic processes, while other compounds of bromine are used in the production of dyes.

Undoubtedly the most familiar use of *iodine* is in tincture of iodine, which is so commonly used as an antiseptic. This consists of a solution of iodine and potassium iodide in alcohol and is one of the most effective antiseptics known. Iodine is used to make many useful compounds including sodium and potassium iodides, which are important laboratory reagents; silver iodide, which (like silver bromide) is used in photography; and numerous organic compounds of iodine which are useful in drugs, in the production of dyes, and so forth.

Hydrogen compounds of the halogens

Each member of the halogen family combines with hydrogen to form a compound of the type HX, where X represents a halogen.

The hydrogen compound of fluorine is exceptional in that it exists in different molecular forms from HF to H_6F_6. For the sake of simplicity, the formula HF is employed consistently in this book.

General methods of preparation.   There are two general methods which, with appropriate modifications, may be used to produce the hydrogen compounds of the halogens.

Direct union of the elements.   The manner in which the halogens react with hydrogen demonstrates admirably the trend in chemical activity within this group of elements. Fluorine and hydrogen combine with explosive violence even in the absence of light and at low temperatures. Chlorine and hydrogen do not react readily in the dark at ordinary temperatures, but under the influence of the actinic rays of a bright light (or sunlight) their union is violent. Bromine and hydrogen combine only upon application of heat, while iodine and hydrogen combine only upon elevation of the temperature in the presence of a catalyst such as spongy platinum. The following thermochemical reactions show the quantities of heat involved and provide evidence of the marked decrease in chemical activity of the halogens with increasing atomic number:

$$H_2 + F_2 \rightarrow 2HF + 128{,}000 \text{ cal}$$

$$H_2 + Cl_2 \rightarrow 2HCl + 44{,}120 \text{ cal}$$

$$H_2 + Br_2 \rightarrow 2HBr + 17{,}300 \text{ cal}$$

$$H_2 + I_2 \rightarrow 2HI - 11{,}820 \text{ cal}$$

For practical purposes, only hydrogen chloride is produced in any appreciable quantity by direct union.

Treatment of a salt with a strong acid. The metathetical reaction between a salt and a strong acid may be used in the preparation of compounds of the type HX, provided the correct reactants are chosen. Almost any salt containing the desired halogen may be employed, but one would, of course, select a cheap and readily available salt. The acid selected must be *concentrated, nonvolatile,* and *nonoxidizing.* The reasons for these requirements become evident from a consideration of a specific case. If, as would be convenient in the laboratory, an apparatus of the boiling flask type were employed, the HX would be formed in the reaction mixture in the retort. If this reaction mixture contained water (i.e., if the acid were not concentrated), the HX would dissolve in the water and hence would not escape at the flask's vapor outlet as desired. If a volatile acid were used, some of this acid would escape at this outlet along with the HX; hence, the material collected would be a mixture and not the anticipated pure HX. Finally, if an acid which acts as an oxidizing agent toward the desired HX were employed, one would again obtain not HX but a mixture of HX and the elemental halogen X_2.

For the preparation of hydrogen fluoride and hydrogen chloride, the following equations represent the use of suitable reactants:

$$CaF_2 + H_2SO_4 \rightarrow CaSO_4 + 2HF$$

and

$$NaCl + H_2SO_4 \rightarrow NaHSO_4 + HCl$$

$$NaCl + NaHSO_4 \rightarrow Na_2SO_4 + HCl$$

Concentrated sulfuric acid is relatively nonvolatile and does not act as an oxidizing agent toward HF and HCl. However, as has already been pointed out, both HBr and HI are oxidized to the respective halogens by concentrated sulfuric acid. Consequently, to prepare pure hydrogen bromide or pure hydrogen iodide, one would use sodium bromide or sodium iodide in reaction with concentrated phosphoric acid, which is also relatively nonvolatile and which does not act as an oxidizing agent toward either HBr or HI.

Other methods of preparation.   For each of the hydrogen compounds of the halogens there are certain specific methods that may be used to advantage in the laboratory. Thus, hydrogen fluoride may be produced by heating potassium hydrogen fluoride:

$$KHF_2 + heat \rightarrow KF + HF$$

Hydrogen bromide and hydrogen iodide may readily be prepared by the hydrolysis of phosphorus tribomide and phosphorus triiodide, respectively:

$$PBr_3 + 3H_2O \rightarrow H_3PO + 3HBr$$

$$PI_3 + 3H_2O \rightarrow H_3PO_3 + 3HI$$

Properties.   All four of the hydrogen halides are colorless gases with pungent irritating odors. These compounds are predominantly covalent in the pure dry condition. However, when they are dissolved in water, all except hydrogen fluoride form solutions (by reaction) which behave as very strong acids:

$$HX + H_2O \rightleftarrows H_3O^+ + X^-$$

With increase in atomic number of the halogen, the hydrogen halides become increasingly unstable and become increasingly strong reducing agents (hence increasingly weak oxidizing agents).

Although hydrofluoric acid is a relatively weak acid, it is unusually reactive in certain other respects. For example, it has an extremely corrosive action on the skin and produces burns which are very painful and slow to heal. Hydrofluoric acid also attacks the silica and silicates present in glass:

$$SiO_2 + 4HF \rightarrow SiF_4 + 2H_2O$$

$$CaSiO_3 + 6HF \rightarrow SiF_4 + CaF_2 + 3H_2O$$

These and similar reactions are involved in etching glass.

Hydrogen fluoride is also used as a catalyst in the production of aviation gasoline. More recently, it has replaced sulfuric acid in the alkylation process for the production of automobile gasoline.

Oxygen compounds of the halogens

Having noted the regularities exhibited by the halogens in their combinations with hydrogen, it is interesting to consider the compounds formed by the halogens and oxygen. Table 11.11 lists the known oxides.

Most of these oxides are unstable, and even at ordinary temperatures, some of them decompose with explosive violence. For the most part, they are very strong oxidizing agents. Of the four halogens, fluorine and bromine show but little tendency to combine with oxygen; in fact, all the known oxides of these elements have been discovered since about 1930.

Ternary oxygen acids of the halogens

Just as the halogens differ with respect to their tendencies toward combination with oxygen and the stability of the resulting oxide, so these elements differ appreciably in regard to the properties of their ternary oxygen acids. Table 11.12 lists the names and formulas of the known acids, together with the names and formulas of the corresponding sodium salts. An inspection of this table makes evident the manner in which the halogens differ in the form of their ternary acids.

It is apparent from Table 11.12 that ternary acids of fluorine are unknown. There is, however, some evidence that salts of hypofluorous and fluoric acids exist in solution.

The following discussion is limited to the acids of chlorine and their salts, since these compounds are by far the most common and useful. The ternary acids of chlorine have certain properties in common. Thus, all four of these acids are rather unstable and are strong oxidizing agents. However, each of them forms salts that are markedly more stable than the corresponding acids.

Hypochlorous acid and its salts. When chlorine gas is dissolved in water, the following equilibrium becomes established:

$$Cl_2 + HOH \rightleftarrows HCl + HClO$$

TABLE 11.11   Known Oxides of the Halogens

Halogen	Known oxides
F	F_2O, F_2O_2
Cl	Cl_2, ClO_2, ClO_3 (or Cl_2O_6), Cl_2O_7, ClO_4
Br	Br_2O, BrO_2, Br_3O_8
I	I_2O_4, I_2O_5, IO_4, I_4O_9

TABLE 11.12   Ternary Acids of the Halogens

Acid		Salt	
Formula	Name	Formula	Name
HClO	Hypochlorous acid	Na^+ClO^-	Sodium hypochlorite
$HClO_2$	Chlorous acid	$Na^+ClO_2^-$	Sodium chlorite
$HClO_3$	Chloric acid	$Na^+ClO_3^-$	Sodium chlorate
$HClO_4$	Perchloric acid	$Na^+ClO_4^-$	Sodium perchlorate
HBrO	Hypobromous acid	Na^+BrO^-	Sodium hypobromite
$HBrO_3$	Bromic acid	$Na^+BrO_3^-$	Sodium bromate
HIO	Hypoiodous acid	Na^+IO^-	Sodium hypoiodite
HIO_3	Iodic acid	$Na^+IO_3^-$	Sodium iodate
HIO_4	Metaperiodic acid	$Na^+IO_4^-$	Sodium metaperiodate
H_5IO_6	Orthoperiodic acid	$Na_2^+H_3IO_6^{-2}$	Disodium orthoperiodate

The quantities of hydrochloric and hypochlorous acids present at equilibrium depend upon the temperature. The concentration of HClO in the equilibrium mixture may be increased by neutralizing the strong acid by the addition of a base such as sodium hydroxide. In practice, the best method for the formation of the hypochlorites and their parent acids is to pass chlorine into a *cold* solution of a strong base such as sodium hydroxide or calcium hydroxide:

$$Cl_2 + 2NaOH \rightarrow NaCl + NaClO + H_2O$$

A solution of hypochlorite having thus been obtained, a solution of hypochlorous acid may be produced by acidification with a strong acid such as sulfuric:

$$NaClO + H_2SO_4 \rightarrow HClO + NaHSO_4$$

Both hypochlorous acid and the hypochlorites are relatively unstable and decompose readily with liberation of oxygen and the formation of chlorides. In fact, the acid is known only in solution. These compounds are excellent oxidizing agents; they are used as bleaching agents, as antiseptics, as disinfectants, and in the production of ethylene glycol. The commercial production of sodium hypochlorite by electrolysis has been described previously.

Bleaching powder.   If, instead of cold sodium hydroxide, cold calcium hydroxide solution is used in reaction with chlorine, one obtains a solution containing calcium chloride and calcium hypochlorite:

$$2Cl_2 + 2Ca(OH)_2 \rightarrow CaCl_2 + Ca(ClO)_2 + 2H_2O$$

If, on the other hand, chlorine gas is allowed to react with dry, solid calcium hydroxide, a compound having the formula $CaOCl_2$ is formed. Commercially, this compound is known as *bleaching powder* and is regarded as being, at the same time, a salt of both hydrochloric and hypochlorous acids. This relationship may be clarified by the following formulas:

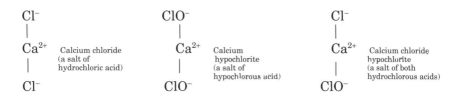

When bleaching powder is treated with a *strong* acid such as sulfuric, both hydrochloric and hypochlorous acids are liberated:

$$Ca(Cl)ClO + H_2SO_4 \rightarrow HCl + HClO + CaSO_4$$

These acids then interact to form chlorine:

$$HCl + HClO \rightarrow Cl_2 + H_2O$$

By the use of a *weak* acid, however, only hypochlorous acid is liberated:

$$Ca(Cl)ClO + H_2CO_3 \rightarrow HClO + Ca(Cl)HCO_3$$

The ability of such solutions to act as bleaching agents is due to the liberated hypochlorous acid or to the oxygen formed by the decomposition of this acid:

$$HClO \rightarrow HCl + [O]$$

The enclosure of the symbol for oxygen implies the liberation of highly reactive oxygen.

Chlorous acid and its salts.   As compared with the other acids of chlorine and their salts, chlorous acid and the chlorites are less well known and have no really important large-scale uses. Chlorous acid and chloric acid are formed together by the reaction between chlorine dioxide and water:

$$2ClO_2 + H_2O \rightarrow HClO_2 + HClO_3$$

Salts of these acids (chlorites and chlorates) may be formed by neutralizing the resulting solution with an appropriate base, and the two salts may then be separated.

Chloric acid and its salts. This acid or its salts may be prepared by heating solutions of hypochlorous acid or hypochlorites:

$$3HClO + heat \rightarrow 2HCl + HClO_3$$

or

$$3NaClO + heat \rightarrow 2NaCl + NaClO_3$$

The chlorates may be prepared directly by passing chlorine into a *hot* solution of a strong base; for example:

$$3Cl_2 + 6NaOH \rightarrow 5NaCl + NaClO_3 + 3H_2O$$

Chloric acid may be produced (in solution) by treating an aqueous solution of a chlorate with a strong acid:

$$NaClO_3 + H_2SO_4 \rightarrow HClO_3 + NaHSO_4$$

Both chloric acid and chlorates are useful as oxidizing agents. Potassium chlorate is used in the manufacture of matches, fireworks, and explosives.

Perchloric acid and its salts. The best method for the preparation of pure perchloric acid is first to prepare a salt of this acid by the action of heat on a chlorate:

$$4NaClO_3 + heat \rightarrow NaCl + 3NaClO_4$$

The sodium perchlorate may be separated and used in the preparation of perchloric acid. It is preferable, however, to use barium perchlorate

$$Ba(ClO_4)_2 + H_2SO_4 \rightarrow BaSO_4 + 2HClO_4$$

since the salt formed as a by-product is insoluble and may be separated by filtration. If the water is evaporated from the resulting filtrate, one obtains a concentrated solution of perchloric acid, but because such solutions are explosive, it is best to allow the acid to remain in more dilute solutions. The uses of perchloric acid and its salts are similar to those given for chloric acid and the chlorates. Perchloric acid is generally used as a reagent in analytical chemistry. Among the perchlorates, potassium perchlorate is of interest because it is one of the very few potassium salts that are relatively insoluble in water.

Commercial Production of Acids and Bases

Among the first chemicals that you encounter in the laboratory are the common acids and bases. Since these materials are readily available and relatively inexpensive, you are seldom called upon to prepare them in the laboratory. Nevertheless, one should become familiar with the manner in which these chemicals are produced commercially and with some of the problems that arise in connection with the conduct of chemical reactions on a large scale.

A few of the common acids and bases, together with certain salts that are discussed, are produced in such very large quantities that they are commonly referred to as the *heavy* chemicals (i.e., heavy in the sense of quantity production). In the discussion to follow, the methods used in the production of the more important of these chemicals are considered briefly.

Development of chemical processes

Any considerable expenditure of effort, time, and money on the development of a commercial chemical process is based on the premise that a market exists for the product (or products) or that the product has properties such that uses for it may be found and a market created. New uses may be found for compounds known to the chemist for many years but not previously produced on a large scale. On the other hand, it is often necessary for the chemist to discover hitherto unknown chemical compounds in order to find a substance which fills an existing need.

Laboratory investigation. New and potentially useful chemical changes are usually carried out first on a small scale in the laboratory. The discovery of such reactions or products may result from the desire to attain a preconceived objective, or it may be a by-product of experiments conducted for a totally different purpose. It is necessary then to determine by experiment the most favorable conditions (temperature, pressure, concentration, catalysis, etc.) for the occurrence of the reactions in question. All of this knowledge must be acquired by the chemist before one can consider seriously the problem of large-scale utilization. If the laboratory investigations indicate probability of success, there arises next the problem of adapting the process to a scale that renders economical the operation of a large chemical plant.

Process development. It is not unusual to find that a reaction which proceeds smoothly when a few grams of the reactants are brought together may behave differently when much larger quantities of the reactants are involved. With drastic change in the quantities of materials, the reaction may occur no differently, more favorably, less favorably, or not at all. Therefore, after the laboratory investigation has been completed, it is necessary to determine whether the reaction occurs under controlled conditions on a larger scale. This phase of the work is usually conducted in a pilot plant or *semiworks,* which is simply a miniature factory wherein the process in question is studied on a scale intermediate between that of the research laboratory and that of the anticipated manufacturing plant. When it is found that the reactions are not affected adversely by increased scale of operation, there arises the problem of the design and construction of all of the equipment that is necessary for the production and purification of the desired product or products in still larger quantities.

Chemical economics. The preceding problems must always be approached in relation to economic considerations. One must consider the availability, location, and cost of raw materials. The cost of equipment and its maintenance, labor, power, and so forth, are only a few of the factors that govern the feasibility of the process. The actual location of the plant must be considered in relation to the cost of transportation of raw materials to the plant and of finished products to the market.

Ideally, a chemical reaction for use in a commercial process should yield only one product and this a pure one. However, this situation is infrequently realized, and one must therefore consider the problem of utilization of by-products. The solution to this problem often makes the difference between success and failure. Finally, the feasibility of putting a chemical process into operation must be studied in relation to the existing or possible future competition, both domestic and foreign. Attention has been called previously to certain competitive chemical processes, and numerous additional cases are cited in the following discussions.

Acids

Although a large number of acids are known to the chemist, only a few of them are produced by the chemical industries on a tonnage basis. Of these, the most important are hydrochloric, nitric, phosphoric, and sulfuric acids.

Hydrochloric acid

At the present time, hydrochloric acid is produced commercially (1) from sodium chloride and sulfuric acid, (2) by the direct union of chlorine and hydrogen, and (3) as a by-product of other chemical processes. Although each of these sources contributes materially to the available supply of hydrochloric acid, the latter source is becoming increasingly important.

Preparation from sodium chloride and sulfuric acid. The chemical changes involved in this procedure are the same as those cited in connection with the discussion of the laboratory preparation of hydrogen chloride; that is:

$$NaCl + H_2SO_4 \rightarrow NaHSO_4 + HCl$$

$$NaCl + NaHSO_4 \rightarrow Na_2SO_4 + HCl$$

The first of these reactions takes place readily at only slightly elevated temperatures. The second, however, requires considerably higher temperatures, which may approach (but should not exceed) the melting temperature of sodium sulfate (884°C). Although the batch NaCl furnace is used commercially, other types of furnaces are also employed. A charge of salt and sulfuric acid is placed in a cast-iron pan, where the first step in the reaction occurs. This charge is then moved through an opening onto the hearth, where, under the influence of higher temperatures, the second reaction occurs. In the meantime, another charge of salt and acid is introduced into the cast-iron pan and is ready for transfer to the hearth as soon as the preceding batch has reacted and the by-product sodium sulfate, known commercially as *salt cake,* has been removed. The hydrogen chloride so produced is removed (under suction) from the furnace through the flues. These gases are then cooled and absorbed in water to form hydrochloric acid solution of any desired concentration.

Preparation by direct union. Some of the chlorine produced as a by-product of the electrolytic production of sodium hydroxide is combined directly with hydrogen. The hydrogen is first fed into a quartz burner where the hydrogen is ignited in air, after which chlorine is introduced, and the hydrogen continues to burn in the atmosphere of chlorine to form substantially pure hydrogen chloride. There remains only the necessity of cooling the hydrogen chloride and dissolving it in water.

The simplicity of this process suggests that it should be used to the exclusion of the more complicated procedure involving the reaction between salt and sulfuric acid. However, producers of sodium hydroxide do not use their by-product chlorine in this manner if there is a sufficient demand for the elemental chlorine as such. Furthermore, if a profitable market for sodium sulfate exists, it may be cheaper to produce hydrochloric acid from salt and sulfuric acid than from the elements. Finally, another competitive aspect to this situation arises—the availability of by-product acid.

By-product hydrochloric acid. Hydrogen chloride is a by-product of the commercial production of many organic chemicals. Thus, in the manufacture of chlorobenzene, hydrogen chloride is liberated:

$$C_6H_6 + Cl_2 \rightarrow C_6H_5Cl + HCl$$
$$\text{Benzene} \qquad\qquad \text{Chlorobenzene}$$

After removal of impurities, the hydrogen chloride is absorbed in water to form hydrochloric acid. Because so many reactions used in the organic chemical industries produce hydrogen chloride as a by-product, this is the single most important source of hydrochloric acid.

Properties and uses of commercial hydrochloric acid.   The crude product obtained from salt and sulfuric acid contains ferric chloride as an impurity. This acid has a light yellow color (owing to the presence of ferric chloride) and is known commercially as *muriatic acid*. The hydrochloric acid produced by the other two methods is colorless and usually of a high degree of purity. The purest commercial hydrochloric acid is that made by direct union of the elements. Large quantities of hydrochloric acid are used in the metallurgical industries, particularly in the pickling of iron used for galvanizing and in the manufacture of wire. The processes of electroplating and etching also require considerable quantities of this acid. Most of the chemical industries use appreciable quantities of hydrochloric acid. The textile industries use this acid in dyeing and printing processes as well as in the manufacture of artificial silks and cotton goods. In the ceramic industries, hydrochloric acid is used in the purification of sand and clays used in the manufacture of glass and pottery. The acid is also used in the manufacture of dextrose, gelatin, glue, soaps, and a multitude of other chemical products.

Nitric acid

So far as domestic production of nitric acid is concerned, only one process need be considered since all but a very small quantity of the nitric acid produced in this country is made by the catalytic oxidation of ammonia. In addition, a little-used method and one obsolete method are discussed briefly.

Ostwald process.   This, the most widely used method for the production of nitric acid, depends on the oxidation of ammonia by atmospheric oxygen in the presence of a catalyst consisting of gauze made of platinum or platinum and rhodium. The reaction

$$4NH_3 + 5O_2 \rightarrow 4NO + 6H_2O$$

is carried out in a reaction chamber of the one-pass type at temperatures ranging from 600 to 900°C. Upon leaving the reaction chamber, the gases are cooled and more atmospheric oxygen is provided to permit the occurrence of the reaction

$$2NO + O_2 \rightarrow 2NO_2$$

The resulting nitrogen dioxide is then transferred to an absorption chamber and allowed to react with warm water:

$$3NO_2 + H_2O \rightarrow 2HNO_3 + NO$$

The nitric oxide liberated in this reaction is recovered and reconverted to nitrogen dioxide, and the nitric acid solution is concentrated and purified by distillation. Ordinarily, the nitric acid produced in this manner is about 50% HNO_3. A more concentrated product may be obtained by distilling mixtures of nitric and sulfuric acids.

Production of nitric acid from sodium nitrate. Although widely used in the past, this method now accounts for less than 10% of the nitric acid produced in the United States. In this process, dried sodium nitrate and sulfuric acid are heated to about 150°C in a retort:

$$NaNO_3 + H_2SO_4 \rightarrow NaHSO_4 + HNO_3$$

The nitric acid vapors are led from the retort through a series of condensers and finally into absorption towers in which the acid is dissolved in water. The by-product sodium hydrogen sulfate is known as *niter cake*. Although it would seem reasonable to expect that this material might be used in a second reaction to produce more nitric acid,

$$NaNO_3 + NaHSO_4 \rightarrow Na_2SO_4 + HNO_3$$

this reaction is not used commercially because the temperature requited for its occurrence is sufficient to decompose the nitric acid thereby formed.

Arc process. The arc or Birkeland-Eyde process was devised in Norway. This process is no longer used extensively and is of interest here only because of certain unusual features. By this method, the endothermal reaction

$$N_2 + O_2 + 43,000 \text{ cal} \rightleftarrows 2NO$$

utilized atmospheric oxygen and nitrogen. By means of a powerful magnetic field, an electric arc was spread into a large disk, and air was blown through the flame at temperatures in excess of 3000°C. Under these conditions, approximately 4% of the atmospheric nitrogen was converted to nitric oxide. After passing through the arc, the gases were cooled to 1000°C, whereupon the nitric oxide reacted with more atmospheric oxygen to form nitrogen dioxide, which was subsequently hydrated to form dilute nitric acid. Despite the fact that no expense is involved in the way of raw materials, the high cost of the electric power required made it impossible for this process to compete with the Ostwald process. However, an abundant supply of cheap hydroelectric power in Norway permitted the arc process to be operated economically for more than 25 years.

Uses of nitric acid. Most of the nitric acid produced is used in the manufacture of explosives. This is done either by converting the acid into its salts, the nitrates, or by using the acid in the nitration certain organic compounds. Some of these nitrated organic substances are used as explosives, while others are used in the manufacture of dyes, pharmaceuticals, and so forth. Nitric acid is used also in the production of fertilizers and many other chemicals.

Sulfuric acid

Much has been written about the role of sulfuric acid as an index of national prosperity. Since this acid is generally considered to be one of the most important heavy chemicals and is used in so many of the industries, it is possible to evaluate trends in economic conditions by following the trends in the consumption of sulfuric acid. This applies not only to industry but also to agriculture because of the use of sulfuric acid in the manufacture of fertilizers.

There are two processes by which sulfuric acid is produced commercially. Because it can be described more simply, the newer process is considered first.

Contact process. This process has been in use for about 100 years and involves the use of a sulfur burner, dust remover, gas cooling exchanger, water scrubber, gas absorber, and contaminant remover. The first step in the process is the formation of sulfur dioxide, either by burning sulfur

$$S + O_2 \rightarrow SO_2$$

or by burning pyrite, galena, sphalerite, or other sulfides. If sulfur is used, the sulfur dioxide requires little purification, but if sulfide ores are employed, extensive and costly purification is necessary in order to remove arsenic and selenium oxides, halogens, and other impurities. After its formation in the sulfur burner, the sulfur dioxide is passed successively through a dust catcher, cooling coils, and scrubbers. The gas is then led into a reaction chamber containing a suitable catalyst and maintained at 350 to 400°C. In this step, the reaction is as follows:

$$2SO_2 + O_2 \rightleftarrows 2SO_3 + 45,200 \text{ cal}$$

The successful operation of the contact process became possible only through a knowledge of chemical equilibrium and the factors that influence such equilibria. Since the reaction is exothermal, the temperature must be controlled carefully in order to avoid favoring the reverse reaction—that is, the decomposition of the desired sulfur trioxide.

Although the forward reaction is favored by increase in pressure, this is not employed in practice since 97 to 99% conversion of sulfur dioxide to sulfur trioxide can be accomplished at the temperature specified here, provided suitable catalysts are used. The first catalyst used for this reaction consisted of finely divided platinum dispersed in asbestos, anhydrous magnesium sulfate, or silica gel. Other catalysts were later discovered. Mixtures of ferric and cupric oxides are useful, but these are less efficient than platinum. Certain mixtures containing vanadium pentoxide (V_2O_5) and other compounds of vanadium appear to be as good as or better than platinum. There has been much controversy over the relative merits of platinum and vanadium catalysts, and only time will provide the answer as to which is best.

The sulfur trioxide is finally led into an absorber, where the gas is dissolved in concentrated sulfuric acid. This is necessary since sulfur trioxide does not dissolve readily in water or in dilute sulfuric acid. However, after the trioxide has been dissolved in concentrated acid and this solution is added to water, the trioxide hydrates to form sulfuric acid:

$$SO_3 + H_2O \rightarrow H_2SO_4$$

Obviously, the concentration of the final product is governed by the quantity of water employed. In most contact plants, the quantity of water used is such that the final product is a concentrated acid (about 96% H_2SO_4).

The chief advantages of the contact process are the high purity of the product and the fact that the product is a concentrated acid. Disadvantages are the high cost of the catalysts and the fact that if sulfides are used as raw materials, costly purification of the sulfur dioxide is necessary, because impurities such as arsenic trioxide and selenium dioxide "poison" the catalyst (i.e., render the catalyst inactive). Platinum catalysts are particularly sensitive to these impurities, while vanadium catalysts are claimed to be free from this disadvantage.

Lead-chamber process.   The essential features of the lead-chamber process are the use of nitric acid to oxidize SO_2 to SO_3 and the absorption of SO_3 by water in large absorber towers. This process has been used for well over 100 years, and the details of equipment and operation are essentially the same today as when the process was first designed.

Sulfur dioxide is produced by burning either sulfur or a sulfide such as pyrite in the furnace. The dioxide is then mixed with the catalyst, NO, which may be prepared in either of two ways. In the older method, nitric acid vapor is formed in the "niter pot," and when this vapor comes into contact with sulfur dioxide, nitric oxide is formed in accordance with the following equation:

$$2HNO_3 + 3SO_2 \rightarrow 2NO + 3SO_3 + H_2O$$

However, the more modern practice is to produce the nitric oxide by the catalytic oxidation of ammonia in the manner already described. The gases are then passed into the Glover tower, where thorough mixing occurs. This tower (25 to 30 ft high and 10 to 20 ft in diameter) is usually constructed of heavy sheet lead or acid-proof stoneware and is filled with quartz or stoneware tower packing. The gaseous mixture leaves the top of the Glover tower and enters the first of the lead-lined chambers from which the name of the process originated.

Three to six of these reaction chambers are ordinarily used. The dimensions vary considerably and may be 50 to 150 ft in length, 15 to 25 ft in height, and 20 to 30 ft in width. Although these may be of the same size, the first chamber is usually larger than the others. All these chambers are lined with sheet lead for the reason that this metal reacts with sulfuric acid to produce a tenacious coating of *insoluble* lead sulfate, which thereafter protects the underlying metallic lead from further attack by the acid.

In addition to the gaseous mixture that enters the chambers from the Glover tower, steam is introduced. Consequently, all the reactants necessary for the formation of sulfuric acid are present.

The reactions that occur in the chambers are rather complex and not fully understood. However, the following equations probably describe the principal changes. Nitrosylsulfuric acid is formed by the following reaction:

$$4SO_2 + 4NO + 2H_2O + 3O_2 \longrightarrow 4 \left(\begin{array}{c} O{=}N{-}O \diagdown \diagup O \\ S \\ HO \diagup \diagdown O \end{array} \right)$$

This acid is converted to sulfuric acid either by hydrolysis

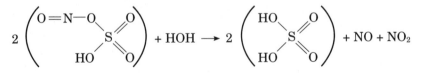

$$2 \left(\begin{array}{c} O{=}N{-}O \diagdown \diagup O \\ S \\ HO \diagup \diagdown O \end{array} \right) + HOH \longrightarrow 2 \left(\begin{array}{c} HO \diagdown \diagup O \\ S \\ HO \diagup \diagdown O \end{array} \right) + NO + NO_2$$

or by reduction with sulfur dioxide

$$2 \left(\begin{array}{c} \text{O}{=}\text{N}{-}\text{O} \diagdown \qquad \diagup \text{O} \\ \qquad \text{S} \diagup \diagdown \\ \text{HO} \diagup \qquad \diagdown \text{O} \end{array} \right) + SO_2 + 2H_2O \longrightarrow 3 \left(\begin{array}{c} \text{HO} \diagdown \qquad \diagup \text{O} \\ \qquad \text{S} \diagup \diagdown \\ \text{HO} \diagup \qquad \diagdown \text{O} \end{array} \right) + 2NO$$

The dilute sulfuric acid is drained from the chambers; it contains 60 to 65% H_2SO_4. This acid is normally used at this concentration. Although there are several inconvenient and somewhat costly processes for concentrating the lead-chamber acid, it is difficult to concentrate this product economically in view of the ease with which concentrated sulfuric acid may be produced directly by the contact process.

From the foregoing reactions, it is seen that nitric oxide is liberated upon formation of sulfuric acid. In order to avoid the loss of this catalyst, the gases that leave the last chamber are led upward through the packed Gay-Lussac tower, against a downward spray of concentrated sulfuric acid. Nitric oxide and nitrogen dioxide are dissolved by virtue of the reformation of nitrosylsulfuric acid:

$$NO + NO_2 + 2H_2SO_4 \longrightarrow 2 \left(\begin{array}{c} \text{O}{=}\text{N}{-}\text{O} \diagdown \qquad \diagup \text{O} \\ \qquad \text{S} \diagup \diagdown \\ \text{HO} \diagup \qquad \diagdown \text{O} \end{array} \right) + H_2O$$

The resulting solution of nitrosylsulfuric acid in concentrated sulfuric acid is withdrawn at the bottom of the Gay-Lussac tower, and this so-called *niter acid* is transferred by means of compressed air to the top of the Glover tower. Here it is allowed to mix with dilute acid or with water, and this dilution, together with the rise in temperature due to both the heat of dilution and the contact with the hot gases streaming upward in the Glover tower, liberates nitric oxide and nitrogen dioxide. The catalyst is again made available and is ready to begin another cycle of operation. This recovery of catalyst is fairly efficient, but some of the oxides of nitrogen are lost, and the required quantity of catalyst must be maintained by frequent introduction of fresh portions of nitric acid.

The lead-chamber process is more economical than the contact process, but it produces a more dilute and less pure product. Thus, the chamber process can compete only in the market that can use a relatively impure and dilute acid. Although chamber-acid plants now in use will undoubtedly be operated for many years to come, it seems probable that all sulfuric acid plants constructed in the future will employ the contact process or some still more efficient process.

Uses of sulfuric acid. It is not feasible here to do more than briefly mention the major uses of this acid. Of these, the use of dilute sulfuric acid in the manufacture of fertilizers (such as ammonium sulfate and primary calcium phosphate) is the most important from the standpoint of quantity of acid consumed. For these purposes, lead-chamber acid is adequate, and some fertilizer companies operate their own lead-chamber plants.

Large quantities of this acid are used in the metallurgical industries in the pickling of iron and steel, in the production of zinc, in the electrolytic refining of copper and other metals, in electroplating operations, and so forth.

In the petroleum industry, sulfuric acid is used in refining processes to remove certain undesired components from crude petroleum. This acid is also employed in large quantities in the manufacture of explosives, paints, pigments, storage batteries, textiles (rayon and other cellulose products), dyes, drugs, and so forth.

In addition to the chemicals previously mentioned, many others are manufactured by processes involving the use of sulfuric acid. Hydrochloric and nitric acids, which have already been discussed, may be cited as examples. Others include phosphoric acid, sodium carbonate, sulfates, and alums.

Phosphoric acid

The old but still useful method for the production of phosphoric acid involves treatment of phosphate rock, bones, or other material containing phosphates with sulfuric acid at elevated temperatures:

$$Ca_3(PO_4)_2 + 3H_2SO_4 \rightarrow 2H_3PO_4 + 3CaSO_4$$

Filtration provides a dilute solution of impure phosphoric acid which is then subjected to procedures designed to provide for both purification and concentration in whatever degree may be necessary.

The more modern process is based upon the previously described method for the production of elemental phosphorus from phosphate rock. Pure white phosphorus is oxidized to phosphorus pentoxide, which is then hydrated to form phosphoric acid. This method is particularly useful where a concentrated product of high purity is sought.

Uses of phosphoric acid.   Although some phosphoric acid is used in the rustproofing of iron and in the manufacture of most high-grade phosphate fertilizers, the major uses for this acid are still concerned with the production of other chemicals. Baking powders, phosphate syrups for use in making soft drinks, and water-softening agents are typical examples.

From phosphoric acid (H_3PO_4) one may produce other acids of phosphorus. Thus, pyrophosphoric acid is formed by the mutual dehydration of two molecules of H_3PO_4 by application of heat:

$$2H_3PO_4 \xrightarrow{225°C} H_4P_2O_7 + H_2O$$

or

Further, an elevation of temperature to 400°C converts pyrophosphoric acid to metaphosphoric acid:

$$H_4P_2O_7 \xrightarrow{400°C} 2HPO_3 + H_2O$$

or

An inspection of the foregoing graphic formulas shows that the oxidation number of phosphorus is the same in all three of these acids—that is, they are *all* phosphoric acids. The prefixes *pyro-* and *meta-* are used in naming $H_4P_2O_7$ and HPO_3, respectively, while ordinary phosphoric acid (H_3PO_4) is more properly designated as *orthophosphoric* acid.

Sodium hexametaphosphate [$(NaPO_3)_6$], which is the sodium salt of metaphosphoric acid, is an important chemical sold under the trade name of Calgon and used in the treatment of industrial water supplies.

Bases

In general, the commercial production of soluble bases does not introduce any problems that have not been considered heretofore. Accordingly, the methods used for the production of these compounds are mentioned only briefly.

Sodium and potassium hydroxides

The production of sodium hydroxide by electrolysis has been described in sufficient detail, and this procedure serves equally well for the production of potassium hydroxide.

The older method for the production of sodium (or potassium hydroxide) is sometimes called the *chemical process* to distinguish it from electrolytic processes. The chemical process involves the treatment of sodium carbonate with calcium hydroxide:

$$Na_2CO_3 + Ca(OH)_2 \rightarrow 2NaOH + CaCO_3$$

Removal of the precipitated calcium carbonate by filtration provides a filtrate consisting of sodium hydroxide solution, which may be used as such or purified and evaporated to supply solid sodium hydroxide.

More than a million tons of sodium hydroxide are produced annually in the United States. This chemical is used in the manufacture of rayon and other textiles, pulp and paper, in the refining of petroleum and vegetable oils, in reclaiming used rubber, and in the production of a wide variety of other chemicals.

Calcium hydroxide

This compound and the corresponding compounds of the related elements barium and strontium are produced by the hydration of their oxides. Thus, calcium hydroxide (slaked lime) is made by addition of water to calcium oxide (quicklime).

$$CaO + H_2O \rightarrow Ca(OH)_2$$

If the quantity of water added is only that required in the reaction, the product is a white powder.

Because it is cheap, calcium hydroxide is used in many chemical processes that require a strong base. It is used in the production of sodium hydroxide, ammonia, bleaching powder, and many other chemicals. Calcium hydroxide is also used as an insecticide in the form of lime-sulfur spray, in water softening, and in the production of numerous materials such as stucco and mortar which are widely used in the building construction industries.

Ammonium hydroxide

The production of ammonium hydroxide (*aqua ammonia*) is accomplished simply by dissolving gaseous ammonia in water. Under atmospheric pressure at 20°C, 710 L of ammonia gas dissolves in 1 L of liquid water. The resulting solution is a weak base because of the occurrence, to a slight extent, of the forward reaction:

$$NH_3 + HOH \rightleftarrows NH_4OH$$

Ammonium hydroxide is an important and useful reagent both in the industries and in the laboratory. Large quantities of dilute aqueous ammonia are used as a cleansing agent.

Commercial Production and Utilization of Salts

A review of the commercial processes discussed thus far serves to illustrate the frequency with which salts serve as raw materials for the chemical industries. The salts so employed may be either naturally occurring materials or the principal products or by-products of other large-scale chemical operations. It is not feasible here to consider these matters either broadly or exhaustively. Only a few of the more common types of salts are considered briefly in the following subsections.

Chlorides

The chlorides are, of course, the salts of hydrochloric acid and include many common and useful substances. Among others, sodium chloride, calcium chloride, zinc chloride, and the chlorides of mercury, are produced in considerable quantities and serve a variety of needs. Of, these, only sodium chloride is considered in detail.

Sodium chloride.   If one were to single out one chemical substance and designate it as being of first rank as a raw material for the chemical industries, that substance would probably be sodium chloride. The annual production of this material amounts to approximately 30 million tons, of which about one-fourth is produced in the United States. Other countries that produce large quantities of salt are Russia, Germany, France, Great Britain, India, and China. Within this country, the chief salt-producing areas are found in the states of Michigan, New York, Ohio, California, Louisiana, Kansas, West Virginia, and Texas. Salt is obtained from both underground deposits and the saltwater of the oceans and salt lakes.

Methods of production.   In any case, the method by which salt is produced from naturally occurring materials is determined by the quality of the product desired. In Louisiana and Texas, salt is produced by mining in a manner essentially the same as for any other underground mining operations. In these regions, a product of high purity (99.8% NaCl) is obtained directly. However, the salt produced in Michigan, New York, and Kansas by this type of mining is relatively less pure.

More commonly, salt is produced by the evaporation of natural brines which are formed by pumping water into salt wells and subsequently pumping the resulting brine to the surface. These brines usually contain sodium chloride as the main solute, together with relatively small quantities of other salts, such as sodium and calcium sulfates and potassium and magnesium chlorides. Upon evaporation of the water, the resulting solid therefore consists largely of sodium chloride, but if a purer product is desired, the brine must be suitably treated to remove at least the major portion of the impurities.

Uses. The most familiar use of salt is in the seasoning of foods. Ordinary table salt is a relatively pure product, yet one that contains some hygroscopic magnesium chloride, which, in humid weather, causes the salt to become moist and to pack. To prevent this, it is common practice to add inert materials such as starch, magnesia, or calcium phosphate, which form a protective coating on the surface of the salt crystals and thus retard the absorption of moisture from the surrounding atmosphere, or to add sodium hydrogen carbonate to convert the magnesium chloride into the nonhygroscopic carbonate. Iodized salt contains a small percentage (usually about 0.0025 to 0.0030%) of potassium iodide.

As already indicated, vast quantities of salt are used in the chemical industries in the production of other chemicals. Including these indirect as well as the direct uses, there are over 1000 more or less distinct uses for this chemical.

Carbonates

Since carbonates are salts of carbonic acid, both the acid salts and the normal salts are included under this heading. Of the many known salts that belong to this class, those most extensively used in the chemical industries are calcium carbonate and the carbonates of sodium.

Calcium carbonate. In the form of ordinary *limestone,* calcium carbonate is a very abundant raw material. In addition to calcium carbonate, limestone contains a variety of impurities, notably compounds of iron, aluminum, magnesium, and silicon. The mineral *dolomite* is the double carbonate of calcium and magnesium ($CaCO_3 \cdot MgCO_3$). Eggshells, oyster shells, coral, and similar materials contain high percentages of calcium carbonate. The mineral *calcite* is nearly pure calcium carbonate. Including all sources, about 100 million tons of limestone are mined annually in the United States. Although this material is produced in practically all the contiguous 48 states, Ohio, Illinois, Michigan, Pennsylvania, and New York are the largest producers.

Uses. In the form of large compact blocks of ordinary limestone or in the form known as *marble,* large quantities of calcium carbonate are used as building stones. Limestone is also used in building roads, in manufacturing cement, in the metallurgical industries, and as a fertilizer on acidic soils. Calcium carbonate in the form of limestone is the source of all of the lime and much of the carbon dioxide produced in this country. When calcium carbonate is heated to about 900°C, it decomposes, as shown by the following equation:

$$CaCO_3 \rightleftarrows CaO + CO_2$$

This reversible reaction proceeds to completion if the gaseous carbon dioxide is progressively removed as the reaction proceeds.

On a commercial scale, lime is produced by heating crushed limestone in large furnaces or *kilns*. The usual lime kiln is about 50 ft in height and is constructed so that heat is supplied by fireboxes located near the bottom of the kiln. A current of air is admitted at the bottom; this air serves to cool the lime near the bottom of the kiln and provides oxygen for burning the fuel used in the fireboxes. Further, this upward flow of gas serves to carry out the carbon dioxide as rapidly as it is liberated. The residual solid lime (*quicklime*) is removed periodically at the bottom, and new charges of limestone are admitted at the top. When quicklime (CaO) is brought into contact with water, calcium hydroxide (*slaked lime*) is formed, and heat is liberated in the process:

$$CaO + H_2O \rightleftarrows Ca(OH)_2$$

Sodium hydrogen carbonate. Although sodium hydrogen carbonate (commonly known as *sodium bicarbonate*) is a component of the naturally occurring double salt $Na_2CO_3 \cdot NaHCO_3 \cdot 2H_2O$, known as *trona,* practically all of the commercial product is manufactured from sodium chloride, ammonia, and carbon dioxide by the Solvay process, which was devised in 1863 by Ernest and Alfred Solvay, two Belgian chemists. It is very widely used and is the source of most of the world's supply of sodium hydrogen carbonate (and, as is shown later, of normal sodium carbonate as well). The essential chemical changes involved in the Solvay process are the following:

$$NH_3 + H_2O \rightarrow NH_4OH$$

$$CO_2 + H_2O \rightarrow H_2CO_3$$

$$NH_4OH + H_2CO_3 \rightarrow H_2O + NH_4HCO_3$$

$$NH_4HCO_3 + NaCl \rightarrow NaHCO_3 + NH_4Cl$$

The precipitation of $NaHCO_3$ from aqueous solutions is carried out below 15°C, under which conditions the acid carbonate is only slightly soluble. The product so obtained is usually quite pure. It may be noted that ammonia used in the Solvay process appears at the end of the preceding series of reactions in the form of the by-product ammonium chloride. The ammonia is recovered by treating the solution (containing the ammonium chloride) with lime:

$$CaO + H_2O \rightarrow Ca(OH)_2$$

$$2NH_4Cl + Ca(OH)_2 \rightarrow 2NH_3 + 2H_2O + CaCl_2$$

The overall efficiency of recovery of ammonia is commonly in the neighborhood of 99%; hence, most of the ammonia is used over and over again in the process. The lime used in the recovery of ammonia is ordinarily produced by the thermal decomposition of limestone,

$$CaCO_3 \rightarrow CaO + CO_2$$

a reaction that also furnishes carbon dioxide, which is one of the raw materials used in the main process.

Sodium carbonate. The normal carbonate (commonly known as *soda*) occurs in nature as *natron* $(Na_2CO_3 \cdot 10H_2O)$, *thermonatrite* $(Na_2CO_3 \cdot H_2O)$, *trona* $(Na_2CO_3 \cdot NaHCO_3 \cdot 2H_2O)$, and as *soda brine*. Deposits of these materials are widely distributed, and some of the more important deposits are found in the United States, Africa, China, Siberia, South America, and Central Europe. In the United States, the most important sources are those of Owens Lake and Searles Lake in California and the Green River deposits in Wyoming. Despite the existence of these minerals in nature, only about 3% of our domestic supply of soda comes from these sources. In fact, the total tonnage of soda produced from all natural sources is not very large when compared to the tonnage produced by the Solvay and Leblanc processes.

Solvay process. Sodium hydrogen carbonate having been produced by the Solvay process as described, the normal carbonate is made by heating the acid salt to 175 to 190°C in a rotary drier:

$$2NaHCO_3 \rightarrow Na_2CO_3 + H_2O + CO_2$$

Of course, the liberated carbon dioxide is returned to the main process and used again to make more of the acid salt.

Leblanc process. In 1775 the French Academy of Sciences offered a prize of 100,000 francs to anyone who could devise a process for the production of soda from raw materials found in France. The prize was claimed (but never received) in 1791 by Nicolas Leblanc, who used ordinary salt, sulfuric acid, coke, and limestone as the starting materials. The essential reactions involved are as follows:

$$NaCl + H_2SO_4 \rightarrow NaHSO_4 + HCl$$

$$NaCl + NaHSO_4 \rightarrow Na_2SO_4 + HCl$$

$$Na_2SO_4 + 2C \rightarrow Na_2S + 2CO_2$$

$$Na_2S + CaCO_3 \rightarrow Na_2CO_3 + CaS$$

The first of these reactions occurs at ordinary temperatures, some heating is required for completion of the second, and the third and fourth are carried out in a furnace at about 1000°C. The sodium carbonate is extracted in water, purified, and sold as *at soda* ($Na_2CO_3 \cdot 10H_2O$), *glassmaker's soda* ($Na_2CO_3 \cdot H_2O$), or the anhydrous salt (Na_2CO_3), which is commonly known as *soda ash*.

Further details of the Leblanc process are omitted here for the reason that it is primarily of historical interest. The Solvay process is so superior in all respects that the Leblanc process is used little, if at all, at the present time and has never been used in the United States. However, it is interesting to point out that while the primary product of the Leblanc process is the normal carbonate, it may be converted to the acid carbonate by treatment with carbon dioxide in aqueous solutions:

$$Na_2CO_3 + H_2O + CO_2 \rightarrow 2NaHCO_3$$

Uses of sodium carbonate. In its various forms, more than 3 million tons of soda are used annually. Approximately one-half of this amount is used in the production of other chemicals, about one-fourth is used in the manufacture of glass, and the remainder is used for a variety of purposes including the production of pulp and paper, soap, textile products, and petroleum. Appreciable quantities of sodium carbonate are also used in water softening.

Chemical treatment of natural waters. Both directly and indirectly, the general problem of purification and treatment of natural waters is related to the chemical and physical properties of the normal and acid salts of carbonic acid. The common impurities in natural waters consist of suspended solid organic and inorganic materials and of certain dissolved salts, particularly the acid carbonates, chlorides, and sulfates of sodium, calcium, and magnesium. The solid matter may be removed by filtration, the presence of limited quantities of sodium salts is not objectionable, and the calcium and magnesium salts are eliminated only through appropriate chemical treatment. The ions that are most objectionable are Ca^{2+}, Mg^{2+}, and HCO_3^-. If calcium and magnesium ions are allowed to remain in water for use in laundries, they react with soap to form insoluble salts. Thus, a considerable quantity of soap may have to be added to soften the water, and no cleansing action is secured until all the calcium and magnesium ions have been precipitated. This results not only in an enormous waste of soap but also in the appearance of an undesirable and unsightly scum of insoluble salts. If Ca^{2+} and Mg^{2+} ions together with HCO_3^- ions are permitted to remain in water for use in steam boilers, hot-water heating pipes, or radiators, conversion to insoluble normal carbonates results in the formation of scale or incrustation known as *boiler scale*. The formation of these deposits not only results in lowered efficiency of the boilers or heating systems but also may lead to serious explosions if the scale deposits become excessive.

The particular chemical treatment to which natural waters are subjected depends upon the use or uses for which the water is intended. If water is intended for human consumption, provision must be made for the elimination of significant quantities of harmful bacteria. This is accomplished in part by the chemical treatment designed to remove hardness, but it is accomplished most effectively by treatment with chlorine, the characteristic odor of which is frequently detectable in municipal water supplies.

Temporary hardness. Probably because of the ease with which, these impurities may be removed, the acid carbonates of calcium and magnesium (and in some cases ferrous iron) are said to constitute *temporary hardness* in water. This type of hardness may be removed by boiling the water, whereupon the soluble acid carbonates are converted to the insoluble normal carbonates, which may be removed by filtration:

$$Ca(HCO_3)_2 \rightarrow CaCO_3 + H_2O + CO_2$$

$$Mg(HCO_3)_2 \rightarrow MgCO_3 + H_2O + CO_2$$

Although this method could be used on a small scale, its use in the treatment of municipal water supplies would involve prohibitive costs. Accordingly, advantage is taken of the fact that the objectionable ions may be removed by treatment with a base such as calcium hydroxide (slaked lime):

$$Ca^{2+}(HCO_3)_2^- + Ca^{2+}(OH)_2^- \rightarrow 2H_2O + 2CaCO_3$$

$$Mg^{2+}(HCO_3)_2^- + Ca^{2+}(OH)_2^- \rightarrow 2H_2O + MgCO_3 + CaCO_3$$

Permanent hardness. Although the use of the adjective *permanent* is inappropriate, *permanent hardness* is the term commonly used to describe the hardness due to the presence of the chlorides and sulfates of calcium and magnesium. This type of hardness may be removed by treatment with sodium carbonate:

$$Ca^{2+}Cl_2^- + Na_2^+CO_3^- \rightarrow CaCO_3 + 2Na^+Cl^-$$

$$Mg^{2+}SO_4^- + Na_2^+CO_3^- \rightarrow MgCO_3 + Na_2^+SO_4^-$$

The sodium carbonate is usually added in the form of soda ash.

Lime-soda process. Combination of the use of slaked lime for the removal of temporary hardness and sodium carbonate for the removal of permanent hardness constitutes the *lime-soda process* for softening natural waters. This method is commonly used in municipal water-treatment plants and is a cheap and yet fairly effective process. If sufficient time can be allowed, the insoluble carbonates may be permitted to settle out in settling basins, or they may be more rapidly removed by means of filters. Frequently, iron or aluminum salts are added, and these hydrolyze to form gelatinous precipitates of ferric or aluminum hydroxides. As these precipitates slowly settle, they carry with them the insoluble normal carbonates, as well as any other suspended matter such as sand, clay, or organic matter which is sometimes slow in settling otherwise.

Zeolite processes. There is a simple and remarkably effective method of removing hardness from water. This scheme utilizes a reversible metathetical reaction between calcium and magnesium salts and substances called *zeolites*.

These materials are natural or artificial sodium aluminum silicates (e.g., $Na_2Al_2SiO_8$), which may be considered as complex salts of orthosilicic acid (H_4SiO_4). When hard water is allowed to flow upward through a bed of granular zeolite in a vertical vessel, reactions of the type illustrated by the following equations result:

$$Ca^{2+}Cl_2^- + Na_2Al_2Si_2O_8 \rightleftarrows 2Na^+Cl^- + CaAl_2Si_2O_8$$

$$Mg^{2+}SO_4^- + Na_2Al_2Si_2O_8 \rightleftarrows Na^{2+}SO_4^- + MgAl_2Si_2O_8$$

$$Ca^{2+}(HCO_3)_2^- + Na_2Al_2Si_2O_8 \rightleftarrows 2Na^+HCO_3^- + CaAl_2Si_2O_8$$

Thus, both temporary and permanent hardness are removed almost completely by an exchange of sodium ions for calcium and magnesium ions. When the sodium in the original zeolite has all been replaced by calcium or magnesium, the sodium zeolite is regenerated by allowing a concentrated salt solution to flow through the bed. This treatment, in accordance with the law of mass action, reverses the reactions, and the zeolite bed is restored to its original condition; by periodic regeneration of this sort, the zeolite may be used repeatedly over long periods of time. Other synthetic materials permit the removal of anionic constituents.

Nitrates

The use of nitrates as raw materials for the production of nitric acid has already been discussed. On a quantity basis, however, this application accounts for only a relatively small proportion of the annual consumption of these salts by the chemical industries.

Production and uses of nitrates.   The chief source of naturally occurring nitrates is a rather sharply limited area along the arid coastal range of Chile. The ore in these deposits is known as *caliche* and consists principally of sodium nitrate, which is commonly known as *Chile saltpeter.* The deposits range in depth from 5 to 15 ft and are up to 200 mi in length. Sodium nitrate is removed from the crude ore (about 60% $NaNO_3$) by leaching with water, after which the resulting solutions are evaporated. The annual production of Chile saltpeter amounts to more than 3 million tons.

Since practically all nitrates are soluble in water, one would not expect to find extensive deposits of these salts in nature excepting in arid or semiarid regions. In addition to the nitrate deposits in Chile, limited quantities of potassium nitrate occur in East Asia. Most of the potassium nitrate used at the present time is made by the metathetical reaction between sodium nitrate and potassium chloride,

$$Na^+NO_3^- + K^+Cl^- \rightarrow K^+NO_3^- + Na^+Cl^-$$

a reaction that may be caused to proceed to completion by taking advantage of the fact that, at the higher temperature, NaCl is much less soluble in water than KNO_3. The bulk of the available supply of nitrates, however, is produced directly or indirectly from nitric acid.

Nitrate fertilizers.   Of the numerous elements that must be present to support the growth of plants in the soil, nitrogen, phosphorus, and potassium are those most commonly added in the form of commercial fertilizers. Nitrates are often used as a source of nitrogen, and the nitrates most used in fertilizer mixtures are those of sodium, calcium, and potassium. The latter has the obvious advantage of supplying both nitrogen and potassium and hence has an unusually high fertilization value on a comparative weight basis.

The use of nitrates in the maintenance of soil fertility is a part of the general problem of nitrogen fixation. Accordingly, any process that accomplishes the conversion of atmospheric nitrogen to ammonia or nitric acid is a potential source of nitrates for use as fertilizers. It is evident that nitrates (either natural or synthetic) for use as fertilizers need not be of a high degree of purity.

Nitrate explosives. The term *explosion* is applied to the effect produced by a sudden change in the pressure of one or more gases. This may be the result of either chemical or physical changes—that is, the sudden liberation or absorption of gases in chemical reactions or the sudden formation of gases from either liquids or solids. Because certain nitrates decompose readily with liberation of gaseous products, these substances are useful in compounding a variety of commercial explosives. Ammonium nitrate is stable under ordinary atmospheric conditions and may be handled safely in small quantities, even at elevated temperatures. When the dry salt is heated, it decomposes with liberation of nitrous oxide and water,

$$NH_4NO_3 \rightarrow N_2O + 2H_2O$$

and this reaction provides probably the most convenient laboratory method for the preparation of nitrous oxide. If solid ammonium nitrate is detonated, the decomposition occurs with explosive violence. A number of high explosives consist of mixtures containing ammonium nitrate (e.g., ammonal and amatol).

Potassium nitrate (together with sulfur and charcoal) is used in the manufacture of gunpowder (*black powder*). When such a mixture is ignited by means of a spark, both gaseous and solid decomposition products are produced in the resulting explosive reaction, which is quite complex. Black powder is used in the manufacture of ammunition for small firearms, in the production of time fuses, and as a blasting powder in mining operations.

Phosphates

Since there are three common acids of phosphorus, the normal and acid salts of which are phosphates, it follows that this is indeed a large group of salts. This fact becomes all the more evident if the salts of the less common acids of pentavalent phosphorus are included. Of this large number of salts, however, only a few may be classed as important commercial chemicals.

Orthophosphates. Aside from their use as reagents in the chemical laboratory, most of the possible metal salts of orthophosphoric acid have not found extensive application in the chemical industries. There are, however, a few notable exceptions. The normal phosphates of the alkali metals are readily soluble in water and are used to a considerable extent as cleansing agents and water softeners.

Primary calcium phosphate [$Ca(H_2O_4)$] is used in the manufacture of certain baking powders. In the presence of moisture, the acid phosphate reacts with sodium hydrogen carbonate (another constituent of such powders) to produce gaseous carbon dioxide, which causes the dough to rise.

Phosphate fertilizers. From the extensive phosphate rock deposits in Tennessee, South Carolina, Florida, Idaho, and Montana, approximately 3 million tons of ore are used annually in the manufacture of fertilizers. The chief phosphate present in this rock is the normal calcium salt [$Ca_3(PO_4)_2$]. Because of its insolubility, this compound is not very useful as a fertilizer. If a phosphorus compound is to provide phosphorus that can be assimilated by growing plants, the compound must be one that is appreciably soluble.

The utilization of tricalcium phosphate therefore requires its conversion into a soluble salt. This is accomplished by treating the normal phosphate with dilute sulfuric acid (lead-chamber acid), thereby producing the soluble primary calcium phosphate:

$$Ca_3(PO_4)_2 + 2H_2SO_4 \rightarrow Ca(H_2PO_4)_2 + 2CaSO_4$$

The resulting *mixture* of calcium dihydrogen phosphate and calcium sulfate is sold as "superphosphate" fertilizer. It is of interest to note that the production of calcium sulfate is avoided by using phosphoric acid rather than sulfuric:

$$Ca_3(PO_4)_2 + 4H_3PO_4 \rightarrow 3Ca(H_2PO_4)_2$$

Although superphosphate fertilizer provides available phosphorus, it is advantageous to provide other needed elements such as nitrogen and potassium at the same time. Thus, diammonium hydrogen phosphate [$(NH_4)_2HPO_4$] provides both nitrogen and phosphorus, while potassium ammonium phosphate [$K(NH_4)HPO_4$] makes available the three elements most needed in the maintenance of soil fertility.

Metaphosphates. The salts of metaphosphoric acid may be produced in a number of ways, but most conveniently by heating the appropriate dihydrogen orthophosphate. Thus,

$$NaH_2PO_4 \rightarrow H_2O + NaPO_3$$

This metaphosphate is used in water treatment in the form of sodium hexametaphosphate [$(NaPO_3)_6$], which is known by the trade name Calgon, to prevent the precipitation of small quantities of calcium and magnesium salts not removed in the lime-soda process and to dissolve scale that has formed in boilers, water pipes, and so forth.

Pyrophosphates. The pyrophosphates may be formed by heating the corresponding monohydrogen orthophosphates:

$$2Na_2HPO_4 \rightarrow H_2O + Na_4P_2O_7$$

Although there are no really large-scale uses for pyrophosphates, small quantities are employed (in solution) to dissolve boiler scale.

Silicates

Because the class of salts known as *silicates* includes so many familiar naturally occurring materials, this subsection gives some attention to the chemical character of these salts. Although the silicates are important raw materials for a number of industries, their commercial applications are mentioned only briefly.

Salts of silicic acids. The silicates may be looked upon as salts of a considerable variety of silicic acids, some of which are far more complex than any acids discussed thus far. The two simplest acids of silicon are *orthosilicic* and *metasilicic* acids. The relationship between these two acids and their anhydride may be shown as follows:

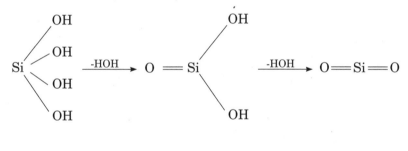

| Orthosilicic | Metasilicic | Silicon |
| acid | acid | dioxide |

Examples of common orthosilicates include *zircon* ($ZrSiO_4$), *mica* [$KH_2Al_2(SiO_4)_3$], and *kaolin* [$H_2Al_2(SiO_4)_2 \cdot H_2O$]. The minerals *beryl* [$Be_3Al_2(SiO_3)_6$] and *asbestos* [$Mg_3Ca(SiO_3)_4$] are examples of common metasilicates.

More complex silicic acids, known only in the form of their salts, may be considered as derived from orthosilicic acid by processes of selective dehydration. Thus, one may represent the formation of series of disilicic acids

$$2H_4SiO_4 \xrightarrow{-HOH} H_6Si_2O_7 \xrightarrow{-HOH} H_4Si_2O_6 \xrightarrow{-HOH} H_2Si_2O_5$$

or trisilicic acids

$$\text{3H}_4\text{SiO}_4 \xrightarrow{-\text{HOH}} \text{H}_{10}\text{Si}_3\text{O}_{11} \xrightarrow{-\text{HOH}} \text{H}_8\text{Si}_3\text{O}_{10} \xrightarrow{-\text{HOH}} \text{H}_6\text{Si}_3\text{O}_9$$

and so forth.

Serpentine ($\text{Mg}_3\text{Si}_2\text{O}_7 \cdot 2\text{H}_2\text{O}$) is a disilicate, and *orthoclase* (KAISi_3O_8) is a trisilicate.

Water glass. A familiar and useful silicate is that known as *water glass.* Although the commercial product consists of a rather complex mixture of silicates, water glass is commonly represented by the formula for the normal sodium salt of metasilicic acid, Na_2SiO_3. This material is produced by heating sand with sodium hydroxide under pressure

$$\text{SiO}_2 + 2\text{NaOH} \rightarrow \text{H}_2\text{O} + \text{Na}_2\text{SiO}_3$$

or by fusing sand with sodium carbonate

$$\text{SiO}_2 + \text{Na}_2\text{CO}_3 \rightarrow \text{Na}_2\text{SiO}_3 + \text{CO}_2$$

Water glass is usually sold in the form of clear syrupy solutions and is used as an egg preservative, as a cement in the manufacture of cardboard boxes, as a fireproofing agent, in the manufacture of soaps and cleaners, and so forth.

Ceramic industries. The chemical and physical properties of silicates are utilized to advantage in a group of industries which collectively may be termed the *ceramic* industries. These are concerned with the manufacture of products such as cement, glass, porcelain, bricks, tile, terra cotta, and enamels. All these industries utilize a wide variety of complex naturally occurring silicates.

Nuclear Chemistry

This section outlines how the study of radioactive decay has contributed to a better understanding of atomic structures. You should recall that the spontaneous disintegration of some of the heavy elements produces not only alpha and beta particles and gamma rays, but also elements of atomic number less than that of the parent radioactive species. Since one element leads to another of different atomic number, it follows that these disintegrations involve changes in the number of protons in the nuclei of the atoms. Accordingly, the phenomenon of radioactivity is of interest not only in connection with the extranuclear structures of atoms but also because of the occurrence of chemical changes that involve alterations in atomic nuclei. These phenomena are of such far-reaching importance that they influence practically every aspect of the physical and biological sciences. It is appropriate, therefore, to present at least a brief discussion of the subject of *nuclear chemistry,* a body of knowledge that is equally of interest to the chemist, the physicist, and the engineer.

Natural radioactivity

The spontaneous disintegration of radium is a typical example of natural radioactivity. The final product of the decay of radium is a stable isotope of lead having an atomic weight of 206 (as compared with the average atomic weight of ordinary lead, 207.21). Between radium and lead having an atomic weight of 206, there are produced as intermediate products two other isotopes of lead (which are also radioactive) and radioactive isotopes of other heavy elements as well. Thus, it appears that radioactive decay may constitute a well-defined sequence of changes in which many isotopes (radioactive and otherwise) may be formed.

Radium is formed by a series of changes which begins with the element uranium. In addition to the *uranium decay series,* there are two other decay series, which have their beginnings in the elements *thorium* and *actinium.* In the study of these more or less complex sequences of nuclear reactions, it is convenient to express the rates of radioactive change in terms of periods of half-life. By a *half-life period* is meant the time required for one-half of a given weight (such as 1 g) of an element to undergo radioactive decay. For example, the half-life period of radium is 1590 years; this means that of an initial quantity of 1 g, 0.5 g would remain unchanged after 1590 years had elapsed, 0.25 g after the next 1590 years, and so on.

Transmutation of the elements

For very many years, the alchemist's dream of changing base metals into gold was ridiculed even by the most reputable of scientists. Although it was known that the nuclei of certain atoms undergo alteration in the course of natural radioactive decay, researchers' inability to exercise any control over the nature or rate of these spontaneous decompositions probably did much to foster the belief that the nucleus of the atom was inviolate. However, in the year 1919 the English physicist Ernest Rutherford accomplished the first transmutation of an element, and this notable discovery was quickly followed by other equally significant developments.

Using radium as a source of alpha particles, Rutherford bombarded nitrogen with these particles and converted the nitrogen to an isotope of oxygen:

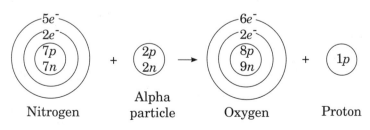

You should recall that alpha particles are helium nuclei bearing two *positive* charges and that the nucleus of the nitrogen atom also bears a positive charge. However, the speed and energy of these particles are such that a few of them can overcome the repulsive force of the nitrogen nuclei and thus strike these nuclei in direct hits. The products of the interaction of nitrogen and alpha particles proved to be oxygen atoms and protons, or in terms of a common scheme of notation used in representing nuclear reactions:

$$^{14}_{7}N + ^{4}_{2}\alpha \rightarrow ^{17}_{8}O + ^{1}_{1}p$$

In this reaction, the subscripts represent nuclear charge, and the superscripts represent the mass. As must be true in any reaction, the sums of the masses and charges on the left of the arrow are equal to those on the right.

Note that the transmutation of nitrogen leads to a kind of oxygen which is different from ordinary oxygen to the extent of having an additional neutron in the nucleus, hence a mass of 17 rather than the more usual 16.

As is evident from the following paragraphs, Rutherford's accomplishment of the first transmutation led to the discovery of many similar transmutations.

Artificial radioactivity

Fifteen years after Rutherford's transmutation of nitrogen, an equally if not more significant experiment was conducted successfully by Frederic and Irene Curie Joliot. They bombarded boron with alpha particles and produced nitrogen in a manner shown by the following reaction:

$$^{10}_{5}B + ^{4}_{2}\alpha \rightarrow ^{13}_{7}N + ^{1}_{0}n$$

The remarkable feature of this transmutation was the fact that the nitrogen isotope produced was *radioactive*. These nitrogen atoms of mass 13 (rather than 14) underwent radioactive decay by emission of positive electrons, or *positrons,* to form stable carbon atoms of mass 13,

$$^{13}_{7}N \rightarrow ^{13}_{6}C + \beta^{+}$$

where β^{+} represents the positron. For the first time, an artificial radioactive isotope had been produced, and its decay was found to occur in a manner wholly analogous to the decay of the naturally radioactive elements.

The discovery of artificial radioactivity stimulated a vast amount of experimental work by both physicists and chemists. That this work has been productive is shown by the fact that several hundred artificial radioactive isotopes are now known, which include isotopes of practically every known element.

Methods of inducing nuclear transformations

Both of the nuclear reactions cited previously were brought about by using alpha particles as "bullets" to bombard the nuclei of nitrogen and boron atoms. This raises the question as to whether other minute projectiles, such as neutrons or protons, might not also be used to induce nuclear transformations. These and other particles have indeed been used in this manner. The nuclei of deuterium atoms are known as *deuterons* (usually represented by d), and these too have been used extensively. Nuclear reactions have also been brought about by the use of gamma rays.

It has already been pointed out that the bombardment of a positively charged nucleus with positive ions, such as the alpha particle, the proton, or the deuteron, suffers the disadvantage inherent in the repulsion of like charges. One means of increasing the percentage of effective hits is by increasing the kinetic energy of these ionic bullets. Of several methods of providing these high-energy projectiles, the *cyclotron* has proved to be the most useful. This instrument was invented by E. O. Lawrence in 1932, and for this invention he was awarded the Nobel Prize in 1940.

The cyclotron is a device in which positive ions are accelerated in a powerful magnetic field until they attain velocities of the order of thousands of miles per second. The use of such an instrument not only makes possible the use of particles other than the alpha particles from naturally radioactive elements but also serves to increase the fraction of effective impacts between ionic bullets and nuclear targets that bear like charges.

Through the use of alpha particles, neutrons, and other particles, many nuclear reactions have been induced since Rutherford's demonstration of the first transmutation. Examples of several different types of nuclear reactions follow, but no effort is made here to illustrate all the types that are known to occur. An additional example of a nuclear reaction involving bombardment by means of alpha particles is found in the production of radioactive carbon of mass 14:

$$^{11}_{5}B + ^{4}_{2}\alpha \rightarrow ^{14}_{6}C + ^{1}_{1}p$$

This isotope has a half life of about 5000 years and is the one most likely to find widespread use in the study of carbon compounds. Other radioactive isotopes of carbon having masses of 10 and 11 have been produced, but their half-lives (8.8 s and 21 min, respectively) are so short as to render them relatively much less useful.

Another example of the use of alpha particles to induce a nuclear reaction is the conversion of aluminum to radioactive phosphorus:

$$^{27}_{13}\text{Al} + ^{4}_{2}\alpha \rightarrow ^{30}_{15}\text{P} + ^{1}_{0}n$$

Nuclear reactions induced by bombardment with protons and by deuterons are illustrated by the following examples:

$$^{12}_{6}\text{C} + ^{1}_{1}p \rightarrow ^{13}_{7}\text{N}$$

$$^{9}_{4}\text{Be} + ^{1}_{1}p \rightarrow ^{9}_{5}\text{B} + ^{1}_{0}n$$

and

$$^{75}_{33}\text{As} + ^{2}_{1}d \rightarrow ^{76}_{33}\text{As} + ^{1}_{1}p$$

$$^{20}_{10}\text{Ne} + ^{2}_{1}d \rightarrow ^{18}_{9}\text{F} + ^{4}_{2}\alpha$$

Of the many known nuclear reactions induced by neutrons, the following serve as examples of typical cases:

$$^{107}_{47}\text{Ag} + ^{1}_{0}n \rightarrow ^{108}_{47}\text{Ag}$$

$$^{32}_{16}\text{S} + ^{1}_{0}n \rightarrow ^{32}_{15}\text{P} + ^{1}_{1}p$$

$$^{27}_{13}\text{Al} + ^{1}_{0}n \rightarrow ^{24}_{11}\text{Na} + ^{4}_{2}\alpha$$

$$^{31}_{15}\text{P} + ^{1}_{0}n \rightarrow ^{30}_{15}\text{P} + 2^{1}_{0}n$$

Other examples of similar reactions are given in the following sections.

Characteristics of nuclear reactions

From the foregoing discussion and illustrations, one might gain the erroneous impression that a given bombardment produces only one or two products. Such is indeed not generally true. Ordinarily, the bombardment of any given element with a particular projectile results in the occurrence of several different nuclear reactions, and it may not be possible to predict which of these competing reactions occurs to the greatest extent. Furthermore, the efficiency of utilization of the projectiles is usually extremely low; only 1 out of 10,000 to 1 million projectiles may suffer an effective impact with a target nucleus. For these and other reasons, a given set of conditions used in a bombardment may result in only an extremely low yield of the desired product, while in other cases the yields are surprisingly high.

A nuclear reaction may involve either the absorption or release of energy; hence, the terms *endothermal* and *exothermal* are applicable.

However, it must be recognized that there is an important difference between the energy changes involved in nuclear reactions and those in ordinary chemical changes. In relation to the masses of matter that undergo transformation, the energy changes in nuclear reactions are enormous in comparison with those which accompany ordinary chemical reactions.

It is a significant fact that the products of nuclear transformations are by no means restricted to isotopes that occur in nature. The element carbon, for example, exists in nature as a mixture of $^{12}_{6}C$ (98.9%) and $^{13}_{6}C$ (1.1%). The production of the artificial radioactive isotopes $^{10}_{6}C$, $^{11}_{6}C$, and $^{14}_{6}C$ has been mentioned previously. Although there are limitations as to the variety of isotopes of a given element which may be produced, it is evident that isotopes are extremely useful tools for solving a wide variety of problems. From the standpoint of practical utilization, it is desirable to have radioactive isotopes of appreciable half-life. If one wished to conduct a prolonged chemical experiment involving a radioactive isotope having a half-life of a few seconds or even a few minutes, the bulk of the original quantity of the isotope employed would have undergone radioactive decay before the experiment could be completed. The advantages of using an isotope having a half life of hours, days, or even years are self-evident.

When the products of nuclear transformations are radioactive, they can be detected and determined quantitatively in terms of the radiations that characterize their radioactive decay. Instruments can measure the quantities of radioactive isotopes present in samples, and these methods are much more rapid and convenient than laborious chemical analyses.

One may rightfully raise the question as to why some products of nuclear reactions are radioactive while others are not. The answer concerns the stability of atomic nuclei. Essentially, any radioactive element, whether artificial or natural, can be considered abnormal. A nucleus that undergoes radioactive decay is in an unstable condition, and the process of decay always leads to stable isotopes. This tendency toward the achievement of stability is illustrated by the stepwise decay of naturally radioactive uranium to form a stable isotope of lead and the formation of stable carbon by the decay of artificial radioactive nitrogen. Although the conditions resulting in the instability of atomic nuclei are fairly well understood, further consideration of these factors is beyond the scope of this discussion.

Tritium

Tritium is an isotope of hydrogen having a mass of 3. This isotope is present in ordinary hydrogen, but at a concentration so low that its isolation presents very great difficulties. It is produced artificially by means of suitable nuclear transformations, making its isolation less difficult than from natural sources. Thus, tritium (T) may be formed by bombardment of deuterium (D) with deuterons:

$$_1^2 D + {_1^2}d \rightarrow {_1^3}T + {_1^1}p$$

or, better, by bombardment of lithium with neutrons:

$$_3^6 Li + {_0^1}n \rightarrow {_1^3}T + {_2^4}\alpha$$

Tritium is also one of the products obtained by bombardment of fluorine, beryllium, antimony, copper, or silver with deuterons, or the bombardment of boron and nitrogen with neutrons. Tritium is the simplest known radioactive isotope. It decays by emission of beta particles to form an isotope of helium and has a half-life of about 12 years.

Transuranium elements

Prior to 1939, elements of atomic number greater than 92 were unknown, and there was considerable doubt as to whether such elements could exist. In 1939, however, Edwin McMillan and Philip Abelson reported the production of the first of the transuranium elements. By bombarding uranium with neutrons, they produced an element of atomic number 93, which was later named *neptunium* (Np).

The formation of neptunium depends on capture of a neutron by an atom of $_{92}^{238}U$ to form $_{92}^{239}U$, which decays by emission of beta particles:

$$_{92}^{238}U + {_0^1}n \rightarrow {_{92}^{239}}U$$

$$_{92}^{239}U \rightarrow {_{93}^{239}}Np + \beta^-$$

McMillan and Abelson predicted that the radioactive element $_{93}^{239}Np$ would decay by emission of beta particles to form an element of atomic number 94. This prediction was confirmed in 1940 by Glenn Seaborg, McMillan, Wahl, and Kennedy, who produced element 94 and made detailed studies of its chemical properties:

$$_{93}^{239}Np \rightarrow {_{94}^{239}}Pu + \beta^-$$

This new element was assigned the name *plutonium* (Pu).

These developments led directly to the discovery of four additional transuranium elements. It should be emphasized that these are products of the laboratory and that none of the transuranium elements exists in nature in any appreciable concentration. Different isotopes of these elements may be formed by several types of nuclear transformations; those given here are simply illustrative.

The synthesis of element 95, *americium* (Am), was accomplished in 1945 by Seaborg, James, and Morgan:

$$^{238}_{92}U + ^4_2\alpha \rightarrow ^{241}_{94}Pu + ^1_0n$$

$$^{241}_{94}Pu \rightarrow ^{241}_{95}Am + \beta^-$$

In the same year, Seaborg, James, and Ghiorso succeeded in producing element 96, *curium* (Cm):

$$^{239}_{94}Pu + ^4_2\alpha \rightarrow ^{242}_{96}Cm + ^1_0n$$

In 1950, Seaborg, Thompson, and Ghiorso reported the discovery of element 97, *berkelium* (Bk):

$$^{242}_{96}Cm + ^2_1d \rightarrow ^{243}_{97}Bk + ^1_0n$$

Seaborg, Street, Thompson, and Ghiorso also produced an isotope of element 98, *californium* (Cf), in 1950:

$$^{242}_{96}Cm + ^4_2\alpha \rightarrow ^{244}_{98}Cf + 2^1_0n$$

Practically all of the work leading to the production of these six transuranic elements was done at the University of California and at the University of Chicago. For the most part, these scientific achievements were the outgrowth of the World War II atomic bomb project. This fact illustrates in a very real way the manner in which notable advances in scientific knowledge often result from work initiated because of a national emergency.

Note that the elements of atomic number 90 to 103 inclusive constitute a second series of rare-earth elements.

Other synthetic elements

In addition to the transuranic elements, the completion of the periodic arrangement of the elements has been possible only as the result of the synthesis of three elements. In all three cases, discovery of these elements by the methods of synthesis cleared up earlier erroneous claims to discovery.

The first element ever to be produced artificially is that of atomic number 43. In 1937, Perrier and Segre isolated minute amounts of a radioactive isotope of this element from a sample of molybdenum that had been bombarded with deuterons in a cyclotron. This element was given the name *technetium* (Tc), which is derived from the Greek word meaning artificial. The bulk of the evidence now available indicates that this element does not occur in nature. Although several isotopes of technetium have been produced by a variety of nuclear reactions, the following is a typical example:

$$^{94}_{42}\text{Mo} + ^{2}_{1}d \rightarrow ^{95}_{43}\text{Tc} + ^{1}_{0}n$$

In 1940, Corson, Mackenzie, and Segre announced the production of an isotope of element 85, *astatine* (At), which is the last member of the halogen family. This synthesis involved bombardment of bismuth with alpha particles in a cyclotron:

$$^{209}_{83}\text{Bi} + ^{4}_{2}\alpha \rightarrow ^{211}_{85}\text{At} + 2^{1}_{0}n$$

The quantity of astatine produced in this manner was extremely small but nevertheless sufficient to permit study of some of the chemical properties of this element and unequivocally establish its identity.

In 1945, Marinsky, Glendenin, and Coryell first identified isotopes of element 61, *promethium* (Pm), which was the last member of the lanthanide series of rare-earth elements to be discovered. Isotopes of this element were obtained both as products of the fission of uranium and as products of several different types of nuclear reactions, most of which involve suitable bombardment of isotopes of neodymium; for example:

$$^{146}_{60}\text{Nd} + ^{1}_{0}n \rightarrow ^{147}_{60}\text{Nd} + \gamma$$

$$^{147}_{60}\text{Nd} \rightarrow ^{147}_{61}\text{Pm} + \beta^{-}$$

The only additional element about which there had been considerable uncertainty until comparatively recent times is element 87, *francium* (Fr), which is the last of the alkali metals in Group I. This element was finally identified by Perey in 1937 as a product of the decay of the naturally occurring actinium isotope of mass 227:

$$^{227}_{89}\text{Ac} \rightarrow ^{223}_{87}\text{Fr} + ^{4}_{2}\alpha$$

Several other isotopes of francium are also known as products of synthesis, but most of these isotopes are of very short half life.

Nuclear fission

The fact that uranium is capable of undergoing a process known as *fission* was discovered as an indirect result of the use of neutrons as projectiles in the production of artificial radioactive isotopes. At the University of Rome, the Italian physicist Enrico Fermi bombarded many different elements with neutrons and thereby produced many new radioactive isotopes. The use of neutrons as projectiles has the distinct advantage that the collision of these uncharged particles with nuclei is not hindered by the repulsion of like charges. Largely for this reason, the use of neutrons has been extremely fruitful in the inducement of nuclear transformations.

Fermi's bombardment of uranium by neutrons led to radioactive products that for a time were believed to be the then-unknown elements 93 and 94 (Np and Pu). In 1939, however, the German chemists Hahn and Strassmann proved that one of the products of the bombardment of uranium with neutrons is an isotope of *barium*. Since the atomic weight of uranium is nearly twice that of barium, it was immediately apparent that the uranium atoms must have undergone a process of cleavage into at least two smaller fragments. The discovery of this "splitting" or fission of uranium led to a vast amount of effort directed toward a better understanding of the nature of the fission process and toward the identification of other fission products of uranium. Many have been identified, and the following are cited merely as a few examples: Kr, Y, Sr, Rh, Zr, Se, La, Sn, Nb, Te, Ce, Mo, and Br.

The occurrence of fission in uranium is due to the presence of $^{235}_{92}$U, which makes up only 0.07% of natural uranium. Other elements including thorium, protactinium, neptunium, and plutonium are also capable of undergoing fission, and fission may be induced not only by neutrons but also by protons, deutrons, alpha particles, and even gamma rays.

Atomic energy

The most striking aspect of nuclear fission is the tremendous release of energy that occurs. For example, each fission in $^{235}_{92}$U is accompanied by the liberation of energy in the amount of 200 MV. To use a now familiar analogy, this means that the energy obtainable from 1 lb of $^{235}_{92}$U is roughly equivalent to that made available by the burning of 10,000 tons of coal. The possibility of utilizing atomic energy became apparent when it was learned that each fission of $^{235}_{92}$U produces not only two lighter elements, but neutrons as well. Thus, one fission produces neutrons, which in turn may induce fission of other atoms of $^{235}_{92}$U and thereby initiate a nuclear chain reaction which would result in the liberation of vast quantities of energy. However, predictions relative to the early utilization of atomic energy were uniformly pessimistic because of the difficulty of separating $^{235}_{92}$U from the much more abundant $^{238}_{92}$U.

During the latter years of World War II, the utilization of atomic energy in the development of atomic bombs became a reality, as the result of two distinct accomplishments. Although it had been freely predicted that the separation of $^{235}_{92}$U would present almost unsurmountable difficulties, ingenious methods for accomplishing this separation and producing usable quantities of this isotope were devised and put into operation at Oak Ridge, Tennessee. Since isotopes are chemically identical, it was necessary to effect this separation by purely physical methods which take advantage of the very small difference in the masses of the uranium isotopes involved. Atomic energy was also made available through the production and use of plutonium, which also undergoes fission with liberation of vast quantities of energy. Plutonium was formed, purified, and isolated in quantity at the Hanford Engineer Works in southeastern Washington. The reactions essential to the formation of plutonium follow here. The nature of the radioactive decay is presented above the arrow, and the half-life of the isotope concerned is presented below the arrow.

$$ {}^{238}_{92}U + {}^{1}_{0}n \rightarrow {}^{239}_{92}U \xrightarrow[\text{23 min}]{\beta^-} {}^{239}_{93}Np \xrightarrow[\text{2.33 days}]{\beta^-} {}^{239}_{94}Pu \xrightarrow[\text{24,300 years}]{{}^{4}_{2}\alpha} {}^{235}_{92}U $$

These reactions are initiated by neutrons from the spontaneous fission of ${}^{235}_{92}U$. It is also necessary to have a *moderator* present, which may consist of hydrogen-containing substances, *heavy water* (deuterium oxide), graphite, and so forth. The moderator must be a substance that does not capture neutrons to any appreciable extent, so that high-energy neutrons may collide with atoms of the moderator and lose energy. Some of the resulting low-energy neutrons react with ${}^{238}_{92}U$ to form plutonium as indicated, and some are lost, while others react to induce fission in other ${}^{235}_{92}U$ atoms. Thus, more neutrons are made available, and a self-sustaining nuclear chain reaction is realized. The chain reaction is allowed to proceed until the concentration of plutonium becomes high enough to warrant its separation. Since plutonium is chemically different from uranium, the separation can be accomplished by chemical rather than by physical methods.

The successful utilization of atomic energy is unquestionably one of the outstanding accomplishments in the history of the physical sciences. The use of the first atomic bomb marked the beginning of a new era, and no one could then foresee the impact of atomic power on the future progress of civilization.

Nuclear fusion

The release of energy that accompanies the process of fission results from the fact that a small fraction of the mass of the atom which undergoes fission is converted into energy. That is, the sum of the masses of the light-element fission products and the neutrons resulting from the fission of any given atom is less than the mass of the original parent atom; the difference in mass is converted to energy in accordance with the Einstein's familiar equation:

$$E = mc^2$$

where E = energy
$\quad m$ = mass
$\quad c$ = velocity of light

Relatively early in the research into means of utilizing atomic energy, it became apparent that still another approach was possible. Just as mass is converted to energy when small atoms are made from very heavy atoms during the fission process, so it appeared likely that a conversion of mass to energy would occur if light atoms could be caused to combine to form heavier atoms. This is the so-called fusion process, that is the principle behind the hydrogen bomb. Possible fusion reactions include the following:

$$2{}_1^2\mathrm{D} \rightarrow {}_2^4\mathrm{He} + E$$

$$_3^6\mathrm{Li} + {}_1^3\mathrm{T} \rightarrow 2{}_2^4\mathrm{He} + {}_0^1n + E$$

$$_1^3\mathrm{T} + {}_1^1\mathrm{H} \rightarrow {}_2^4\mathrm{He} + E$$

The occurrence of these and similar fusion reactions requires extremely high temperatures on the order of several million degrees— that is, temperatures of a magnitude comparable to those made possible only by the energy release characteristic of nuclear reactions, including fission.

Uses of radioactive isotopes

Wholly aside from questions regarding the use of atomic energy, artificial radioactive isotopes constitute a most useful tool for the study of problems in practically all branches of the physical and biological sciences. Important facts have been brought to light through the use of radioactive isotopes of elements such as hydrogen, oxygen, sodium, iron, carbon, and phosphorus. The use of radioactive isotopes as tracers is advantageous because the radiation emitted by such isotopes can easily be detected. Thus, if one wishes to perform an experiment employing ${}_{15}^{32}\mathrm{P}$, which decays by emission of beta particles, one can detect the presence of a given quantity of this isotope by counting the number of beta particles emitted. Another advantage lies in the fact that radiation detection instruments are so sensitive that usually only very small quantities of radioactive isotopes are needed. A small amount of radioactive isotope mixed with a much larger amount of the corresponding ordinary nonradioactive element serves to trace the behavior of the latter.

For example, it was demonstrated by such methods that phosphorus assimilated from the soil by tomato plants tends to concentrate in the stems and certain parts of the leaves. Similarly, radioactive zinc was used to show that this element localizes in the seeds of tomatoes. The rate of absorption of iodine by the thyroid gland was established by the use of radioactive iodine; this and related work did much to add to the understanding of the diagnosis and treatment of goiter. These and many similar uses of radioactive isotopes show that these substances have been of inestimable value in the study of the mechanism of chemical reactions, problems relating to plant and animal metabolism, and the diagnosis and treatment of diseases.

Appendix

A

LMTD Correction Factors

TABLE A.1 Fouling Factors

Temperature of heating medium	≦240°F		240–400°F*	
Temperature of water	≦125°F		>125°F	
Water	Water velocity, ft/s		Water velocity, ft/s	
	≦3 ft	>3 ft	≦3 ft	>3 ft
Sea water	0.0005	0.0005	0.001	0.001
Brackish water	0.002	0.001	0.003	0.002
Cooling tower and artificial spray pond				
Treated makeup	0.001	0.001	0.002	0.002
Untreated	0.003	0.003	0.005	0.004
City or well water (such as Great Lakes)	0.001	0.001	0.002	0.002
Great Lakes	0.001	0.001	0.002	0.002
River water				
Minimum	0.002	0.001	0.003	0.022
Mississippi	0.003	0.002	0.004	0.003
Delaware, Schuylkill	0.003	0.002	0.004	0.003
East River and New York Bay	0.003	0.002	0.004	0.003
Chicago sanitary canal	0.008	0.006	0.010	0.008
Muddy or silty	0.003	0.002	0.004	0.003
Hard (>15 gr/gal)	0.003	0.003	0.005	0.005
Engine jacket	0.001	0.001	0.001	0.001
Distilled	0.0005	0.0005	0.0005	0.0005
Treated boiler feedwater	0.001	0.0005	0.001	0.001
Boiler blowdown	0.002	0.002	0.002	0.002

* Ratings in the last two columns are based on a temperature of the heating medium of 240 to 400°F. If the heating medium temperature is over 400°F, and the cooling medium is known to scale, these ratings should be modified accordingly.

TABLE A.1   Fouling Factors *(Continued)*

Petroleum Fractions			
Oils (industrial)		Cracking units	
Fuel oil	0.005	Gas oil feed	
Clean recirculating oil	0.001	<500°F	0.002
Machinery and transformer oils	0.001	≧500°F	0.003
Quenching oil	0.004	Naphtha feed:	
Vegetable oils	0.003	<500°F	0.002
Gases, vapors (industrial)		≧500°F	0.004
Coke-oven gas, manufactured gas	0.01	Separator vapors (vapors from	
Diesel-engine exhaust gas	0.01	separator, flash pot, and	
Organic vapors	0.0005	vaporizer)	0.006
Steam (non-oil bearing)	0.0	Bubble-tower vapors	0.002
Alcohol vapors	0.0	Residuum	0.010
Steam, exhaust (oil bearing from		Absorption units	
reciprocating engines)	0.001	Gas	0.002
Refrigerating vapors (condensing		Fat oil	0.002
from reciprocating compressors)	0.002	Lean oil	0.002
Air	0.002	Overhead vapors	0.001
Liquids (industrial)		Gasoline	0.0005
Organic	0.001	Debutanizer, depropanizer,	
Refrigerating liquids, heating,		depentanizer, and alkylation	
cooling, or evaporating	0.001	units	
Brine (cooling)	0.001	Feed	0.001
Atmospheric distillation units		Overhead vapors	0.001
Residual bottoms, <25°API	0.005	Product coolers	0.001
Distillate bottoms, ≧25°API	0.002	Product reboilers	0.002
Atmospheric distillation units		Reactor feed	0.002
Overhead untreated vapors	0.0013	Lube treating units	
Overhead treated vapors	0.003	Solvent oil mixed feed	0.002
Side-stream cuts	0.0013	Overhead vapors	0.001
Vacuum distillation units		Refined oil	0.001
Overhead vapors to oil		Refined oil heat exchangers,	
From bubble tower (partial		water cooled[†]	0.003
condenser)	0.001	Gums and tars	
From flash pot (no appreciable		Oil-cooled and steam	
reflux)	0.003	generators	0.005
Overhead vapors in water-cooled		Water-cooled	0.003
condensers		Solvent	0.001
From bubble tower (final		Deasphaltizing units	
condenser)	0.001	Feed oil	0.002
From flash pot	0.04	Solvent	0.001
Side stream		Asphalt and resin	
To oil	0.001	Oil-cooled and steam	
To water	0.002	generators	0.005
Residual bottoms, <20°API	0.005	Water-cooled	0.003
Distillate bottoms, >20°API	0.002	Solvent vapors	0.001
Natural gasoline stabilizer units		Refined oil	0.001
Feed	0.0005	Refined oil water cooled	0.003
Overhead vapors	0.0005	Dewaxing units	
Product coolers and exchangers	0.0005	Lube oil	0.001
Product reboilers	0.001	Solvent	0.001
H_2S removal units		Oil wax mix heating	0.001
For overhead vapors	0.001	Oil wax mix cooling[†]	0.003
Solution exchanger coolers	0.0016		
Reboiler	0.0016		

[†] Precautions must be taken against deposition of wax.

TABLE A.1 **Fouling Factors** *(Continued)*

	Crude Oil Streams											
	0–199°F			200–299°F			300–499°F			≧500°F		
	Velocity, ft/s											
	<2 ft	2–4 ft	≧4 ft	<2 ft	2–4 ft	≧4 ft	<2 ft	2–4 ft	≧4 ft	<2 ft	2–4 ft	≧4 ft
Dry	0.003	0.002	0.002	0.003	0.002	0.002	0.004	0.003	0.002	0.005	0.004	0.003
Salt‡	0.003	0.002	0.002	0.005	0.004	0.004	0.006	0.005	0.004	0.007	0.006	0.005

‡ Refers to a wet crude—any crude that has not been dehydrated.

SOURCE: *Standards of Tubular Exchanger Manufacturers Association,* 2d ed., New York, 1949. Reprinted courtesy of McGraw-Hill from D. Q. Kern, *Heat Process Transfer,* McGraw-Hill, New York, 1950.

TABLE A.2   Approximate Overall Design Coefficients

Values include total dirt factors of 0.003 and allowable
pressure drops of 5 to 10 psi on the controlling stream

Hot fluid	Cold fluid	Overall U_D
	Coolers	
Water	Water	250–500[a]
Methanol	Water	250–500[a]
Ammonia	Water	250–500[a]
Aqueous solutions	Water	250–500[a]
Light organics[b]	Water	75–150
Medium organics[c]	Water	50–125
Heavy organics[d]	Water	5–75[e]
Gases	Water	2–50[f]
Water	Brine	100–200
Light organics	Brine	40–100
	Heaters	
Steam	Water	200–700[a]
Steam	Methanol	200–700[a]
Steam	Ammonia	200–700[a]
Steam	Aqueous solutions:	
Steam	<2.0 cP	200–700
Steam	>2.0 cP	100–500[a]
Steam	Light organics	100–200
Steam	Medium organics	50–100
Steam	Heavy organics	6–60
Steam	Gases	5–50[f]
	Exchangers	
Water	Water	250–500[a]
Aqueous solutions	Aqueous solutions	250–500[a]
Light organics	Light organics	40–75
Medium organics	Medium organics	20–60
Heavy organics	Heavy organics	10–40
Heavy organics	Light organics	30–60
Light organics	Heavy organics	10–40

[a] Dirt factor 0.001.

[b] *Light organics* are fluids with viscosities of less than 0.5
cP and include benzene, toluene, acetone, ethanol, methyl
ethyl ketone, gasoline, light kerosene, and naphtha.

[c] *Medium organics* have viscosities of 0.5 to 1.0 cP and
include kerosene, straw oil, hot gas oil, hot absorber oil, and
some crudes.

[d] *Heavy organics* have viscosities above 1.0 cP and
include cold gas oil, lube oils, fuel oils, reduced crude oils,
tars, and asphalts.

[e] Pressure drop 20 to 30 psi.

[f] These rates are greatly influenced by the operating pressure.

SOURCE: Reprinted courtesy of McGraw-Hill from D. Q.
Kern, *Heat Process Transfer*, McGraw-Hill, New York, 1950.

TABLE A.3   **Heat Exchanger and Condenser Tube Data**

Tube OD, in	BWG	Wall thickness, in	ID, in	Flow area per tube, in²	Surface per lin ft, ft²		Weight per lin ft, lb steel
					Outside	Inside	
½	12	0.109	0.282	0.0625	0.1309	0.0748	0.493
	14	0.083	0.334	0.0876		0.0874	0.403
	16	0.065	0.370	0.1076		0.0969	0.329
	18	0.049	0.402	0.127		0.1052	0.258
	20	0.035	0.430	0.145		0.1125	0.190
¾	10	0.134	0.482	0.182	0.1963	0.1263	0.965
	11	0.120	0.510	0.204		0.1335	0.884
	12	0.109	0.532	0.223		0.1393	0.817
	13	0.095	0.560	0.247		0.1466	0.727
	14	0.083	0.584	0.268		0.1529	0.647
	15	0.072	0.606	0.289		0.1587	0.571
	16	0.065	0.620	0.302		0.1623	0.520
	17	0.058	0.634	0.314		0.1660	0.469
	18	0.049	0.652	0.334		0.1707	0.401
1	8	0.165	0.670	0.355	0.2618	0.1754	1.61
	9	0.148	0.704	0.389		0.1843	1.47
	10	0.134	0.732	0.421		0.1916	1.36
	11	0.120	0.760	0.455		0.1990	1.23
	12	0.109	0.782	0.479		0.2048	1.14
	13	0.095	0.810	0.515		0.2121	1.00
	14	0.083	0.834	0.546		0.2183	0.890
	15	0.072	0.856	0.576		0.2241	0.781
	16	0.065	0.870	0.594		0.2277	0.710
	17	0.058	0.884	0.613		0.2314	0.639
	18	0.049	0.902	0.639		0.2361	0.545
1¼	8	0.165	0.920	0.665	0.3271	0.2409	2.09
	9	0.148	0.954	0.714		0.2498	1.91
	10	0.134	0.982	0.757		0.2572	1.75
	11	0.120	1.01	0.800		0.2644	1.58
	12	0.109	1.03	0.836		0.2701	1.45
	13	0.095	1.06	0.884		0.2775	1.28
	14	0.083	1.08	0.923		0.2839	1.13
	15	0.072	1.11	0.960		0.2896	0.991
	16	0.065	1.12	0.985		0.2932	0.900
	17	0.058	1.13	1.01		0.2969	0.808
	18	0.049	1.15	1.04		0.3015	0.688
1½	8	0.165	1.17	1.075	0.3925	0.3063	2.57
	9	0.148	1.20	1.14		0.3152	2.34
	10	0.134	1.23	1.19		0.3225	2.14
	11	0.120	1.26	1.25		0.3299	1.98
	12	0.109	1.28	1.29		0.3356	1.77
	13	0.095	1.31	1.35		0.3430	1.56
	14	0.083	1.33	1.40		0.3492	1.37
	15	0.072	1.36	1.44		0.3555	1.20
	16	0.065	1.37	1.47		0.3587	1.09
	17	0.058	1.38	1.50		0.3623	0.978
	18	0.049	1.40	1.54		0.3670	0.831

SOURCE: Reprinted courtesy of McGraw-Hill from D. Q. Kern, *Heat Process Transfer,* McGraw-Hill, New York, 1950.

TABLE A.4   Tube-sheet Layouts (Tube Counts)

Square Pitch

¾-in-OD tubes on 1-in square pitch					1-in-OD tubes on 1¼-in square pitch				
Shell ID, in	1-P	2-P	4-P	6-P	8-P	Shell ID, in	1-P	2-P	4-P	6-P	8-P

Shell ID, in	1-P	2-P	4-P	6-P	8-P	Shell ID, in	1-P	2-P	4-P	6-P	8-P
8	32	26	20	20		8	21	16	14		
10	52	52	40	36		10	32	32	26	24	
12	81	76	68	68	60	12	48	45	40	38	36
13¼	97	90	82	76	70	13¼	61	56	52	48	44
15¼	137	124	116	108	108	15¼	81	76	68	68	64
17¼	177	166	158	150	142	17¼	112	112	96	90	82
19¼	224	220	204	192	188	19¼	138	132	128	122	116
21¼	277	270	246	240	234	21¼	177	166	158	152	148
23¼	341	324	308	302	292	23¼	213	208	192	184	184
25	413	394	370	356	346	25	260	252	238	226	222
27	481	460	432	420	408	27	300	288	278	268	260
29	553	526	480	468	456	29	341	326	300	294	286
31	657	640	600	580	560	31	406	398	380	368	358
33	749	718	688	676	648	33	465	460	432	420	414
35	845	824	780	766	748	35	522	518	488	484	472
37	934	914	886	866	838	37	596	574	562	544	532
39	1049	1024	982	968	948	39	665	644	624	612	600

1¼-in-OD tubes on 1⁹⁄₁₆-in square pitch					1½-in-OD tubes on 1⅞-in square pitch				
10	16	12	10							
12	30	24	22	16	16	12	16	16	12	12	
13¼	32	30	30	22	22	13¼	22	22	16	16	
15¼	44	40	37	35	31	15¼	29	29	25	24	22
17¼	56	53	51	48	44	17¼	39	39	34	32	29
19¼	78	73	71	64	56	19¼	50	48	45	43	39
21¼	96	90	86	82	78	21¼	62	60	57	54	50
23¼	127	112	106	102	96	23¼	78	74	70	66	62
25	140	135	127	123	115	25	94	90	86	84	78
27	166	160	151	146	140	27	112	108	102	98	94
29	193	188	178	174	166	29	131	127	120	116	112
31	226	220	209	202	193	31	151	146	141	138	131
33	258	252	244	238	226	33	176	170	164	160	151
35	293	287	275	268	258	35	202	196	188	182	176
37	334	322	311	304	293	37	224	220	217	210	202
39	370	362	348	342	336	39	252	246	237	230	224

Triangular Pitch

¾-in-OD tubes on ¹⁵⁄₁₆-in triangular pitch					¾-in-OD tubes on 1-in triangular pitch				
Shell ID, in	1-P	2-P	4-P	6-P	8-P	Shell ID, in	1-P	2-P	4-P	6-P	8-P

Shell ID, in	1-P	2-P	4-P	6-P	8-P	Shell ID, in	1-P	2-P	4-P	6-P	8-P
8	36	32	26	24	18	8	37	30	24	24	
10	62	56	47	42	36	10	61	52	40	36	
12	109	98	86	82	78	12	92	82	76	74	70
13¼	127	114	96	90	86	13¼	109	106	86	82	74
15¼	170	160	140	136	128	15¼	151	138	122	118	110
17¼	239	224	194	188	178	17¼	203	196	178	172	166

TABLE A.4 Tube-sheet Layouts (Tube Counts) (*Continued*)

1-in-OD tubes on 1¼-in triangular pitch					1¼-in-OD tubes on 1⁹⁄₁₆-in triangular pitch				
Shell ID, in	1-P	2-P	4-P	6-P	8-P	Shell ID, in	1-P	2-P	4-P	6-P	8-P

Shell ID, in	1-P	2-P	4-P	6-P	8-P	Shell ID, in	1-P	2-P	4-P	6-P	8-P
19¼	301	282	252	244	234	19¼	262	250	226	216	210
21¼	361	342	314	306	290	21¼	316	302	278	272	260
23¼	442	420	386	378	364	23¼	384	376	352	342	328
25	532	506	468	446	434	25	470	452	422	394	382
27	637	602	550	536	524	27	559	534	488	474	464
29	721	692	640	620	594	29	630	604	556	538	508
31	847	822	766	722	720	31	745	728	678	666	640
33	974	938	878	852	826	33	856	830	774	760	732
35	1102	1068	1004	988	958	35	970	938	882	864	848
37	1240	1200	1144	1104	1072	37	1074	1044	1012	986	870
39	1377	1330	1258	1248	1212	39	1206	1176	1128	1100	1078

1-in-OD tubes on 1¼-in square pitch					1¼-in-OD tubes on 1⁹⁄₁₆-in square pitch				
8	21	16	16	14						
10	32	32	26	24		10	20	18	14		
12	55	52	48	46	44	12	32	30	26	22	20
13¼	68	66	58	54	50	13¼	38	36	32	28	26
15¼	91	86	80	74	72	15¼	54	51	45	42	38
17¼	131	118	106	104	94	17¼	69	66	62	58	54
19¼	163	152	140	136	128	19¼	95	91	86	78	69
21¼	199	188	170	164	160	21¼	117	112	105	101	95
23¼	241	232	212	212	202	23¼	140	136	130	123	117
25	294	282	256	252	242	25	170	164	155	150	140
27	349	334	302	296	286	27	202	196	185	179	170
29	397	376	338	334	316	29	235	228	217	212	202
31	472	454	430	424	400	31	275	270	255	245	235
33	538	522	486	470	454	33	315	305	297	288	275
35	608	592	562	546	532	35	357	348	335	327	315
37	674	664	632	614	598	37	407	390	380	374	357
39	766	736	700	688	672	39	449	436	425	419	407

1½-in-OD tubes on 1⅞-in triangular pitch				
12	18	14	14	12	12
13¼	27	22	18	16	14
15¼	36	34	32	30	27
17¼	48	44	42	38	36
19¼	61	58	55	51	48
21¼	76	72	70	66	61
23¼	95	91	86	80	76
25	115	110	105	98	95
27	136	131	125	118	115
29	160	154	147	141	136
31	184	177	172	165	160
33	215	206	200	190	184
35	246	238	230	220	215
37	275	268	260	252	246
39	307	299	290	284	275

SOURCE: Reprinted courtesy of McGraw-Hill from D. Q. Kern, *Heat Process Transfer,* McGraw-Hill, New York, 1950.

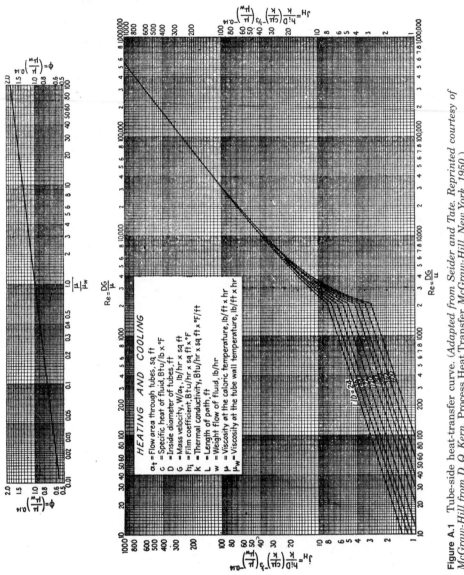

Figure A.1 Tube-side heat-transfer curve. (*Adapted from Seider and Tate. Reprinted courtesy of McGraw-Hill from D. Q. Kern, Process Heat Transfer, McGraw-Hill, New York, 1950.*)

HEATING AND COOLING

a_t = Flow area through tubes, sq ft
c = Specific heat of fluid, Btu/lb × °F
D = Inside diameter of tubes, ft
G = Mass velocity, W/a_t, lb/hr × sq ft
h_i = Film coefficient, Btu/hr × sq ft × °F
k = Thermal conductivity, Btu/hr × sq ft × °F/ft
L = Length of path, ft
w = Weight flow of fluid, lb/hr
μ = Viscosity at the caloric temperature, lb/ft × hr
μ_w = Viscosity at the tube wall temperature, lb/ft × hr

$\phi = \left(\dfrac{\mu}{\mu_w}\right)^{0.14}$

$Re = \dfrac{DG}{\mu}$

$j_H = \dfrac{h_i D}{k}\left(\dfrac{c\mu}{k}\right)^{-1/3}\left(\dfrac{\mu}{\mu_w}\right)^{-0.14}$

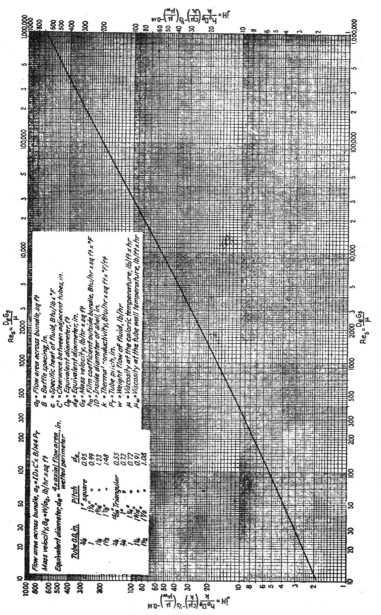

Figure A.2 Shell-side heat-transfer curve for bundles with 25% cut segmental baffles. (*Reprinted courtesy of McGraw-Hill from D. Q. Kern, Process Heat Transfer, McGraw-Hill, New York, 1950.*)

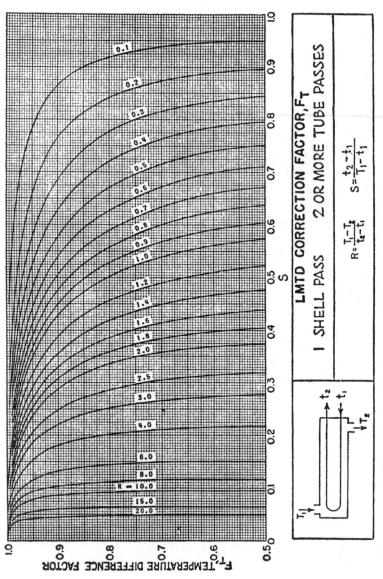

Figure A.3 LMTD correction factors for 1–2 exchangers. (Standards of Tubular Exchanger Manufacturers Association, 2d ed., New York, 1949. Reprinted courtesy of McGraw-Hill from D. Q. Kern, Process Heat Transfer, McGraw-Hill, New York, 1950.)

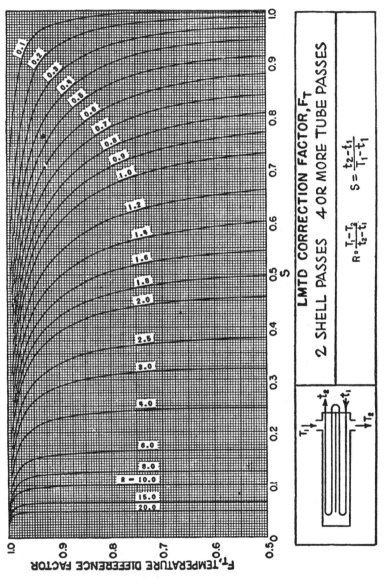

Figure A.4 LMTD correction factors for 2–4 exchangers. (Standards of Tubular Exchanger Manufacturers Association, 2d ed., New York, 1949. Reprinted courtesy of McGraw-Hill from D. Q. Kern, *Process Heat Transfer*, McGraw-Hill, New York, 1950.)

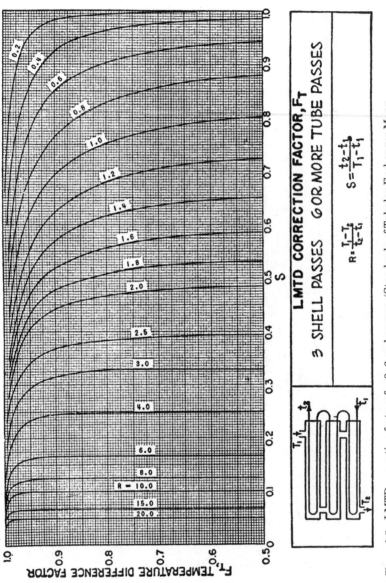

Figure A.5 LMTD correction factors for 3–6 exchangers. (Standards of Tubular Exchanger Manufacturers Association, 2d ed., New York, 1949. Reprinted courtesy of McGraw-Hill from D. Q. Kern, Process Heat Transfer, McGraw-Hill, New York, 1950.)

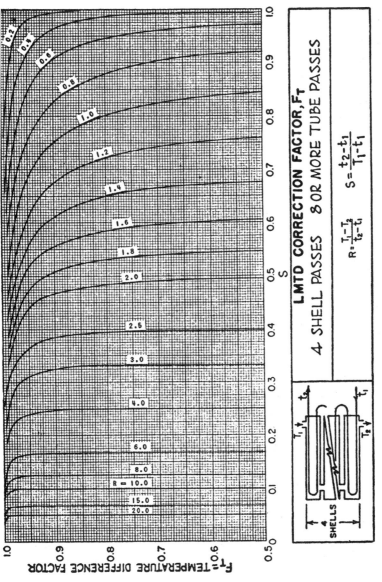

Figure A.6 LMTD correction factors for 4–8 exchangers. (Standards of Tubular Exchanger Manufacturers Association, 2d ed., New York, 1949. Reprinted courtesy of McGraw-Hill from D. Q. Kern, Process Heat Transfer, McGraw-Hill, New York, 1950.)

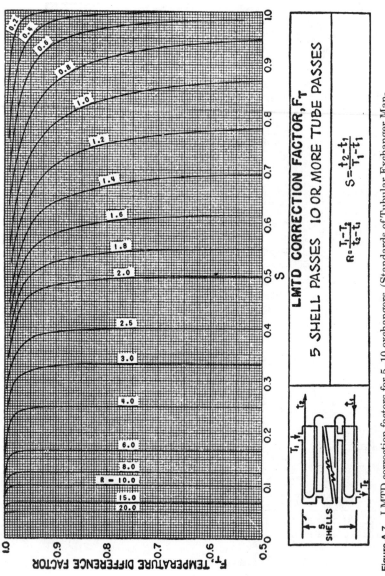

Figure A.7 LMTD correction factors for 5–10 exchangers. (Standards of Tubular Exchanger Manufacturers Association, 2d ed., New York, 1949. Reprinted courtesy of McGraw-Hill from D. Q. Kern, Process Heat Transfer, McGraw-Hill, New York, 1950.)

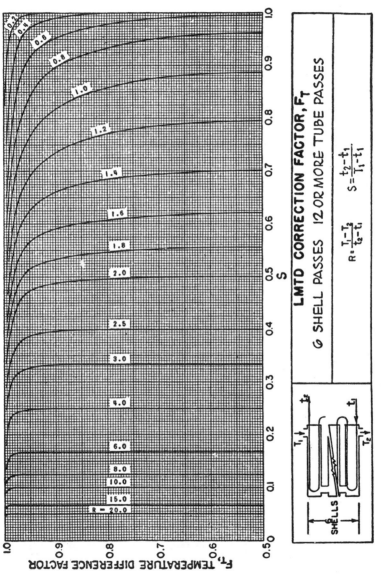

Figure A.8 LMTD correction factors for 6–12 exchangers. (Standards of Tubular Exchanger Manufacturers Association, 2d ed., New York, 1949. Reprinted courtesy of McGraw-Hill from D. Q. Kern, *Process Heat Transfer*, McGraw-Hill, New York, 1950.)

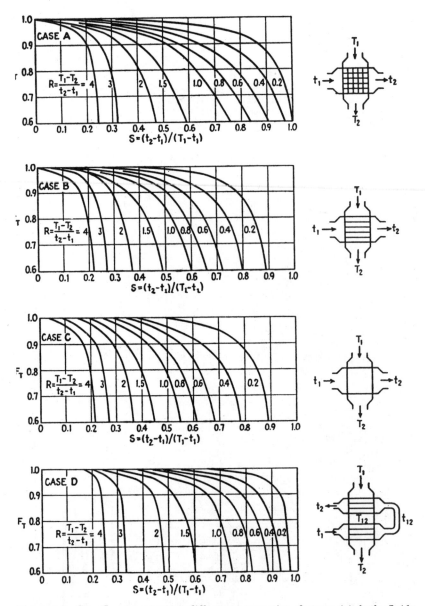

Figure A.9 Crossflow-temperature-difference correction factors: (*a*) both fluids unmixed; (*b*) one fluid mixed, other fluid unmixed; (*c*) both fluids mixed; and (*d*) two-pass counterflow, shell fluid mixed, tube fluid unmixed. (*Bowman, Mueller, and Nagle,* Transactions of the ASME. *Reprinted courtesy of McGraw-Hill from D. Q. Kern,* Process Heat Transfer, *McGraw-Hill, New York, 1950.*)

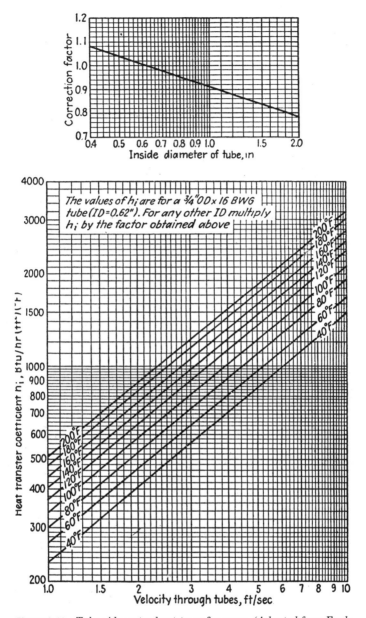

Figure A.10   Tube-side water-heat-transfer curve. (*Adapted from Eagle and Ferguson,* Proc. Roy. Soc. *A127:540, 1930. Reprinted courtesy of McGraw-Hill from D. Q. Kern,* Process Heat Transfer, *McGraw-Hill, New York, 1950.*)

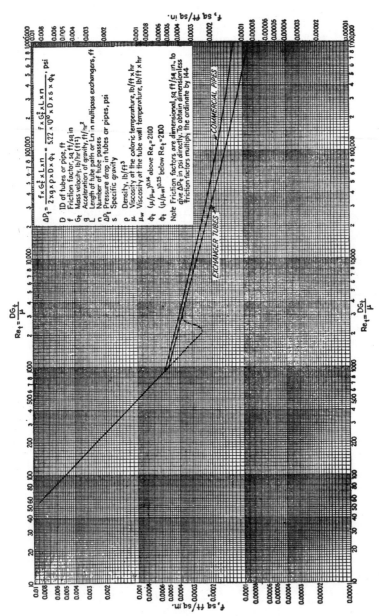

Figure A.11 Tube-side friction factors. (Standards of Tubular Exchanger Manufacturers Association, 2d ed., New York, 1949. Reprinted courtesy of McGraw-Hill from D. Q. Kern, Process Heat Transfer, McGraw-Hill, New York, 1950.)

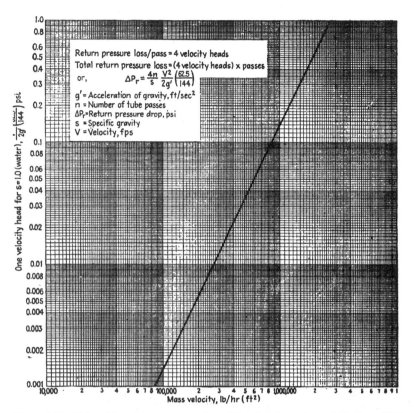

The y-axis is labeled: One velocity head for s=1.0 (water), $\frac{V^2}{2g'}\left(\frac{1}{144}\right)$ psi

Inside the plot box:

Return pressure loss/pass = 4 velocity heads

Total return pressure loss = (4 velocity heads) x passes

or, $\Delta P_r = \dfrac{4n}{s}\dfrac{V^2}{2g'}\left(\dfrac{62.5}{144}\right)$

g' = Acceleration of gravity, ft/sec^2
n = Number of tube passes
ΔP_r = Return pressure drop, psi
s = Specific gravity
V = Velocity, fps

x-axis: Mass velocity, lb/hr (ft^2)

Figure A.12   Tube-side return pressure loss. (*Reprinted courtesy of McGraw-Hill from D. Q. Kern,* Process Heat Transfer, *McGraw-Hill, New York, 1950.*)

Software Program Exhibits

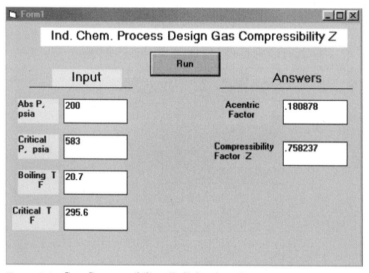

Figure B.1 Gas Compressibility Z dialog box. Insert your input in the squares on the left. Use the mouse to locate the input square. Left-click Run and your answers appear in the squares on the right.

Please note that you may obtain the same values simply by clicking on Run before you make any inputs. These are the defaulted values made in the program code. To bring up any program, simply double-click on the Z.exe file in the CD accompanying this book. Please note that you may load this file to any file location desired. *Please be advised that you must also load Vbrun300.dll to the same file location for any Visual Basic .exe program to run.*

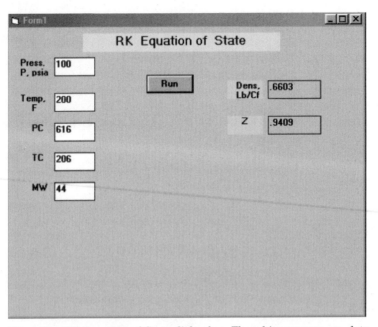

Figure B.2 RK Equation of State dialog box. The white squares are data input squares. This program is RK.exe, and all the same notes apply to it as for Fig. B.1 and all other .exe programs provided with this book.

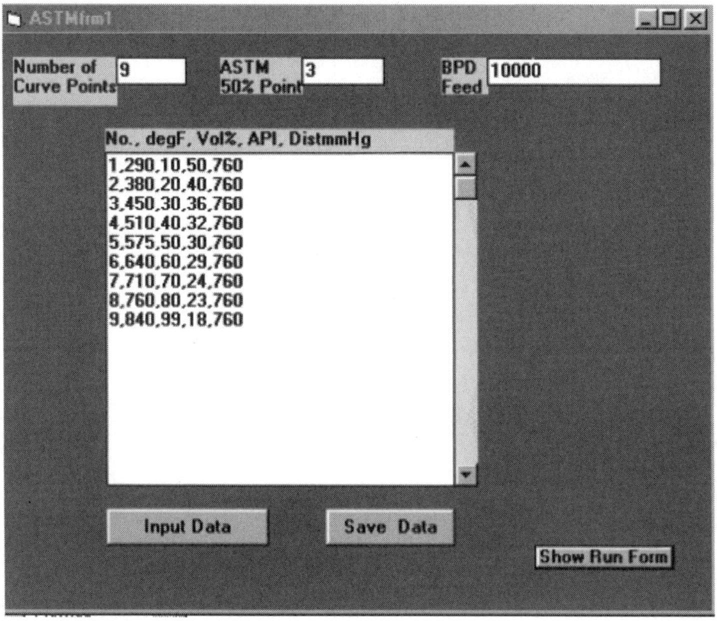

Figure B.3a Crude oil characterization.

1. Install the number of ASTM curve points you have.
2. Install the 50% number point of your data, or a best estimate.
3. Next install the point number, the ASTM °F value, the corresponding API gravity value, and the pressure value of the ASTM distillation as shown on each of the lines in the large center input square. Please note that each value is separated by a comma.
4. Click on Input Data, then click on Save Data, and last click on Show Run Form.

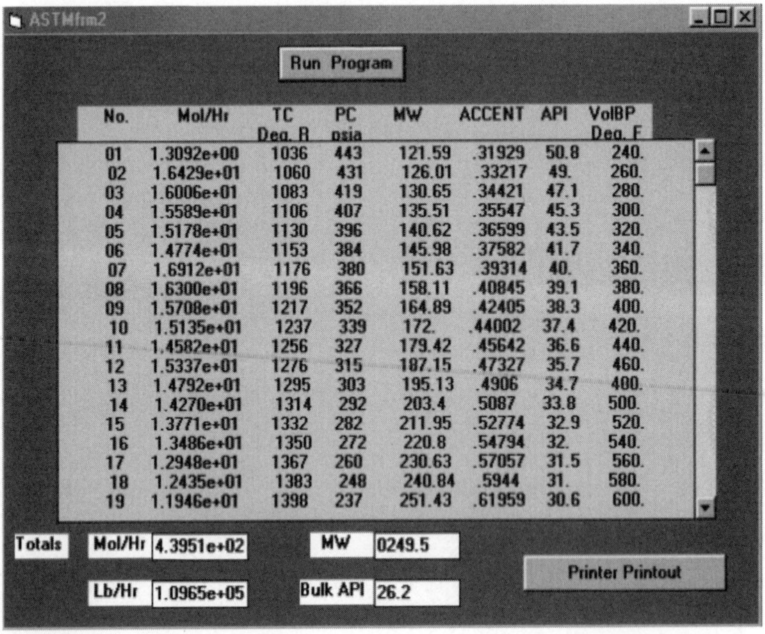

No.	Mol/Hr	TC Deg. R	PC psia	MW	ACCENT	API	VolBP Deg. F
01	1.3092e+00	1036	443	121.59	.31929	50.8	240.
02	1.6429e+01	1060	431	126.01	.33217	49.	260.
03	1.6006e+01	1083	419	130.65	.34421	47.1	280.
04	1.5589e+01	1106	407	135.51	.35547	45.3	300.
05	1.5178e+01	1130	396	140.62	.36599	43.5	320.
06	1.4774e+01	1153	384	145.98	.37582	41.7	340.
07	1.6912e+01	1176	380	151.63	.39314	40.	360.
08	1.6300e+01	1196	366	158.11	.40845	39.1	380.
09	1.5708e+01	1217	352	164.89	.42405	38.3	400.
10	1.5135e+01	1237	339	172.	.44002	37.4	420.
11	1.4582e+01	1256	327	179.42	.45642	36.6	440.
12	1.5337e+01	1276	315	187.15	.47327	35.7	460.
13	1.4792e+01	1295	303	195.13	.4906	34.7	400.
14	1.4270e+01	1314	292	203.4	.5087	33.8	500.
15	1.3771e+01	1332	282	211.95	.52774	32.9	520.
16	1.3486e+01	1350	272	220.8	.54794	32.	540.
17	1.2948e+01	1367	260	230.63	.57057	31.5	560.
18	1.2435e+01	1383	248	240.84	.5944	31.	580.
19	1.1946e+01	1398	237	251.43	.61959	30.6	600.

Totals			
Mol/Hr	4.3951e+02	MW	0249.5
Lb/Hr	1.0965e+05	Bulk API	26.2

Figure B.3*b* Crude oil characterization.

5. Click on Run Program, and the data calculations are displayed as shown.
6. You may use the defaulted values simply by first clicking on Input Data before doing anything else on the ASTMfrm1 (Fig. B.3*a*). Then click on Save Data and Show Run Form. Last, click on Run Program. By following these steps, you should then see the defaulted screens, Figs. B.3*a* and B.3*b*.
7. Be sure that you also copy astm4dat and Vbrun300.dll with ASTM4.exe to any drive or folder desired. The program will then run as shown in Figs. B.3*a* and B.3*b*.

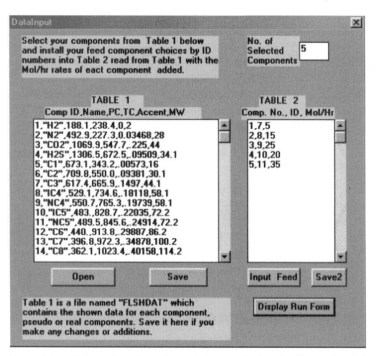

Figure B.4a   Hydrocarbon flash, dew point, and bubble point. To run the Ref-Flsh.exe program, double-click it in the directory file in which you placed it. Click first on Open. Second, click on Save and then Input Feed. You should then see the defaulted values shown. Next click on Save2 and then click on Display Run Form. Make sure you copy the files FlshDat, FeedDat, and tmsf1 to the drive and folder from which you run RefFlsh.exe.

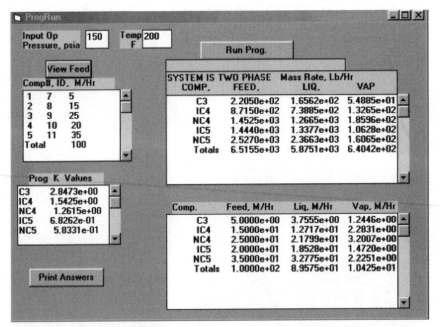

Figure B.4b Hydrocarbon flash, dew point, and bubble point. Click on Run Prog. to bring up the answers calculated in Fig. B.4a here. Please note that you may install any database desired in Fig. B.4a. You must, however, follow the convention of data input in Table 1 as shown. Note that you can change this Table 1 data with your keyboard and save the data. Save the files you want to keep in another folder. Please note that you can also swap other compatible files with the FlshDat file.

Change the figure's defaulted temperature from 200 to 250 and then click on Run Prog. again. Now observe the notation given, "All Vapor, not two phase."

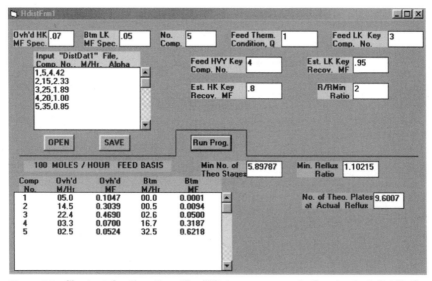

Figure B.5 Shortcut fractionation. The Hdist.exe program is the shortcut distillation program. To display the defaulted values, first click on OPEN and next click on Run Prog. The defaulted values will then be displayed on your PC screen. You may then change any input value and click again on the Run Prog. button. Note that the yellow tabbed values are the input values, and the purple tabs are the output answers.

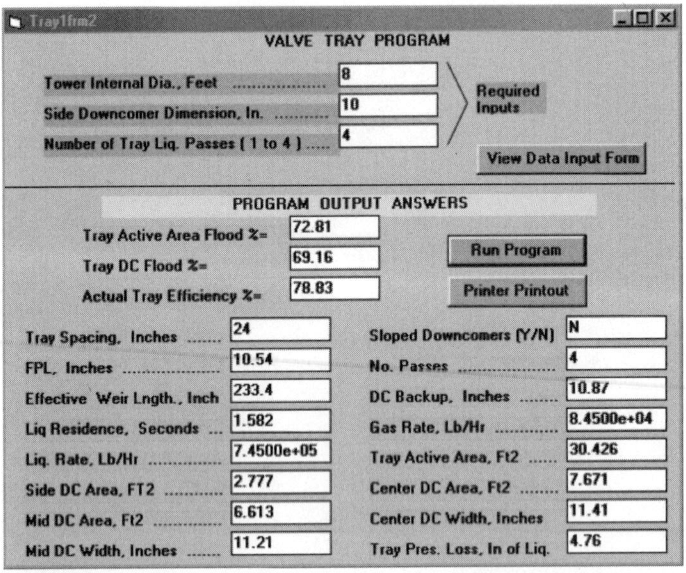

Figure B.6a Fractionation tray design and rating. Please take special note that you must input values in the three squares at the top of the window before the program will run. Next click on Run Program to display all the default calculated values shown. Calculated values are shown with purple tabs.

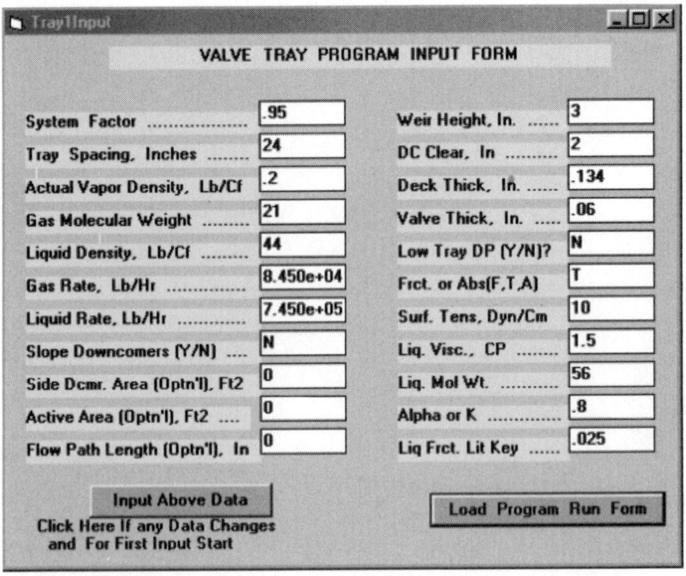

Figure B.6b Fractionation tray design and rating. Double-click the tray10.exe file to display the program on your PC screen. You may click on the Input Above Data button to display all of the default values shown.

Figure B.7*a* Fractionation tray design and rating. Double-click on the tray11.exe file for the program running as shown. Click on Input Above Data for the defaulted inputs shown.

Figure B.7*b* Fractionation tray design and rating. You must input the Required Inputs at the top of the window before this program will run. If you do not have a good guess, use the shown values. See Chap. 3 for more details on the program and its operation.

BUBBLE CAP TYPE TRAY

System Factor	.95	Weir Height, In.	2
Tray Spacing, Inches	24	Cap Dia., Inches	5.875
Actual Vapor Density, Lb/Cf	.2	Riser Dia. , In.	4.125
Gas Molecular Weight	21	Cap Pitch, S/T	T
Liquid Density, Lb/Cf	44	Cap Pitch, Inches	8.25
Liq. Mol Wt.	56	Slot Ht., Inches	.75
Gas Rate, Lb/Hr	8.450e+04	Slot Width, In.	.25
Liquid Rate, Lb/Hr	7.450e+05	No. Slots/Cap	40
Slope Downcomers (Y/N)	N	Cap Slot Seal, In.	.5
Side Dcmr. Area (Optn'l), Ft2	0	Surf. Tens, Dyn/Cm	10
Active Area (Optn'l), Ft2	0	Liq. Visc., CP	1.5
Flow Path Length (Optn'l), In	0	Frct. or Abs(F,T,A)	T
DC Clear, In	2	Alpha or K	.81
		Liq Frct. Lit Key	.025

Input Above Data   Load Program Run Form

Figure B.8a Fractionation tray design and rating. Double-click on the tray12.exe file for the program running as shown in Fig. B.8(a). Click on Input Above Data for the defaulted inputs shown.

BUBBLE CAP TYPE TRAY PROGRAM OUTPUT

Tower Internal Dia., Feet	10
Side Downcomer Dimension, In.	8
Number of Tray Liq. Passes (1 to 4)	4

Required Inputs

View Data Input Form

PROGRAM OUTPUT ANSWERS

Tray Active Area Flood %=	90.65
Tray DC Flood %=	89.92
Actual Tray Efficiency %=	75.11

Run Program

Printer Printout

Tray Spacing, Inches	24	Sloped Downcomers (Y/N)	N
FPL, Inches	20.52	No. Passes	4
Effective Weir Lngth., Inch	282.03	DC Backup, Inches	7.11
Liq Residence, Seconds	2.372	Gas Rate, Lb/Hr	8.4500e+04
Liq. Rate, Lb/Hr	7.4500e+05	Tray Active Area, Ft2	57.48
Side DC Area, FT2	2.249	Center DC Area, Ft2	6.211
Mid DC Area, Ft2	5.354	Center DC Width, Inches	7.39
Mid DC Width, Inches	7.26	Tray Pres. Loss, In of Liq.	2.64

Figure B.8b Fractionation tray design and rating. You must input the Required Inputs at the top of the window before this program will run. If you do not have a good guess, use the shown values. See Chap. 3 for more details on the program and its operation.

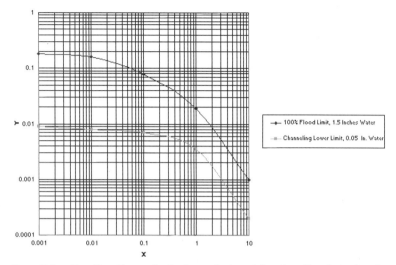

Figure B.9a Fractionation packed-column design and rating. Absorb.exe is a fractionation packed-column program in Fig. B.9(a). For any packed-column vapor-liquid loading, the X and Y coordinate points must be between the two curves shown. A point outside of these two curves is either in severe weeping-channeling zone or in the severe flood blockage zone.

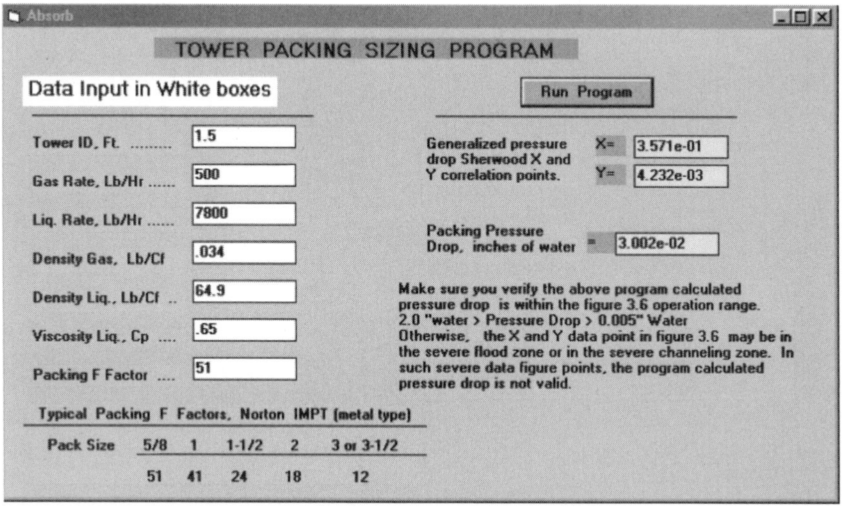

Figure B.9b Fractionation packed column design and rating. Double-click on the Absorb.exe file. Next click on Run Program, and the default inputs come up, along with the calculated answers, as shown. The answers are shown with a purple background color. The X and Y values refer to Fig. B.9a.

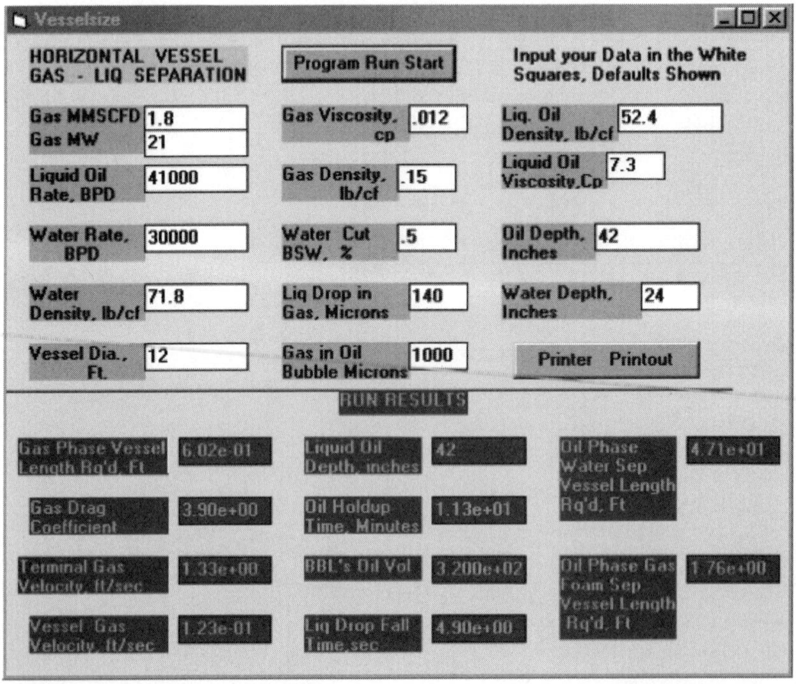

Figure B.10 Horizontal production oil, gas, and water separator. Double-click on Vessize.exe to bring up the display. Then click on Program Run Start to display the default values shown. The gray tabs are user input data and the blue tabs are the program output answers.

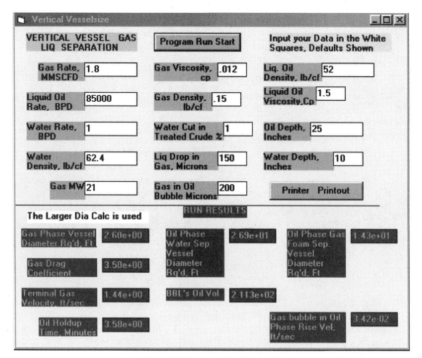

Figure B.11   Vertical production oil, gas, and water separator. Double-click on Vessize V.exe to bring up the display. Then click on Program Run Start to display the default values shown. The gray tabs are user input data and the blue tabs are the program output answers.

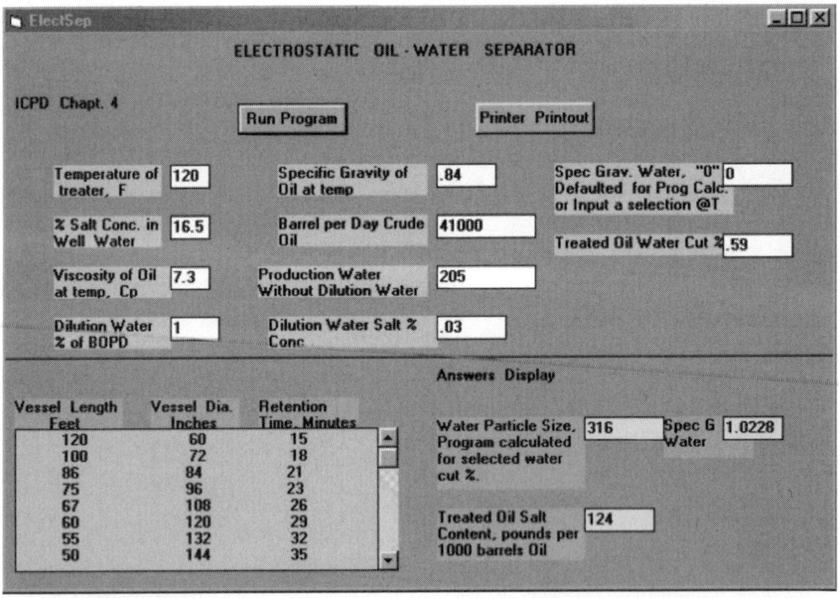

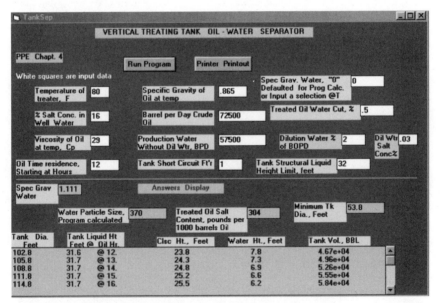

Figure B.12 Electric production oil and water dehydrator and desalter. Double-click on Elec2.exe to bring up the display. Then click on Run Program to display the default values shown. The yellow tabs are user input data and the orange tabs are the program output answers.

Figure B.13 Storage tank production oil and water dehydrator and desalter. Double-click on tksep.exe to bring up the display. Then click on Run Program to display the default values shown. The violet tabs are user input data and the blue tabs are the program output answers.

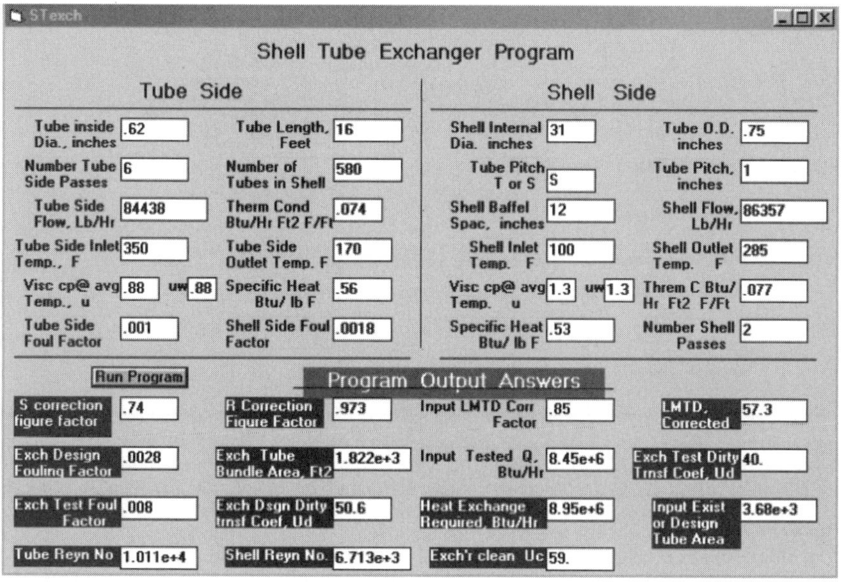

Figure B.14 Shell/tube exchanger program. Double-click on STEXCH.exe to bring up the display. Then click on Run Program to display the default values shown. The blue-green tabs are user input data and the blue tabs are the program output answers.

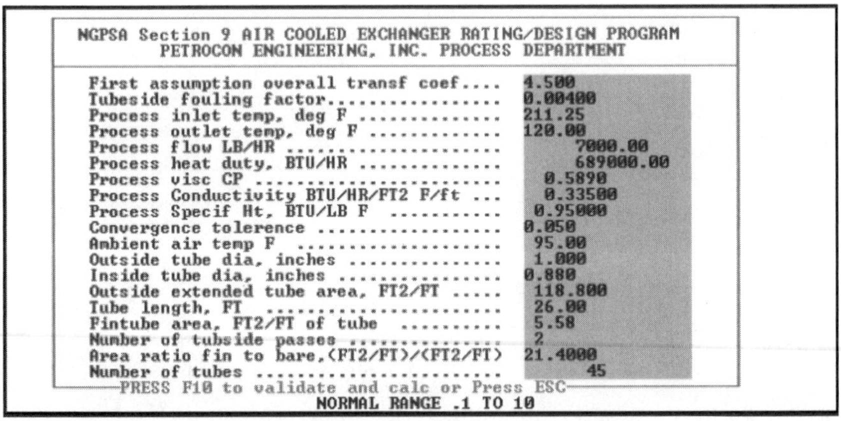

Figure B.15a Air-cooler design and rating, AirClr8 program. Double-click on AirClr8.exe to bring up this DOS program. This is an executable program for any single-phase fluid to be cooled in an air cooler. A specific heat is to be given, which the program will use to determine the process heat exchanged. Make certain you have the applications file Brun20.exe in the same directory as AirClr8.exe.

This program is also used for steam condensation. A given Q fixes heat transfer. A steam condensation tubeside film coefficient of 1500 Btu/h · ft^2 · °F is assumed. This condensing film coefficient is set by inputting a specific heat value of 0. The program appears with the default input values as shown.

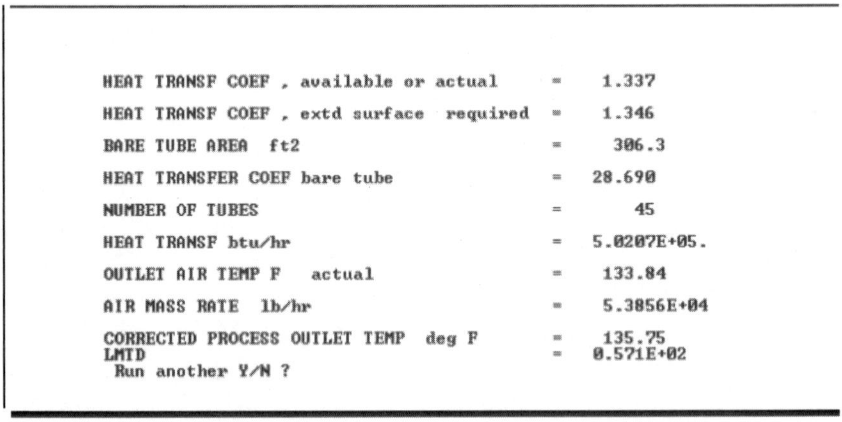

Figure B.15b Air-cooler design and rating, AirClr8 program, answers form.

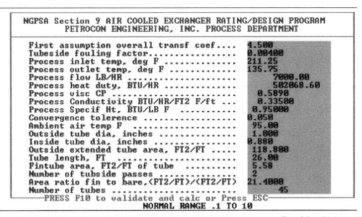

```
NGPSA Section 9 AIR COOLED EXCHANGER RATING/DESIGN PROGRAM
      PETROCON ENGINEERING, INC. PROCESS DEPARTMENT

First assumption overall transf coef....  4.500
Tubeside fouling factor.................  0.00400
Process inlet temp, deg F ..............  211.25
Process outlet temp, deg F .............  135.75
Process flow LB/HR .....................       7000.00
Process heat duty, BTU/HR ..............     502068.60
Process visc CP ........................  0.5890
Process Conductivity BTU/HR/FT2 F/ft ...  0.33500
Process Specif Ht, BTU/LB F  ...........  0.95000
Convergence tolerence ..................  0.050
Ambient air temp F  ....................  95.00
Outside tube dia, inches ...............  1.000
Inside tube dia, inches ................  0.880
Outside extended tube area, FT2/FT .....  118.800
Tube length, FT ........................  26.00
Fintube area, FT2/FT of tube ...........  5.58
Number of tubside passes ...............  2
Area ratio fin to bare,<FT2/FT>/<FT2/FT> 21.4000
Number of tubes ........................           45
   PRESS F10 to validate and calc or Press ESC
                NORMAL RANGE .1 TO 10
```

Figure B.16a   Air-cooler design and rating, AirClr9 program. Double-click on AirClr9.exe to bring up this DOS program. This is also an executable program for any condensing fluid. The input Q will be the heat duty transferred. All prompted inputs of variable data for the condensing film must be entered. The program appears with the default input values as shown.

```
HEAT TRANSF COEF , available or actual    =    1.295

HEAT TRANSF COEF , extd surface  required =    1.304

BARE TUBE AREA  ft2                       =    306.3

HEAT TRANSFER COEF bare tube              =    27.803

NUMBER OF TUBES                           =       45

HEAT TRANSF btu/hr                        =    4.8478E+05.

OUTLET AIR TEMP F    actual               =    138.18

AIR MASS RATE   lb/hr                     =    4.6785E+04

CORRECTED PROCESS OUTLET TEMP  deg F      =    138.35
LMTD                                      =    0.569E+02
   Run another Y/N ?
```

Figure B.16b   Air-cooler design/rating, AirClr9 program, answers form.

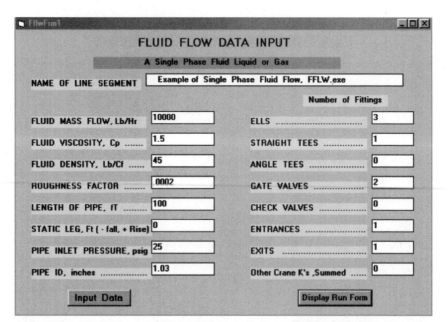

Figure B.17a Single-phase fluid flow, FFLW.exe program. Click on Input Data to get the default example run. You may input any desired data in the data input boxes. Simply move the mouse pointer to the desired box and left-click. To display the answers, simply click on Display Run Form. It is important that you click on the Input Data box. *Note that failure to click on this box first will cause the program to fail.* Also, be advised that you may go back to this first display countless times to change input data, but when you do so, you must click on Input Data again. Clicking on the Input Data box inputs the data changes you made.

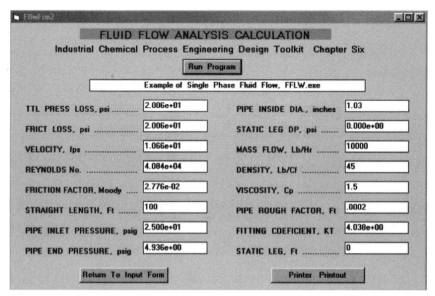

Figure B.17*b*   Single-phase flow program, FFLW.exe, answers form. Click on Run Program and all the inputs you installed on the FflwFrm1 screen display will be used to calculate the answers now shown. If you installed none and instead used the default values, the default answers will be shown. If you installed new data in FflwFrm1, you must again click on Run Program to obtain the new results. This procedure may be used any number of times. An example is making repeated trial-and-error runs to find desired results in fluid flow piping design. However, the first display—the FflwFrm1 display—must receive input before you click on Run Program to get new answers.

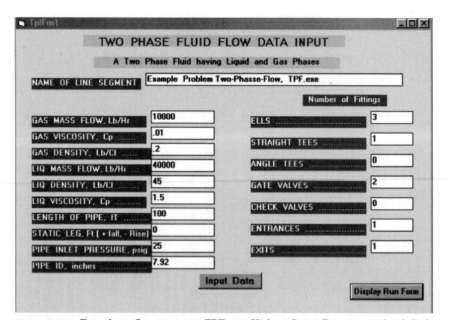

Figure B.18a Two-phase flow program, TPF.exe. Click on Input Data to get the default example run. You may input any desired data in the data input boxes. Simply move the mouse pointer to the desired box and left-click. You may write over or clear any data input box for new data input. To display the answers, simply click on Display Run Form. It is important that you click on the Input Data box before going to the Display Run Form. *Note that failure to click on this box first will cause the program to fail.* Also, be advised that you may go back to this first display countless times to change input data, but when you do so, you must click on Input Data again. Clicking on the Input Data box inputs the data changes you made.

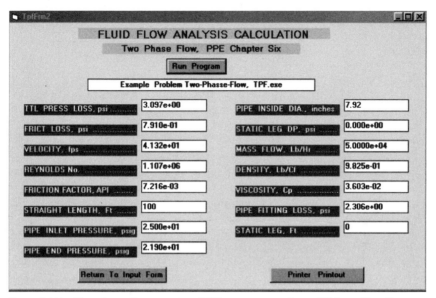

Figure B.18b Two-phase flow program, TPF.exe, answers form. Click on Run Program and all the inputs you installed on the TpfFrm1 screen display will be used to calculate the answers now shown. If you installed none and instead used the default values, the default answers will be shown. If you installed new data in TpfFrm1, you must again click on Run Program to obtain the new results. This procedure may be used any number of times. An example is repeated trial-and-error runs to find desired results in fluid flow piping design. However, the first display—the TpfFrm1 display—must receive input before you click on Run Program to get new answers.

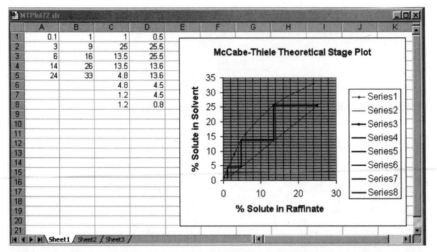

Figure B.19 McCabe-Thiele liquid/liquid extraction, Excel spreadsheet. Open this program, MTPlot72.xls, in the Excel spreadsheet. Use it as a template to make any liquid-liquid extraction simulation desired. This is the conventional McCabe-Thiele fractionation methodology.

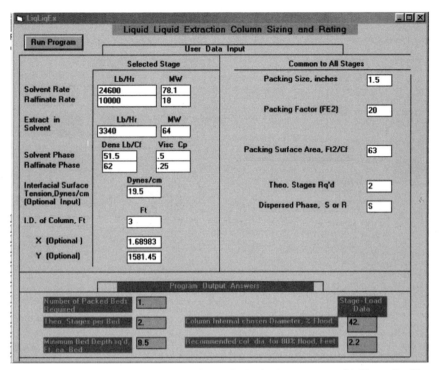

Figure B.20 Liquid/liquid extraction column design/rating program, LiqX.exe. Double-click on LiqX.exe to bring up the display. Then click on Run Program to display the default values shown. The gray tabs are user input data and the blue tabs are the program output answers.

PROJECT:	CC5 Recovery				28-Jul-01	
PROJECT SUMMARY						
EQUIPMENT TYPE					NEW EQUIPMENT COST	
FRACTIONATION COLUMNS					$1,721,000	
PUMPS					$459,300	
SHELL TUBE HEAT EXCHANGERS					$357,000	
HORIZONTAL PRESSURE VESSELS					$140,500	
AIR FINFAN COOLERS					$642,000	
MISCELLANEOUS EQUIPMENT						

TOTAL NEW EQUIPMENT COST					$3,319,800	
USED EQUIPMENT CREDIT					$0	
ASSOCIATED COSTS:						
MATERIALS						
PIPING					$1,522,894	
CONCRETE					$250,353	
STEEL					$151,832	
INSTRUMENTS					$303,689	
ELECTRICAL					$300,312	
INSULATION					$169,339	
PAINT					$36,110	

TOTAL ASSOCIATED COSTS, MATERIAL					$2,734,529	
TOTAL MATERIAL COSTS					$6,054,329	
LABOR					$2,568,500	
LABOR ITEMS ESTIMATED SEPARATELY					$0	
					==========	
TOTAL DIRECT COSTS					$8,622,829	
TOTAL INDIRECT COSTS					$3,132,177	
SALES TAX AND FREIGHT		10.0%	OF MATERIAL		$605,433	
CLIENT SUPERVISION		5.0%	OF TOTAL		$846,605	
OUTSIDE ENGINEERING		12.0%	OF TOTAL		$2,031,853	
CONTINGENCY		10.0%	OF TOTAL		$1,693,211	
					==========	
TOTAL PROJECT INSTALLED COST					$16,932,107	

Figure B.21 Cost summary, Excel spreadsheet EquipTable.xls. This is an Excel spreadsheet program; it shows the data from Table 8.38. There are five other spreadsheets in the EquipTableCost.xls file. Use these spreadsheets as templates for estimating the cost of modular equipment.

Raoult's Law

Raoult's law is a generalized statement that "the equilibrium vapor pressure which is exerted by a component in a solution is proportional to the mole fraction of that component in the solution" [1]. The term *solution* here designates a liquid phase having two or more components which each exert vapor pressure. Thus the following equation may be established:

$$p_a = P_a \frac{n_a}{n_a + n_b + n_c + \cdots} = N_a P_a$$

where $n_a, n_b, n_c \cdots$ = moles of components $a, b, c, \cdots$
N_a = mole fraction of component a in liquid solution
p_a = vapor pressure of component a in solution with components $b, c, \cdots$
P_a = vapor pressure of component a in pure state (*Lange's Handbook of Chemistry* [2] or other data sources)

Application to Liquid-Liquid Extraction

Having now established a methodology to find the solute a vapor pressure in both the feed phase and the solvent phase of liquid-liquid extraction, the Henry's law constants H_f and H_s may be calculated independently by the following equations:

$$\text{Feed phase } H_f = \frac{p_a}{x_m}$$

$$\text{Solvent phase } H_s = \frac{p_a}{y_m}$$

where x_m = mole fraction of component a (the solute) in the feed
y_m = mole fraction of component a (the solute) in the solvent

Vapor pressure calculations for component a are to be made independently. More simply, calculate the feed vapor pressure a disregarding any effect of partial pressures exerted by the solvent phase. The same holds true for the solvent phase. If the true vapor pressure of the solute is available for either the solvent phase or the feed phase, then use this value. Otherwise, the vapor pressure of the pure solute must be used.

For conditions where actual vapor pressure data is available for the solute, these calculations of H_f and H_s are to be repeated to find equilibrium points where $H_f = H_s$. Hold the mole fraction value of a constant, say x_m, and vary y_m until $H_f = H_s$. Repeat until several equilibrium points are calculated.

For conditions where only the pure solute vapor pressure is known in either solvent or feed phase, note that when $H_f = H_s$ becomes true, $x_m = y_m$ is also true. Therefore, $K = 1.0$ for this condition. Hundreds of feed, solute, and solvent combinations are given in Table C.1. Each K value shown in Table C.1 has been taken from the referenced data source, and each is the lowest K value given in the source work. It is also important to note that every K value in Table C.1 tends to approach unity as the solute concentration y_m is increased in the solvent phase. In many cases, K becomes unity as y_m reflects a concentration of 20 to 30%. A conservative assumption here is to assume that $K = 1.0$ when $y_m = 0.15$. For a better assumption in any particular case, refer to the Table C.1 reference sources [3]. Thus using the table and assuming that $K = 1.0$ at a 15% concentration of solute, a straight equilibrium line may be drawn for any K value given.

H values may be given in units of bars, mmHg, or psi. All three have been used and may be applied here as long as the units are kept consistent. Although Table C.1 uses weight ratios for y and x values, you should make transforming calculations to molar values when applying the Henry's law rule.

Although these calculations are not exacting for every case, such as electrolytic solutions, for the general majority of liquid-liquid extraction cases this equilibrium data is believed to be accurate to within $\pm 15\%$. Replacing this empirical method with laboratory test data is of course encouraged in every case. Please also consider that this Henry's law application is good only for dilute solute component concentrations ($\leq 15\%$ by volume) and low pressures (≤ 10 bar). Temperature does have a considerable effect, as seen in the vapor pressure relationships.

References

1. Hougen, O. A., and Watson, K. M., *Chemical Process Principles,* Part 1, John Wiley & Sons, New York, 1953, p. 80.
2. Dean, J. A., and Lange, N. A., *Lange's Handbook of Chemistry & Physics,* McGraw-Hill, New York, 1998.
3. Perry, R. H., Green, D. W., and Maloney, J. O., *Perry's Chemical Engineer's Handbook,* McGraw-Hill, New York, 1997, Table 15-5, p. 15-10.

TABLE C.1 Selected List of Ternary Systems

Component a = feed solvent, component b = solute, and component s = extraction solvent. K_1 is the partition ratio in weight fraction solute y/x for the tie line of lowest solute concentration reported. Ordinarily, K will approach unity as the solute concentration is increased.

Component a	Component b	Component s	Temperature, °C	K_1	Reference
Centane					
	Benzene	Aniline	25	1.290	47
	n-Heptane		25	0.0784	47
Cottonseed oil					
	Oleic acid	Propane	85	0.150	46
			98.5	0.1272	
Cyclohexane					
	Benzene	Furfural	25	0.680	44
	Benzene	Nitromethane	25	0.397	127
Docosane					
	1,6-Diphenylhexane	Furfural	45	0.980	11
			80	1.1	11
Dodecane					
	Methylnaphthalene	β,β'-Iminodipropionitrile	$\approx$25	0.625	92
		β,β'-Oxydipropionitrile	$\approx$25	0.377	92
Ethylbenzene					
	Styrene	Ethlyene glycol	25	0.190	10
Ethylene glycol					
	Acetone	Amyl acetate	31	1.838	86
		n-Butyl acetate	31	1.940	86
		Cyclohexane	27	0.508	86
		Ethyl acetate	31	1.850	86
		Ethyl butyrate	31	1.903	86
		Ethyl propionate	31	2.32	86
Furfural					
	Trilinolein	n-Heptane	30	47.5	15
			50	21.4	15
			70	19.5	15

TABLE C.1 Selected List of Ternary Systems (Continued)

Component a	Component b	Component s	Temperature, °C	K_1	Reference
Glycerol	Triolein	n=Heptane	30	95	15
			50	108	15
			70	41.5	15
n-Heptane	Ethanol	Benzene	25	0.159	62
		Carbon tetrachloride	25	0.0667	63
	Benzene	Ethylene glycol	25	0.300	50
			125	0.316	50
		β,β'-Thiodipropionitrile	25	0.350	92
		Triethylene glycol	25	0.351	89
	Cyclohexane	Aniline	25	0.0815	47
		Benzyl alcohol	0	0.107	29
			15	0.267	29
		Dimethylformamide	20	0.1320	28
		Furfural	30	0.0635	78
	Ethylbenzene	Dipropylene glycol	25	0.329	90
		β,β'-Oxydipropionitrile	25	0.180	101
		β,β'-Thiodipropionitrile	25	0.100	101
		Triethylene glycol	25	0.140	89
	Methylcyclohexane	Aniline	25	0.087	116
	Toluene	Aniline	0	0.577	27
			13	0.477	27
			20	0.457	27
			40	0.425	27
		Benzyl alcohol	0	0.694	29
		Dimethylformamide	0	0.667	28
			20	0.514	28
		Dipropylene glycol	25	0.331	90
		Ethylene glycol	25	0.150	101
		Propylene carbonate	20	0.732	39
		β,β'-Thiodipropionitrile	25	0.150	101
		Triethylene glycol	25	0.289	89
	m-Xylene	β,β'-Thiodipropionitrile	25	0.050	101

n-Hexane	o-Xylene	β,β'-Thiodipropionitrile	25	0.150	101
	p-Xylene	β,β'-Thiodipropionitrile	25	0.080	101
Neohexane	Benzene	Ethylenediamine	20	4.14	23
Methylcyclohexane	Cyclopentane	Aniline	15	0.1259	96
		Aniline	25	0.311	96
Isooctane	Toluene	Methylperfluorooctanoate	10	0.1297	58
		Methylperfluorooctanoate	25	0.200	58
Perfluoroheptane	Benzene	Furfural	25	0.833	44
	Cyclohexane	Furfural	25	0.1076	44
	n=Hexane	Furfural	30	0.083	78
Perfluoro-n-hexane	Perfluorocyclic oxide	Carbon tetrachloride	30	0.1370	58
	n-Heptane	n-Heptane	30	0.329	58
Perfluorotri-n-butylamine	Benzene	Benzene	30	6.22	80
	Carbon disulfide	Carbon disulfide	25	6.5	80
Toluene	Isooctane	Nitroethane	25	3.59	119
		Nitroethane	31.5	2.36	119
		Nitroethane	33.7	4.56	119
Triethylene glycol	Acetone	Ethylene glycol	0	0.286	100
	α-Picoline	Ethylene glycol	24	0.326	100
	Methylcyclohexane		20	3.87	14
	Diisobutylene		20	0.445	14
	Mixed heptanes		20	0.317	14
Triolein	Propane	Oleic acid	85	0.138	46
Water	n-Amyl alcohol	Actaldehyde	18	1.43	74
	Benzene	Actaldehyde	18	1.119	74
	Furfural	Actaldehyde	16	0.967	74

TABLE C.1 Selected List of Ternary Systems (Continued)

Component a	Component b	Component s	Temperature, °C	K_1	Reference
		Toulene	17	0.478	74
		Vinyl acetate	20	0.560	81
	Acetic acid	Benzene	25	0.0328	43
			30	0.0984	38
			40	0.1022	38
			50	0.0558	38
			60	0.0637	38
Acetic acid		1-Butanol	26.7	1.613	102
		Butyl acetate	30	0.705	45
				0.391	67
		Caproic acid	25	0.349	73
		Carbon tetrachloride	27	0.1920	91
			27.5	0.0549	54
		Chloroform	≈25	0.178	70
			25	0.0865	72
			56.8	0.1573	17
		Creosote oil	34	0.706	91
		Cyclohexanol	26.7	1.325	102
		Diisobutyl ketone	25–26	0.284	75
		Din-butyl ketone	25–26	0.379	75
		Diisoproply carbinol	25–26	0.800	75
		Ethyl acetate	30	0.907	30
		2-Ethylbutyric acid	25	0.323	73
			25	0.286	73
		Ethylidene diacetate	25	0.85	104
		Ethyl propionate	28	0.510	87
		Fenchone	25–26	0.310	75
		Furfural	26.7	0.787	102
		Heptadecanol	25	0.312	114
			50	0.1623	114
		3-Heptanol	25	0.828	76
		Hexalin acetate	25–26	0.520	75
		Hexane	31	0.0167	85
		Isoamyl acetate	25–26	0.343	75

Substance	Temp, °C	Value	Ref
Isophorone	25–26	0.858	75
Isopropyl ether	20	0.248	31
	25–26	0.429	75
Methyl acetate		1.273	67
Methyl butyrate	30	0.690	66
Methylcyclohexanone	25–26	0.930	75
Methylisobutyl carbinol	30	1.058	83
Methylisobutyl ketone	25	0.657	97
	25–26	0.755	75
Monochlorobenzene	25	0.0435	77
Octyl acetate	25–26	0.1805	75
n-Propyl acetate		0.638	67
Toluene	25	0.0644	131
Trichloroethylene	27	0.140	91
	30	0.0549	54
Vinyl acetate	28	0.294	103
Amyl acetate	30	1.228	117
Benzene	15	0.940	11
	30	0.862	11
	45	0.725	11
n-Butyl acetate	30	1.127	67
Carbon tetrachloride	25	0.238	12
Chloroform	25	1.830	43
		1.720	3
Dibutyl ether	25–26	1.941	75
Diethyl ether	30	1.00	54
Ethyl acetate	30	1.500	117
Ethyl butyrate	30	1.278	117
Ethyl propionate	30	1.385	117
n=Heptane	25	0.274	112
n=Hexane	25	0.343	114
Methyl acetate	30	1.153	117
Methylisobutyl ketone	25–26	1.910	75
Monochlorobenzene	25–26	1.000	75
Propyl acetate	30	0.243	117
Tetrachloroethane	25–26	2.37	57
Tetrachloroethylene	30	0.237	88

Acetone

TABLE C.1 Selected List of Ternary Systems (Continued)

Component a	Component b	Component s	Temperature, °C	K_1	Reference
		1,1,2-Trichloroethane	25	1.467	113
		Toluene	25–26	0.835	75
		Vinyl acetate	20	1.237	81
			25	3.63	104
		Xylene	25–26	0.659	75
	Allyl alcohol	Diallyl ether	22	0.572	32
	Aniline	Benzene	25	14.40	40
			50	15.5	40
		n-Heptane	25	1.425	40
			50	2.20	40
		Methylcyclohexane	25	2.05	40
			50	3.41	40
		Nitrobenzene	25	18.89	108
		Toluene	25	12.91	107
	Aniline hydrochloride	Aniline	25	0.0540	98
	Benzoic acid	Methylisobutyl ketone	26.7	76.9*	49
	Isobutanol	Benzene	25	0.989	1
		1,1,2,2-Tetrachloroethane	25	1.80	36
		Tetrachloroethylene	25	0.0460	7
	n-Butanol	Benzene	25	1.263	126
			35	2.12	126
		Toluene	30	1.176	37
	tert-Butanol	Benzene	25	0.401	99
		tert-Butyl hypocholorite	0	0.1393	130
			20	0.1487	130
			40	0.200	129
			60	0.539	129
	2-Butoxyethanol	Ethyl acetate	20	1.74	5
		Methyl ethyl ketone	25	3.05	68
	2,3-Butylene glycol	n-Butanol	26	0.597	71
			50	0.893	71
		Butyl acetate	26	0.0222	71
			50	0.0326	71
		Butylene glycol diacetate	26	0.1328	71

Solute	Solvent	Temperature, °C	Value	Reference
	Methylvinyl carbinol acetate	75	0.565	71
n-Butylamine	Monochlorobenzene	26	0.237	71
1-Butyraldehyde	Ethyl acetate	50	0.351	71
Butyric acid	Methyl butyrate	75	0.247	71
	Methylisobutyl carbinol	37.8	1.391	77
Cobaltous chloride	Dioxane	30	41.3	52
Cupric sulfate	n-Butanol	30	6.75	66
	sec-Butanol	25	12.12	83
	Mixed pentanols	30	0.0052	93
p-Cresol	Methylnaphthalene	30	0.000501	9
Diacetone alcohol	Ethylbenzene	30	0.00702	9
	Styrene	35	0.000225	9
	Monochlorobenzene	25	9.89	82
Dichloroacetic acid	Benzene	25	0.335	22
1,4-Dioxane	n-Amyl alcohol	25	0.445	22
	Benzene	25–26	0.0690	77
Ethanol	n-Butanol	25	1.020	8
	Cyclohexane	25	0.598	75
	Cyclohexane	20	0.1191	13
	Dibutyl ether	25	0.0536	115
	Di-n-propyl ketone	25	3.00	26
	Ethyl acetate	25–26	0.0157	118
	Ethylisovalerate	25–26	0.0244	124
	Heptadecanol	0	0.1458	75
	n-Heptane	20	0.592	75
	3-Heptanol	70	0.0263	5
	n-Hexane	25	0.500	5
	n-Hexanol	25	0.455	41
	sec-Octanol	30	0.392	13
	Toluene	25	0.270	114
	Trichloroethylene	25	0.274	94
		28	0.783	76
		28	0.00212	111
		25	1.00	56
		25	0.825	56
			0.01816	122
			0.0682	16

TABLE C.1 Selected List of Ternary Systems (Continued)

Component a	Component b	Component s	Temperature, °C	K_1	Reference
	Ethylene glycol	n-Amyl alcohol	20	0.1159	59
		n-Butanol	27	0.412	85
		Furfural	25	0.315	18
		n-Hexanol	20	0.275	59
		Methyl ethyl ketone	30	0.0527	85
	Formic acid	Chloroform	25	0.00445	72
			56.9	0.0192	17
		Methylisobutyl carbinol	30	1.218	83
	Furfural	n-Butane	51.5	0.712	42
			79.5	0.930	42
		Methylisobutyl ketone	25	7.10	19
		Toluene	25	5.64	53
	Hydrogen chloride	Isoamyl alcohol	25	0.170	21
		2,6-Dimethyl-4-heptanol	25	0.266	21
		2-Ethyl-1-butanol	25	0.534	21
		Ethyl butyl ketone	25	0.01515	79
		3-Heptanol	25	0.0250	21
		1-Hexanol	25	0.345	21
		2-Methyl-1-butanol	25	0.470	21
		Methylisobutyl ketone	25	0.0273	79
		2-Methyl-1-pentanol	25	0.502	21
		2-Methyl-2-pentanol	25	0.411	21
		Methylisopropyl ketone	25	0.0814	79
		1-Octanol	25	0.424	21
		2-Octanol	25	0.380	21
		1-Pentanol	25	0.257	21
		Pentanols (mixed)	25	0.271	21
	Hydrogen fluoride	Methylisobutyl ketone	25	0.370	79
	Lactic acid	Isoamyl alcohol	25	0.352	128
	Methanol	Benzene	0	0.01022	4
		n-Butanol	15	0.600	65
			30	0.479	65
			45	0.51	65
				1.26	65

	Solvent	Temp, °C		
	p-Cresol	60	0.682	65
	Cyclohexane	35	0.313	82
	Cyclohexene	25	0.0156	125
	Ethyl acetate	0	0.01043	124
Methanol	n-Hexanol	20	0.0589	5
	Methylnaphthalene	28	0.238	5
	sec-Octanol	25	0.565	55
	Phenol	35	0.025	82
	Toluene	28	0.0223	82
	Trichloroethylene	25	0.584	55
Methyl-n-butyl ketone	n-Butanol	27.5	1.333	82
Methyl ethyl ketone	Cyclohexane	37.8	0.0099	60
	Gasoline	25	0.0167	54
	n-Heptane	30	53.4	52
	n-Hexane	25	1.775	48
	2-Methyl furan	25	3.60	85
	Monochlorobenzene	37.8	1.686	64
	Naphtha	25	1.548	112
	1,1,2-Trimethylpentane	25	1.775	112
	Trichloroethylene	26.7	2.22	52
	2,2,4-Trimethylpentane	25	84.0	109
Nickelous chloride	Dioxane	25	2.36	68
		26.7	0.885†	6
		25	3.44	68
		25	3.27	68
		25	1.572	64
		25	0.0017	93
Nicotine	Carbon tetrachloride	25	9.50	34
Phenol	Methylnaphthalene	25	7.06	82
α-Picoline	Benzene	20	8.75	14
	Diisobutylene	20	1.360	14
	Heptanes (mixed)	20	1.378	14
	Methylcyclohexane	20	1.00	14
Isopropanol	Benzene	25	0.276	69
	Carbon tetrachloride	20	1.405	25
	Cyclohexane	25	0.0282	123
	Cyclohexene	15	0.0583	124

TABLE C.1 Selected List of Ternary Systems (Continued)

Component a	Component b	Component s	Temperature, °C	K_1	Reference
	n-Propanol	Diisopropyl ether	25	0.0682	124
			35	0.1875	124
		Ethyl acetate	25	0.406	35
		Tetrachloroethylene	0	0.200	5
			20	1.205	5
		Toluene	25	0.388	7
		Isoamyl alcohol	25	0.1296	121
		Benzene	25	3.34	20
		n-Butanol	37.8	0.650	61
			37.8	3.61	61
		Cyclohexane	25	0.1553	123
			35	0.1775	123
		Ethyl acetate	0	1.419	5
			20	1.542	5
		n-Heptane	37.8	0.540	61
		n-Hexane	37.8	0.326	61
		n-Propyl acetate	20	1.55	106
			35	2.14	106
	Propionic acid	Toluene	30	0.299	2
		Benzene	30	0.598	57
		Cyclohexane	31	0.1955	84
		Cyclohexene	31	0.303	84
		Ethyl acetate	30	2.77	87
		Ethyl butyrate	26	1.470	87
		Ethyl propionate	28	0.510	87
		Hexanes (mixed)	31	0.186	84
		Methyl butyrate	30	2.15	66
		Methylisobutyl carbinol	30	3.52	83
		Methylisobutyl ketone	26.7	1.949*	49
		Monochlorobenzene	30	0.513	57
		Tetrachloroethylene	31	0.167	84
		Toluene	31	0.515	84
		Trichlorethylene	30	0.496	57
	Pyridine	Benzene	15	2.19	110

Sodium chloride	Monochlorobenzene	25	3.00	105
	Toluene	25	2.73	120
	Xylene	45	2.49	110
	Isobutanol	60	2.10	110
		25	2.10	77
		25	1.900	120
		25	1.260	120
	n-Ethyl-sec-butyl amine	25	0.0182	36
	n-Ethyl-tert-butyl amine	32	0.0563	24
	2-Ethylhexyl amine	40	0.1792	24
	1-Methyldiethyl amine	30	0.187	24
	1-Methyldodecyl amine	39.1	0.0597	24
	n-Methyl-1,3-dimethylbutyl amine	30	0.693	24
		30	0.0537	24
1-Methyloctyl amine		0.589	24	
	tert-Nonyl amine	30	0.0318	24
	1,1,3,3-Tetramethylbutyl amine	30	0.072	24
Succinic acid	Isobutanol	25	0.00857	36
	Dioxane	15	0.0246	95
	Ethyl ether	20	0.220	33
		25	0.198	33
Trimethyl amine	Benzene	25	0.1805	33
		25	0.857	51
		70	2.36	51

* Concentrations in lb · mol/ft³.
† Concentrations in volume fraction.

Source: Perry, R. H., Green, D. W., and Maloney, J. O., *Perry's Chemical Engineer's Handbook*, McGraw-Hill, New York, 1997, p. 15-10. Reprinted courtesy of McGraw-Hill.

References

1. Alberty and Washburn, *J. Phys. Chem.* 49:4, 1945.
2. Baker, *J. Phys. Chem.* 59:1182, 1955.
3. Bancroft and Hubard, *J. Am. Chem. Soc.* 64:347, 1942.
4. Barbaudy, *Compt. Rend.* 182:1279, 1926.
5. Beech and Glasstone, *J. Chem. Soc.* 67: 1938.
6. Berg, Manders, and Switzer, *Chem. Eng. Prog.* 47:11, 1951.
7. Bergelin, Lockhart, and Brown, *Trans. Am. Inst. Chem. Engrs.* 39:173, 1943.
8. Berndt and Lynch, *J. Am. Chem. Soc.* 66:282, 1944.
9. Blumberg, Cejtlin, and Fuchs, *J. Appl. Chem.* 10:407, 1960.
10. Boobar, et al., *Ind. Eng. Chem.* 43:2922, 1951.
11. Briggs and Comings, *Ind. Eng. Chem.* 35:411, 1943.
12. Buchanan, *Ind. Eng. Chem.* 44:2449, 1952.
13. Chang and Moulton, *Ind. Eng. Chem.* 45:2350, 1953.
14. Charles and Morton, *J. Appl. Chem.* 7:39, 1957.
15. Church and Briggs, *J. Chem. Eng. Data* 9:207, 1964.
16. Colburn and Phillips, *Trans. Am. Chem. Engrs.* 40:333, 1944.
17. Conti, Othmer, and Gilmont, *J. Chem. Eng. Data* 5:301, 1960.
18. Conway and Norton, *Ind. Eng. Chem.* 43:1433, 1951.
19. Conway and Phillips, *Ind. Eng. Chem.* 46:1474, 1954.
20. Coull and Hope, *J. Phys. Chem.* 39:967, 1935.
21. Crittenden and Hixson, *Ind. Eng. Chem.* 46:265, 1954.
22. Crook and Van Winkle, *Ind. Eng. Chem.* 46:1474, 1954.
23. Cumming and Morton, *J. Appl. Chem.* 3:358, 1953.
24. Davison, Smith, and Hood, *J. Chem. Eng. Data* 11:304, 1966.
25. Denzler, *J. Phys. Chem.* 49:358, 1945.
26. Drouillon, *J. Chem. Phys.* 22:149, 1925.
27. Durandet and Gladel, *Rev. Inst. Franc. Petrole* 9:296, 1954.
28. Durandet and Gladel, *Rev. Inst. Franc. Petrole* 11:811, 1956.
29. Durandet, Gladel, and Graaziani, *Rev. Inst. Franc. Petrole* 10:585, 1955.
30. Eaglesfield, Kelly, and Short, *Ind. Chem.* 29:147, 243, 1953.
31. Elgin and Browning, *Trans. Am. Inst. Chem. Engrs.* 31:639, 1935.
32. Fairburn, Cheney, and Chernovsky, *Chem. Eng. Prog.* 43:280, 1947.
33. Forbes and Coolidge, *J. Am. Chem. Soc.* 41:150, 1919.
34. Fowler and Noble, *J. Appl. Chem.* 4:546, 1954.
35. Frere, *Ind. Eng. Chem.* 41:2365, 1949.
36. Fritzsche and Stockton, *Ind. Eng. Chem.* 38:737, 1946.
37. Fuoss, *J. Am. Chem. Soc.* 62:3183, 1940.
38. Garner, Ellis, and Roy, *Chem. Eng. Sci.* 2:14, 1953.
39. Gladel and Lablaude, *Rev. Inst. Franc. Petrole* 12:1236, 1957.
40. Griswold, Chew, and Klecka, *Ind. Eng. Chem.* 42:1246, 1950.
41. Griswold, Chu, and Winsauer, *Ind. Eng. Chem.* 41:2352, 1949.
42. Griswold, Klecka, and West, *Chem. Eng. Prog.* 44:839, 1948.
43. Hand, *J. Phys. Chem.* 34:1961, 1930.
44. Henty, McManamey, and Price, *J. Appl. Chem.* 14:148, 1964.
45. Hirata and Hirose, *Kagaku Kogaku* 27:407, 1963.
46. Hixon and Bockelmann, *Trans. Am. Inst. Chem. Engrs.* 38:891, 1942.
47. Hunter and Brown, *Ind. Eng. Chem.* 39:1343, 1947.
48. Jeffreys, *J. Chem. Eng. Data* 8:320, 1963.
49. Johnson and Bliss, *Trans. Am. Inst. Chem. Engrs.* 42:331, 1946.
50. Johnson and Francis, *Ind. Eng. Chem.* 46:1662, 1954.
51. Jones and Grigsby, *Ind. Eng. Chem.* 44:378, 1952.
52. Jones and McCants, *Ind. Eng. Chem.* 46:1956, 1954.
53. Knight, *Trans. Am. Inst. Chem. Engrs.* 39:439, 1943.
54. Krishnamurty, Murti, and Rao, *J. Sci. Ind. Res.* 12B:583, 1953.
55. Krishnamurty and Rao, *J. Sci. Ind. Res.* 14B:614, 1955.
56. Krishnamurty and Rao, *Trans. Indian Inst. Chem. Engrs.* 6:153, 1954.
57. Krishnamurty, Rao, and Rao, *Trans. Indian Inst. Chem. Engrs.* 6:161, 1954.

58. Kyle and Reed, *J. Chem. Eng. Data* 5:266, 1960.
59. Laddha and Smith, *Ind. Eng. Chem.* 40:494, 1948.
60. Mason and Washburn, *J. Am. Chem. Soc.* 59:2076, 1937.
61. McCants, Jones, and Hopson, *Ind. Eng. Chem.* 45:454, 1953.
62. McDonald, *J. Am. Chem. Soc.* 62:3183, 1940.
63. McDonald, Kluender, and Lane, *J. Phys. Chem.* 46:946, 1942.
64. Moulton and Walkey, *Trans. Am. Inst. Chem. Engrs.* 40:695, 1944.
65. Mueller, Pugsley, and Ferguson, *J. Phys. Chem.* 35:1314, 1931.
66. Murty, Murty, and Subrahmanyam, *J. Chem. Eng. Data* 11:335, 1966.
67. Murti, Venkataratnam, and Rao, *J. Sci. Ind. Res.* 13B:392, 1954.
68. Newman, Hayworth, and Treybal, *Ind. Eng. Chem.* 41:2039, 1949.
69. Olsen and Washburn, *J. Am. Chem. Soc.* 57:303, 1935.
70. Othmer, *Chem. Met. Eng.* 43:325, 1936.
71. Othmer, Bergen, Schlechter, and Bruins, *Ind. Eng. Chem.* 37:890, 1945.
72. Othmer and Ku, *J. Chem. Eng. Data* 4:42, 1959.
73. Othmer and Serrano, *Ind. Eng. Chem.* 41:1030, 1949.
74. Othmer and Tobias, *Ind. Eng. Chem.* 34:690, 1942.
75. Othmer, White, and Treuger, *Ind. Eng. Chem.* 33:1240, 1941.
76. Oualline and Van Winkle, *Ind. Eng. Chem.* 44:1668, 1952.
77. Peake and Thompson, *Ind. Eng. Chem.* 44:2439, 1952.
78. Pennington and Marwill, *Ind. Eng. Chem.* 45:1371, 1953.
79. Pilloton, *ASTM Spec. Tech. Publ.* 238:5, 1958.
80. Pilskin and Treybal, *J. Chem. Eng. Data* 11:49, 1966.
81. Pratt and Glover, *Trans. Inst. Chem. Engrs.* (London) 24:52, 1946.
82. Prutton, Walsh, and Desar, *Ind. Eng. Chem.* 42:1210, 1950.
83. Rao, Ramamurty, and Rao, *Chem. Eng. Sci.* 8:265, 1958.
84. Rao and Rao, *J. Appl. Chem.* 6:270, 1956.
85. Rao and Rao, *J. Appl. Chem.* 7:659, 1957.
86. Rao and Rao, *J. Sci Ind. Res.* 14B:204, 1955.
87. Rao and Rao, *J. Sci. Ind. Res.* 14B:244, 1955.
88. Rao and Rao, *Trans. Indian Inst. Chem. Engrs.* 7:78, 1954–1955.
89. Rifai, *Riv. Combust.* 11:811, 1957.
90. Rifai, *Riv. Combust.* 11:829, 1957.
91. Saletore, Mene, and Warhadpande, *Trans. Indian Inst. Chem. Engrs.* 2:16, 1950.
92. Saunders, *Ind. Eng. Chem.* 43:121, 1951.
93. Schott and Lyncy, *J. Chem. Eng. Data* 11:215, 1966.
94. Schweppe and Lorah, *Ind. Eng. Chem.* 46:2391, 1954.
95. Selikson and Ricci, *J. Am. Chem. Soc.* 64:2474, 1942.
96. Serjian, Spurr, and Gibbons, *J. Am. Chem. Soc.* 68:1763, 1946.
97. Sherwood, Evans, and Longcor, *Ind. Eng. Chem.* 31:1144, 1939.
98. Sidgwick, Pickford, and Wilson, *J. Chem. Soc.* 99:1122, 1911.
99. Simonsen and Washburn, *J. Am. Chem. Soc.* 68:235, 1946.
100. Sims and Bolme, *J. Chem. Eng. Data* 10:111, 1965.
101. Skinner, *Ind. Eng. Chem.* 47:222, 1955.
102. Skrzec and Murphy, *Ind. Eng. Chem.* 46:2245, 1954.
103. Smith, *J. Phys. Chem.* 45:1301, 1941.
104. Smith, *J. Phys. Chem.* 46:229, 1942.
105. Smith, *J. Phys. Chem.* 46:376, 1942.
106. Smith and Bonner, *Ind. Eng. Chem.* 42:896, 1950.
107. Smith and Drexel, *Ind. Eng. Chem.* 37:601, 1945.
108. Smith, Foecking, and Barber, *Ind. Eng. Chem.* 41:2289, 1949.
109. Smith and La Bonte, *Ind. Eng. Chem.* 44:3740, 1952.
110. Smith, Stibolt, and Day, *Ind. Eng. Chem.* 43:190, 1951.
111. Taresenkow and Paulsen, *J. Gen. Chem.* (USSR) 7:2143, 1937.
112. Treybal and Vondrak, *Ind. Eng. Chem.* 414:1761, 1949.
113. Treybal, Weber, and Daley, *Ind. Eng. Chem.* 38:817, 1946.
114. Upchurch and Van Winkle, *Ind. Eng. Chem.* 44:618, 1952.
115. Varteressian and Fenske, *Ind. Eng. Chem.* 28:928, 1936.
116. Varteressian and Fenske, *Ind. Eng. Chem.* 29:270, 1937.

117. Venkataratnam, Rao, and Rao, *Chem. Eng. Sci.* 7:102, 1957.
118. Vold and Washburn, *J. Am. Chem. Soc.* 54:4217, 1932.
119. Vreeland and Dunlap, *J. Phys. Chem.* 61:329, 1957.
120. Vriens and Medcalf, *Ind. Eng. Chem.* 45:1098, 1953.
121. Washburn and Beguin, *J. Am. Chem. Soc.* 62:579, 1940.
122. Washburn, Beguin, and Beckord, *J. Am. Chem. Soc.* 61:1694, 1939.
123. Washburn, Brockway, Graham, and Deming, *J. Am. Chem. Soc.* 64:1886, 1942.
124. Washburn, Graham, Arnold, and Transue, *J. Am. Chem. Soc.* 62:1454, 1940.
125. Washburn and Spencer, *J. Am. Chem. Soc.* 56:361, 1934.
126. Washburn and Strandskov, *J. Phys. Chem.* 48:241, 1944.
127. Weck and Hund, *Ind. Eng. Chem.* 46:2521, 1954.
128. Weiser and Geankoplis, *Ind. Eng. Chem.* 47:858, 1955.
129. Westwater, *Ind. Eng. Chem.* 47:451, 1955.
130. Westwater and Audrieth, *Ind. Eng. Chem.* 46:1281, 1954.
131. Woodman, *J. Phys. Chem.* 30:1283, 1926.

Index

ABOUT THE AUTHOR

Douglas L. Erwin, P.E., is a practicing process engineer for engineering design and construction companies, major oil- and gas-producing companies, major refinery-operating companies, and major chemical product companies. His project assignments have ranged from daily, routine technical services in refineries through the conception of new, complex, first-of-their-type patented processes. His chief quest in engineering practice has been to extend our knowledge beyond commercial process-simulation software, covering the severe technology gaps these packages miss. Process unit operation equipment, design and rating of fractionation equipment, heat-exchange equipment, rotating equipment, liquid-liquid extraction equipment, three-phase separation equipment, packaged plant cost analysis, and equipment emergency pressure relief have been his main fields of accomplishment. Since graduating from Lamar University with a BS in chemical engineering in 1956, he has achieved these accomplishments in such locations worldwide as Korea, Saudi Arabia, Indonesia, Peru, Singapore, Canada, Alaska, the United Kingdom, Europe, Africa, Egypt, and the United States. He is a registered Professional Engineer in Texas by written examination. He is accounted by clients worldwide to be an expert consultant on code safety practice, the design and operation of oil and gas production and gas processing plants, and refinery liquid-liquid extraction design and operation.